PLAYING VIDEO GAMES

Motives, Responses, and Consequences

LEA'S COMMUNICATION SERIES
Jennings Bryant/Dolf Zillmann, General Editors

For a complete list of titles in LEA's Communication Series, please contact Lawrence Erlbaum Associates, Publishers, at www.erlbaum.com.

PLAYING VIDEO GAMES

Motives, Responses, and Consequences

Edited by

Peter Vorderer
University of Southern California

Jennings Bryant
University of Alabama

LEA
2006

LAWRENCE ERLBAUM ASSOCIATES, PUBLISHERS
Mahwah, New Jersey   London

Senior Acquisitions Editor: Linda Bathgate
Assistant Editor: Karin Wittig Bates
Cover Design: Stacey Spiegel
Cover Layout: Tomai Maridou
Full-Service Compositor: TechBooks
Text and Cover Printer: Hamilton Printing Company

This book was typeset in 10/12 pt. Times, Italic, Bold, and Bold Italic.
The heads were typeset in GillSans, Bold, Italics, and Bold Italics.

Lawrence Erlbaum Associates, Inc., Publishers
10 Industrial Avenue
Mahwah, New Jersey 07430
www.erlbaum.com

Library of Congress Cataloging-in-Publication Data

Playing video games : motives, responses, and consequences / edited by
   Peter Vorderer [and] Jennings Bryant.
      p. cm.—(LEA's communication series)
   Includes bibliographical references and index.
   ISBN 0-8058-5321-9 (cloth : alk. paper)—ISBN 0-8058-5322-7 (pbk. : alk. paper)
   1. Video games–Psychological aspects. I. Vorderer, Peter. II. Bryant, Jennings. III. Series.
   GV1469.3.P484   2006
   794.8—dc22

                                                              2005029416

Books published by Lawrence Erlbaum Associates are printed on
acid-free paper, and their bindings are chosen for strength and durability.

Printed in the United States of America
10 9 8 7 6 5 4 3 2 1

Contents

Foreword

When Palladas, the Greek poet who flourished in the 4th century A. C. E., said that life is but a game, he hardly could have imagined how pervasive games would become in every aspect of our modern lives. From security training simulations to war games and role-playing games, from sports games to gambling, playing video games has become a social phenomena and the increasing number of players that cross gender, culture, and age is on a dramatic trajectory.

Game play—and by that I mean simply computer-based game play—has become a driving economic force that is now giving shape to the technology landscape that supports it. With game play producing revenues in excess of $30 billion worldwide, it is not hard to imagine that a cutting-edge 3-D game can push the lagging revenues of a graphic chip manufacturer into soaring profitability. And so it goes that the fantastic developments in low-cost graphics capability feed the demand for more realism, simulation, and complex game play, which in turn require an even more sophisticated graphics capability in order to play the more demanding game.

This market thrust of pushing development by interweaving content and technology is the dominant feature of today's game play. The latest releases of popular games such as *DOOM 3, HALO 2,* and *World of Warcraft* outperform the commercial success of Hollywood movies and have an extraordinary level of realism that often can not be fully experienced without players upgrading their graphics hardware to the latest specifications. Game experience drives development even in the dedicated high-end game platforms that are another competitive solution for gamer players. Each of the majors—Sony, Microsoft, and Nintendo—regularly releases proprietary hardware that support new levels of graphic capability, but each release is short-lived only to be replaced by yet another, newer technology innovation.

Four years ago E3, the world's largest trade show for game players, distributors, and developers, featured a panel on massively multiplayer online games (MMOGs). The panel was made up of badly bruised and battered developers of these games, each of whom had lost money in the realization of their product and now faced another tough sentence: The audience was unrepentant and vocal in their belief that there would never be a market for people playing together in networked game environments.

In only a few short years, the world for online gamers has shifted completely. Online game play is now considered one of the highest growth opportunities for the commercial future of games. In North America we recently saw the U.S. military host a game developers' conference called "Serious Games" that focused on game-based team training, while the entertainment industry released the hugely popular MMOG called World of Warcraft. These uniquely distinct events share a common thread: They firmly establish the growth of social computing.

With all this frenzied development, what has not been rapidly upgraded or easily replaced is the thinking and academic research about game play itself. This anthology, providing an in-depth review and analysis of playing video games based on study of motives, responses and consequences, is long overdue. By deconstructing the topic into products, motivation and selection, reception and reaction processes, and effects and consequences, the editors have established a foundation in the understanding of what playing video games is all about.

Editors Vorderer and Bryant establish the playing field, focusing their extensive knowledge in entertainment theory to tackle challenging question and putting them into a context of academic research on entertainment. Unlike movies, however, games by their very nature set up the player as the director, with the action taking place in real time. Games, therefore, seem to be the real entertainment of our times, but playing them at the same time is so much different from traditional entertainment.

A total of 27 chapters, written by authors from all around the world, deal with every issue that is most pressing and urgent for our understanding of the more specific nature of playing these games. Overall, this seminal text arrives at a pivotal time in the history of video game development. The relevance of these writings will be equally meaningful to teachers, academics, and parents as to the established commercial game industry and experimental new indie producers.

—Stacey Spiegel, President and CEO
Immersion Studios, Inc., Toronto

Preface

This book has two ancestors: One of them is another book; the other one is a misfit. The other book is currently in press with Lawrence Erlbaum Associates under the title *Psychology of Entertainment*. Here is the backstory: When Jennings Bryant tired of never having adequate materials to teach a doctoral seminar on entertainment theory and research, he asked Peter Vorderer to join him in editing a scholarly volume that would have the advanced content and perspectives needed for such a course. While they were contemplating potential chapters and contributors for such a book, it became obvious that video games have become one of the—if not *the*—most important means of entertainment, at least for the younger generation. Moreover, when they taught entertainment at their various schools or presented papers on entertainment at conferences in the United States and abroad, it became equally obvious that the greatest interest in entertainment theory and research often came from junior faculty and from graduate students who were particularly intrigued by the newest and latest ways of becoming entertained. Often their preferred sources of entertainment were not traditional media, such as television, movies, or music. The delivery format for such contemporary entertainment was something that has been called digital or interactive entertainment. Although such media seemed to have lots of similarities with traditional entertainment, significant and substantial differences were also abundantly present. Many of these young scholars frequently complained that, despite the emerging status of entertainment theory in psychology and communication, there have been few attempts to apply the theory to the playing of video games and to thereby systematize this new field of research. After being hit over the head with such persuasive arguments several times, your responsive editors finally realized that a book that would bring together all these young (typically, but not always) scholars from around the world could enlighten our understanding of what happens when people play video and computer games.

At about the same time, the two editors also submitted a panel to the annual conference of the International Communication Association (ICA). This panel also was supposed to bring together scholars from around the world who would demonstrate and share their findings and insights on electronic games with their peers. The panel proposal, however, was rejected by the Mass Comm division of ICA on the grounds that the reviewers did not think it would fit into this particular branch of the discipline. So here is the misfit—one that puzzled and irritated your editors to such an extent that it finally motivated them to pull together a book that would be able to reach across divisions of a discipline. But, there was a hitch. This other book on the *Psychology of Entertainment* was already in the works, and who would be crazy enough to do two books at the same time?

Sure, our spouses dissuaded us from doing it, and our students and friends suggested we do one after the other, but we could not help developing both projects at the same time. One reason for our unwillingness to defer either idea was because we thought each project was so timely that if either were deferred, the discipline would suffer from the delay. Personally, neither of us has seen another topic within our discipline that has received more attention, more concern, and a greater need for understanding and explanation in such a short period of time as this burgeoning research area of entertainment theory. In particular, the video game industry is growing faster than any other entertainment industry, and electronic games have infiltrated and already changed our lives as much if not more than any other medium. Moreover, public opinion is highly alert to several facets of video game uses and effects, and universities around the world have started to put together programs, curricula, and research teams to better understand what playing electronic games may do to us. We simply cannot afford to postpone a more systematic and empirical study of playing video games.

We also found that most of the university programs that have been put together so far have focused on the creative side of video games, that is, on the art of storytelling and on the production of games. While this aesthetic and engineering perspective is crucial for developing and producing games, equally important are programs that take the perspective of examining their uses and effects. Such perspectives ask how people play games, why they play, and which games they play under what conditions and reasons, and what these games do to users as well as what gamers do with their games, both in the short term and in the longer term. Naturally, these are the questions scholars in media psychology and communication are interested in and are capable of answering, which is why we thought it was essential to put together this volume— if we could only obtain the commitment of these extremely busy scholars. Fortunately, our contributors were able and willing to meet the challenge.

No doubt, such a project with 51 contributors from the United States, Europe, Asia, and Australia would not have been possible without a publishing company that has supported us from the very beginning, and one that is as reliable and competent as we have always found them to be. We are extremely thankful to Linda Bathgate and to her team at Erlbaum for all their advice and for the patience they have had with us. We are also grateful to our contributors, who not only came from different places around the globe but also from different disciplines within the academic world and various gaming industries. We offer special thanks to Stacey Spiegel, who not only penned the foreword to the volume, but who also had his company's best designers create a cover for it.

The book is aimed at students, young and old, who would like to understand how, why, and with which consequences people play video games. It may be used inside and outside of classrooms for communication and media studies, in psychology, in human development, and in education both as an introductory reference resource as well as a textbook. It brings together an extremely talented group of international scholars who recognize—indeed, insist on—the relevance of video games in our lives.

—Peter Vorderer
—Jennings Bryant

Contributors

Craig A. Anderson
Iowa State University

Ann-Sofie Axelsson
Chalmers University of Technology
 Göteborg, Sweden

Kaysee Baker
Florida State University

Jennings Bryant
University of Alabama

Katherine E. Buckley
Iowa State University

Elaine Chan
University of Southern California

John Davies
University of North Florida

Mari Siân Davies
University of California, Los Angeles

Kevin Durkin
University of Strathclyde

Bradley S. Greenberg
Michigan State University

Patricia M. Greenfield
University of California, Los Angeles

Jeffrey T. Hancock
Cornell University

Tilo Hartmann
Hannover University of Music and Drama

Seung-A Jin
University of Southern California

Christoph Klimmt
Hannover University of Music and Drama

G. Christopher Klug
Carnegie Mellon University

Anna Kostygina
University of Southern California

Astrid Kristen
Freie Universität Berlin (Germany)

Ken Lachlan
Boston College

Kwan Min Lee
University of Southern California

Debra A. Lieberman
University of California, Santa Barbara

Scott D. Lipscomb
Northwestern University

Henry Lowood
Stanford University

Kristen Lucas
Purdue University

Gerhild Nieding
University of Würzburg

Peter Ohler
University of Technology Chemnitz

Caroline Oppl
Freie Universität Berlin (Germany)

Namkee Park
University of Southern California

Jorge Peña-Herborn
Cornell University

Katherine M. Pieper
University of Southern California

Wei Peng
University of Southern California

Arthur A. Raney
Florida State University

Tim Regan
Microsoft Research Cambridge, UK

Ute Ritterfeld
University of Southern California

Maria von Salisch
Universität Lüneburg (Germany)

Jesse Schell
Carnegie Mellon University

Michael Sellers
Online Alchemy, Inc.

Michael A. Shapiro
Cornell University

John L. Sherry
Michigan State University

Paul Skalski
University of Minnesota, Duluth

Jason K. Smith
Florida State University

Stacy L. Smith
University of Southern California

Barry P. Smith
University of Alabama

Francis F. Steen
University of California, Los Angeles

Ron Tamborini
Michigan State University

Brendesha Tynes
University of California, Los Angeles

Peter Vorderer
University of Southern California

René Weber
Michigan State University

Dmitri Williams
University of Illinois at Urbana-Champaign

Sean M. Zehnder
Northwestern University

PLAYING VIDEO GAMES

Motives, Responses, and Consequences

1

Playing Video Games as Entertainment

Peter Vorderer
University of Southern California

Jennings Bryant
University of Alabama

Katherine M. Pieper
University of Southern California

René Weber
Michigan State University

Video games have surpassed the designation of "fad" or "new technology" to become a staple of contemporary entertainment. In 2003, computer and video game software sales totaled $7 billion in the United States—more than 239 million units—which is nearly enough for every American household to have purchased two games (Entertainment Software Association, 2004a). In the year 2000, "the demand for computer and video games created a $10.5 billion market" for the game industry, including such subsets as transporting and wholesaling (IDSA, 2001, p. 4). Clearly, entertainment needs that can be satisfied by game playing can generate quite a bit of revenue.

Increasing game sales are not the only indication that gaming has found a place in American—and international—homes. The Entertainment Software Association (ESA), which represents the computer and video game industry in the United States, tracks the demographics of game players. According to the ESA, 50% of all Americans play games, and the average age of a game player is 29 years old (ESA, 2004b). Interestingly, the ESA states that 39% of all gamers are female, dispelling the popular notion that games are a totally male-dominated pastime. Perhaps most importantly, the increases in players and revenue show no signs of stopping or even slowing; according to the ESA, more than 50% of gamers predict that in 10 years they will play as much or more than they currently play (ESA, 2004a).

A 2003 report by the Kaiser Family Foundation shows that even the youngest children have experience with video games; nearly half of all children (49%) ages 0–6 have a video game player in their home, and 10% have a video game console in their bedroom. Thirty percent of young children have played video games, including 3% of children younger than 2. Although game playing is less common among children this age than using other media, 50% of children ages 4–6 play video games, and on a typical day, 16% play for a little more than an hour (1:04). Among boys this age, 9% play games every day, but only 2% of girls ages 4–6 play games this often (Kaiser Family Foundation, 2003). Clearly, video games are popular with younger members of society.

Industry members and nonprofit organizations are not the only groups interested in the pervasiveness of video game play. Academicians who study computer and video games have formed research groups, such as the Center for Computer Games Research Copenhagen, MIT's Comparative Media Studies program, and other similar groups across the globe. One such group at the University of Southern California, the Annenberg Studies on Computer Games Group (ASC Games), recently conducted an exploratory survey to determine the prevalence of game play and other factors associated with play. A total of 314 individuals completed the online survey; of these, 94% ($n = 297$) responded that they play video games, and 75% of respondents said that they played video games every day.

Players' favorite genres were shooter (57.5%), role-playing (54.8%), adventure (48.6%), and strategy/puzzle (48.3%) games. Examples of these games include the popular *Max Payne* and *Lara Croft: Tomb Raider* series, and even PC-based games such as solitaire. Game players expect these genres to remain their favorite for some time as well. Ninety-four percent responded that they expect to be playing games in 10 years, with little variation expected in their favorite genres: role-playing (55.8%), shooter (53.8%), strategy/puzzle (49.8%), and adventure (45%). These findings are similar to the ESA's list of top-selling game genres. In 2003, the best selling console genres were action (27.1%), sports (17.6%), racing (15.7%), and role-playing (8.7%), and for computer games were strategy (27.1%), children's (14.5%), shooter (13.5%) and family entertainment (9.5%) games (ESA, 2004b).

Video and computer games are quite obviously *entertaining* to those people who play them. Respondents overwhelmingly indicated that when forced to choose between video games and other media, they would rather give up television (73.6%) or movies (69.3%). However, does this qualify them as *entertainment*? Is it possible to use theories that explain "traditional" entertainment products (i.e., television, films) to describe what happens when people play games?

THE BOUNDARIES OF TRADITIONAL ENTERTAINMENT THEORY

Vorderer (2000) has pointed out that interactive entertainment poses special challenges to theories of entertainment, which rely on the assumption that users are receptive to content and process what is given to them. Interactivity, however, assumes that content evolves as the user participates with the medium. Computer and video games, unlike television or films in general, contain content that is modified by the user and may change as play develops.

Interactivity poses a unique question for individuals seeking to understand what it is that drives players to use a particular game or even games in general. Respondents to the ASC Games Group survey stated that "competition" was the most important factor (31%), followed by "challenge" (21.4%), for their enjoyment of game play. Respondents also said that they are most likely to purchase a game "because the game will be challenging" (55%). Despite these results, very few tests determine what factors of "competition" or "challenge" are particularly stimulating to players. Other studies (Sherry, de Souza, Greenberg, & Lachlan, n.d.) also demonstrate that the rewarding nature of a challenge or competition drives individuals to use games. These gratifications hold among children, adolescents, and college students (Sherry et al., n.d.). The most intricate process of establishing new gratifications of video game play, however, comes in defining what it is about "challenge" that is motivating. Researchers have already contemplated the defining characteristics of such gratifications as "information," "diversion," or even "entertainment," but have yet to clearly delineate what "challenge" and "competition" mean for video game players and why they are so appealing.

Continually evolving content requires special approaches to the narrative structure of video games. Grodal (2000) discussed three dimensions of entertainment that are experienced fundamentally differently in video games than in more traditional entertainment genres. In a situation in which the viewer or player takes an active role in constructing content, our academic understanding of the role or structure of curiosity, surprise, and suspense needs adjustment. These three aspects of narrative create arousal in viewers and players that will govern their emotional experience while using a film or video game. Interactivity allows for multiple unique interactions with a given entertainment product, which changes the function of curiosity, surprise, and suspense. According to Grodal, especially in video games, "the experience of given situations will change over time, due to learning processes that will change arousal and will change the cognitive labeling of the arousal" (2000, p. 207). Instead of experiencing surprise only one time throughout the game, users experience continual surprises as they encounter new challenges, battles, or characters in a game (Grodal, 2000). Suspense, an important storytelling tool in film (Vorderer, Wulff, & Friedrichsen, 1996), changes drastically when applied to games. A user's experience with a game revolves around the use of the avatar, which they must guide through a series of increasingly difficult tasks. Thus, although players may feel suspense about outcomes of the games, they are at the same time in control of those outcomes (to some extent). This fusion of narrative and interactivity results in a much different emotional experience than that of traditional entertainment (Grodal, 2000; Vorderer, 2000).

CONVERGENCE AND DIVERGENCE: VIDEO GAMES AND TRADITIONAL ENTERTAINMENT THEORY

Video games have altered the public conception of entertainment, but it follows that they must also transform the way that academics research entertainment. The differences between video games and traditional entertainment should force researchers to question an established base of research and demand not only assimilation into older theoretical traditions, but also the construction of new theories in the discipline.

There is a clear convention in entertainment research to investigate several different elements of the viewing experience, namely motivation, selection, experience, and effects (Vorderer, Klimmt, & Ritterfeld, 2004). These different phases of viewing capture the unique and variable emotional states that may guide or result from the use of entertainment. This practice in entertainment research has spawned numerous studies and research projects; so many, in fact, that it is impossible to cover them all in the space of this chapter. Instead, two of these theories will be discussed, alongside the problems that arise when video games become the object of investigation.

Mood Management Theory

One way to explain the motivation of individuals to use entertainment products is given by mood management theory (Zillmann, 1988a, 1988b; Zillmann & Bryant, 1985). The theory considers individuals as hedonically motivated to place themselves in situations in which they amplify pleasure while ameliorating pain. Although the underlying conception of humans as beings who enjoy pleasurable experiences translates to situations other than entertainment, mass media situations represent a common practice individuals may employ to regulate mood states that are noxious and to maintain those that are satisfying (Oliver, 2003). Mood management theory asserts that one of the fundamental states that requires modification is physiological arousal. Overly stimulated or bored individuals may seek to use entertainment to reduce their arousal or stimulate their interest. Highly absorbing entertainment fare that is unrelated to an individual's current affective state can reduce stressful arousal levels because it is "likely to disrupt rehearsal processes that would perpetuate states of elevated arousal associated with negative affective experiences" (Zillmann, 1991, p. 109). Similarly, when faced with a choice, under-stimulated individuals turn to entertainment options that will increase their level of arousal to a "normal" degree (Bryant & Zillmann, 1984; Zillmann, 1991). Additionally, affective states can be regulated by media exposure. Individuals experiencing a negative mood turn to entertainment stimuli that will alleviate these feelings of sadness or upset, and instead provide feelings of joy or cheer (Oliver, 2003).

Video games, however, come as a challenge to mood management theory, because they provide a much different form of entertainment than traditional film or television. Games are very arousing, highly involving, and require the user to participate in the action instead of providing a distraction. Research on exactly which games are more stimulating to individuals and why is certainly needed. Also, perhaps certain other attributes of games—challenge, inter-personal activity while playing, or fast-paced action—have implications for how individuals can regulate their positive or negative affective states. Correspondingly, users may be forced to not only choose which content best suits their needs, but also which media will best modify their arousal level. Individuals seeking a reduction or increase in arousal have a host of options, and some might forego film or television in favor of games, or eschew their favorite games in favor of less stimulating material.

Affective Disposition Theory

Affective disposition theory (Zillmann, 1980, 1983, 1996; Raney, 2003; Raney & Bryant, 2002) represents a second element of the entertainment discipline, which is focused on the experience a viewer has as he or she is entertained. As a narrative unfolds, its central characters (are forced to) make choices. Disposition theory regards each of these choices as an opportunity for viewers to assess the moral valence of the characters—whether or not the characters' judgments are in line with an individual's own attitudes and beliefs. When characters act in a manner that is in line with a viewer's own opinions, the viewer hopes for positive outcomes and fears negative events for this character. Characters whose actions oppose the viewpoints of the individual are resented or disliked, and the viewer hopes for negative outcomes but fears that these characters will experience positive ends. At the end of the presentation, the resolution will be enjoyable if the desired outcomes are achieved (Zillmann, 2000).

Disposition theories, like mood management theory, are difficult to apply to video and computer games. According to Klimmt and Vorderer (2003), disposition theory considers the audience "to be passive witnesses of the ongoing events" (p. 351). However, as stated earlier, the active audience is an essential component of video and computer games—which require not only action, but also *interaction* with the medium for successful results. Additionally, the lynchpin of disposition theory is the moral judgments individuals form about the actions of the characters. The avatars in nearly all computer games may be evaluated much differently than traditional characters, as the player controls them, rather than viewing their progress as a "moral monitor" (Zillmann, 2000, p. 38). Moral judgments about these characters may not apply because they are behaving in user-controlled patterns. Although these characters may invite users to empathize with their situations and form affective connections, thereby improving the sense of presence an individual feels (Klimmt & Vorderer, 2003), avatars are a problematic application of the basic tenets of disposition theory. Disposition theory, in its current state, is challenged when it is expected to explain the process of enjoyment that users feel as they play a game because it does not account for the unscripted nature of the medium (see, however, Bryant & Davies, chap. 13 in this volume).

Overall, from an academic point of view, the situation has become most interesting: Al-though there has been an established body of research on traditional entertainment, including well-supported theories and a pallet of empirical findings, many questions involving "new media" such as video and computer games remain open. How will entertainment theory deal with issues of interactivity? How do video games regulate mood states or arousal levels? What kinds of judgments or dispositions do game players form with their avatars, and what impact does this have not only on enjoyment of the game, but also on feelings, cognitions, attitudes, and behaviors? These are just a few of the many questions that will need to be dealt with as video games continue to infiltrate our lives.

IN THIS VOLUME

This book brings together scholars from various disciplines and from different countries around the globe to provide answers to questions like those just mentioned. It is structured into four segments that deal with the games themselves, that is, their content and their history, users' motivation and selection processes, their responses to these games, and the consequences that playing them may have on the users.

Before dealing with these aspects in detail, we first approach computer games twice from a business perspective: In his foreword, Stacey Spiegel contemplates the scope of this volume from his background as a CEO and president of Immersion Studios, Inc., a company that develops games and other multimedia products. Michael Sellers, straddling the typically separate worlds of academia and game design, looks at the process of designing a game from an industry standpoint.

With respect to the games themselves, four chapters introduce and describe their most important features: Henry Lowood traces the history of computer games by providing a brief, but nevertheless most comprehensive, biography of computer games. The following two chapters content-analyze the most popular current games, that is, systematically describe what may be found in these. Barry Smith does this with a rather broad scope that serves very well as a general introduction. Stacy Smith is particularly interested in negative content patterns and character portrayals, such as perps, pimps, and provocative clothing. By analyzing their "weight," she addresses the eligibility of many public concerns expressed in recent years. The final chapter in this section deals with so-called massively multiplayer online games, which Elaine Chan and Peter Vorderer introduce as a new and quite different spin on computer games.

Following a rather established allocation of different phases of the entertainment process, section 2 of this book deals with the processes of motivation and selection, essentially asking, "Why do people play games?" Again, a number of very different perspectives, academic disciplines, theories, and paradigms are put together to illuminate these topics in all their complexity: Jesse Schell and Chris Klug lead off this section by providing answers that are dominant within the game industry. Very much in contrast to this approach, Peter Ohler and Gerhild Nieding take an evolutionary perspective on game play and selection, a stance that has become prominent in contemporary psychology. The following two chapters then use personality theory, motivational psychology, and communication theory to address the question of selection: In the first chapter by Tilo Hartmann and Christoph Klimmt, the role that personality factors play in the selection of computer games is addressed. In the subsequent chapter by Christoph Klimmt and Tilo Hartmann, effectance motivation and self-efficacy, in particular, account for the motivation and selection of specific games. Two further chapters in this section examine players of different ages: Maria von Salisch, Caroline Oppl, and Astrid Kristen explore why children, arguably the most vulnerable group of players, are attracted to games. And Arthur Raney, Jason Smith, and Kaysee Baker focus on adolescents, for whom games have become such an important factor in life. Jennings Bryant and John Davies unite these age variables and others to integrate them in their explanation of selective exposure to computer games.

While the preceding section deals with motivation and selection—that is, processes that occur before the actual entertainment experience—section 3 of this book focuses on reception and reaction processes, such that develop in a phase when somebody is in fact engaged in play and, hopefully, entertained. The section opens with a chapter by Dmitri Williams, who provides a social history of game play, asking, "How has the way we have played computer games changed over the years?" John Sherry, Kristen Lucas, Bradley Greenberg, and Ken Lachlan then take on a very popular research paradigm in communication to summarize what we know about the uses and gratifications of computer games. The four following chapters deal with specific aspects (i.e., features) of the entertainment experience that occur as a particular response to playing computer games: Ron Tamborini and Paul Skalski examine the role of presence (as a "sense of being there") in playing. Sean Zehnder and Scott Lipscomb do this in respect to the role of music in games. Kwan Min Lee, Namkee Park, and Seung-A Jin discuss the importance of narrative and of interactivity in games, and Michael Shapiro, Jorge Peña-Herborn, and Jeff Hancock summarize what is known about the relevance of realism and imagination in computer games. The final two chapters in this section address the playing of online games: Ann-Sofie Axelsson and Tim Regan discuss what it means to play online by examining *Asheron's Call*, and Francis Steen, Patricia Greenfield, Mari Siân Davies, and Brendesha Tynes pick another game, *The Sims Online*, to examine why, in contrast to its offline version (*The Sims*), the online version has failed so dramatically.

The final section of the book is devoted to the various effects and consequences playing computer games can have on their users. Again, the section starts with an overview of what is known in respect to the topic: Kwan Min Lee and Wei Peng summarize the social and psychological effects of computer games. Their chapter is followed by René Weber, Ute Ritterfeld, and Anna Kostygina, who introduce current theoretical positions on the effects of violent games, discuss various methodologies to investigate the short-term and long-term impacts of game playing, and report empirical findings. Katherine Buckley and Craig Anderson then elaborate and expand the most prominent theoretical model about the effects of violent games, namely the general aggression model (GAM), in their chapter. Whereas these chapters primarily focus on negative effects, the final chapters turn the question around and ask what can be and what is actually learned by playing games: Debra Lieberman gives an overview of various studies that show the educational potential of playing. Ute Ritterfeld and René Weber look particularly at the potential of interactivity for enjoyment and the enhancement of developmental processes while elaborating paradigms of entertainment–education. Finally, Kevin Durkin claims that adolescent users are "at risk" if they do not grow up with games.

This sheer quantity of ideas, assumptions, perspectives, theses, and research results is admittedly quite a reading load for anyone who is interested in the study of this new medium. Up to now, there have been no canonized research programs, undisputed theories, or conclusive findings concerning the uses, enjoyment, and consequences of playing electronic games. The field grows rapidly and diversely, driven primarily by the energy of junior scholars who try to come to terms with the entertainment medium that still persists from their childhood. No doubt, this field is still very much work in progress. The only thing that may already be stated is that computer games have become extremely important for people of different ages and cultures, and gender alike. Games are just about to become the most important entertainment product that people use for leisure. This book solicits, examines, and further encourages systematic research on playing computer games by approaching it from different disciplines and research paradigms, and from scholars around the globe, in order to account for the great complexity of this new phenomenon.

REFERENCES

Bryant, J., & Zillmann, D. (1984). Using television to alleviate boredom and stress: Selective exposure as a function of induced excitational states. *Journal of Broadcasting, 28*, 1–20.

Entertainment Software Association. (2004a). *Top ten industry facts.* Retrieved August 23, 2004 from http://www.theesa.com/pressroom.html.

Entertainment Software Association. (2004b). *Essential facts about the computer and video game industry.* Retrieved September 2, 2004 from http://www.theesa.com/pressroom.html.

Grodal, T. (2000). Video games and the pleasures of control. In D. Zillmann & P. Vorderer (Eds.), *Media entertainment: The psychology of its appeal* (pp. 197–214). Mahwah, NJ: Lawrence Erlbaum Associates.

International Digital Software Association. (2001). *Economic impacts of the demand for playing interactive entertainment software.* Washington, DC: Author.

Kaiser Family Foundation. (2003, Fall). *Zero to six: Electronic media in the lives of infants, toddlers and preschoolers.* Menlo Park, CA: Author.

Klimmt, C., & Vorderer, P. (2003). Media psychology "is not yet there": Introducing theories on media entertainment to the presence debate. *Presence, 12*, 346–359.

Oliver, M. B. (2003). Mood management and selective exposure. In J. Bryant, D. Roskos-Ewoldsen, & J. Cantor (Eds.), *Communication and emotion* (pp. 85–116). Mahwah, NJ: Lawrence Erlbaum Associates.

Raney, A. A. (2003). Disposition-based theories of enjoyment. In J. Bryant, D. Roskos-Ewoldsen, & J. Cantor (Eds.), *Communication and emotion* (pp. 61–84). Mahwah, NJ: Lawrence Erlbaum Associates.

Raney, A. A., & Bryant, J. (2002). Moral judgment and crime drama: An integrated theory of enjoyment. *Journal of Communication, 52*, 402–415.

Sherry, J., de Souza, R., Greenberg, B. S., & Lachlan, K. (n.d.). *Why do adolescents play video games? Developmental stages predict video game uses and gratifications, game preference, and amount of time spent in play.* Unpublished manuscript.

Vorderer, P. (2000). Interactive entertainment and beyond. In D. Zillmann & P. Vorderer (Eds.), *Media entertainment: The psychology of its appeal* (pp. 21–36). Mahwah, NJ: Lawrence Erlbaum Associates.

Vorderer, P., Klimmt, C., & Ritterfeld, U. (2004). Enjoyment: At the heart of media entertainment. *Communication Theory, 14*, 388–408.

Vorderer, P., Wulff, H. J., & Friedrichsen, M. (1996). *Suspense: Conceptualizations, theoretical analyses, and empirical explorations.* Mahwah, NJ: Lawrence Erlbaum Associates.

Zillmann, D. (1980). Anatomy of suspense. In P. H. Tannenbaum (Ed.), *The entertainment functions of television* (pp. 133–163). Hillsdale, NJ: Lawrence Erlbaum Associates.

Zillmann, D. (1983). Disparagement humor. In P. E. McGhee & J. H. Goldstein (Eds.), *Handbook of humor research: Vol. 1, basic issues* (pp. 85–107). New York: Springer-Verlag.

Zillmann, D. (1988a). Mood management: Using entertainment to full advantage. In L. Donohew, H. E. Sypher, & E. T. Higgins (Eds.), *Communication, social cognition, and affect* (pp. 147–171). Hillsdale, NJ: Lawrence Erlbaum Associates.

Zillmann, D. (1988b). Mood management through communication choices. *American Behavioral Scientist, 31*, 327–340.

Zillmann, D. (1991). Television viewing and physiological arousal. In J. Bryant & D. Zillmann (Eds.), *Responding to the screen: Reception and reaction processes* (pp. 103–133). Hillsdale, NJ: Lawrence Erlbaum Associates.

Zillmann, D. (1996). The psychology of suspense in dramatic exposition. In P. Vorderer, H. J. Wulff, & M. Friedrichsen (Eds.), *Suspense: Conceptualizations, theoretical analyses, and empirical explorations* (pp. 199–231). Mahwah, NJ: Lawrence Erlbaum Associates.

Zillmann, D. (2000). Humor and comedy. In D. Zillmann & P. Vorderer (Eds.), *Media entertainment: The psychology of its appeal* (pp. 37–58). Mahwah, NJ: Lawrence Erlbaum Associates.

Zillmann, D., & Bryant, J. (1985). Affect, mood, and emotion as determinants of selective exposure. In D. Zillmann & J. Bryant (Eds.), *Selective exposure to communication* (pp. 157–190). Hillsdale, NJ: Lawrence Erlbaum Associates.

2

Designing the Experience of Interactive Play

Michael Sellers

Online Alchemy, Inc.

Games occupy a unique place in human society. They are almost infinite in form and yet possess a singular quality that makes them immediately recognizable. They serve no clear practical purpose and yet are found in every culture and era. As computers have developed in our technological society, games have evolved right along with them—first reiterating classical games (chess, checkers, etc.) and then enabling new games that could not be played in any other form (from *Hunt the Wumpus* [1972] to *The Sims* [2000] and *Grand Theft Auto* [1997]). Computer games have changed as technology has advanced, but the game developer's essential task of designing the game play experience has remained fundamentally the same.

HOW GAMES ARE DIFFERENT FROM OTHER SOFTWARE

Unlike almost all other kinds of software such as enterprise or office products, presentation or tax software, games are peculiar in two important ways. The first of these is that no one has to play a game. An office worker faced with learning a new database program typically has little choice in the matter; but if a game fails to grab a person's attention, it fails as a product. Games are not *required* for anyone, and must therefore succeed on their own engaging qualities to attract and retain a person's interest.

This condition leads to a highly competitive landscape, where games compete in a Darwinian fashion for the attention of potential players—a scarce resource, especially as the number of games expands faster than the pool of likely players. Game developers today are driven by risk-averse financial necessity to design games that are similar to games that have been successful in the recent past, and by creative necessity to throw in innovations along the way. This creates a highly mutative, highly competitive environment where game developers are constantly trying out new ideas (but not *too* new) to get people to buy and play their games.

There is intense pressure in game development to create products that have more usable user interfaces, better graphics and sound, and overall provide a more appealing and satisfying game play experience than their competitors. This pressure exceeds that found in any other field of software development, because the feedback loop (revenue) is so directly linked to these attributes.

The second key difference between games and other forms of software is that in games, there is no set of user goals and tasks to model. In contrast, other forms of software exist to streamline, expand on, or augment a preexisting external set of these goals. In user-centered design (a prominent methodology for software development), identifying and understanding the users' goals and tasks is the central aspect around which all the rest of the product's development revolves. In game development, however, there is no such set of predefined goals or tasks. Instead, game developers are faced with coming up with a set of goals and tasks that is comprehensible, novel, appealing, and—most elusively of all—fun. The developers must create goals that the players understand and find sufficiently meaningful, and provide them the ability to navigate a conceptual landscape with appropriate controls that lead to some goal-state.

As predicted by the classic "inverted-U" Yerkes-Dodson arousal curve (Yerkes & Dodson, 1908; explored more fully in Silverman, 2001), this must be done in a way that maintains the user's psychological arousal in the area between boredom and overstimulation, between mechanistic tedium and the paralyzing anxiety and frustration of having too many options. As discussed below, creating the player's tasks effectively to maintain this level of arousal and engagement is the heart of designing the interactive play experience.

GAMES LEAD OTHER FORMS OF SOFTWARE DEVELOPMENT

The combination of these factors—the nature of games as discretionary items, having no external task to design to, and the highly competitive design environment—results in games emerging at the forefront of consumer product development. Games often lead the way in user interface design, artificial intelligence, asset and database organization, and the use of consumer computer hardware (e.g., graphics cards). New developments in industrial and other consumer software are often first seen in games several years earlier.

Despite (or perhaps because of) this leading position in many aspects of consumer product development, the actual game design process—by which the developers decide what the game play experience will be—remains intuitive and ad hoc, with little agreement between designers or teams on necessary and sufficient methods. This has begun to change in the past few years as developers have begun to gather a corpus of knowledge about what works in games, but still remains highly fragmented and idiosyncratic. In the remainder of this chapter, I review common aspects of game design and the development process and explore possible unifying abstractions that focus on defining interactivity from a psychosocial point of view.

COMMONALITIES IN COMPUTER GAME DESIGN

Despite the lack of a de facto accepted methodology for designing game play experiences, there are important commonalities that can be found across the computer games industry. These include distinct genres that have evolved and that now both inform and constrain game development, as well as high-level abstractions concerning the development of interactive play.

Understanding computer game genres will help to frame the discussion of design principles and the game development process.

Computer game genres are styles of game play and are important to understand as part of the design process. Each of these genres is fluid but has identifiable hallmarks. Much as with movie genres like "comedy" or "buddy picture," these classifications form useful abstractions for everyone from the initial designers to the product distributors and consumers. As a result, each genre also has its typical dynamics and design constraints. These inform and often limit everything from user interface conventions (e.g., key bindings in first-person shooters) to game play dynamics and aesthetics (e.g., the feeling of heroic accomplishment in role-playing games). The following discussion of genre types is not exclusive, but represents the majority of computer games developed over the past 20 years.

Puzzle games are typically the simplest games to pick up and play, and are aimed at the most casual players. These games present the player with one or more puzzles to be solved, often under time pressure or increasing complexity. These are often visually attractive, easy to learn, and require little time commitment. At their best, these games enable players to "surf" in a psychologically aroused state, often described as being "in the zone" or as the experience of *flow* (Csikszentmihalyi, 1990), a highly attractive state to which people will return again and again. Examples include *Tetris*, *Myst*, and *Bejeweled*.

Shooters (also known as "first-person shooters" or FPS games) have emerged in the last decade as one of the most exciting if violent types of games; examples include *Doom*, *Quake*, and *Unreal*. The typical interaction is to shoot at and kill multiple enemies in pursuit of some goal. The quality of the graphics in FPS games is of paramount importance, as the game play experience is primarily perceptually driven.

Role-playing games (RPGs) allow the player to take on the role of a hero on a quest to right a wrong or achieve a great destiny. These games are typically played from a graphical third-person perspective, as if looking down on the people represented in the game. The player's character fights monsters, retrieves treasure, and from time to time talks with computer-run characters (typically through predefined menus onscreen). Graphical quality is important in these games from an atmospheric point of view (as shown in *Diablo II* and *Dungeon Siege*, among many others), but does not trump story and game play in importance to the players.

An important but nearly vanished subgenre of the RPG is that of *interactive fiction*. In contrast to the other types of games described here, these games use only text as a display, often using sophisticated language parsers to make sense of the user's typed input. These games first became popular on college mainframe computers in the late 1970s and moved to personal computers in the 1980s. There is still a small but persistent community of players and developers devoted to this area, but it is no longer a major presence in the computer game industry. This type of game does illustrate the primacy of game play over graphics in this genre, however.

Sports games simulate known activities ranging from realistic professional baseball to car racing, fly-fishing, or snowboarding taken to the point of fantasy. These games are known for their attention to detail, scrupulous following of the rules (as set down in the sport itself), and faithful real-world physics.

Simulation games also attempt to faithfully simulate some real-world activity or an extension of some activity. Some, such as fighting flight simulators (e.g., *Crimson Skies*), are hybrids of this and other genres. Others such as the popular city-building game *SimCity*, the Tycoon games (*Rollercoaster Tycoon*, *Zoo Tycoon*, etc.), and Sim games (*SimTown*, *SimGolf*, *The Sims*) enable players to do something they could not do otherwise through the simulation of a familiar series of events in a compressed time and physical context. These games, while brightly graphical, succeed primarily on the cognitive engagement provided by the game play rather than on the perceptual elements of the graphics themselves.

Strategy games fall into two categories—turn-based and real-time strategy games. The latter now dominates the market and these games are often referred to simply as RTS games. The *Heroes of Might and Magic* series are some of the best-selling turn-based strategy games, while the *WarCraft* and *StarCraft* games represent some of the most financially successful real-time strategy games. The game play most often consists of using many units to explore the landscape, expand a controlled area, exploit resources to create new units, and exterminate enemies (for this reason, they are often called "4X" games—eXplore, eXpand, eXploit, eXterminate). As with simulation games, the primary game play elements here are cognitive and sometimes social; the graphics are important, but some of the most celebrated games in this genre (e.g., *Civilization*) have had minimal graphics and animation when compared to other games.

The last major genre of games today are *massively multiplayer games*, often abbreviated MMP games (or even with the unwieldy MMORPG—massively multiplayer online role-playing game). Examples include *Lineage*, *World of Warcraft*, and *Dark Age of Camelot*, which range from having about two hundred thousand to several million players each. These games are discussed in greater detail in chapter 6.

MMP games present the player with a persistent world that remains in place (and in play) after any individual player has ended his or her play session. Thousands of individual players may be found in this online world at any given time. The game play typically focuses on killing monsters, gathering treasures, completing (often formulaic) quests, and gaining power within the game. However, players also spend a great deal of time socializing with each other, crafting in-game items, creating communities within and outside the game, and exploring the game landscape.

The presence of many players in a persistent world sets the stage for social dynamics not seen in any other type of game or online community. This results in an entirely different experience for the developers and the players; developing an MMP game poses a series of design and production challenges that differ greatly from those found in other genres. These games more than any others combine the perceptual elements of FPS games, the cognitive aspects of strategy and simulation games, and provide a greater social component than do games in other genres.

Genres of computer games are not static, but do represent fundamental abstractions of forms of game play. It is significant that these genres have evolved empirically rather than from design principles: They exist because they have been shown to create games that people will play. As a result, while existing game genres are not a canon, they nevertheless drive and constrain game development from the beginning of conceptual design to the game's final polishing and testing. Additional information on game genres can be found in chapter 3 of this book.

COMMON DESIGN ABSTRACTIONS

Beyond aspects such as genre, point of view, and game play style, there are design abstractions that increasingly act as touchstones during the design and development process. While game developers differ in the formality or intuitiveness of their approach to game play design, a common vocabulary is beginning to emerge centered on these abstractions.

Among those gaining acceptance at a fundamental level is LeBlanc's "Mechanics–Dynamics–Aesthetics" or MDA framework (LeBlanc, 2004; Hunicke, 2004). In brief, this is a method for conceptualizing the different components of any game design into three different but interrelated parts.

The first of these, *mechanics*, refers to the specifics of what can happen in a game—what the pieces or actors in a game do. For example, the mechanics of chess defines how each piece can move: Rooks move only in straight lines, bishops only diagonally on their color, and so forth.

In a computer game, the mechanics define the rate of a vehicle's acceleration or the damage done by a sword.

A game's *dynamics* describes how the specific mechanics interact. In chess, two knights can form a pincer to trap another piece; and in *Tetris* the player can create higher-scoring combinations of pieces by placing the individual blocks carefully. The mechanics of a game are always defined explicitly and are thus entirely predictable. The dynamics rely on the effects of multiple pieces acting together, and so often result in unpredictable systemic behavior. When the dynamics are unforeseeable at first but obvious in retrospect; when the player can learn and manipulate them easily via the game's mechanics; and when they create perceptually pleasing, cognitively nonlinear, or socially valuable results, then the game play tends to be describable as engaging and fun. The tentative "tends to" disclaimer is necessary as even these factors may not always describe a sufficient set for creating enjoyment; there is still a large element of unexplored and unarticulated art in designing game play dynamics.

Finally, the *aesthetics* describe the reactions in the player evoked by the unpredictability arising from the dynamic interactions of specific objects in the game. A particularly subtle set of moves in chess can create a sense of wonder (or dismay for the opponent); pulling off a difficult trick and winning first place in a snowboard race in *SSX* can create the feeling of *fiero*, or personal triumph. Ultimately, game developers are concerned with the aesthetics their game creates, the psychological arousal and emotions it evokes in the player. But, just as movie directors work with scripts, actors, lights, camera angles, and so on to engender the reactions in the audience they are after, game developers must work through their game's mechanics and dynamics to create the desired aesthetic. For example, in the highly successful game *Prince of Persia: The Sands of Time*, a key mechanic is that the Prince can rewind time using his magical dagger. This creates the dynamic of enabling the player to avoid mistakes by seeing what does not work—albeit at a cost, since the dagger has limited uses. This and other dynamics (including how objects in the world can be used) help set up in the player a sense of excitement, wonder, foreboding, exploration, and even regret when he or she is unable to use the Prince's dagger in a difficult situation.

LINEARITY, INTERACTIVITY, AND EMERGENCE

Along with the MDA conceptual framework, the qualities of interactivity, linearity, and emergence must also be considered. These can be thought of as types of dynamics often considered by developers during the design process.

One common definition of interactivity used in game development states that a computer program (or any other device) can be said to be interactive if it:

- presents state information to the user,
- enables the user to take actions indirectly related to that state,
- changes state based on the user's action, and
- displays that new state.

Together the human user(s) and the device or program form an interactive system, each altering the others' behavior. For example, a thermostat, furnace, and person in combination can be considered to have rudimentary interactivity: The thermostat displays state (temperature), enables the user to change its setting, which indirectly turns on or off the furnace, which then in turn eventually changes the thermostat's displayed state. Setting a thermostat is not a game, however, as it lacks the enjoyment that comes from effective dynamics. Different types of interactivity and their accompanying aesthetics will be considered later in this chapter.

In a game, the user can typically view various aspects of the game state, make decisions about what to do next to achieve some goal, take actions via the user interface (mouse, keyboard, console controller, etc.) to enact those decisions, and, after the game processes this input, view the new state. Whether this is more enjoyable than adjusting a thermostat moves back into the realm of dynamics and aesthetics, and is what sets successful game designs apart from unsuccessful ones. In general, a game needs to provide the players with not just the ability to make and act on decisions but must provide *meaningful* decisions. Meaningfulness occurs in several contexts (described below), but in games is always tied to the player's perceived ability to interact with the game, to change an outcome by making decisions within the context of the game.

Interactivity can take many forms. In many games, the most basic of kind of interactivity is an almost completely linear experience: As in a book or movie, the game proceeds from A to B to C and the player is powerless to change this. Game developers speak of a game being "on rails" (like a train) to describe the fact that the game mechanics do not allow the player to deviate from a predefined path. While making a game completely linear in some aspects can work well (for example, in *Medal of Honor: Allied Assault*, the missions are presented in a predetermined order), too much linearity eliminates interactivity. Whenever an experience is linear, the player is robbed of any ability to make decisions; all the player can do is view the game's state. This leaves the player with mechanics (e.g., this gun shoots 10 rounds per second), but without any dynamics from which the desired aesthetics can be derived.

This highlights the essential difference between games and books or movies. Games are often compared to books or (especially) movies as parallel forms of popular entertainment, but they differ at a fundamental level: Books and movies are noninteractive in their content—nothing the reader or viewer does will change how the story ends. These are still compelling to us psychologically and socially, and our knowledge that the end has been predetermined does not reduce our enjoyment along the way. Games, on the other hand, are necessarily interactive experiences; those in which the outcome is perceived to be predetermined and in which the player's decisions have no effect quickly lose their appeal, no matter how strong the writing or the narrative arcs within them. This difference affects every aspect of the design of the interactive experience.

As a way to maintain some semblance of traditional story and to straddle the line between linearity and interactivity, many games have branching–linear plotlines. In these, the player is given a few opportunities to make meaningful decisions that affect the overall game, but the possible choices and their consequences have all been mapped out during design and development. The advantage of this design method is that it enables the developer to enforce a narrative arc on the game (albeit at the price of to some degree reducing the player to the status of passive viewer) and does not require subtlety in automated characters in the game; their responses can be entirely predetermined. Branching linearity can be successful if the game is sufficiently interactive at some level within the plot structure, so that the player maintains the feeling of having been able to make sufficiently meaningful decisions between forced branch-points.

The difficulties of linearity, even branching linearity where a number of branch points are presented to the player, make classical narrative storytelling difficult in games. This has set up an enduring conundrum for game developers: People enjoy stories, but stories are linear; linearity destroys game dynamics, and people typically do not enjoy linear games or interactive stories. Creating meaningful narrative structures with increasing tension, climax, and resolution within a satisfyingly interactive context—in which the player retains the ability to truly alter the course of events—remains an unsolved problem.

Beyond linearity (branching or otherwise) lies emergent game play. Emergence is to games what narrative is to fiction. Rather than propelling the game forward on a single track via artificial or prescripted structures, the developer provides the building blocks—the game mechanics—that the player can use to create meaningful dynamics with compelling and unexpected-but-welcome aesthetic effects. When such effects arise because of indirect mechanic or dynamic connections, this is called emergent game play. Emergence requires locally specified conditions and consequences, but not an overall plot or narrative direction (though in the best cases, the individual mechanics and dynamics combine to create an emergent directional flow, similar to the narrative flow in a story). Thus, persons playing *SimCity* will make decisions that, acting on the game mechanics and dynamics, determine the growth of their city. Each time they play, the city that emerges will be different. Similarly, players of *Grand Theft Auto* may have entirely different experiences based on their decisions and interactions each time they play the game. In either case, the game developers can predict only how individual pieces of the game will act and interact (the mechanics and dynamics), but not ultimately what the players will make of this (the aesthetics). The sequence of events that emerges becomes the narrative for that particular game within the player's mind, though it may not follow the path of a traditional linear plotline.

It is important to understand that these emergent narratives are often as satisfying for the individual as is a well-crafted, predetermined linear story. This suggests that the more we can embed nascent narrative structures within the mechanics and dynamics of a game without destroying its interactive and emergent properties, the more satisfying the aesthetic experience will be for the player.

It is worth noting that in massively multiplayer games the presence of hundreds or thousands of people in the same game space adds a new element to emergence—that of the other people and the social context. In effect, the players become part of the dynamics of the game for each other. This means that the degree of unexpected emergent consequence in MMP games far exceeds than in any other type of game; it is common for two or more innocuous mechanics to combine to create a dynamic with unforeseeable—sometimes undesirable—aesthetic experience for the players.

For example, in *Ultima Online* it was at one time possible to learn a skill simply by standing near someone who was using the skill: If you watched someone cook, you would become a better cook, too. And in this game, characters had a limited pool of skill they could attain: At some point their skill-gain became zero-sum, and they could not gain a skill without losing from one someplace else. The unfortunate consequence of these two mechanics is that a player, via his or her character, could *harm* another character by reducing their highest skill. That is, character A comes up near character B and begins cooking. Character B's cooking skill begins to rise due to the observation effect. But, because of the zero-sum nature of the skill cap, another skill—character B's highest and thus most desired skill—was reduced. So by the simple act of cooking near someone, you could reduce their ability to fight or work with magic in the game. This is only one of many similar dynamics based on underlying mechanics that did not result in the desired aesthetics within the players.

THE INTERACTIVE PLAY DESIGN PROCESS

To create new games, developers may take existing experiences and turn them into games (e.g., table tennis in *Pong* or developing a city in *SimCity*), or they may come up with entirely novel experiences that have no direct analog outside of the game (e.g., "dropping blocks" games such as *Tetris*). In either case, they must devise ways to make a set of tasks immediately attractive and appealing enough to hold the player's attention. Moreover, developers of online subscription-based games face an additional challenge: Along with creating a play experience that is novel and appealing, it must be one that retains its attractive quality over a period of months or years rather than the tens of hours typically considered for a game experience. This extension for online games adds an entirely new layer to game play design.

Idea and Concept

Game development typically starts with a specific idea. This may be a setting, a visual style, a particular game event the designer sees in his or her mind's eye, or an emotion or reaction the developer hopes to instill in the player. Alternatively, the design process may start with a licensed property to be developed as a game, or with various technological or business constraints that drive the developers in a particular design direction. However it begins, the idea of the game is only a seed. Although people become attached to their ideas, experienced designers know that great ideas are to be found everywhere; coming up with the core idea for a game is by far the easiest part of the development process.

Once the developers believe they have a viable seed for the game, they—typically a single designer or a small group working together—develop the high-level game concept. This often takes the form of a design treatment or overview, sometimes accompanied by early graphical mockups of what the designer envisions for the final product. This brief document outlines the type or genre (RTS, FPS, etc.) of the game, which specifies the overall style of play; the market opportunity (such commercial constraints are never far from the surface in professional game development); a few sketches of the user interface or conceptual graphics, including the player's visual perspective on the game (whether it is viewed from above or from a first-person perspective); and often a brief vignette or two outlining the kind of experience the player will have within the game. Ultimately it is the player's internal psychological experience that determines the success of a game: If this experience is satisfying, people will continue playing and will tell their friends about the game; if not, they will stop playing and tell their friends that, too. No matter the marketing budget or the strength of the license, a game that is simply not enjoyable is quickly lost under the tide of new games ready to wash over it.

From Concept to Design

After the overall concept is designed and approved by a publisher or studio executives, the developers begin work on the mechanics and dynamics of the game—the objects and actions in the game that will support the aesthetic outlined in the high-level concept document (note that many developers do not think explicitly in terms of mechanics, dynamics, and aesthetics, though this is still indicative of the thought process). Some game developers spend a great deal of time in the early macro/concept phase searching for the right aesthetic, the right experience that they want to communicate via the game. This is often considered in terms of the player's experience, phrased as the question, "What is the game?" or "What's fun about that?" Other developers rush right to the detailed micro design of the mechanics and use this as a way to determine the eventual aesthetic, trusting in their ability to create sufficiently interesting pieces that the aggregate will also be interesting to players. Still other developers take their intuitive feeling for how the pieces should interact—the game's dynamics—and use this to create both the mechanics and the aesthetics in the eventual game. No single path is accepted overall in the game industry, and none can be said to have a greater success rate than another. Depending on the strengths of the developers (as individuals and as teams), one of these approaches will be more likely to lead to an optimal set of mechanics and dynamics that create a satisfying, engaging, fun aesthetic for the player.

The risk inherent in this process, no matter how it is approached, is that the developer will not be able to converge on a viable game, but will instead cycle interminably, searching for the right combination of mechanics, dynamics, and aesthetics without ever finding it. This is an unfortunate but common fate for many game projects. At the present stage of evolution of the game development process, there are few tools for evaluating actual progress in the design process. The necessary but difficult evaluations of progress in a game's development typically still reduce to intuitive gut-checks by management or publishers who have to weigh the game developer's creativity and vision against their concomitant optimism.

From Design to Production and Deployment

Development of the game proceeds from the early idea and concept stage through preproduction and production. The first of these typically involves completing the specifics of the game mechanics and testing out a few crucial dynamics. Ideally in preproduction the team comes to understand what they are going to build, and in production they actually build it. The reality is typically far from this ideal, however, as the building of the game exposes previously unknown problems and design insufficiencies and may lead to better ideas than were conceived early on. Currently there is no single accepted game development process, though phases such as concept, preproduction, and production are used as general heuristics throughout the industry (Sellers, 2002).

Once a game has been developed to the point that most of its internal objects—including programmed mechanics and graphical representations—are in place, and when the developers are confident that most of the dynamics are behaving as they intend, the game is made available to a limited audience for beta testing. During this phase, nondeveloper players get their first chance to try out the game and the developers react as best they can to fill in holes, balance out-of-kilter mechanics, and generally hone the game so that it provides the desired aesthetic experience. Once the game is sufficiently playable (it provides the desired aesthetic experience)—or, alternatively, when the schedule or financing demands it—the game is released to the public.

EXISTING PRINCIPLES FOR INTERACTIVE GAME DESIGN

The current game development process does not rely on any theory of game play or game aesthetics. However, while there are few overarching principles of the design of the interactive game play experience, there are hints and outlines that have emerged from many different developers and researchers. Placing these in a psychosocial context leads to a deeper, more explanatory framework for understanding and developing games.

Caillois (1967) defined four categories or types of games: *alea*, games of chance; *agôn*, games of competition; *ilinx*, games of physical or sensory pleasure; and *mimicry*, games of fantasy and make-believe. In terms of the mechanics–dynamics–aesthetics framework discussed earlier, the first two of these relate to the game's mechanics and dynamics, while the latter two refer more significantly to the game's aesthetics. Caillois also included another dimension of *paidia*, or informal play, to *ludus*, or formal game. Again, these refer to both the mechanics— the formalisms involved in a particular game—and the aesthetic created in the player. For example, Frasca (1991) noted that *SimCity* is an example of *paidia* in that it is a game without a goal. While it has formal rules, it has no "victory condition" typically made explicit in *ludic* games. Costikyan (1994) echoed this in differentiating between software toys (Caillois'*paidia*) like *SimCity* and games (*ludus*) that have stated goals. The heart of Costikyan's definition of game play is that it consists of players making meaningful decisions in pursuit of a goal. If the game is trivial or linear or if there is no opposition, there are no decisions; and if there are no goals, the decisions are meaningless. This is a strongly cognitive (and, as Yee, 2002 pointed out, strongly male) view of game play, but it provides a necessary piece of the game play puzzle. In a complementary vein, Huizinga (1968) offered an extensive definition of games, a key part of which is the aesthetic view that a game involves "a feeling of tension, joy, and the consciousness that it is different from ordinary life."

In reference to online games, Bartle (1996) posited that there are four broad types of players in MUDs (Multi-User Dungeons), the text-based predecessors of current massively multiplayer online games: *killers*, *achievers*, *socializers*, and *explorers*. While these four types are broad, they provide explanatory archetypes that have helped drive and clarify the design of many online game designs. However, Yee (2002) found in a survey of over 6,700 players of several popular MMP games that Bartle's typology was insufficient to entirely explain actual player motivations. Yee added *leadership* as a prime motivation along with a more qualitative desire for *immersion* in the game world.

Two other contributors to a psychosocial view of game design have already been discussed: the Yerkes-Dodson concept of rising and then falling performance (and interest) as psychological arousal increases, and LeBlanc's mechanics–dynamics–aesthetics framework. In his framework, LeBlanc (2004) also posited "eight kinds of fun" as specific types of aesthetics: sensory pleasure; make-believe or fantasy; drama or narrative; challenges to be overcome; fellowship with others; exploration and discovery; self-discovery and expression; and submission, or a way to pass the time. These types of fun are not exclusive; for example, a competitive game (Caillois' *agôn*) offers a combination of challenge and fellowship experiences.

It is significant that the MDA framework enables game descriptions in terms of the player's experience. It is ultimately not the game mechanics or dynamics—the formality or informality, for example—that determine whether a game is enjoyable (and, from a commercial perspective, successful). These are the tools the game developer uses, but the players' enjoyment depends entirely on their internal psychological and emotional experience created or evoked by the game. At the same time, a range of psychological aspects beyond emotions or aesthetics— such as cognitive problem solving and goal orientation—must be included in any complete description of game play.

A PSYCHOSOCIAL DESCRIPTION OF INTERACTIVITY FOR GAME DESIGN

By returning to the view discussed earlier of the player and game as complementary parts of a whole system, with interactivity describing the communication between them, we can place the game play descriptors used above into a single psychosocial context. Considering game play as the combination of physical/perceptual, cognitive, social, and cultural experiences, each with its own aesthetic and emotional components, provides a psychosocial framework that illuminates an understanding of game play and supports the game play design process.

Perceptual and Physical Interactivity

As humans, we are evolutionarily built to react to physical stimuli and action, especially visual and aural perceptions. We are attracted to bright colors, flashes, moving images, rhythmic or explosive sounds, and to specific proportions in form and color. We are also predisposed to memorize and perform rhythmic or sequential physical actions such as dance steps or songs— the behavioral counterpart to rhythmic perceptions. These predilections are not a matter of conscious choice but of psychophysics; in this we are like cats with string or dogs with thrown balls: We cannot help but be attracted to stimuli or actions deemed relevant by our evolution. Each of these stimuli thus represents a perceptual hook that can be used as part of the game play experience. Repetitive or rhythmic actions create a similar action-oriented hook as seen in everything from rhythmic clapping games to specific memorized keyboard or mouse click sequences in many computer games.

It should be no surprise that many games make use of bright flashing colors, explosions and the like, as these are, in the right proportions and doses, almost irresistible to the human perceptual system (consider too noninteractive applications of these principles we tend to find enjoyable, such as the bright flashes of fireworks, often accompanied by music). Other games such as the popular *Myst* series offer a quieter experience of visual form and sound design that are pleasing to the eye and the ear. Finally, many games rely on repetitive or rhythmic action—usually in the hands in the case of computer games, though some such as *Dance Dance Revolution* involve the whole body.

The common aspect to all perceptual and physical-level interactivity is that of a short time horizon: These are pleasing only in the psychological present tense. This immediacy is related to the nature of perception and physical action itself. Cognitive interaction includes planning for the future and memories of the past, but perceptual and physical enjoyment lasts only so long as the bright color, musical tone, or proportional form is perceived or as long as the dance or song lasts. This is nevertheless one of the most accessible—and thus the most often used—forms of interactivity in computer games. The entire first-person shooter genre depends on players finding enjoyment in the fast-changing graphics, bright colors and explosions, and fast-beat musical soundtracks. In LeBlanc's (2004) terms, perceptually oriented game play creates the aesthetic of sense-pleasure; this also refers to Caillois' (1967) *ilinx* type of game play.

Short-Term Cognitive Interactivity

Adjacent in time to perceptual and physical interactivity is short-term cognitive activity. This incorporates attentional aspects such as short-term memory and accomplishment of a proximate task. In game terms, this includes solving puzzles and (in game-military terms) the short-term tactics that combine to form longer-term strategies. The common aspect here is that of attentional focus: The player must be able to comprehend and consider the entire puzzle or task at once in his or her mind, or must be able to follow a known sequence that exceeds the short-term store (as, for example, in completing a known series of dance steps or musical notes). The link to perceptual and physical interactivity can be seen in this dual aspect of short-term cognition: The player must be able to cognitively perceive the entire puzzle or task at once or must be able to follow the cognitive dance steps (literal or metaphorical) through a known sequence. As with the sensory and perceptual pleasure found in bright colors, finding enjoyment in such activities is not so much a matter of choice as cognitive ability fused with arousal and emotion. If a puzzle is not too tedious (too low arousal) or frustrating (requiring too high arousal or more cognitive resources than are available), it is almost certain to be perceived with aesthetic descriptors like challenging, dynamic, or compelling—each a type of enjoyment or, more simply, fun.

Long-Term Cognitive Interactivity

Beyond task-oriented short-term cognition lies goal-oriented longer-term cognition. This is where planning, strategy, considered decisions, and memory all come into play; for many this is the true *ludic* form of play. Achieving a goal may require stringing together many short-term tasks, each the product of a cognitively intensive decision. As with short-term cognitive interaction, if the planning and decision making for long-term cognitive interactivity is not too simple or repetitive (that is, too low arousal) or too complex and taxing of the user's cognitive resources (too high arousal), and in particular if the goals themselves are meaningful to the player in an emotional sense, then the player will likely find enjoyment in both the pursuit and accomplishment of the goals. In LeBlanc's (2004) terms, this enjoyment typically takes the form of satisfaction in goal accomplishment, overcoming a challenge, discovery, and self-expression.

Prior to this level of interactivity, a game may be seen as enjoyable in itself without reference to any external context; at the perceptual and short-term cognitive levels emotional attachment (the sense of enjoyment) does not require a significant cognitive or reflective component. Once the player's long-term cognition and goal selection is part of the game, meaningfulness (as determined by the player) is more important. Once significant planning, balancing, and strategizing are involved, the question of "why am I doing this" becomes more germane. This desire for a meaningful context may be created in many ways, such as with an elaborate background fiction, characters the player cares about, or by making reference to a real-world situation that is important to the player. This lattermost context may include the player's own capabilities; thus, games of skill, especially intellectual or cognitive skill, become ends in themselves. The cognitive nature of the game enables the player to increase his or her own cognitive abilities.

Most computer games draw on a combination of perceptual, short-term, and long-term cognitive interactivity for their game play. For example, *Myst* is perceptually beautiful and uses a linear series of linked task-puzzles to lead the player toward the accomplishment of the game's goal. In contrast, chess depends heavily on short- and long-term cognition, but typically does not rely on the perceptual component of the shape of the pieces or board to increase the player's enjoyment. Real-time strategy games such as *Age of Empires* and *Starcraft* present the player with a lush visual and aural landscape onto which is built a set a series of short- and long-term cognitive decisions. While there is one overall goal to such games (eliminate all competitors), there are innumerable approaches to this goal, making for a cognitively diverse— and thus more enjoyable—game play landscape than is found in a linear sequence. Finally, sandbox games such as *SimCity* that do not direct the player to any particular goal also provide rich perceptual and cognitive landscapes. Some players complain that the undirected nature of these games detracts from their enjoyment, as they do not provide a sufficiently goal-oriented framework. This may be viewed as the result of individual differences in preferred arousal level: It may be that players who require less cognitive arousal may be happier with an open, undirected experience, while those who thrive in higher-arousal environments may find greater enjoyment in contending with externally imposed challenges.

Social Interactivity

Historically, most computer games have been created for a single player interacting with the computer or game console, but an increasing number also include a social component. This enables a new form of interactivity that leads from the purely internal and psychological to the interpersonal and psychosocial, as players are now able to interact not only with the game but also with each other around and as part of the game. The social component becomes especially important in games where the players have a persistent identity and the ability to affect the game state together, as in massively multiplayer games. The persistence of identity enables long-term relationships to form, greatly enhancing social enjoyment ("fellowship" in LeBlanc's terms). Social game play often acts as a form of meta-game, surrounding and enhancing other perceptual and cognitive game play elements. It also creates new game dynamics and aesthetics as the other players become both part of the game play landscape (e.g., as competitors) and allies with complementary goals.

By providing "social scaffolding" (Kim, 1998) such as persistent identity, the ability for players to communicate easily with each other (both synchronously and asynchronously—via chat and message boards, for example), and the ability to form their own ad hoc and permanent groups, game developers greatly increase the players' potential for fellowship and building social enjoyment (Sellers, 2002). This is a particularly strong form of game-related enjoyment for many people, especially women (Yee, 2002).

Unlike perceptual or cognitive interactivity, social interactivity takes place on a time scale of hours, days, weeks, or even years. The short-term interactivity may involve cooperation or competition in a goal-context, but the longer baseline interactivity involves relationships built over longer periods of time. Games that adequately support these longer-term relationships are both more commercially successful and provide greater social enjoyment (fellowship, self-expression, and new forms of group-related narrative) than do those that limit socialization to the in-the-moment social interactions.

Cultural Interactivity

On the far end of the spectrum of interactivity is the largely unexplored area of cultural interactivity. This form of game play involves giving the player a new historical or cultural perspective or articulating previously tacit cultural knowledge. Cultural interactivity has the longest baseline in terms of time, typically occurring over the course of many hours of play. It leads to subtler forms of enjoyment than the other types of interactivity—self-discovery, discovery of the world, and a combined reconsideration of one's place in history and culture. In many ways this can be the most powerful form of enjoyment, the kind of fun that does not fade as quickly as perceptual or short-term puzzle-based interactivity. The aesthetics associated with cultural interactivity might be achieved via combinations of the other types of game play that provide a previously unknown perspective, such as in the life of a family in *The Sims*, a city in *SimCity*, a group of pioneers in *Oregon Trail*, or an ancient civilization in *Age of Empires*. Our knowledge of how to enable the aesthetics of wonder and discovery in the player is still rudimentary. As our understanding of these grows we should see more games evoking these emotions and the sense of fulfilling enjoyment that goes with them.

CONCLUSION

The heart and soul of interactive play is the human experience while playing. Games fill no practical need, but we play them anyway. As discussed in this chapter, the enjoyment from a game may be the ephemeral but visceral perceptual enjoyment of bright colors or fast movement; the cognitive thrill of overcoming obstacles and being victorious over a challenge (or challenger); the sense of community and fellowship derived from sharing experiences with others; or the sense of wonder and discovery at learning more about one's place in the larger scheme of things.

While many think of games primarily in terms of solving puzzles or reaching goals, the various forms of enjoyment inevitably touch us emotionally. Designing the interactive game play experience requires successfully evoking the desired psychosocial experiences through the careful application of game mechanics and dynamics. By creating a conceptual landscape of emotionally and aesthetically meaningful decisions rather than relying on an unchangeable predetermined narrative line, the game developer is able to create an enjoyable, psychologically satisfying experience unlike any other.

REFERENCES

Bartle, R. (1996). *Hearts, clubs, diamonds, spades: Players who suit MUDs*. Retrieved May 5, 2004 from http://www.mud.co.uk/richard/hcds.htm.

Caillois, R. (1967). *Les jeux et les hommes. Le masque et le vertige*. Paris: Gallimard.

Costikyan, G. (1994). I have no words and I must design. *Interactive Fantasy #2*. Retrieved May 5, 2004 from http://www.costik.com/nowords.html.

Csikszentmihalyi, M. (1990). *Flow: The psychology of optimal experience*. New York: Harper & Row.

Frasca, G. (1991). *Videogames of the oppressed: Videogames as a means for critical thinking and debate*. Masters thesis, Georgia Institute of Technology. Retrieved May 5, 2004 from http://www.ludology.org/articles/thesis/FrascaThesisVideogames.pdf.

Hunicke, R., LeBlanc, M., & Zubek, R. (2004). MDA: A formal approach to game design and game research. *Proceedings of the AAAI Workshop on Challenges in Game AI*. Tech Report WW-04-04. Menlo Park, CA: AAAI Press.

Huizinga, J. (1968). *Homo Ludens*. Buenos Aires: Emecé Editores.

Kim, A. J. (1998). 9 *Timeless principles for building community*, Retrieved August 8, 2005 from http://www.Webtechniques.com/archives/1998/01/kim/

LeBlanc, M. (2004). *Game design and tuning workshop*, Game Developers Conference 2004. Retrieved May 5, 2004 from http://algorithmancy.8kindsoffun.com/GDC2004/.

Sellers, M. (2002, September). *Creating effective groups and group roles in MMP games*. Retrieved May 5, 2004 from http://www.gamasutra.com/resource_guide/20020916/sellers_01.htm.

Sellers, M. (2003). The stages of game development. In F. Laramee (Ed.), *Secrets of the game business* (section 4.1). Hingham, MA: Charles River Media.

Silverman, B. (2001). *More realistic human behavior models for agents in virtual worlds: Emotion, stress, and value ontologies* (Technical Report). Retrieved May 5, 2004 from http://www.seas.upenn.edu/~barryg/TechRpt.pdf.

Yee, N. (2002). *Facets: 5 motivation factors for why people play MMORPGs*. Retrieved May 5, 2004 from http://www.nickyee.com/facets/home.html.

Yerkes, R., & Dodson, J. (1908). The relation of strength of stimulus to rapidity of habit-formation. *Journal of Comparative Neurology and Psychology, 18*, 459–482.

THE PRODUCT

3

A Brief Biography
of Computer Games

Henry Lowood
Stanford University

Biography provides an interesting metaphor for a survey of computer game history. After all, the computer game was born in the early 1960s, so it has only just entered middle age. Even if we located the origins of the computer game near the invention of electronic computing, its entire history would easily fit within a human life span.

This biography of the computer game is rooted in an observation made by Loftus and Loftus more than 20 years ago: "Video games are fundamentally different from all other games in history because of the computer technology that underlies them" (Loftus & Loftus, 1983, pp. ix–x). They referred to the nature of games as interactive software. In framing the history of the computer game as a history of technology, it is nevertheless crucial to consider the roles of game culture and design; game design is discussed in greater detail in chapter 2. The modifiability of game software provides a historical thread that encompasses these themes, leading from early games such as *Spacewar!* to "mods" such as *Half-Life: Counterstrike*.

THE DOOM REVOLUTION

This is the first game to really exploit the power of LANs and modems to their full potential. In 1993, we fully expect to be the number one cause of decreased productivity in businesses around the world. (id Software, 1993)

Released in December 1993, *DOOM* immediately left its imprint on almost every aspect of computer gaming, from graphics and networking technology to styles of play, notions of authorship, and public scrutiny of content. The development team, led by John Romero and John Carmack, had designed games for *Softdisk* magazine and Apogee Software, notably the *Commander Keen* series. Carmack's first significant programming achievement, a demonstration of a smooth, side-scrolling platform game called "Dangerous Dave in Copyright Infringement," led to the formation of id Software in February 1991. Other than the protagonist, "Danger-ous Dave" reproduced the first level of Nintendo's flagship game, *Super Mario Brothers 3*. More than homage, it demonstrated that personal computers could meet or possibly exceed the graphics produced by video game consoles.

Indeed, during the 1990s, personal computers, not proprietary game consoles, paced the progress of graphical game engines. Id's *Catacomb 3-D*, completed in 1991, provided another milestone as their first texture-mapped action game, and in 1992, *Wolfenstein 3-D,* another action title published by Apogee as shareware, showcased the improvement of id's graphics technology. (Origin's *Ultima Underworld* brought 3-D graphics technology to role-playing games at about the same time.) *Wolfenstein 3-D* set the stage for *DOOM* to define the game genre now known as the first-person shooter (FPS). *DOOM* added improvements such as a superior graphics engine, fast peer-to-peer networking for multiplayer gaming, a modular design that encouraged independent authors to create new levels, and a new mode of competitive play that Romero called "death match." *DOOM* established competitive multiplayer gaming as the leading-edge category of PC games (Kushner, 2003; *Book of id,* 1996). Its subject matter (slaughtering demons in outer space), moody graphics and audio, and vocabulary ("shooters," "death match") also called public attention to the levels of violence depicted in computer games.

DOOM stimulated a new model of game development. According to its own corporate history:

> The team of innovators also made DOOM's source code available to their fan base, encouraging would-be game designers to modify the game and create their own levels, or "mods." Fans were free to distribute their mods of the game, as long as the updates were offered free of charge to other enthusiasts. The mod community took off, giving the game seemingly eternal life on the Internet. (id Software, n.d.)

Inspired by programming hacks that altered games, such as the "Barney patch" for Silas Warner's original *Castle Wolfenstein* (Muse Software, 1981), Carmack had often altered games he played. He built modifiability into *DOOM* by simplifying the process. He did this by separating the core "game engine" from the code for specific "levels" of the game defined by maps, objects, monsters, graphics, sound, and so on. Level-specific information was captured in wad files, which were loaded separately into the game to play these levels; editing or creating wad files changed a game's content without hacking at the game engine. This mechanism spawned independent and third-party level design, and encouraged the development of software tools to make this new content.[1] Manovich suggested that *DOOM* spawned a new "cultural economy" of game design, in which "hacking and adding to the game became its essential part" (Manovich, 1998). Games become design tools as much as finished designs, as independent level, scenario, and mod designers begin creating their own modified versions ("mods") almost as soon as the original game ships. The popularity of *Counter-Strike,* a competitive, multiplayer modification of *Half-Life* (itself based on the *Quake* engine), shows how mainstream the mod-based economy of game design has become. Manovich contrasted modifiable games to the more traditional characteristics of a game like *Myst,* "more similar to a traditional artwork than to a piece of software: something to behold and admire, rather than take apart and modify" (Manovich, 1998). This "letting go of authorial control" (Packer, 1998) is a distinguishing characteristic of modern computer games, especially in comparison to other artistic or entertainment media.

It is tempting to date the birth of the modern computer game at *DOOM's* release from id's womb. It established modes of authorship, production, and distribution for computer games, at least on PC platforms. Certainly, id embraced this notion of *DOOM's* historical impact, proclaiming "on December 10, 1993, id unleashed *DOOM* on the world. A technically stunning opus of heart-stopping action, unspeakable horror and pure gaming bliss, *DOOM* heralded a paradigm shift in video games" (id Software, n.d.). As developers of games defined by player actions rather than narrative content, the loss of authorial control hardly troubled Carmack and Co.; rather, *DOOM* established technology as id's foundation for game development; they encouraged the player community to modify their games. Romero viewed *DOOM* as necessarily an "open" game during its development, "because of Wolf3D [*Wolfenstein 3-D*]— people figured out how to make maps for it without our help, plus change all the graphics, etc. and we were so impressed that we knew that *DOOM* just 'had' to be modifyable [sic]. That is the real reason" (Romero, 1997). Scott Miller of Apogee (and 3D Realms) cited *Wolfenstein 3-D* and his own *Duke Nukem* (1991) as games that introduced the "hacking and proliferation of user levels," thus making it imperative that *DOOM* be "easy to modify at the start." For Miller, later mods such as *Counter-Strike* merely raised "the importance to a new level" (Miller, 2001). Carmack's position is even more pointed: "There is not a hell of a lot of difference between what the best designer in the world produces, and what quite a few reasonably clued in players would produce at this point" (Carmack, 2002a).

Id followed *Doom* with *Quake,* released in June 1996; it preserved *DOOM*'s modes of competitive play, thus establishing the FPS as a genre. *Quake* was a technological tour-de-force. Its built-in client/server networking stimulated the popularity of Internet-based multiplayer games, and it offered Carmack's first genuinely 3-D graphics engine, optimized by Michael Abrash. The complexity of creating fully 3-dimensional game levels might have daunted many enthusiasts, but id provided information and tools to help them. During development of the game, Carmack described a new editing tool, QuakeEd, and disseminated some source code (Carmack, 1996). A Usenet discussion group, rec.games.computer.Quake.editing, was launched in January 1996, and the participants filled it with discussion of how *Quake*'s levels could be edited and the game modified. The full retail version included the QuakeC script programming language, along with its source code. Access to QuakeC, QuakeEd, and other software tools created by *Quake* players and programmers made it easier to create new "skins" or textures, program "bots" (robot opponents controlled by computer AI), design new levels, and even develop new games (such as "team fortress" and "capture the flag" modes of play). *Quake*, though far more complex than *DOOM*, became even more accessible as an arena for demonstrating programming as well as playing skills.

DOOM and *Quake* introduced a new culture of computer game design. "Communities of networked gamers" competed online, influenced the development of games, and acquired a linked-up life as virtual communities. The impact of *DOOM* and *Quake* has been cast in various lights. King and Borland wrote, "the networked age of gamers had begun in earnest" (2003, p. 116). Miklaucic asserted that "*DOOM* redefined gaming, virtually overnight," by spawning "an entire subculture" (Miklaucic, 2002). According to Kushner, the independently developed *DOOM* Editor Utility was "a watershed in the evolution of the participatory culture of mod making." He added that the player community, by making tools like the DEU, transformed "game players into game makers" and this culture of community authorship, beginning with *DOOM,* offered a "radical idea not only for games but, really, for any type of media" (Kushner, 2002, pp. 71–72). Raymond, writing for the Open Source software community, concluded that, "*Doom*'s life cycle... may be coming to typify that of applications software in today's code ecology" (Raymond, 1999, 10.3–10.4). Schleiner suggested that *DOOM* and *Quake* realized the "many to many" notions of cultural production that have influenced cable television and other media since the 1960s (Schleiner, 1999).

Did the modern technology and culture of computer game design begin with id's first-person shooters? Similar historical questions could be asked about cyberculture and new media generally. For example, "virtual communities" have transformed the computer from the mainframe of cultural repression into a symbol for wired freedom of expression, experimentation, and open collaboration. Turner has argued that their ideologies and practices echoed "habits of mind" established decades earlier (Turner, 2002). The biography of the modern computer game, including an assessment of id's status as "one of [its] fathers" (id Software, n.d.), also requires a closer look at its history.

FAMILY HISTORY: PLAYING GAMES ON COMPUTERS

> Of all the toys that are machines and that work by themselves and can be enjoyed in solitude for endless periods of time, the apotheosis is undoubtedly today's video game. The "video game" is an automaton that might have made Descartes shout with delight. (Sutton-Smith, 1986, pp. 61–62)

The idea of playing games on computers is about as old as the computer itself because games often inform the study of computation and computer technology. For example, Shannon proposed in 1950 that computers could be programmed to play chess in order to explore whether a machine "would be capable of 'thinking.'" His goal was not the chess-playing machine—"of no importance in itself"—but "techniques that can be used for more practical applications" (Shannon, 1950, p. 48). His essay stimulated decades of research on chess- and checkers-playing programs in the field of artificial intelligence. Shannon's notion of the machine as player could be traced back to Wolfgang von Kempelen (1734–1804) and his "Turk," a chess-playing contrivance presented as a life-sized mechanical model in Turkish garb (Windisch, 1783; Standage, 2001). Before World War I, the Spanish engineer Leonardo Torres y Quevedo (1852–1936) constructed an electromechanical automaton that could play a perfect endgame ("Torres . . . ", 1915; Randell, 1982). Following Shannon's essay, research through the 1960s considered strategy games such as chess and checkers as laboratories for investigations of weightier computer science problems (Newell et al., 1958; Samuel, 1959). These projects encouraged tolerance for play with a "serious purpose" (Shannon, 1950, p. 48) in computer science laboratories. The first generation of games were thus born and incubated largely at universities or in industrial laboratories.

Anthropologists and philosophers of play have described games as expressing an "interdependence of games and culture." (Caillois, 1961, p. 82). Huizinga (1938) depicted this "play-element" as historically and culturally specific. McLuhan traced this social element of play to the "simultaneous participation of many people in some significant pattern of their own corporate lives" (McLuhan, 1964, p. 245). Sutton-Smith has put games in the historical context of providing tools for rehearsing adaptive problems faced by a society. He applied this framework of analysis to the topic of electronic and computer games by reasoning that, for contemporary society, the "adaptive problem to which the video game is a response is the computer" (Sutton-Smith, 1986, p. 64). For Sutton-Smith, computer games function as a rehearsal for life and work in a world dominated by information technology and demanding "training to understand how to read a computer and even more to understand how to make it give the answers one wants." Game play overcomes the anxiety associated with achieving mastery over the computer as a "machine which in many respects is more intelligent than human beings" (Sutton-Smith, 1986, p. 65). In these critical writings, play as real-world mastery of the machine became something more than a game in a wider social context, reinforcing the respectability of computer games in laboratories that associated them with programs of computer research.

The project often considered the first interactive, electronic game was assembled at one of the centers of Cold War technology. William ("Willy") Higinbotham, head of the Brookhaven

National Laboratory's Instrumentation Division, created *Tennis for Two* for the laboratory's open house exhibit in 1958 from an analog computer, control boxes, and an oscilloscope display. Higinbotham was a physicist who had worked in the Manhattan Project during World War II and at the national laboratory in Los Alamos. Having witnessed the first test detonation of the atomic bomb and founded Brookhaven's own nuclear safeguards group in the 1960s, he sympathized with anxiety about nuclear research and probably created *Tennis for Two* to suggest an alternative, perhaps less threatening realm for the work of laboratory scientists (Schwarz, 1990). He felt that "people were not much interested in static exhibits," and that a "hands-on display" such as an interactive game would provide an enjoyable alternative ("Video Games," n.d.). David Ahl, who later founded *Creative Computing* magazine, remembered that "hundreds of students saw it and went away with the idea that in addition to doing thousands of statistical calculations in a remarkably short time, computers could also be fun" (Ahl, 1983). Yet, despite this enthusiasm, the technology of *Tennis for Two* was never patented and, dismissed as a "lab curiosity" by later inventors such as Ralph Baer (1996?), apparently had little or no impact on future work.

PARENTS: EARLY COMPUTER GAMES

Spacewar!

> Not surprisingly, the first computer wargames were developed unofficially by students at universities specializing in computer research.... Space War was the ancestor of all PC based wargames. (Dunnigan, 2000, p. 237)

> If, when walking down the halls of MIT, you should happen to hear strange cries of "No! No! Turn! Fire! ARRRGGGHHH!!," do not be alarmed. Another western is not being filmed— MIT students and others are merely participating in a new sport, SPACEWAR! (Edwards & Graetz, 1962, p. 2)

Steve Russell, Alan Kotok, J. Martin Graetz, and others at the Massachusetts Institute of Technology created *Spacewar!* as a demonstration program in 1962. M.I.T.'s new PDP-1 minicomputer and the Precision CRT Display Type 30, donated by Digital Equipment Corporation (DEC), appealed to the "hacker" culture of M.I.T.'s Tech Model Railroad Club (TMRC), including the *Spacewar!* authors. They were unimpressed by previous "little pattern-generating programs" that were "not a very good demonstration." So Russell's group reasoned that with this computing power and display technology, they could make a "two-dimensional maneuvering sort of thing." They concluded that "naturally the obvious thing to do was spaceships" (Brand, quoting Russell, 1972). They believed that a good demo program "should involve the onlooker in a pleasurable and active way—in short, it should be a game" (Graetz, 1983). Playful programs had been written in the lab before *Spacewar!*, such as Tic-Tac-Toe or Prof. Marvin Minsky's "Tri-Pos: Three Position Display" for the PDP-1, better known as the Minskytron. The *Spacewar!* collaborators wrote software and built control boxes so that players could maneuver virtual spaceships and fire at opponents against a background of black, empty space and a few bright stars shown on the CRT.

Spacewar! demonstrated the technical mastery of programmers and hardware hackers by producing a popular and competitive game available in any U.S. computer science laboratory of the 1960s and 1970s. It drew upon the popularity of the emerging genre of science fiction, especially the serialized novels of Edward Elmer ("Doc") Smith's *Lensman* series, reprinted in book form by Fantasy Press in the late 1940s and through the 1950s. This was the Space Age, and it is not surprising that a fan like Russell would place his game in a setting reminiscent of these novels. Smith excelled at action, with spaceships blasting away at each other, so Russell's homage became a fast-paced shoot-'em-up game. Setting the game in outer space meant that a black visual backdrop with a few flickering stars sufficed, easy to program and render graphically. As Russell noted, "by picking a world which people weren't familiar with, we could alter a number of parameters of the world in the interests of making a good game and of making it possible to get it onto a computer" (Brand, 1972). Dan Edwards' gravity calculations were a realistic feature, but the game's "photon torpedoes" ignored its attraction (thus easing the computational task) and "hyperspace" jumps allowed players to move instantly to a random location.

The game was modified soon and often. Peter Samson coded "Expensive Planetarium" to portray stars in the night sky more accurately; Edwards improved gravity algorithms; and so on. The game superbly showcased the lab's new computer, stimulating experimentation with its graphics, I/O, and display technology. Edwards told the emerging PDP users community shortly after *Spacewar!* was unveiled that the "use of switches to control apparent motion of displayed objects amply demonstrates the real-time capabilities of the PDP-1." He could "verify an excellent performance" (Edwards & Graetz, 1962). A new configuration of real-time processing power, control hardware, and graphical display had been assembled for the game.

A community of programmers and players formed around *Spacewar!* In his reportage of the 1972 *Spacewar!* Olympics competition at Stanford University, Brand described players with sharp competitive skills, "brandishing control buttons in triumph" (Brand, 1972, photo caption) after winning the tournament and achieving renown. Public competition signaled the dawn of the cyberathlete, but at the same time, *Spacewar!* grew out of an unstructured development process, a vast collaboration of programmers who added significant elements to the game or merely tweaked settings and controls—"within weeks of its invention *Spacewar* was spreading across the country to other computer research centers, who began adding their own wrinkles" (Brand, 1972). Performance was not limited to game play, but included displays of technical mastery, such as a superior programming trick or impressive feature. *Spacewar!* established computer game performance as a convergence of competitive skill, programming wizardry, and the formation of player communities.

Adventure

Nelson (2001) has called computer games emerging during the 1970s from research laboratories "university games." This term encompasses both the technical interests embedded in these games and the institutional setting of their creators. Like *Spacewar!*, Willv Crowther's *ADVENT*[2] (henceforth: *Adventure*) sprang from the network—both technical and institutional—of computer science and engineering. As part of the software team at Bolt Beranek & Newman (BBN) that in 1969 had built the first packet-switching Interface Message Processor (IMP) under a contract from the Advanced Research Projects Agency, Crowther figured prominently in laying down a fundamental piece of the ARPANET infrastructure. Like other members of BBN's carefully assembled group, he was a crack programmer. The setting was a "kind of hybrid version of Harvard and M.I.T." and among corporate labs, it was "the cognac of the research business, very distilled" (Abbate, 1999, quoting Bob Kahn, p. 57).

Programmers at BBN shared an intellectual and engineering culture with the M.I.T. hackers that joined programming skill, laboratory life, and enthusiasm for games and fantasy worlds. As with *Spacewar!*, a fantasy setting inspired Crowther's game. This was the fantasy role-playing game, *Dungeons and Dragons (D&D)*. Crowther played in a regular *D&D* game with several BBN and Boston-area computer programmers, starting soon after these "rules for fantastic medieval wargames campaigns" were issued (Gygax & Arneson, 1974). In this long-running game, the players role-played a series of adventures inspired by J. R. R. Tolkien's *Lord of the*

Rings trilogy. The game's leader, or Dungeon Master, was Eric Roberts, a Harvard student recruited to the IMP team by another player, Dave Walden. Roberts had "dungeonmastered up a dungeon and a bunch of us from the project team got sucked into playing" (Koster, 2002, quoting Sandy Morton). Roberts carefully chronicled the proceedings (Roberts, 1977). His account of their *Mirkwood Tales* describes a distinctive style of play, with the group diverging from *D&D's* origins in historical miniatures or "hack-and-slash" adventuring based on combat. They preferred to solve imaginative plot puzzles and role-play cleverly with the objects and locations in the game.

Game studies have paid little attention to the ways in which "paper gamers, as they would be known after the rise of the computer age, served very much as prototypes for the kinds of digital communities that would come later" (King & Borland, 2003, p. 5). Even though Crowther's *Adventure* was not a role-playing game, many of its elements can be tracked back to the *Mirkwood Tales,* an example of how games "playable with paper and pencil" (Gygax & Arneson, 1974, title) inspired game programmers. Crowther began work on his computer game after the group "had been playing *D&D* for a few months" (Koster, 2002, quoting Sandy Morton). The original version has apparently been lost. He wrote the FORTRAN program on a DEC PDP-10 and most likely completed it in 1975, releasing *Adventure* to the small ARPANET community by early 1976 (Montfort, 2003; Nelson, 2001; Hafner & Lyon, 1996, p. 206).

Adventure was not an action game; unlike *Spacewar!*, it did not require fast graphics displays or special controllers. Rather, it became the prototype for a narrative genre of games. Players revealed scripted story lines by typing responses to text generated and displayed by the computer program; hence, these games were called text adventures. They moved through the virtual game space by reading descriptions of "rooms" they occupied, then typing simple instructions at the keyboard that could be understood by the software "parser," the set of routines that translated phrases such as "go north" or "pick up lantern." Descriptive details about the locations were often based on Crowther's personal experiences as a caver, adding to the delight of exploration the knowledge that players could match specific rooms in *Adventure*'s Colossal Cave to real locations in the Mammoth Cave system. *Adventure* was played by completing specific tasks through movement, actions, and puzzle-solving, always expressed as text. As a "vehicle for the delivery of fictional texts" (Atkins, 2003, p. 7), the tension between "fixity" of narrative and freedom of movement through the game world was carried by another term often applied to such games: *interactive* fiction.

Adventure, born in the development lab of the ARPANET, grew alongside it. Arguably the preeminent coder at BBN, Crowther's work on the IMP software contributed significantly to the success of the early ARPA network (Hafner & Lyon, 1996, pp. 108–114). ARPANET-connected researchers, graduate students, and programmers distributed and played *Adventure*. In 1976, Don Woods, a graduate student in the Stanford Artificial Intelligence Laboratory (SAIL), found the source code on the lab's computer. He thoroughly revised the game, adding elements that reoriented the cave crawl into a magical fantasy world. Wood's popular revision became the canonical version of the game and its variants were distributed as an open guest account on the SAIL computer or as free software distributed by the DEC Users Group (DECUS). *Adventure* became ubiquitous on the growing network of university-based computer laboratories:

> I remember being fascinated by this game when John McCarthy showed it to me in 1977. I started with no clues about the purpose of the game or what I should do; just the computer's comment that I was at the end of a forest road facing a small brick building. Little by little, the game revealed its secrets, just as its designers had cleverly plotted. What a thrill it was when I first got past the green snake! Clearly the game was potentially addictive, so I forced myself to stop playing— reasoning that it was great fun, sure, but traditional computer science research is great fun too, possibly even more so. (Knuth, 2002)

Some programmers and players, like Woods, sought to improve on *Adventure*. One group, which included players from Roberts' *D&D* group, "played ADVENT, liked it, wished it were better, and tried to do a 'better' one" (Koster, 2002, quoting Mark Blank). They wrote *Zork* for the PDP-10 in 1977 while still at M.I.T. This group, with several M.I.T. professors and students, founded Infocom 2 years later to market a version of *Zork* for home computers; this company became one of the leading computer game developers of the early 1980s. *Adventure* and other text adventures such as Infocom's wildly successful *Zork* series reached the height of their popularity by the mid-1980s.

The linkage between game authorship and the network of computer science laboratories was not limited to M.I.T., the Boston area (with BBN and M.I.T.), Stanford University, or even the ARPANET. Other active centers of game development could be found in the PLATO (Programmed Logic for Automatic Teaching Operations) Project founded in 1962 and head-quartered at the University of Illinois, where multiplayer, social and graphics-enhanced games such as *Empire* and *Mines of Moria* were developed (Woolley, 1994), as well as several English universities, such as the University of Essex, home of the first Multi-User Dungeon (MUD).

While it emerged from a similar environment for computing, the impact of *Adventure* on game design differed from that of *Spacewar!* The difference was more than the contrast of competitive action versus narrative. *Adventure* was also significant in presenting a richer virtual world than *Spacewar!*, one that could be explored in different ways determined by players. Also unlike *Spacewar!*, the keyboard and text interface of *Adventure* was identical to that used for communicating in new computer networks such as the ARPANET. Perhaps partly for this reason it inspired dial-up and networked games that combined *Adventure*-like exploration of virtual spaces with social interaction. Persistent virtual worlds such as the original MUD, developed by Bartle and Trubshaw at the University of Essex in late 1978 and early 1979 (Bartle, 2003), along with its many offshoots, created environments for performative play within a community of player–actors. Made possible by the modem, hundreds of themed multiplayer MUDs, MOOs (MUDS, Object-Oriented), and BBS-based games were written and played throughout the 1980s and 1990s. These games underwent constant and extensive modification, as players attained the required status and developed skills for creating intricate rooms that could be showed off to other players (Dibbell, 1998). *Adventure* defined a kind of computer game flexible enough to redefine the computer program itself as a game *space*, both for play and as something to be studied, taken apart, and changed by the player community.

Families of computer games descended from *Spacewar!* and *Adventure*. Both games grew out of institutions that defined networked computing during the 1960s and 1970s, thus establishing a contextual relationship between exploratory work in computer science and the emergence of computer games. Computer games rose out of the very institutions that defined the networked computer—M.I.T., BBN, the University of Utah, and Stanford. The networked and graphics-based games of the PLATO Project, close ties between Infocom and M.I.T., and programming projects by graduate students and researchers such as SHRDLU and Eliza, refined and extended this relationship—one that would never again be so close. By the mid-1970s, the emergence of the video game console from television technology and consumer-oriented industries focused on commercial exploitation transformed the institutional matrix for game development. Nonetheless, the connection between the development of computer technology and the first computer games, particularly as demonstrations of the computer's capabilities, recalls Sutton-Smith's claim that games are fundamentally "problems in adaptation," specifically, that computer games address the adaptive problem that "is the computer" (Sutton-Smith, 1986, p. 64).

CHAPTER 2. BIRTH: EXPANSION, COMPETITION, CRASH, CONTROL

And there was another Atari logo, and another, and still more, a brand-new complex of over a dozen buildings, large lawns, fresh instant landscaping, discreetly small logos, street numbers, and that crystalline silicon sunlight. Laidback, sophisticated, nouveau riche residences for the PAC-MAN family, sleek red and terra-cotta buildings with lots of glass and pitched tile roofs, retinas themselves for all I knew, snazzy industrial homes for Space Invaders, Asteroids, Tempest, and their brothers. (Sudnow, 1983, p. 90)

Like computing generally, computer games during the 1970s broke out of the laboratory and computer center into the living room and study. Technical, social, and cultural factors that launched the home and personal computer also produced game machines and software. Advances in microelectronics and component miniaturization fostered expectations of affordable computers and electronic devices. Innovation in software design such as prototypes of graphical user interfaces and other productivity applications redefined computers from calculating machines to technologies of personal productivity, communication, and entertainment. The MITS Altair 8800, generally considered the first microcomputer, made its debut as the cover story of *Popular Electronics* in January 1975. Hobbyists responded by sharing information about the new technology in groups such as the Homebrew Computer Club in Palo Alto, California. Its members were "were intensely interested in getting computers into their homes to study, to play with, to create with," and they were willing to build the hardware to do it (Levy, 1984, p. 202).

Ted Nelson's *Computer Lib/Dream Machines* (1974) had already envisioned that "computer liberation" would bring computing power to the masses who "can and must understand computers NOW" (Nelson, 1974, title page). He recognized that these "versatile gizmos" had been "turned to any purpose, in any style," including games, noting that "wherever there are graphic displays, there is usually a version of the game *Spacewar*." Noting that "games with computer programs are universally enjoyed in the computer community, Nelson insisted that "computers belong to all mankind." He foresaw that wider access to computer games was a means to achieving this vision (Nelson, 1974, pp. 3, 48). His favorite examples included John Horton Conway's popular *Game of Life* and BASIC recreational and educational games published by the People's Computer Company to bring programming power to the people.

The transition from games available within the research network of computing to an accessible entertainment and educational medium did not happen all at once. Revolutionary computer hardware that could be sited in the home was not enough; the community of programmers and players needed to grow dramatically. In the mid-1970s, an unprecedented number of hobbyist programmers were introduced to easily mastered programming languages, particularly BASIC, and they usually honed their skills by programming games. The popular and influential *Hunt the Wumpus* emerged from this scene for game design and illustrates its importance. Originally programmed by Gregory Yob, it appeared in the inaugural issue of *People's Computer Company* (1973), founded by Robert Albrecht to promote use of computers by hobbyists, children, and others who might benefit from its educational and recreational potential. John Kemeny and Thomas Kurtz had created BASIC (Beginner's All-Purpose Symbolic Instruction Code) in 1964 as a general-purpose, high-level programming language whose relative ease of use would encourage students to program without making them learn all the intricacies of computers. Albrecht realized that games attracted the previously uninitiated to programming, and BASIC provided a satisfactory beginner's language. He realized that games made good demos of programming power, and every issue of *PCC* included the source code for BASIC games.

Albrecht also organized a walk-in computing center in Menlo Park, California, not far from the Stanford University campus. In 1973, Yob happened by the center and noticed that several simple BASIC games featured the same topology, a flat 10×10 grid. He wondered about the possibility of a "topological computer game," that is, a more imaginative layout of the virtual game space. His solution was a simple explore-and-hunt game taking place in a "squashed dodecahedron." Players explored this space—an arrangement of "rooms"—by inputting simple commands such as room numbers from their keyboards. He programmed it immediately for free use at the PCC center. About a month later, while attending a Synergy conference on the Stanford campus, "where many of the far-out folk were gathered to share their visions of improving the world," Yob realized that *Hunt the Wumpus* was being played on every computer monitor in the room. He had "spawned a hit computer game!!!" (Yob, 1976). Yob's *Hunt the Wumpus* introduced countless programmers to the notion of defining a virtual space by coordinates or simple room numbers. Games like this could easily be shared, modified, and extended by programmers, resulting in a great variety of games based on similar designs. Variants could be found for any computer system of the 1970s. For example, Kenneth Thompson, developer of UNIX at the Bell Telephone Laboratories, wrote one version in C shortly after he had used Dennis Ritchie's new programming language to rewrite the operating system in 1973 (Ritchie, 2001), and the *UNIX Programmer's Manual* included a listing for WUMP, while noting under bugs that "it will never replace Space War" (Bell Telephone Laboratories, 1979, section 6, n.p.). Even so, *Hunt the Wumpus* spearheaded a generation of simple games that brought game programming to the people.

Just as the people's computing movement was getting underway, the microcomputer revolution opened up access to computing technology. Dozens of companies emerged between 1975 and 1977 to manufacture microcomputers. Apple Computer was the most significant with respect to the future of computer gaming. Steve Jobs and Stephen Wozniak founded Apple in 1976 to sell Wozniak's elegantly designed Apple I microcomputer, and they launched its successor, the Apple II, at the first West Coast Computer Fair in 1977. Wozniak had previously designed a successful game, *Breakout,* for Atari. Apple's home computer was nothing less than a *Breakout* machine, with features such as color graphics, sound, and paddle support built into the Apple II; Wozniak acknowledged that many of the features "that made the Apple II stand out in its day came from a game" and the "fun features . . . were built in . . . only to do one pet project, which was to program a BASIC version of *Breakout* and show it off to the [Homebrew Computer] Club" (quoted in Connick, 1986, p. 24). The Apple II became a leading platform for grassroots game programming through the early 1980s; its alumni developed commercial games and founded game publishers. Scott Adams, for example, popularized "adventure games" for microcomputers such as the Apple II after learning about *Adventure* at the Canadian Computer Conference in 1977 (Adams, 1979). Home computers provided a significant market for commercial game software through the 1980s.

Emergence of the Video Game Console

It is tempting to conclude from *Spacewar!* and *Adventure* that computer games emerged whole out of laboratories and research centers. This conclusion ignores, however, the significant role of consumer product development in the definition and development of the dedicated game console. The technical efforts of television engineers established proprietary console designs as delivery mechanisms for location- and home-based interactive entertainment, not as openly accessible software programs, but as closed boxes for proprietary computer games. This is not to say that the early development of arcade and home console systems was disconnected from research spaces. As we have seen, *Spacewar!* was available in virtually any computer science laboratory of the 1970s, such as the University of Utah, home of a strong program in computer graphics. Nolan Bushnell began to play *Spacewar!* as a student in electrical engineering there. After graduating, he moved to California to work for Ampex Corporation, near Stanford University and the active *Spacewar!* players in the Stanford Artificial Intelligence Laboratory (SAIL), the same lab in which Don Woods would discover *Adventure*. Bushnell had worked in an amusement park as a student, and he recognized that *Spacewar!* presaged a new form of entertainment arcade filled with computer games. He may have known of *The Galaxy Game*, undertaken by a recent SAIL graduate, Bill Pitts, and his high school friend, Hugh Tuck. This coin-operated game was a version of *Spacewar!* for the PDP-11 installed in a custom cabinet. Constructed in 1971, this *Spacewar!* console was installed in the Stanford student union, and a later version remained there until 1979 (Pitts, 1997).

Bushnell, joined by Ampex coworker Ted Dabney, drew upon his knowledge of arcade machines in designing *Computer Space* (1971), essentially a coin-operated version of *Spacewar!* Nutting Associates manufactured 1,500 of the consoles in wildly futuristic fiberglass cabinets. Due to the complexity of its interface and game play, *Computer Space* was a commercial failure, yet Bushnell had established a design format and configuration for video game arcade consoles. Bushnell and Dabney soon severed their relationship with Nutting and founded a new company, incorporated as Atari Corporation in June 1972. Al Alcorn, another talented engineer from Ampex, joined them. Bushnell assigned him the task of designing a simple electronic game based on Ping-Pong. Alcorn rapidly produced a prototype from an off-the-shelf television set, a homemade cabinet, and several tricks from his bag of analog and television engineering tricks. Unable to attract interest from manufacturers of pinball games such as Bally, Bushnell and Alcorn designed their own coin-operated version of the game, named *Pong*. Installed in in a local bar, it was an immediate success. In order to begin volume production of *Pong*, however, Atari first cleared legal hurdles caused by Magnavox's hold on Ralph Baer's video game patent (about which more below) and Bushnell's attendance at a demonstration of the new home system based on this patent.

Bushnell and Alcorn exemplified the intersection of computer science and television engineering as converging technical paths to the modern computer game. Alongside computers and coin-operated arcades, television also stimulated development of computer games. Ralph Baer, a television engineer and manager of consumer product development at the military electronics firm Sanders Associates, personified its influence. Since the 1950s, he had been intrigued by the idea of interactive television as a way of increasing its educational value. By September 1966, he had work out several ideas for a technology he described as "low cost data entry devices" enabling an operator to "communicate with a monochrome or color TV set of standard, commercial unmodified type" (Baer, 1996). He designed circuitry to display and control moving dots on a television screen, followed by the simple chase game *Fox and Hounds* in 1967. Following this success, Sanders management gave Baer permission and funding to assemble a small development team, the TV Game Project. Within a year, he established fundamental design parameters for home video game consoles and demonstrated several rudimentary games. By early 1969, Baer's group completed the Brown Box, a solid-state prototype for a video game console. Two years later, Baer applied for the U.S. patent on a "television gaming apparatus" that would be granted in 1973, with rights assigned to Sanders. Magnavox Corporation licensed the technology from Sanders and in May 1972 began production of the first home video console, the "Odyssey Home Entertainment System."

Takeoff and Crash of the Video Game Industry

Pong and the Odyssey inspired numerous imitators, both arcade and home systems. Atari led by creating Atari *Pong*, a home version designed by Alcorn, Harold Lee, and Bob Brown. It was released in 1975 and sold by Sears under its Tele-Games brand. The popularity of *Pong* home consoles established an important synergy between arcade and home systems for Atari. Its phenomenal success also led to brutal competition as more companies entered the market and released new home and arcade systems. Some manufacturers followed the Odyssey's model by offering flexibility in the choice of games. Unlike *Pong*, these new consoles were platforms for playing multiple cartridge-based games. Atari released its 2600 VCS (Video Computer System) in 1977. The coin-operated arcade business depended on exclusive distribution of hardwired games playable only on dedicated machines; home consoles such as the market-leading VCS were programmable in that software contained in a game cartridge's read-only memory (ROM) could be read after insertion into special slots and then executed by the system's processing unit. The separation of game development from hardware manufacturing symbolized by the game cartridge stimulated a boom in demand for new games through the early 1980s. Activision, founded in 1979 by former Atari game designers, became the first third-party game publisher, followed by a rush of competitors.

Rudimentary action games dominated the title lists of arcade and home consoles circa 1980. Display technologies, microprocessors, and other components of the time limited designers, but quick, repetitive games also swallowed more quarters or, in the case of home consoles, could be manufactured cheaply and run reliably on underpowered hardware. While the designs of unqualified hits such as *Breakout* (1976) or Taito's *Space Invaders* (1978) were elegantly streamlined, most of the early console games offered little in terms of strategic and narrative depth. By 1983, competition, overreliance on knockoff imitations of proven hits, and a flood of weak game titles depressed the arcade and home console markets. The disastrous Christmas 1982 release of Atari's *ET* for the 2600 was the beginning of the crash and shakeout. Companies such as Mattel, Coleco, and Magnavox dropped out of the industry; Atari began a long decline, never again leading the industry. Software manufacturers also suffered; as Chris Crawford, then at Atari, put it, "The dozens of opportunistic cartridge publishers that had sprouted like weeds in 1982 died just as quickly in 1983" (Crawford, 1991). The details of the industry crash are well documented (Herman, 2001, pp. 89–98; Kent, 2001, pp. 234–240, 252–255; Sheff, 1999, pp. 150–157). Their significance lies in the lessons learned by the survivors.

Those survivors learned not only from the demise of the Atari generation, but also from its successes. The greatest was arguably *Pac-Man*, released by Japanese arcade game manufacturer Namco in 1980. The lead designer was Toru Iwatani, eager to find an alternative to the implied violence of earlier hits such as *Space Invaders*. By careful attention to concept, design, and color, Namco tried to create an arcade game that appealed to women and girls. The concept was inspired by food and eating, not shooting as in most arcade games. Players used joysticks to maneuver through a simple maze, gobbling colored dots—and occasionally devouring or being devoured by a gang of colored "ghosts"—until all were devoured and a fresh maze appeared. In Japanese slang, *paku paku* describes a mouth snapping open and shut, and thus the central character was given the name *Pac-Man*. It became the most popular arcade game in terms of unit sales, with more than 100,000 consoles sold in the United States alone. Its impact on popular culture was unprecedented, due largely to Iwatani's innovative design. Players discovered that the ghosts moved in patterns, became obsessed with devising precise routes for Pac-Man to follow, but were thwarted in this quest by the vast number of game levels. Guidebooks to playing *Pac-Man* became best-sellers, followed by popular songs, cartoon shows, merchandise, and magazine articles as well as versions or imitations for every electronic gaming platform. *Pac-Man* represented the game as tightly controlled intellectual property, the product of a closed industrial design studio, a toy to be played—not toyed with.[3] Given the failure of most crashed hardware manufacturers to control the quality of software cartridges playable on their machines, the next generation of companies would carefully guard their technology platforms and intellectual property.

The Nintendo Generation: Control of the Product

By 1985, this new generation was led from Japan. On the heels of the commercial collapse of the Atari generation, Nintendo released its video console, the Famicom (Family Computer), in Japan in 1983, followed by the Nintendo Entertainment System, a U.S. version of this system, in 1985. Its notable features included improved graphics processing supplied by Nintendo's MMC (Memory Map Controller) chip and the provision of battery-powered backup storage in the game cartridge. The NES and its followers, such as the Sega Genesis (1989), equaled or exceeded contemporary home or personal computers as game machines. Above all, Nintendo had learned how to control its platform and product, deploying technical, legal, and business measures that restrained access by independent software developers to its cartridge-based console. For example, Nintendo vigorously protected its patent on the cartridge device to restrict game software developers from publishing compatible cartridges without its explicit permission. It also insisted on a high level of quality control, both for titles developed in-house, such as Shigeru Miyamoto's *Super Mario Brothers* (1985) and *The Legend of Zelda* (1986, U.S. version, 1987), and third-party titles.

Published by Nintendo for the Japanese market in 1986 and in the United States one year later, *The Legend of Zelda* justified heightened expectations for video games. It exemplified the new technology, design aspirations, and business culture. Miyamoto was already Nintendo's star designer, having produced *Donkey Kong* and the *Mario Bros.* series. Now he added open-ended exploration by giving players a large fantasy world in which to find their own path for the story and main character, Link. It exploited several capabilities of the NES, such as the graphical rendering of a navigable, 2-dimensional world and the ability to use backup storage as one progressed through the lengthy game. Miyamoto also added interface elements such as screens that were activated to manage the hero's items and abilities, much like the pull-down menus then beginning to appear in business software applications. Finally, Miyamoto paid equally careful attention to the pacing and complexity of the game, so that players attained requisite abilities before progressing to increasingly difficult challenges. Miyamoto raised expectations for the narrative scope and game mechanics of a new generation of video games, qualities that encouraged comparisons to other narrative media such as cinema.

Nintendo was not the only game studio built on high production qualities, carefully guarded intellectual property, and *auteur* designers. LucasArts, founded in 1982 as LucasFilm Games by *Star Wars* filmmaker George Lucas, added interactive media such as game software to his multifaceted vision for the future of entertainment technology. Beginning with *Ballblazer* and *Rescue on Factalus!* in 1984, LucasArts established itself as a leading publisher of adventure games with strong stories, memorable characters, and vivid worlds, such as *Maniac Mansion* (1987), the *Secret of Monkey Island* (1987) and *Grim Fandango* (1998), as well as games from the worlds of *Star Wars* and *Indiana Jones*. Electronic Arts, founded in 1982, was inspired by United Artists Pictures, the Hollywood "company built by the stars." Just as UA had promoted independent production, Electronic Arts initially left game development to established designers and programmers while gaining control over publishing of its games in a manner that "deliberately [emulated] the music recording industry in producing and marketing its computer software" (Duberman, 1983, p. 63). The elements of its business success during the 1980s included strong branding, sports licensing, distinctive packaging and marketing, and control of the distribution network. Electronic Arts became strong enough to challenge the strict licensing requirements of Nintendo and Sega for access to their consoles, most notably by reverse engineering the 16-bit Sega Genesis and facing down Sega management to secure more favorable terms. Like Sega and Nintendo, its formula for success depended on control of its product.

CONCLUSION

Bigger critters than Atari have bitten the dust; bigger industries than ours have shriveled and died. Size and past success are no guarantee of permanence. We need substantive reasons for confidence in the future rather than simple extrapolations of past history. I am convinced that substantive reasons for optimism exist. . . . For now let me say that computer games satisfy a fundamental desire for active recreation, and as such are assured of a bright future. (Crawford, 1982 p. 76).

Convergences of technology between the latest video game consoles, such as the Sony Playstation 2 (2000) and Microsoft Xbox (2001), and personal computers with high-end graphics cards and peripherals are misleading. These convergences mask historical differences in the development and business cultures of these platforms for computer games, one built on the Nintendo model of corporate control and the other on the *DOOM* economy of mod-makers and player-generated content.

John Carmack has noted that *DOOM* was a "really significant inflection point for things, because all of a sudden the world was vivid enough that normal people could look at a game and understand what computer gamers were excited about" (Carmack, 2002b). This chapter began with *DOOM* as the beginning of the modern computer game, not only as a technical achievement, but also as the springboard for networked player communities, modifiable content, and other characteristics associated with PC-based computer games since the mid-1990s. The parents of the first generation of computer games—*Spacewar!* and *Adventure*—expressed similar qualities in a rather different milieu. The success of these open "university games" led to commercialization, rapid expansion, and cutthroat competition, particularly in the arcade and home console businesses. After the Atari generation crashed, savvy successors such as Nintendo, Sega, and Electronic Arts applied lessons learned about technology, marketing, and content. The result was an expanded market for a tightly controlled product, but with the loss of close connections with the open community based in research laboratories or the "people's computing" movement.

In this framework of technology, culture, and business, the relationship of the new culture of computer gaming after *DOOM* to its past shape-shifts between two dominant images: the offspring that reveals the characteristics of its ancestors (*Spacewar!, Adventure*) and the petulant child that breaks with its parents (the Nintendo/Electronic Arts generation). These rather different relationships between present and past in post-*DOOM* game culture are not exclusive. It is possible to be part of something new and yet share a bond with the past; the history of modern computing offers examples. In *Hackers* (1984), Levy proposed described a loose set of values he called the Hacker Ethic, which he traced historically in three phases from M.I.T. through People's Computing and the Microcomputer Revolution, and then to the computer game industry of the early 1980s. Hacker culture emphasized sharing, openness, decentralization, and other qualities similar to post-*DOOM* game culture, such as "access to computers should be unlimited and total" or "you can create art and beauty on a computer" (Levy, 1984, pp. 27–33). Carmack, author of the Nintendo homage that transformed computer game culture, knew about Levy's book: "I read that as a teenager. At that third section I was like 'Goddammit, I should be here!' Then about 10 years later, I thought back about it: 'You know, if there was a fourth section in that book, maybe I would be in there!' That's a nice thought" (Carmack, 2000).

NOTES

[1] However, the original version of *DOOM* was not released as open-source software and, in fact, id's initial stance toward editing of the game code was not quite as encouraging as it has often been depicted. Id issued a "Data Utility

License" that allowed modification of the game software under strictly defined conditions. With the release of *DOOM II* in 1994, Romero released more information about the game program. Carmack released the *DOOM* source code as a Christmas present to the player community in December 1997.

[2]ADVENT was the file name for the program. It was more often called *The Colossal Cave Adventure* or simply, *Adventure*.

[3]It was possible in principle to alter an arcade game like *Pac-Man*, through daughter-boards, for example, as the history of *Ms. Pac-Man* demonstrates.

REFERENCES

Abbate, J. (1999). *Inventing the Internet*. Cambridge, MA: MIT Press.

Adams, S. (1979). An adventure in small computer game simulation. *Creative Computing*, August, 90–97.

Ahl, D. H. (1983). Editorial. *Creative Computing: Video & Arcade Games, 1*(1). Retrieved March 2004 from Phaze's Classic Videogame Magazine Museum Web site, at http://cvmm.vg-network.com/vag1.htm.

Atkins, B. (2003). *More than a game: The computer game as fictional form*. Manchester, UK and New York: Manchester University Press.

Baer, R. H. (1966). Conceptual material—Conceptual, TV gaming display. Notes and typescript, dated September 2, 1966, retrieved April 2004 from Ralph Baer, "Video Game History," at http://www.ralphbaer.com/video_game_history.htm.

Baer, R. H. (1996?). Who did it first? E-mail contribution to Pong-Story Web site. Retrieved May 2004 from http://www.pong-story.com/inventor.htm.

Bartle, R. A. (2003). MUD1 collection notes. Papers of Richard A. Bartle, Department of Special Collections, Stanford University Libraries.

Bell Telephone Laboratories. (1979). UNIX[TM] time-sharing system. *UNIX programmer's manual* (7th ed., vol. 1). Murray Hill, NJ: Bell Telephone Laboratories.

The Book of id. (1996). Mesquite, Texas: id Software. This booklet was issued as part of the "id Anthology."

Brand, S. (1972). SPACEWAR: Fanatic life and symbolic death among the computer bums. *Rolling Stone*, December 7, 1972. Retrieved May 2004 from http://www.wheels.org/spacewar/stone/rolling_stone.html.

Caillois, R. (1961). *Man, play, and games*. New York: Free Press of Glencoe. Trans. Meyer Barash, from the French ed. of 1958.

Carmack, J. (1996). *Quake* editing tools information. E-mail dated April 5, 1996 to Bernd Kreimeier. Retrieved December 2003, at http://www.gamers.org/dEngine/quake/QuakeEd/qedit_info.html.

Carmack, J. (2000). John Carmack interview, February 9, 2000. Retrieved June 2004 from FiringSquad Web site, at http://www.firingsquad.com/features/carmack/page3.asp.

Carmack, J. (2002a). Re: Definitions of terms. Discussion post to Slashdot, January 2, 2002. Retrieved January 2004, at http://slashdot.org/comments.pl?sid=25551&cid=2775698.

Carmack, J. (2002b). "DOOM 3: The Legacy." Video released at the Electronic Entertainment Exposition, 2002. Transcript retrieved June 2004 from the New *DOOM* Web site, at http://www.newdoom.com/interviews.php?i=d3video.

Connick, J. (1986). And then there was Apple. Call-A.P.P.L.E., October, p. 24. Pt. 2 of the transcript for Steve Wozniak's speech to the Apple World meeting, San Francisco, January 1986.

Crawford, C. (1982). The art of computer game design. On-line version retrieved January 2004, at http://www.mindsim.com/MindSim/corporate/artCGD.pdf.

Crawford, C. (1991) The history of computer games: The Atari years. *The Journal of Computer Game Design, 5*. Online version retrieved May 2004, at http://www.erasmatazz.com/library/JCGD_Volume_5/The_Atari_Years.html.

Dibbell, J. (1998). *My tiny life: Crime and passion in a virtual world*. New York: Henry Holt.

Duberman, D. (1983). Artistry in electronic Gaming: The "star" system pays off. *Antic, 2*(6), September, 63.

Dunnigan, J. F. (2000). *The complete wargames handbook: How to play and design commercial and professional wargames*. (3rd ed.). San Jose, CA: Writers Club Press.

Edwards, D. J., & Graetz, J. M. (1962). PDP-1 Plays at Spacewar. *Decuscope, 1*(1) 2–4.

Graetz, J. M. (1983). The origin of Spacewar! *Creative Computing, 1*(1), 78–85.

Gygax, G., & Arneson, D. (1974). *Dungeons and Dragons: Rules for fantastic medieval wargames campaigns playable with paper and pencil and miniature figures*. Lake Geneva, WI: Tactical Studies Rules.

Hafner, K., & Lyon, M. (1996). *Where wizards stay up late: The origins of the Internet*. New York: Simon & Schuster.

Herman, L. (2001). *Phoenix: The fall and rise of videogames* (3rd ed). Springfield, NJ: Rolenta Press.

Huizinga, J. (1938). *Homo Ludens: Proeve eener bepaling van het spelelement der cultuur*. Haarlem: Willink.

Id Software. (n.d.). Id Software Backgrounder. (n.d.). Retrieved February 2004 from id Software Web site, at http://www.idsoftware.com/business/home/history/.

Id Software (1993). DOOM press release. Retrieved December 2003 from "Lee Killough's Legendary DOOM Archive," preserved by John Romero as part of the "Planet Rome.ro" Web site, at http://www.rome.ro/lee_killough/history/*DOOM*pr3.shtml.

Kent, S. L. (2001). *The ultimate history of video games: From Pong to Pokémon and beyond: The story behind the craze that touched our lives and changed the world.* Roseville, CA: Prima.

King, B., & Borland J. (2003). *Dungeons and dreamers: The rise of computer game culture from geek to chic.* New York: McGraw-Hill/Osborne.

Knuth, D. (2002). Adventure. Retrieved March 2004 from Literate Programming Web site, at http://www.literate-programming.com/fexamples.html.

Koster, R. (2002). Online world timeline. 2002 update. Retrieved March 2004, at http://www.legendmud.org/raph/gaming/mudtimeline.html.

Kushner, D. (2002). The mod squad. *Popular Science, 261*(2), 68–72.

Kushner, D. (2003). *Masters of DOOM: How two guys created an empire and transformed pop culture.* New York: Random House.

Levy, S. (1984). *Hackers: Heroes of the computer revolution.* Garden City, NY: Anchor Press/Doubleday.

Loftus, G. R., & Loftus, E. F. (1983). *Mind at play: The psychology of video games.* New York: Basic Books.

Manovich, L. (1998). Navigable space. Retrieved June 2004 from Lev Manovich's Cultural Software Web site, at http://www.manovich.net/DOCS/navigable_space.doc.

McLuhan, M. (1964). *Understanding media: The extensions of man.* New York: New American Library; London: Routledge, 1994.

Miklaucic, S. (2002). Gaming. In Steve Jones (Ed.), *Encyclopedia of new media: An essential reference to communication and technology* (pp. 197–200). Thousand Oaks, CA: Sage.

Miller, S. (2001). Scott Miller of 3D Realms—part one. Interview by "Sander." Retrieved April 2004 from Eurogamer.net, at http://www.eurogamer.net/article.php?article_id=1261.

Montfort, N. (2003). *Twisty little passages: An approach to interactive fiction.* Cambridge, MA: MIT Press.

Nelson, G. (2001). A short history of interactive fiction. Chap. 46 in *The inform designer's manual* (4th ed.). St. Charles, IL: Interactive Fiction Library. Retrieved March 2004, at http://www.inform-fiction.org/manual /html/s46.html.

Nelson, T. H. (1974). *Computer lib: You can and must understand computers now/Dream machines: New freedoms through computer screens, a minority report.* Chicago: Hugo's Book Service. 9th printing, 1983.

Newell, A., Shaw, J. C., & Simon, H. A. (1958). Chess-playing programs and the problem of complexity. *IBM Journal of Research and Development, 2,* 320–335.

Packer, R. (1998). Net art as theater of the senses: A hypertour of Jodi and Grammatron. Retrieved April 2004 from the Beyond Interface Web site, at http://www.archimuse.com/mw98/beyond_interface/bi_fr.html.

Pitts, B. (1997). The Galaxy Game. Retrieved March 2004 from Computer History Exhibits Web site (Stanford University), at http://www-db.stanford.edu/pub/voy/museum/galaxy.html.

Randell, B. (1982). From analytical engine to electronic digital computer: The contributions of Ludgate, Torres, and Bush. *Annals of the History of Computing, 4,* 327–341.

Raymond, E. S. (1999). The Magic Cauldron. Retrieved June 2004, at http://www.catb.org/~esr/writings/magic-cauldron/magic-cauldron.html#toc10.

Ritchie, D. (2001). Ken, UNIX, and games. *International Computer Chess Association Journal, 24*(2). Online version retrieved June 2004, at http://cm.bell-labs.com/cm/cs/who/dmr/ken-games.html.

Roberts, E. S. (1977). The Mirkwood Tales. Manuscript. Stanford University Archives, Dept. of Special Collections, Stanford University Libraries.

Romero, J. (1997). E-mail dated December 11, 1997. Retrieved January 2004 from "*DOOM* Editing History," a collection of e-mail documentation on this topic. URL: http://www.johnromero.com/lee_killough/history/edhist.shtml.

Samuel, A. L. (1959). Some studies in machine learning using the game of checkers. *IBM Journal of Research and Development, 3,* 211–229.

Schleiner, A.-M. (1999). Parasitic interventions: Game patches and hacker art. Retrieved May 2004 from opensorcery.net Web site, at http://www.opensorcery.net/patchnew.html. Originally prepared for *Mariosophia* (1999), ed. Erkki Huhtamo.

Schwarz, F. D. (1990). The patriarch of Pong. *American Heritage of Invention & Technology, 6*(2), 64.

Shannon, C. E. (1950). A chess-playing machine. *Scientific American, 182*(2), 48–51.

Sheff, D. (1999) *Game over: How Nintendo conquered the world* (2nd ed.). Wilton, CT: Gamepress.

Standage, T. (2001). *The Turk: The life and times of the famous eighteenth-century chess-playing machine.* New York: Walker & Co.

Sudnow, D. (1983). *Pilgrim in the micro-world: Eye, mind, and the essence of video skill.* New York: Warner Books.

Sutton-Smith, B. (1986). *Toys as culture.* New York and London: Gardner Press.

Torres and his remarkable automatic devices. (1915). *Scientific American, 80,* suppl., 296–298.

Turner, F. (2002). From counterculture to cyberculture: How the Whole Earth Catalog brought us "virtual community." PhD dissertation, University of California, San Diego. A revised version is forthcoming as *Counterculture into cyberculture: How Stewart Brand and the Whole Earth Network transformed the politics of information*, University of Chicago Press.

Video games—Did they begin at Brookhaven? (n.d.). Retrieved April 2004 from DOE R&D Accomplishments Web site, at http://www.osti.gov/accomplishments/videogame.html.

Windisch, K. G. (1783). *Lettres de M. Charles Gottlieb de Windisch sur le joueur d' echecs de M. de Kempelen*. Basel.

Woolley, D. R. (1994). PLATO: The emergence of online community. Retrieved April 2004, at http://www.thinkofit.com/plato/dwplato.htm. An earlier version of this article appeared in *Matrix News* (January 1994).

Yob, G. (1976). Hunt the Wumpus. In David Ahl (Ed.), *The best of creative computing*, vol. 1 (pp. 247–250). Morristown, NJ: Creative Computing Press. Organization appeared in *Creative Computing* (September–October 1975), with an earlier version in *People's Computer Company, 2*(2), November 1973.

4

The (Computer) Games People Play: An Overview of Popular Game Content

Barry P. Smith
University of Alabama

Computer games are big business. At a time when the overall U.S. economy has been sluggish and the U.S. technology sector has not recaptured the glory days of the Internet boom, sales of entertainment software such as computer games have grown steadily. According to the NPD Group (2004), game software sales totaled $7 billion in the United States in 2003. Gaming hardware generated another $4.2 billion in U.S. sales during the same year. This $11.2 billion total reflects a 4% decrease in sales from 2002, but this decrease is largely the result of hardware sales entering the downward swing of a multiyear cyclical sales pattern. According to industry group Entertainment Software Association (formerly the Interactive Digital Software Association), game software sales have grown steadily over the past 9 years, from $3.2 billion in 1995 to 2003's $7 billion (2004a). A lot of money is being spent on these games, but what are individuals actually getting for their money?

The entertainment content that individuals get when they buy and play computer games is quite different from the content they get when buying a movie ticket, DVD, or CD. Although games share many aesthetic features with movies and music, games add the element of control (Grodal, 2000) that is largely missing from other media. Control alters the pace, flow, and outcome of the programming that is a computer game. The element of control in a game can also cause a user to identify with a mediated character to a greater degree than is possible with characters portrayed in other media because the user, to some degree, actually is the protagonist in the game. With control comes the possibility of interaction with a virtual environment and the denizens of that environment (see Vorderer, 2000).

This chapter first explores some general aesthetic features of computer games. Second, content ratings and gaming hardware are briefly discussed. Third, game genres are addressed with an eye toward comparing various classification schemes. Finally, specific content features of some of the most popular games of the last few years are examined.

AESTHETICS

Visual Features

One of the areas of game content that receives a great deal of attention in effects research is the area of visual content. Gone are the days when games, such as *Pong*, basically featured squares moving around on a screen. Whereas such early games provided little more in the way of visual aesthetics than board games, recent games provide images that begin to approach the level of realism associated with movies or television.

The first version of *Grand Theft Auto* (released in 1997) gave the player a bird's-eye view of the game environment. The player was presented with a relatively flat, cartoonlike environment in which to steal cars and create mayhem. Cars and characters lacked details—cars were more or less rectangular and characters were more or less bipedal, but cars did not have defined bumpers, and characters did not have faces. A more recent version of the game, *Grand Theft Auto: Vice City*, places the player on the ground in a 3-dimensional environment. Both cars and characters are clearly distinguishable at the individual feature level, down to details such as bent fenders on cars and unique items of (often scanty) clothing on characters.

Grodal's (2000) assertion is correct that it was the interactive capabilities—not the primitive graphics—of early games that attracted users. However, advances in graphics technology have made the visual experience of playing a game aesthetically appealing in its own right. Recent best-selling titles such as *Madden NFL 2004*, *Need for Speed: Underground*, and *Age of Mythology* each have visual features that are appealing both visually and artistically. Users are presented with visual effects such as fairly realistic football players complete with accompanying shadows (that vary with the virtual lighting), motion-distorted neon signs along a midtown racetrack, and flames and ocean waves that faithfully reproduce the movements of their real-world counterparts. Users are encouraged not only to observe these elements as part of a visual landscape but to also consider them as artistic achievements. Advertisements and product packaging for these games tout visual realism as a selling point and bring attention to special visual details within the game (see Shapiro, Pena-Herborn, & Hancock, chap. 19, this volume, on realism).

Auditory Features

Games have developed a great deal in the area of sound as well. Early games featured a lot of *beeps* and *bloops* as the only auditory stimuli for the player. As visual features of games began to more closely resemble nonmediated reality, however, game developers made efforts to provide more realistic sounds to match. The earliest successful attempts at realistic sounds were reproductions of mechanical sounds, such as the sounds associated with cars and guns. Because the sound of a gunshot or a revving engine is not very complex harmonically, these sounds are fairly easy to approximate with some degree of success.

After basic sound effects, game developers provided players with musical accompaniment—much akin to movie scores. Movies without a musical score are relatively rare, and the absence of a score is typically used to create a mood (such as the tension in *Fail-Safe*). Although early games, like the earliest films, did not feature a musical score, music has now achieved the same level of ubiquity in games that it achieved in films decades ago. Current games feature both original scores (some of which are available as stand-alone CD releases) and popular recorded music. *Grand Theft Auto: Vice City*, for example, features a number of popular hits from the 1980s. This inclusion of popular music is an important content feature of the game because *Grand Theft Auto* is modeled after (and designed partly to appeal to fans of) a popular television show of the 1980s, *Miami Vice*. If a positive experience stamps in preference (Zillmann, 1988), then inclusion of fondly remembered popular music should contribute to the creation and management of a positive mood and increase the enjoyment of a game (see Zehnder & Lipscomb, chap. 17, this volume, on music in games).

One of the more recent advances in game development has been the inclusion of realistic voices. Early games featured crudely synthesized or sampled voices on rare occasion that voices were used at all. Recent audio sampling and compression techniques allow for the presentation of character voices in CD-quality (or better) sound. Reeves and Nass (1996) have compiled a body of research indicating that individuals often respond to a voice coming from a computer in ways that are very similar to responses during interpersonal vocal communication. The research presented by Reeves and Nass demonstrates that electronically mediated voices can trigger affective responses. This can be an important factor in the entertainment appeal of computer games, because the existence of affective responses indicates the generation of arousal. Through excitation transfer (Zillmann, 1971, 1978), the arousal generated by the mediated voices of game characters may enhance or heighten the level of arousal generated by subsequent visual elements of a game and the user's interaction with those elements (Zillmann, 1991b), thus making the game more exciting in the player's perception.

Tactile Features

Whereas films, television, and radio feature very little tactile interaction for the user, almost all computer games feature tactile interaction as a major part of the entertainment experience. Players use keyboards, mice, joysticks, trackballs, movement-sensing gloves, footpads, and many other devices to provide continual tactile input. Aside from the standard tactile stimulation created by the player pressing a button or moving a controller, many games now also provide tactile feedback through vibration devices embedded within game controllers. When a player brings his or her character ashore in *Medal of Honor: Frontline*, for example, the player experiences hand-jarring vibrations through the game controller to go along with the barrage of images and sounds of D-Day explosions.

The Value of Realism

The most important aspect of these aesthetic features is their ability to make a computer game seem more real to the player. In every entertainment medium, content producers attempt to draw consumers into an alternate, mediated reality for a period of time. Coleridge's "willing suspension of disbelief" is made easier if fewer noticeably nonreal artifacts attract attention. A more realistic portrayal of a virtual environment can make the illusion of immersion in that environment more complete.

In addition, reality is data rich. The individual as an active information processor is presented with an infinite stream of input information through the five senses—a stream that must be filtered down with various physiological and cognitive mechanisms to make understanding one's environment possible. Given this state of affairs, an individual sitting down to play a relatively data-poor game may not find his or her attention fully engaged by the experience. The production of more realistic, that is, data-rich, games provides players with a form of entertainment that requires more attention and thus provides greater escape from reality and engagement with the game (see Shapiro et al., chap. 19, this volume).

GAME RATINGS

Computer games have been categorized in a number of different ways for purposes of quantifying their content. Age rating, hardware platform, and genre are the three most commonly used criteria for categorizing games. Hardware platform categorizations are typically of most interest to market analysts. Genre categorizations are typically of interest to entertainment scholars and game players. Age ratings, however, seem to be the favored categorization framework for parents and policymakers (see Bushman & Cantor, 2003). The Entertainment Software Rating Board (ESRB) is the most well-known provider of age ratings and content descriptors for computer games.

According to the ESRB, 54% of games sold in the United States in 2003 were rated *E* for Everyone, 30.5% were rated *T* for Teen, and 11.9% were rated *M* for Mature (Entertainment Software Association, 2004b). The small remaining percentage was a mixture of *EC* for Early Childhood, *AO* for Adults Only, and *RP* for Rating Pending.

Of the top 20 best-selling games for the various home video game consoles in 2003, 70% fell into the *E* or *T* categories. Of the top 20 best-selling games for the personal computer in 2003, 90% fell into the *E* or *T* categories (Entertainment Software Association, 2004b). (This is an interesting comparison, given that users playing on home video game consoles tend to be younger than users playing on personal computers on average.)

The ESRB also provides content descriptors to go along with its age ratings. These descriptors are designed to provide notification to parents about specific content (such as violence, blood, sexual themes, etc.) that may be considered objectionable for children. Many parents rely on media ratings (Bushman & Cantor, 2003), however, ESRB ratings and descriptors may miss some potentially objectionable violent and sexual content (Haninger & Thompson, 2004), thus limiting their usefulness to parents and researchers (see S. Smith, chap. 5, this volume, on negative content).

HARDWARE

Categorizing games by their hardware platform can provide some interesting context when sales figures for games are used as a proxy for entertainment value obtained. For example, some home video game consoles are handheld, but others require connection to a television set. Examination of these two different types of hardware systems could reveal differences in how individuals obtain entertainment value in different settings. A handheld unit might be used as a diversion from the dull reality of riding in a car for an extended period of time. A standard-size home video game console connected to a television set might be used by multiple players simultaneously as a form of social activity.

Some variation in game contents may follow from a player's preference for, or use of, one platform over another. Game title availability, for example, can vary by hardware platform. If *Grand Theft Auto: Vice City* is only available for Sony Playstation2, and *Halo* is only available for Microsoft's Xbox, then purchasers of various hardware systems may buy a particular platform so that they can play a particular game (or line of games in a series). In addition, the same game title released across hardware platforms may vary depending on the capabilities of each hardware system or platform—particularly when handheld systems are contrasted with other hardware systems.

Hardware platform preferences also show up over time. Aside from the expected emigration from older video game consoles to newer systems, some noticeable brand preference trends are also in evidence from year to year. Game titles released for Sony PlayStation accounted for 17 of the top 50 games in 2000. The Nintendo 64 accounted for 13 of the top 50 games that same year. By 2002, however, the next-generation PlayStation2 accounted for 27 of the top 50 games (and three more of the top games were PlayStation titles). Nintendo's next-generation GameCube only accounted for 4 of the top 50 games in 2002. On the other hand, Nintendo's series of GameBoy handheld systems (GameBoy, GameBoy Color, GameBoy Advance) have so dominated the handheld system sector of the U.S. gaming market for the past 5 years (1999–2003) that no other handheld system shows up in market data for best-selling games.

Market data indicate some variation in sales of different game genres across hardware platforms. This shows up particularly when a comparison is made between individuals using the personal computer versus individuals using home video game consoles. The top-selling genre for games played on home video game consoles is action, but the top-selling genre for personal computer games is strategy (Entertainment Software Association, 2004b). This difference may be related to age differences in gamers using personal computers versus consoles.

GAME GENRES

Although computer games, like novels or films, are generally considered to fall into differing genres, establishing exactly what these genres are can be a difficult task. Popular industry Web site Gamespot.com classifies games into eight genres. The Academy of Interactive Arts and Sciences (AIAS) gave awards to game developers in 14 different genres. As Table 4.1 indicates, various scholars have found still other numbers of genres.

The AIAS categories make distinctions between games designed for home video game consoles and games designed for personal computers. However, this is a recognition of differences between game development processes for different systems rather than a distinction between the actual entertainment content of games designed for these different systems. For example, *Madden NFL 2004* is available for a number of different hardware platforms and could thus be cross-classified. Although these platforms do differ slightly in the way they present content

TABLE 4.1
Genres Used to Classify Computer Games

Source	Genres Used
GameSpot.com	Action, adventure, driving, puzzle, role-playing, simulation, sports, strategy
Academy of Interactive Arts and Sciences, 6th Annual Awards Nomination Categories	Console action/adventure, console fighting, console first-person action, console platform action/adventure, console racing, console role-playing game, console sports, PC action/adventure, PC first-person action, PC massively multiplayer/persistent world, PC role-playing game, PC simulation, PC sports, PC strategy
Lachlan, Smith, & Tamborini (2000)	Adventure, flight simulator, fighting, music, role-playing, racing, shooter, sports, strategy/puzzle
Sherry, Lucas, Rechtsteiner, Brooks, & Wilson (2001)	Strategy, puzzle, fantasy/role-playing, action–adventure, sports, sims, racing/speed, shooters, fighters, arcade, card/dice, quiz/trivia, classic board games, kids' games
Haninger & Thompson (2004)	Action, adventure, fighting, racing, role-playing, shooting, simulation, sports, strategy, trivia

for the user, the differences are for the most part less significant than factors such as the size of the screen being used.

Seven genres appear fairly consistently from one classification scheme to another: sports, driving (or racing), simulation, strategy, role-playing, shooting (or shooter), and fighting (or fighter). Action, adventure, action–adventure, first-person action, and platform action (or action–adventure) appear in various forms and combinations across classification schemes with little indication of what criteria are used to place a given title in a given genre. Other genres, such as music, puzzle, board games, and trivia, are less consistently used.

Sports

The genre of sports games is somewhat unusual among game genres in its reliance on real-world figures as characters. The most popular titles in this genre prominently feature current athletes, sports celebrities, and sports teams toward which game players are likely to have a preexisting positive disposition. These positive dispositions can be quite intense and lasting (Bryant & Raney, 2000; Zillmann, Bryant, & Sapolsky, 1989). Game publisher Electronic Arts (EA) has achieved near total market dominance in this genre by building on these existing dispositions. One of EA's most successful franchises is the *Madden NFL* series. As the title indicates, this series of games prominently features popular U.S. sports commentator John Madden. Individuals who have been fans of NFL professional football are likely to have viewed and formed a disposition toward Madden while watching televised games. Each year's entry in the series also features actual teams and athlete rosters from the NFL. Game players with a positive disposition toward a particular team or athlete can compete as that team or athlete. Conversely, game players are given the opportunity to play against teams and athletes toward which the players may have a negative disposition.

In addition to team sports such as football, baseball, soccer, hockey, and basketball, a number of individual-athlete sports have attracted the attention (and money) of game players. Probably the most successful franchise in this group is the *Tony Hawk's Pro Skater* series. Individuals familiar with the sport of professional skateboarding in the United States are likely to be familiar with Tony Hawk prior to purchasing or playing these games. In the first *Tony Hawk* game, players could play as Tony Hawk or one of nine other professional skaters. In the second, third, and fourth games, players could play as Tony Hawk, 1 of 12 other professional skaters, or their own, individualized character. Although players can play as other skaters or themselves, the series is centered around the titular Tony Hawk and depends on players' disposition toward Hawk to generate some portion of the enjoyment derived from purchasing and playing the game. Interestingly, the fifth title in this series, *Tony Hawk's Underground*, departs from this strategy. This game has the player playing as him- or herself rather than as a real-world professional skater. In addition, the player is given the ability (for the first time in the franchise) to leave the skateboard and simply move around to explore the virtual environment. While customizing their own characters, players are given the options of creating their own unique skateboarding moves, or tricks, and even downloading images of their own faces for use in the game. Although the title of the game still carries Tony Hawk's name, the actual game play centers around the players' opportunities for exploration and action rather than the players' identification with Hawk.

Although players may be encouraged to initially purchase and play a particular sports game based on a preexisting disposition toward the individuals or teams represented in the game, continued enjoyment of the game will depend on a number of other factors as well. As with all computer games, continued enjoyment of a sports game will depend in part on a player's abilities, self-efficacy, and outcomes achieved during game play. The most popular sports games allow players to overcome negative experiences in these areas by providing a narrative cycle that extends beyond the single contest. Best-selling title *NCAA Football 2003* (another of EA's popular franchises), for example, allows players to go beyond the narrative cycle of a single virtual football game. Players can play full seasons of football. In the game's *Legacy* mode, players can even guide their favorite team, such as the University of Alabama Crimson Tide through several decades of seasons. By providing such an expanded narrative cycle, the game allows players to look beyond negative outcomes at the single-contest level. A player faced with a loss at the single-contest level could experience a negative mood that would reduce enjoyment derived from playing the game and thus potentially reduce the actual playing of that game (Zillmann, 1988). However, if that same player is presented with other available goals—a national championship or a winning legacy for his or her team—then that player may be more likely to continue playing and enjoying the game with the prospect of positive outcomes.

Driving

Driving, or racing, games are centered around the maneuvering of some sort of vehicle—typically, but not always, in a race setting. (Demolition derby games such as *Destruction Derby* do not really emphasize a race, but the game play is focused on driving nonetheless.) Some driving games feature real-world characters, such as the NASCAR drivers featured in the *NASCAR Thunder* series, but most do not. The vehicles, which are often exotic or unusual, are typically featured more prominently than the characters driving the vehicles.

Driving games are exciting to play because of speed and competition. The perception of speed is created by fast-paced movements of objects on-screen. Quick movement in the visual field is arousing, whether the movement is real or virtual (Lang, Bolls, Potter, & Kawahara, 1999; Reeves & Thorson, 1986). In addition, players realize the need for quick reactions to obstacles in the virtual environment and therefore actively pay greater attention to the screen (Anderson & Lorch, 1983).

Although the range of vehicles represented in driving games varies greatly (e.g., all-terrain vehicles in *ATV Off-Road Fury,* tractor-trailers in *Rebel Trucker: Cajun Blood Money*, various exotic racing cars in *Gran Turismo 3: A-Spec*), almost all driving games feature the element of competition against other players or nonplayer characters. Competition is a big appeal for computer game players (Sherry, Lucas, Rechtsteiner, Brooks, & Wilson, 2001), and head-to-head striving among near equal opponents for the same goal can be one of the most intense forms of competition (see Hartmann & Klimmt, chap. 9, this volume, on personality factors and games choice).

Simulation

Simulation games, or simulators, share some characteristics with driving games. Typically, game play is focused around the maneuvering of some type of vehicle, but the vehicle is most often some type of aircraft instead of a land-based vehicle. Whereas simulators may or may not feature a competitive element, the main focus of the game is the realistic piloting required by the virtual aircraft. The focus on realism has led to cross-pollination between the domains of gaming and real-world training. Many professional and military pilots have used computer-based simulators to practice skills in a low-risk setting.

One of the most successful simulator games has been Microsoft's *Flight Simulator* series, which has been around for more than 20 years. Like many aircraft simulations, titles in the *Flight Simulator* series come with a large training manual. Much like learning to pilot a real aircraft, players are required to learn a great deal of technical jargon and mechanical intricacies in order to fly their virtual craft smoothly in the "expert" play mode. Most simulators also feature an easier playing mode to allow new players to experience virtual flight immediately.

Whereas *Flight Simulator* usually features civilian aircraft, other game publishers focus on military aircraft simulators. International intelligence provider Jane's Information Group has used its information to release a number of complex, realistic military aircraft simulators. *Jane's F-15*, for example comes with a manual of more than 300 pages that contains all sorts of information for players with an interest in fighter jets. Most individuals will never have the opportunity to fly a military aircraft, but flight simulators allow players to feel like they have actually been in the cockpit. One of the appeals of computer games is this ability to temporarily transcend the limitations of society and physical abilities (see the chapters in this volume on motivation and selection for greater detail on the appeals of games).

Strategy

As previously mentioned, strategy is the most popular game genre for users playing on personal computers. Strategy games focus on the acquisition or utilization of resources to achieve some type of long-term goal or advancement. The player's character may be an individual or a whole society, and may or may not face an active antagonistic element.

Microsoft's *Age of Empires* and *Age of Mythology* are typical examples of the strategy game. In these games, the player controls one of two or more societies in a virtual environment. The player's society must compete or cooperate with the other societies to acquire resources, such as food and building materials. As materials are collected, the player uses these resources to build and advance his or her society. The game usually ends when one society (or cooperative group of societies) vanquishes any competing societies. Strategy games require a great deal of cognitive activity, as players must simultaneously acquire and use resources, attack and defend against competitive societies, and advance their societies technologically.

The Sims is the all-time best-selling game for the personal computer. The game has also spawned nine commercially successful expansion packs and add-ons thus far. What sets *The Sims* apart from other strategy games is the rather mundane nature of the game. Instead of creating a world-conquering empire, players of *The Sims* create characters who live in a neighborhood, work at a job, and acquire material possessions. The created characters have needs (such as food and social interaction) that must be met for the characters to remain healthy and happy. Much of the game play is centered around having the characters earn money in order to buy things that will satisfy their needs. Rather than providing a fantastic escape from real life, *The Sims* provides players with the opportunity to control the everyday lives of virtual people. The popularity of a game about such mundane activities may further highlight the importance of control as an element of entertainment in computer games (Grodal, 2000; Vorderer, 2000).

Role-playing

Although almost all computer games (with the possible exception of puzzle games such as *Tetris*) involve some degree of role-playing, some games involve little else. Players may or may not face an antagonistic element in role-playing games. Instead, these games are centered around the development of the character's persona and the character's interaction with the virtual environment.

Role-playing games in general, and online role-playing games in particular, typically take a great deal of time for the creation and development of a player's virtual persona. The action of assuming the role of a virtual character may require a player to spend hours collecting virtual resources and using or bartering those resources for virtual food, shelter, protection, and so on. Because of the amount of real-world time involved in collecting these virtual resources, some players earn real income by acquiring and selling virtual resources. Some players of *Ultima Online*, for example, earn a full-time real-world income by selling virtual possessions on auction site eBay (Dibbell, 2003). (*Ultima Online* is an example of a massively multiplayer online game, or MMOG, a subgenre of role-playing games discussed in further detail by Chan & Vorderer, chap. 6, this volume.)

Role-playing games are somewhat unusual in that they may not necessarily have an ultimate goal or ending point. Most games in other genres have an ending point at which a player can say that he or she has won the game or defeated all foes. Although some role-playing games do feature such an outcome, many allow nearly infinite exploration of the virtual environment.

Shooting

Shooting games, or shooters, are among the most graphically violent games on the market. Shooters typically provide the player with a first-person perspective in a 3-dimensional virtual environment. As the name implies, shooters place the player's character among hosts of antagonists who must be shot (or otherwise eliminated). Although most shooters do have some type of overall plot structure, game play is focused more on killing enemies than any other activity.

One of the earliest first-person, 3-dimensional shooters was *Wolfenstein 3-D*, released in 1992. In this game, the player's character is an Allied soldier trapped in a Nazi prison during World War II. The player's character is besieged by armed antagonists as he makes his way to freedom. The player's weapons, such as handguns, rifles, and automatic weapons, are obtained and upgraded by shooting antagonists and taking whatever they have.

Later games in this genre have followed the same general pattern. The player's character is the protagonist, for whom the player is expected to feel a strong, positive disposition following Zillmann's (1991a) theorizing about television protagonists. The player's character is absurdly outnumbered by armed antagonists for whom the player is expected to feel an equally strong, but negative, disposition. The antagonists must be killed to allow the player's character to achieve some important goal, such as freedom or the saving of the world. Placed in this virtual situation, players are given the opportunity to collect a variety of weapons to use on enemies during the game. The killing of enemies is portrayed as justified, and the player is rewarded with points or "kills percentage" numbers at the end of each level of the game. In addition to the presentation of justified, rewarded violence, shooters often depict the killing in explicit and graphic detail— that is, blood and gore is portrayed up close (Lachlan, Smith, & Tamborini, 2000).

The introduction of *Quake* in 1996 allowed players to shoot either non–player characters or other players' characters via an Internet connection. (Previous multi–player games were more limited in their connectivity and were not as popular.) The introduction of other players into the shooter game changes the affective nature of game play. Acquaintances whom a player knows exist in real life are different from computer-generated opponents who only exist to antagonize and harm a player's character. Whereas non-player characters are killed because of their antagonistic nature, the killing of other players' characters is more akin to a sports competition between friends. In the latter case, the number of times one player can virtually kill another assumes a similar function as the keeping of score in a basketball game.

Fighting

Fighting games, or fighters, feature the player's character and one or more antagonists in timed (or otherwise limited) pugilistic rounds. The games are reminiscent of boxing or wrestling matches in general format, although more exotic fighting styles and weapons are typically used in fighters. Game play is usually presented on-screen from a side point of view. The player's character and the player's opponent approach each other from opposite sides of the screen to engage in fighting.

In fighting games, the object is typically not to kill the opponent but to inflict a certain amount of physical damage in each round. Each character's level of vitality is depicted visually in the form of a gauge or meter somewhere on-screen. When one character's level drops to a certain point (as a result of the other player's hits, kicks, etc.), then that character loses that round. Fighting games most often feature a best-of-three rounds format. Upon vanquishing one foe in two or three rounds, the winner advances to face a more difficult opponent. Most games in this genre also feature options for head-to-head competition between player characters.

Popular fighting series *Mortal Kombat* is typical of many games in this genre. In these games, a player is allowed to choose his or her character from a pool of selections. Each character has particular strengths and weaknesses and features some unique fighting technique, weapon, or ability. In the *Mortal Kombat* series (like others in this genre), the fighting actions of characters are a mixture of Eastern martial arts and a supernatural element. Kicks and punches are supplemented with lightning and fireballs.

Action–Adventure

As mentioned previously, games classified in the action genre by one individual or group may be classified in the adventure genre by others, and still other classifiers may combine the two into one genre. When attempting to determine the genre for a given game, some scholars look to the trade press or Web sites for a classification. When these sources do not agree, scholars are forced to follow some clarifying rule such as following the majority of sources cited (Haninger & Thompson, 2004).

The best-selling games of 2001 and 2002 were the latest release in the *Grand Theft Auto* series each year—*Grand Theft Auto III* and *Grand Theft Auto: Vice City*, respectively. Both of these games are alternately classified as action and adventure (in their releases for different hardware platforms) by Gamespot.com. Given the lack of definitive delineation by scholars or the industry, action and adventure should probably be combined into one genre for the present. Action–adventure games feature the player as a protagonist moving through a series of obstacles or challenges to reach some specified goal. Typically, these games also feature an antagonist element that actively attempts to thwart the player's character.

Action–adventure games are distinct from role-playing games in that the character is largely laid out for the player in an action–adventure game, but the player assumes greater responsibility for fleshing out the character in a role-playing game. The action–adventure game also features a definitive goal. Shooting games may be considered a subgenre of the action–adventure genre, with their chief distinction being the almost exclusive focus on killing enemies during the quest for the goal.

In the *Grand Theft Auto* series of games, the player plays a small-time criminal presented with various missions to perform for organized-crime bosses. *Grand Theft Auto* is unusual in that the player's character is not one that players would be expected to feel a positive disposition toward when first coming to the game. Part of the appeal of the series is the player's ability to act out various antisocial behaviors in a setting that rewards, rather than punishes, such actions. The desire to escape the limitations of an orderly life and society has been posited as one of the appeals of mediated entertainment (Mendelsohn, 1966).

Other Genres

Aside from the most common game genres already mentioned, a number of other genres have been proposed. Puzzle games, such as *Tetris*, often feature abstract geometrical shapes which must be manipulated into some specified pattern according to specified rules. Game shows such as *Jeopardy!* and *Who Wants to Be a Millionaire?* have spawned a number of moderately successful quiz games. Other games are simply computerized versions or variations of popular board games or card games (i.e., *Monopoly, FreeCell,* and computerized chess).

Games falling into these other genres are typically underrepresented in lists of each year's best-selling games. One reason may be that these games do not have the same level of affective appeal for game players. Although a favored athlete or a compelling hero may receive a positive affective reaction, variations on solitaire games have no such dispositional focal point. It should be noted, however, that reliance on sales figures alone may not always indicate the relative popularity of, or entertainment value received from, a given game or genre of games. The vast majority of personal computers worldwide run some version of Microsoft's *Windows* operating system. With each copy of that operating system comes at least one version of solitaire (and some later versions of *Windows* come with multiple solitaire card games). That comprises a very large installed base of computer games which do not show up in computer game market data.

POPULAR CONTENTS

Previous sections in this chapter have looked at general aspects of computer games and their content. This section focuses on specific content of some of the best-selling games of the past few years. The focus is on best-selling games because, in general and with other factors being equal, individuals are more likely purchase and consume media fare that is entertaining than media fare that is not entertaining. The most popular games, therefore, are those that have succeeded in providing the greatest amount of entertainment. Of course, other factors are at work as well, such as the price of a given game. Game titles for handheld gaming platforms typically cost 25% to 50% less than titles for video game consoles or personal computers. For this reason, games for handheld platforms are omitted from the following discussion.

During a survey of the top 50 best-selling games each year for 2000, 2001, and 2002, the first item of note was the importance of being a sequel or having had some prior media exposure. Eliminating entries that appeared in multiple years yielded 124 unique game titles from the 150 titles surveyed. (Several add-ons or expansion packs were also eliminated from the survey because they are not actually stand-alone games.) Out of these 124 unique titles, 95 (76.6%) were sequels. Sequels were defined as games that either kept the same title as a previous game but added a number, such as *Driver* and *Driver 2*, or games that did not share the same title but featured a character that had been featured in a previous game, such as Link from the various *Legend of Zelda* games.

In addition to the popularity of sequels, 83 (66.9%) of the games featured either a title or characters from a prior media source, such as a cartoon, a television show, a film, or some other type of game such as a board game or trading cards. Cartoon character Spongebob Squarepants, novel and film character James Bond, and the various Pokemon trading-card characters all appeared in computer-game form during the period examined.

Also, 33 (26.6%) of the games were either endorsed by or featured a celebrity personality. Among the celebrities were: John Madden and various NFL athletes, WWE (formerly WWF— World Wrestling Federation) personalities, and well-known actors (such as Pierce Brosnan). As mentioned previously, a prior positive disposition toward a celebrity can be brought into a game to increase the appeal of playing that game.

Some genres were not well represented in the top 50 games for each year. Role-playing games for example, accounted for only 4 unique titles out of the total 124, and only one of the four, *Disney's Kingdom Hearts,* appeared in the top 10 in any year. (In addition to these four unique titles, there was one add-on for one of the unique titles.) Two of these role-playing games were sequels in the long-running *Final Fantasy* franchise—*Final Fantasy IX* and *Final Fantasy X*, specifically.

Another underrepresented genre among the top-selling games was strategy. There were only 3 unique strategy titles among the total 124. Although strategy is the most popular genre among players using personal computers, these players represent much less of the overall gaming market than players using video game consoles. In 2003, for example, sales of games for personal computers totaled $1.2 billion, but sales of games for video game consoles totaled $5.8 billion (NPD Group, 2004). It is almost impossible, however, to determine how many illegal copies of any particular game may be in circulation. Games distributed for the personal computer are generally easier to illegally copy, so sales figures may not exactly equal playing figures.

Puzzle and quiz games were all but nonexistent among the top 50 games for the years examined. Computer game versions of *Who Wants to Be a Millionaire?* (1 and 2) appeared on the list of top games from 2000 but faded away with the waning popularity of the television game show. *Mario Party 2* appeared on the top 50 list in 2000, and *Mario Party 3* was the only puzzle game to appear on the list from 2001. This diminutive Italian plumber's persistent appeal across decades, genres, and gaming platforms is indicative of the same type of positive disposition toward a character that has kept film characters such as James Bond popular for so many years.

Analysis of a subsample of the surveyed games revealed results similar to those obtained previously by other researchers. Haninger and Thompson (2004) analyzed T-rated games for home video game consoles. In their stratified (by genre), random sample, they found 98% of T-rated games involved intentional violence, and 27% of these same games depicted sexual themes (defined as nudity or behaviors or dialogue related to sex). Lachlan et al. (2000) found violence featured in 90% of sampled games rated T or M. Games rated for general audiences (E or the retired $K - A$) featured violence in 57% of the sample cases.

In the subsample of popular games from 2000, 2001, and 2002, violence was featured in 57% of E-rated games and 80% of T- or M-rated games. The relatively lower rate among games with a T or M designation, as compared to previous research, was partially the result of the popularity of games in the *Tony Hawk's Pro Skater* series, which received T ratings (except for the original, which was rated E). Multiple titles in this series have been among the best-selling games since 2000.

For this analysis, each game sampled was played until basic proficiency with the controls was established. After achieving control proficiency, each game was played at the hardest difficulty setting offered. Following the protocol used by Lachlan et al. (2000), each game was videotaped during play, and the first 10 minutes of each videotaped game were analyzed.

Violence was quantified as discrete acts during the sampled time frame. Hostility was quantified as discrete hostile statements or exclamations. It was also noted whether humor was associated with the observed violence.

Table 4.2 presents some figures regarding violent and hostile content observed in the subsample. Only a single entry from any popular series of games is presented because entries within a series tend to be very similar in regard to featured violent content. Over time, however, violent content in a series is portrayed more realistically with the progress of game development.

Interestingly, T- and M-rated games which featured violence offered a greater variety of weapons with which to perpetrate that violence. *Grand Theft Auto* in particular offers a wide selection of conventional and nonconventional weapons, including fire arms, screwdrivers, baseball bats, and cars. Military-themed games such as *Metal Gear* and *Medal of Honor* feature a more conventional variety of firearms and ordnance.

TABLE 4.2
Violent Content in Popular Games

Title	ESRB Rating	Discrete acts of Violence	Instances of Hostility	Presence of Humor
ATV Off-road Fury	E	5	0	No
Disney's Kingdom Hearts	E	15	0	No
Gran Turismo 3: A-Spec	E	0	0	
Grand Theft Auto: Vice City	M	50	25	Yes
Madden NFL 2002	E	0	0	
Medal of Honor: Frontline	T	*	0	No
Metal Gear Solid 2	M	7	1	No
NCAA Football 2003	E	0	0	
Spec Ops	T	28	0	No
Spider-Man: The Movie	E	13	5	No
Spyro the Dragon	E	16	0	Yes
Tony Hawk's Pro Skater 2	T	0	0	

* *Medal of Honor* begins with a D-Day invasion. The game's montage of continuous explosions and deaths made accurate quantification all but impossible.

CONCLUSION

The future development of computer games looks bright aesthetically but questionable socially. Although game developers are providing ever more realistic escape or engagement for players, many games feature virtual environments in which antisocial behaviors have few, if any, negative consequences. Further research into the appeals and effects of computer games gives rise to hopes, however, that future games can be both entertaining and pro-social.

REFERENCES

Anderson, D. R. & Lorch E. P. (1983). Looking at television: Action or reaction? In J. Bryant & D. R. Anderson (Eds.), *Children's understanding of TV: Research on attention and comprehension* (pp. 1–33). New York: Academic Press.

Bryant, J., & Raney, A. A. (2000). Sports on the screen. In D. Zillmann & P. Vorderer (Eds.), *Media entertainment: The psychology of its appeal* (pp. 153–174). Mahwah, NJ: Lawrence Erlbaum Associates.

Bushman, B., & Cantor, J. (2003). Media ratings for violence and sex: Implications for policymakers and parents. *American Psychologist, 58,* 130–141.

Dibbell, J. (2003, January). The unreal estate boom. *Wired, 11*(1), 106–113.

Entertainment Software Association. (2004a). *Computer and video game software sales break $7 billion in 2003.* Retrieved April 15, 2004, from http://www.theesa.com/1_26_2004.html.

Entertainment Software Association. (2004b). *Industry sales and economic data.* Retrieved April 10, 2004, from http://www.theesa.com/industrysales.html.

Grodal, T. (2000). Video games and the pleasures of control. In D. Zillmann & P. Vorderer (Eds.), *Media entertainment: The psychology of its appeal* (pp. 197–213). Mahwah, NJ: Lawrence Erlbaum Associates.

Haninger, K., & Thompson, K. (2004). Content and ratings of teen-rated video games. *Journal of the American Medical Association, 291,* 856–865.

Lang, A., Bolls, P., Potter, R., & Kawahara, K. (1999). The effects of production pacing and arousing content on the information processing of television messages. *Journal of Broadcasting and Electronic Media, 43,* 451–475.

Lachlan, K., Smith, S., & Tamborini, R. (2000, November). *Popular video games: Assessing the amount and context of violence.* Paper presented at the annual meeting of the National Communication Association in Seattle, WA.

Mendelsohn, H. (1966). *Mass Entertainment.* New Haven, CT: College & University Press.

NPD Group. (2004, January 26). *The NPD Group reports annual 2003 U.S. video game industry driven by console software sales.* Retrieved April 14, 2004, from http://www.npd.com/press/releases/press_040126a.htm.

Reeves, B., & Nass, C. (1996). *The media equation: How people treat computers, television and new media like real people and places.* New York: Cambridge University Press.

Reeves, B., & Thorson, E. (1986). Watching television: Experiments on the viewing process. *Communication Research, 13,* 343–361.

Sherry, J., Lucas, K., Rechtsteiner, S., Brooks, C., & Wilson, B. (2001, May). *Video game uses and gratifications as predictors of use and game preference.* Paper presented at the International Communication Association convention, Washington, DC.

Vorderer, P. (2000). Interactive entertainment and beyond. In D. Zillmann & P. Vorderer (Eds.), *Media entertainment: The psychology of its appeal* (pp. 21–36). Mahwah, NJ: Lawrence Erlbaum Associates.

Zillmann, D. (1971). Excitation transfer in communication-mediated aggressive behavior. *Journal of Experimental Social Psychology, 7,* 419–434.

Zillmann, D. (1978). Attribution and misattribution of excitatory reactions. In J. H. Harvey, W. J. Ickes, & R. F. Kidd (Eds.), *New directions in attribution research* (vol. 2, pp. 335–368). Hillsdale, NJ: Lawrence Erlbaum Associates.

Zillmann, D. (1988). Mood management: Using entertainment to full advantage. In L. Donohew, H. E. Sypher, & E. T. Higgins (Eds.), *Communication, social cognition, and affect* (pp. 147–171). Hillsdale, NJ: Lawrence Erlbaum Associates.

Zillmann, D. (1991a). Empathy: Affect from bearing witness to the emotions of others. In J. Bryant & D. Zillmann (Eds.), *Responding to the screen: Reception and reaction processes* (pp. 135–167). Hillsdale, NJ: Lawrence Erlbaum Associates.

Zillmann, D. (1991b). Television viewing and physiological arousal. In J. Bryant & D. Zillmann (Eds.), *Responding to the screen: Reception and reaction processes* (pp. 103–133). Hillsdale, NJ: Lawrence Erlbaum Associates.

Zillmann, D., Bryant, J., & Sapolsky, B. S. (1989). Enjoyment from sports spectatorship. In J. H. Goldstein (Ed.), *Sports, games, and play: Social and psychological viewpoints* (2nd ed., pp. 241–278). Hillsdale, NJ: Lawrence Erlbaum Associates.

5

Perps, Pimps, and Provocative Clothing: Examining Negative Content Patterns in Video Games

Stacy L. Smith
University of Southern California

The societal war against video games is well underway. The battle cry of parents, policymakers, and educators is that game-related violence is partly responsible for some of the horrific school-yard shootings across the United States and abroad (Murphy, 2002; Woodruff & Schneider, 1999). Commenting on the content found in *Duke Nukem, Carmageddon,* and *Grand Theft Auto 2,* U.S. Senator Sam Brownback (2000) stated, "Defenders of these games say that they are mere fantasy, and harmless role-playing. But is it really the best thing for our children to play the role of a murderous psychopath? Is it all just good fun to positively reinforce virtual slaughter? Is it truly harmless to simulate mass murder?" (p. 2). Congressional concern such as this has caused the crafting of 16 anti-video game bills across the country (Pereira, 2003).

How much objectionable content is featured in video games? The purpose of this chapter is to answer that question. To this end, the discussion is divided into four major sections. The first section examines the marketing of video games to youth. Point-of-purchase advertising is one important content feature that may initially lure youngsters into buying certain games. All of the extant content analyses of video games are reviewed in the second session. Examining the distribution of features such as violence, sex, profanity, and other potentially objectionable content is the first step toward understanding the risk that video games may be posing to youth. Yet some have questioned the validity of conducting content analyses on video games. Therefore, the third section looks at three methodological challenges associated with carrying out this type of research. In the final section, rating systems designed to forewarn parents about any potentially problematic content in video games are discussed. Specifically, the current systems used in North America, Europe, and Australia are reviewed.

A few caveats must be noted before beginning, however. Video games can be played across a variety of platforms—from handheld devices to high-speed computers. Furthermore, gamers can play against the computer, a few other opponents physically present in the same room, or thousands of virtual individuals around the world. Although massive, multiplayer online role-playing games (i.e., MMOGs) such as *Lineage II, Everquest II,* and *Asheron's Call* are exploding in popularity, there is currently no systematic, empirical research assessing the *content patterns* of such games. Because of this, MMOGs are not addressed in this chapter (for discussion of this topic, see Chan & Vorderer, chap. 6, this volume).

The second caveat is that the focus of this chapter is largely on content patterns, not theory. Theories of media processes and effects should always inform the design of any content analysis, even in the most exploratory case. Because many of the other chapters in this book are devoted squarely to explaining the underlying processes of using computer games and their effects from different theoretical perspectives, this chapter remains largely silent on explanatory mechanisms.

MARKETING OF OBJECTIONABLE CONTENT

Ads for video games may appear in many locations such as television, "gaming" magazines, or on Internet Web sites. Another persuasive tool is point-of-purchase (POP) advertising or the packaging (i.e., jacket covers) of video games. The use of appealing visuals and graphic displays to highlight different attributes of game play on box art may be particularly influential when young consumers are alone or with their parents in retail stores contemplating a purchase.

What is typically featured on the jacket covers of video games? Only a few content studies have attempted to answer this question (Brand & Knight, 2002). Using Goffman's work as his theoretical foundation, Provenzo (1991) content analyzed gender portrayals on the box art associated with 47 top-rated Nintendo games. In terms of sheer prevalence, men outnumbered women by a ratio of 13 to 1 (115 males to 9 females)! Of these characters, men were substantially more likely to be shown initiating or dominating the action (e.g., using weapon, leading group) than were women (20 vs. 0, respectively).

Nearly 10 years later, Children Now (2000) examined the packaging of 24 top-selling games across three platforms. Of those games (54%) with female characters, only two featured females on the covers and both presented them in a provocative light. Surely, the results from these studies suggest that females have been grossly underrepresented on the box art of games.

Encompassing aspects of gender portrayals as well as several other content attributes, the Advertising Review Council (ARC) of the Entertainment Software Review Board has advanced a set of guidelines for responsible advertising practices when marketing video games in the United States (ARC, 2001). Both visual portrayals and printed text (i.e., ad copy) on jacket covers are identified as part of a "qualifying advertisement" that falls under the ARC's jurisdiction (ARC, 2001, p. 4). Based on the ARC guidelines (2001, pp. 5–6), publishers are to be "sensitive" when portraying five key attributes: violence, sex, alcohol/drugs, offensive verbal or bodily expression, and beliefs.

For this chapter, Smith, Pieper, and Choueiti (2004) examined whether packaging and ad copy on video game covers are in compliance with the ARC Guidelines. We assessed the packaging of 74 top-selling games for Nintendo GameCube, Sony Playstation2, and Xbox. Using definitions from the *National Television Violence Study* (Smith et al., 1998), we examined the prevalence, type, graphicness, and targets of violence on the front and back covers of the games. Other variables relating to sexuality (e.g., nudity, revealing clothing), gender representation (e.g., presence of males, females), and substance use (e.g., use of alcohol, drugs, smoking) were also measured.

In terms of ad copy, the front and back text of jacket covers was examined for references to violence. Although the ARC does not have specific guidelines regarding ad copy, all violent content is supposed to be sensitive to graphic and/or excessive portrayals of aggression, weapon use, and blood/gore (ARC, 2001, p. 6). A total of 31% ($n = 23$) of the games contained violent ad copy. Table 5.1 shows examples of ad copy explicitly or implicitly referring to violence,

TABLE 5.1

Examples of Violent Ad Copy on Jacket Covers of Video Games

Game Title	Platform	Rating	Cover	Examples of Ad Copy
Ikaruga	NGC	E	Back	"…the lone survivor of a ravaged people. He must battle fiercely…in a new war—one of frighteningly intense firepower and hypnotic beauty."
The Legend of Zelda—The Wind Waker	NGC	E	Back	"Take on hordes of incredibly animated enemies…"…take the fight to spectacular, massive bosses."
Warioworld	NGC	E	Back	"…beat it out of 'em!…I'm ready to brawl! I'll be throwing punches, charging through crowds, and piledriving ugly mugs right and left…do I have to give you a knuckle sandwich too?" "Battle an endless tide…" "Pummel huge-bosses…"
Crash Bandicoot—The Wrath of Cortex	PS2	E	Back	"…demolish all that lies before him"
ATV 2—Quad Power Racing	XBOX	E	Back	"Kick it! Stomp the comp with an arsenal of vicious fighting moves" "Punish opponents…"
Hulk[1,3]	PS2	T	Back	"…a terrifying villain intent on unleashing a relentless army…on the world." "…as you sneak and smash your way through…" "…if you can see it—you can smash it."
Tom Clancy's Splinter Cell—Stealth Action	PS2	T	Back	"You have the right to spy, steal, destroy, and assassinate…" "Force enemies to cooperate or use them as human shields." "Stalk from the shadows. Strike with precision." "…gut-wrenching stealth action."
Wolverine's Revenge[3]	PS2	T	Back	"…you were implanted with a deadly virus that will kill you…" "Time is running out. And so is your life." "Unsheathe your razor-sharp claws, unleash lethal combo attacks…as you tear into intense action"
Resident Evil Zero	NGC	M	Back	"…investigate a series of grisly murders…" "An overturned military transport riddled with corpses"
Grand Theft Auto—Vice City	PS2	M	Back	"Most of Vice City seems to want Tommy dead. His only answer is to fight back and take over the city himself."
Return to Castle Wolfenstein—Operation Resurrection	PS2	M	Back	"…specialist in…heavy weapons and assassination…twisting the science and the occult into an army capable of annihilating the Allies"
Tenchu—Wrath of Heaven	PS2	M	Back	"Live by honor. Kill by stealth." "…amassing an army of ninjas and lords of darkness…" "…you—a cunning stealth assassin"

TABLE 5.1
(Continued)

Game Title	Platform	Rating	Cover	Examples of Ad Copy
Brute Force	XBOX	M	Front	"Dangerous Alone" "Deadly Together" "Alone each of you is deadly. United, you are an unstoppable
			Back	war machine." "Take on an entire race of alien predators . . ."
Halo	XBOX	M	Back	"Bent on Humankind's extermination . . . wiping out Earth's fledging interstellar empire . . . the other surviving defenders of a devastated colony . . . Shot down and marooned . . . Fight for humanity against an alien onslaught . . ." "Attack on all fronts" "Bred for combat, built for war, you are a master of any weapon"
Max Payne	XBOX	M	Back	". . . man with nothing to lose in the violent, cold, urban night" "framed for murder, hunted by cops and the mob" "painstaking attention to detail—even the bullets are modeled accurately"
Tao Feng—Fist of the Lotus	XBOX	M	Back	"the pain is real" ". . . the most brutal and lifelike hand-to-hand fighting game . . . Clothes will tear, blood will spill, and bones will break" "LIMB DAMAGE: Feel every torn muscle and broken limb as targeted strikes shatter bones" "See every cut, gash, and bruise in real-time." "Throw your opponent through glass . . ." "Build your Chi to unleash bone-crushing devastation on your enemies, or use it to heal your own battered body."
Tom Clancy's Ghost Recon[2]	XBOX	M	Back	". . . an elite handful of specially-trained Green Berets, armed with the latest technology and the deadliest weapons . . ."

Note: If a game was featured on more than one platform and had violent ad copy, then it is illustrated by a superscript. 1 = Game cube, 2 = PS2, 3 = Xbox.

weapons, and/or consequences. Perhaps the most extreme instance comes from Xbox's, *Tao Feng: Fist of the Lotus*. Outside of this example, most ad copy does not appear to be graphic or excessive in nature, thereby complying with both the spirit and the letter of the ARC guidelines.

In terms of box art, a full 70% of the games depicted visually one or more instances of violence (see Table 5.2). The most frequent types of violence were credible threats and behavioral acts. Despite the ARC's rules that jacket covers should be sensitive to "excessive" portrayals of violence, some games had up to 12 credible threats and 9 behavioral acts. Adding up these two types of acts yielded a range of violent instances from 1 to 17 on the jacket covers, with 11.7% featuring only 1 instance, 50.9% featuring 2–5, 27.4% featuring 6–10, and 10%

TABLE 5.2

Prevelence of Violence, Gender, and Sexual Content on Vedio Game Jacket Covers

	% of Games
Violence	
Prevalence	
% of jacket covers with violence	70.2% (n = 52)
% of credible threats (range)	85% (n = 44; range 1–12)
% of behavioral acts (range)	75% (n = 39; range 1–9)
% of harmful consequences of unseen violence	11.5 (n = 6)
Explicitness & Graphicness	
% w/ any explicit violence (20% or more)	14% (n = 8)
% w/ graphic violence (any blood/gore)	2% (n = 1)
Weapons	
% w/ any type of weapon shown	69% (n = 51)
% w/ natural means	56% (n = 29)
range of natural means shown	1 to 7 instances
% w/ unconventional means	7.7% (n = 4)
ranges of means shown	1 instance
% w/ conventional weapons, non-firearms	28.8% (n = 15)
range of conventional weapons shown	1 to 15 instances
% w/ guns	56% (n = 29)
range of guns shown	1–14 instances
% w/ heavy weaponry	9.6% (n = 5)
range of heavy weaponry shown	1 to 6 instances
Targets of Violence	
% of games w/ women as targets of violence	12% (n = 9)
% of games w/ children as targets of violence	0
% of games w/ public figures as targets of violence	3% (n = 2)
Gender & Sexual Content	
Prevalence of Men & Women	
% of games with one or more human men on cover	88% (n = 65)
range of human men on jacket covers	1 to 26 men
% of games with one or more human women on cover	38% (n = 28)
range of human women on jacket covers	1 to 12 women
Sexuality	
% of games w/ partial nudity*	43.2% (n = 32)
% w/ women partial nudity*	32% (n = 24)
% w/ men partial nudity *	20% (n = 15)
% w/ sexual behavior	3% (n = 2)
% w/ sexual violence, necrophilia, or STDs	0

Note: * = includes human and anthropomorphized characters.

featuring 10 or more. If six or more instances of violence constitute "excessive," then more than a third (37.4%) of the box art may be in violation of the ARC guidelines.

Graphic or violent depictions involving weapons are also the subject of ARC guidelines. Table 5.2 illustrates that weapons are routinely portrayed (69% of jacket covers), especially guns and natural means. The ARC also suggests that violence should not be shown graphically. Only one video game depicted blood, gore, or viscera, the definition of graphicness used by the NTVS. Furthermore, the ARC guidelines indicate that publishers should be conscientious with portrayals of violence targeting women, public figures, and children. These variables occurred infrequently (12%, 3%, 0%, respectively).

Game publishers seem to be in compliance with portrayals of sexuality and substance use. Smith et al. (2004) observed no instances of full nudity (partial nudity[1] appeared in 43% of games), sexual behavior[2] in only two games (i.e., man holding a woman), and no portrayals of sexual violence, necrophilia, or information about sexually transmitted diseases (STDs). Substances are nonexistent in games (i.e., smoking = 0%, drug use = 0%). However, three games depicted women holding presumably alcoholic beverages (i.e., *Grand Theft Auto*, female holding a martini glass with a green olive; *The Sims* which appears twice in the sample, shows a female in a hot tub clutching a wineglass).

The findings from the Smith et al. (2004) study suggest that the video game industry is doing a fair to good job complying with the ARC standards, especially in the areas of sexual content and substance use. Of course, the portrayal of violence and brandishing of weapons is one area where improvement could be shown.

CONTENT OF GAMES

A recent report released by the Kaiser Family Foundation (1999) revealed that 8- to 18-year-olds spend 27 minutes a day playing video games. What types of portrayals are children likely to see in this time frame? To date, a total of 13 studies could be found that have examined the content of video games. Table 5.3 features a complete list of these investigations. All but three were conducted in the United States and almost all of the studies focused to varying degrees on different types of negative content. Research examining violence, gender representation, profanity, and substance use will be reviewed below.

Violence

Violence has attracted the most empirical attention, presumably due to the potential harm it may cause. Perhaps the most controversial games come from the *Grand Theft Auto* series, in which players are told they can engage in "murder, drug busts . . . road rage, extortion . . . pimping, petty thievery and double parking!" (as cited in Parvaz, 1999, p. 1). The concern over such contents may be warranted, as studies show that playing violent video games is a positive and significant predictor of aggression (Anderson & Bushman, 2001; see this volume, Weber, Ritterfeld, & Kostygina, chap. 24; Lee & Peng, chap. 22).

How much violence exists in video games? Examining popular console games, Dietz (1998) found that 79% contained physical aggression. With a larger sample of platform, handheld, and PC games, Children Now (2001) found that 89% depicted violence as a content attribute. Internationally, Schierbeck and Carstens (2000) conducted a content study in Denmark and found that 53% of PC and platform-based games depicted violence. Shibuya and Sakamoto (2004) found that among games rated highly by fifth graders in Japan, 85% featured violence.

Rating and genre also seem to exert an influence on whether violence is present or not in games (see also B. Smith, chap. 4, this volume). Studies show anywhere from 57% to 79% of

TABLE 5.3
Summary of Video Game Research

Author & Year	N	Sample	ESRB Rating	Type of Sample	Console	Unit of Analysis	Duration of Game Play	# of Game Players	Skill of Game Players	% of Violence
Provenzo (1991)	47 games	Console games	n/a	Top-rated games	Nintendo	Game covers, game theme	n/a	n/a	n/a	85%
Braun & Giroux (1989)	21 games	Arcade games	n/a	Popular games	n/a	Game	n/a	n/a	n/a	71%
Dietz (1998)	33 games	Console games	not in article	Most popular games	Sega Genesis, Nintendo	Game	Not in article	Not in article	Not in article	79%
Children Now (2000)	24 games	Console games	E, T, M	Top-selling games	Sony PS, Sega DC, Nintendo 64	Game cover, character, game	1st level	Not in article	Not in article	46%
Children Now (2001)	70 games, 1,716 characters	Console games, handhelds, PC games	E, T, M	Top-selling games	Sega DC, Sony PS & PS2, Nintendo 64	Character, game	1st level	2	Not in article	89%
Thompson & Haninger (2001)	55 games	Console games	E	Convinence sample	Sega DC, Sony PS & PS2, Nintendo 64	Violent incident, game	90 min/ end of game	1	Expert	64%
Schierbeck & Carstens (2000)	338 games	Console games, PC games	n/a	All games available '98	Playstation, Nintendo 64	Game	Not in article	Not in article	Not in article	53%
Beasley & Standley (2002)	47 games, 597 characters	Console games	E, T, M, RP	Random sample	Sony PS, Nintendo 64	Character, game	20 min	Not in article	Not in article	n/a
Brand & Knight (2002)	130 games	Console games, handhelds, PC games	n/a	Top-selling games	Sony PS2, Xbox, Game Cube	Game covers, two lead characters, game	10 min	4	Had played games or became familiar	
Smith et al. (2003)	60 games 1,389 interactions	Console games	E, K-A, T, M	Top-selling games	Sony PS, Sega DC, Nintendo 64	Violent interaction, game	10 min	3	3–5 yrs gaming experience	68%
Shibuya & Sakamoto (2004)	41 games	Not specified	n/a	Most popular by 5th-grade children	Not in article	Game	Up to 2 hours	11	Not in article	85%
Haninger & Thompson (2004b)	81 games	Console games	T	Random sample of T games	Sega DC, Sony PS & PS2, Nintendo 64	1 sec. epoch, game	1 hour	1	Expert	98%
Haninger et al. (2004)	9 new games (81 games from H & T '04)	Console games	T	Not in article	Xbox, Game Cube	1 sec. epoch, game	1 hour	2	Played games several hours	89%

games for younger audiences (i.e., "E" or "K-A") feature physical aggression (Children Now, 2001; Smith et al., 2003; Thompson & Haninger, 2001) whereas 90% to 98% of those for more mature audiences do (i.e., "T" or "M"; Haninger & Thompson, 2004a; Smith et al., 2003). In Denmark, Schierbeck and Carstens (2000) found the most violent genre to be role-playing (100%), followed by strategy (89%) and action (86%).

Whereas the reviewed research tells us about the prevalence of violence, it reveals little about the importance or saturation of physical aggression in games. In terms of importance, Children Now (2001) found that violence was essential for reaching game-related objectives in 41% of the games and was the central activity in 17%. In terms of saturation, Thompson and Haninger (Haninger & Thompson, 2004a; Thompson & Haninger, 2001) have found that, on average, roughly one third of play in "E" and "T" rated violent games features physical aggression (31%, 36% of game play, respectively). Smith et al. (2003) examined violence in the first 10 minutes of play time across 60 games. The researchers found that "T" and "M" rated games featured 4.59 violent interactions per minute whereas "E" and "K-A" rated games featured 1.17. Because 8- to 18-year-old boys, on average, in the United States play games just over 40 minutes per day (Kaiser Family Foundation, 1999), those choosing "T" or "M" rated interactive media during this time will be exposed to roughly 180 interactions of violence a day, 5,400 per month, or 64,800 a year! Such repeated exposure may contribute to the development and reinforcement of aggressive scripts for solving interpersonal conflicts (Huesmann, 1988; see also Lee & Peng, chap. 22, this volume).

Scholars also have examined the way in which violence is presented in games. From empirical work in the media effects arena, we know that the nature of violence influences how viewers interpret and respond to televised or cinematic acts of aggression (Smith & Donnerstein, 1998). Studies show that violence involving justification (Paik & Comstock, 1994), perpetrators with whom the viewer identifies (Huesmann, Moise-Titus, Podolski, & Eron, 2003; Perry & Perry, 1976), humor (Baron, 1978; Berkowitz, 1970), rewards or no punishments (Bandura, 1965; Bandura, Ross, & Ross, 1963), excessiveness (Huesmann et al., 2003), guns (Berkowitz & LePage, 1967), minimal consequences (Baron 1971a, 1971b), and realism (Huesmann et al., 2003; Thomas & Tell, 1974) all *increase* the risk of aggression.

Assuming that these variables may also be important in players' reactions to video game content, how often do such contextual elements appear across the landscape of popular games? Although some studies have examined the nature of violence in games (Children Now, 2001; Haninger et al., 2004; Haninger & Thompson, 2004a; Shibuya & Sakamoto, 2004; Thompson & Haninger, 2001), only one will be focused on here. Smith et al. (2003) assessed the above contextual features using the framework of the *National Television Violence Study*. As such, all of the definitions of the context features are taken from operationalizations in the experimental, survey, and longitudinal television-based literature demonstrating effects.

Table 5.4 features the overall results, as well as by rating, from the Smith et al. (2003) investigation. Several trends emerge from the findings. First, violence seems to be enacted by humans who are White and male. Second, violent interactions seem to be socially sanctioned or justified, which is somewhat consistent with Children Now's (2001) findings (91% of player-controlled actions are justified). Third, violence does not seem to be sanitized. Although the majority of interactions feature unrealistically low levels of harm, games usually feature the consequences of violence, which can reduce the probability of aggression (Baron, 1971a, 1971b). However, consequences in video games might function as a reward for injuring or killing, thereby heightening the likelihood of aggressive responding. Fourth, a majority of the games reward (56%) but do not punish (98%) characters who engage in violence. Fifth, 14% of all violent interactions are from a first-person perspective.

One interesting contextual feature is the presence of guns. Smith et al. (2003) found that only 8% ($n = 116$) of all violent interactions in their sample featured gun use. All of these

TABLE 5.4
Percentage of Contextual Variables as a Function of Game Rating

	All Audiences	Mature Audiences	Overall
Violence			
% w/physical aggression	57%	90%	68%
rate per 10-minute segment	1.17	4.59	2.315
Perpetrator Qualities			
% of vio interactions w/ male perpetrators	90% ($n = 230$)	75%($n = 618$)	79% ($n = 848$)
% of vio interactions w/ female perpetrators	10% ($n = 25$)	25% ($n = 202$)	21% ($n = 227$)
% of vio interactions w/ White perpetrators	78% ($n = 106$)	69% ($n = 457$)	71% ($n = 563$)
% of vio interactions w/ human perpetrators	40% ($n = 186$)	87% ($n = 802$)	71% ($n = 988$)
% of vio interactions w/ robots	39% ($n = 181$)	7% ($n = 63$)	18% ($n = 244$)
Reason for Violence			
% of vio interactions that are justified	56% ($n = 264$)	77% ($n = 711$)	70% ($n = 975$)
Weapons Used			
% of vio interactions w/ natural means	79% ($n = 369$)	42% ($n = 390$)	55% ($n = 759$)
% of vio interations w/ unconventional weapons	6% ($n = 28$)	20% ($n = 185$)	15% ($n = 213$)
% of vio interactions w/ guns	0%	13% ($n = 116$)	8% ($n = 116$)
Extent			
% of vio interactions involving repeated violence	27% ($n = 125$)	60% ($n = 539$)	49%($n = 664$)
% of vio interactions w/ lethal violence	73% ($n = 345$)	80% ($n = 735$)	78% ($n = 1080$)
Consequences of Violence			
% of vio interactions w/ no harm/pain	6% ($n = 28$)	3% ($n = 24$)	4% ($n = 52$)
% of vio interactions w/ unreal harm/pain	55% ($n = 261$)	52% ($n = 481$)	53% ($n = 742$)
Explicitness/Graphicness			
% of segments w/ up-close violence	74% ($n = 17$)	83% ($n = 15$)	78% ($n = 32$)
% of segments w/ any blood or gore	4% ($n = 1$)	56% ($n = 10$)	27% ($n = 11$)
Rewards & Punishments			
% of segments w/ rewards for violence	57% ($n = 13$)	56% ($n = 10$)	56% ($n = 23$)
% of segments w/ no punishments	100% ($n = 23$)	94% ($n = 17$)	98% ($n = 40$)
Humor			
% of segments w/ funny comment or action	48% ($n = 11$)	33% ($n = 6$)	41% ($n = 17$)
Visual Perspective			
% of vio interactions from 1st-person perspective	0%	21% ($n = 198$)	14% ($n = 198$)

Note: Adapted from Smith, S. L., Lachlan, K., & Tamborini, R. (2003). Popular video games: Quantifying the presentation of violence and its context. *Journal of Broadcasting and Electronic Media, 47*, 58–76. Vio interactions refer to aggressive exchanges between a perpetrator engaging in a particular type of act against a target. Segments refer to 10 minutes of game play.

incidents were featured in "T" and "M" rated games. Other studies have found higher estimates. Thompson and Haninger's (2001) results show that 24% of the "E" games in their sample depicted guns. Gun use was found in 57% of the "T" rated games in Haninger et al.'s (2004) study. Some of the variability in these findings is presumably due to differences in methods employed across investigations, which will be discussed later in this chapter.

The research reviewed above suggests that violence is commonplace in video games. The context is characterized by human aggression that is justified, not sanitized, and rewarded in nature. Experiments are just starting to examine the impact of contextual features in games on aggression (Cope-Farrar, Krcmar, & Nowak, 2004; Lee & Peng, chap. 22, this volume). Thus, the meaning of these findings must be interpreted with caution.

Gender Representation and Sexuality

Another content feature receiving empirical attention is the representation of gender (see Cassell & Jenkins, 1999). It is well known that girls spend less time playing video games than do boys (Kaiser Family Foundation, 1999). One reason for this may be that females are not featured as often as males in this medium. The content analytic research seems to support this argument.

Braun and Giroux (1989) found that almost 60% of arcade games featured males whereas roughly 2% featured females. Similar disparity was observed with male and female voices in games ($\sim$ 20% vs. 2%, respectively). Dietz' (1998) results revealed that 41% of the video games with characters in her sample did not portray any females. Examining 1,716 characters, Children Now (2001) found that males appeared almost four times as frequently as females (64% vs. 17% of characters, respectively). Focusing on player-controlled characters only increases this gender gap (73% vs. 12%). Similar findings were revealed by Beasley and Standley's (2002) analysis, where men (71.5%) were substantially more prevalent than women (13.7%).

Not only are females underrepresented, but they also tend to be depicted in a stereotypical light. Dietz (1998) found that females were presented as the "damsel in distress" (i.e., *Teenage Mutant Ninja Turtles*, *Double Dragon*) in roughly one fifth of the games (21%) in her sample. Children Now's (2001) results show that female characters were more likely than male characters in games to scream, be nurturing, and help/share.

In addition to gender stereotyping, females' bodies are sometimes featured in a hypersexualized light. Femme fatales such as Lara Croft in the *Tomb Raider* series, Ailish in *Sudeki,* and Monica in *Dark Cloud 2* are portrayed with unrealistically large breasts and disproportionately small waists. How common are such body images in games? Research reveals that they occur with some frequency. Children Now (2000) looked at body types in games with female lead characters. In such games, Children Now (2000) found that over a third of females (38%) had large breasts and approaching half (46%) had undersized waists. In their follow-up study, Children Now's (2001) results showed that 11% of females had "very large breasts and a very small waist" (p. 13).

Scantily clad clothing and partial nudity are staples of video games (i.e., *Extreme Volleyball*), especially when such attire is completely unsuited for the character's mission or quest. For instance, *ATV: Off Road Fury* is a game where players can race through the desert. Males are often shown clothed in protective gear from head to toe, while some females are shown in bikini tops and shorts. Such portrayals are not uncommon in games. Almost half of the females in Beasley and Standley's (2002, p. 287) sample were shown sleeveless, with many wearing halter tops, tank tops, and bathing suits during game play. Of the games with females in their sample, Children Now (2000) found that 23% of females uncovered their cleavage, 31% modeled their midriffs, and 15% brandished their buttocks (p. 4). Similar patterns were observed a year later (Children Now, 2001, p. 14) with 21% of women showing partially exposed breasts (7% completely exposed) and 13% showing somewhat exposed behinds (8% completely exposed). Finally, females were twice as likely as males to be shown in revealing clothing.

Although we know very little about sexual content in games on new platforms such as Xbox and GameCube, Haninger et al. (2004) content analyzed nine games from these consoles for sexual content. Three examples, all from "T" rated games, were noted for featuring such behavior. Here is the researchers' verbatim summary of what they found in two of the games (Table 6, ¶1, 3):

> *Tony Hawk's Pro Skater 4* features adult film star Jenna Jameson, who provided the voice and model for Daisy (1 of 2 playable female skaters in the game). Daisy performs provocative tricks on her skateboard and suggestively says, "Oh yeah, now it's time

for some X-X-X-treme action!" Behind-the-scenes film clips in the game show Jenna Jameson fully clothed, although they include shots of her raising and lowering her skirt very quickly, posing provocatively with a skateboard, and lifting her shirt to the camera operator, who exclaims, "Damn, they're bigger than I thought they were!"

WWF SmackDown! contains a 21-second film clip of wrestler Val Venis with 2 women in a hot tub in a scene that depicts sexual images and an ecstatic expression on Val Venis's face that is suggestive of oral sex.

These examples reveal that more research is needed to examine whether such comments and behaviors are typical or idiosyncratic representations of content found in "T" rated games on Xbox and GameCube platforms. Together, the research reviewed above suggests that women are featured substantially less often in video games than are men. Given that same sex role models in the media may be important socializing agents to youth (see Jose & Brewer,1984; Maccoby & Wilson, 1957), young girls have fewer characters to attend to and identify with in games than do boys. The types of role models that are available to girls are also troubling. The hypersexualized and disproportionately thin females may be teaching young girls ideals about beauty and thinness that are damaging to their socioemotional or physical health. Indeed, some studies in the television arena show that females who are attracted to thin characters (Harrison, 1997) as well as those who compare themselves to media figures while viewing (Botta, 1999) score higher on eating disorder-related variables. Such depictions may also affect boys' social learning about women and their attitudes toward the opposite sex.

Profanity

Only two studies have examined the prevalence of profanity in games. Thompson and Haninger (2001) found two instances of mild language in their sample of "E" rated games. The only portrayal of profanity came from *Goemon's Great Adventure*, which featured in text on-screen the word "damn." Looking at "T" rated games, Haninger and Thompson's (2004a) results revealed that 27% featured profanity in conversation, visual text, nonverbal actions, or on the musical soundtrack. The worst offenders were *WWF Smackdown!* with 12 instances per hour, *Starlancer* with 10, and *Ogre Battle 64: Person of the Lordly Caliber* with 9.

Substance Use

Thompson and Haninger (2001) found that none of the "E" rated games in their sample featured substance use. However, one game (*Harvest Moon 64*) gave players the option to "purchase and consume beer, wine, or liquor resulting in a red face and a fall to the floor" (p. 596). Haninger and Thompson (2004a) found that 15% of "T" rated games in their sample depicted substances. Of these, use was depicted in nine games, which either showed the characters drinking alcohol (e.g., *The Simpson's Wrestling*) or smoking cigarettes (e.g., *Alundra 2*).

Summarizing the above research, video games seem to feature quite a bit of violence but very little profanity and substance use. In terms of character representation, males outnumber females by a ratio of roughly 4 to 1. Furthermore, hypersexuality and partial nudity seems to have found a niche in this interactive medium. After attending the unveiling of new games at the Electronic Entertainment Exposition (E3) in Los Angeles in May of 2004, one reporter's headline read, "Video games get raunchy: Animated sex? Naked women? Streakers? They're all in this year's crop of upcoming titles" (Morris, 2004, p. 1). More research is needed to not only monitor sexual content in newly released games but also to assess the effects that exposure to such portrayals may have on youth.

METHODOLOGICAL CHALLENGES TO CONTENT STUDIES OF GAMES

Just like any mass medium, the study of content patterns in video games possesses some unique challenges to social scientists. Such challenges could threaten the validity of the current research findings. I want to focus on only three main issues here: sampling of time frames, player skill, and characters chosen for play. These issues are important, especially in light of the legal and societal concern surrounding this popular media technology.

Time Frames

Table 5.3 features a list of time devoted to game play across the studies, ranging from 10 minutes to 2 hours. Several criticisms could be leveled against the studies only sampling 10 to 20 minutes of play, with the most blatant being that time begets more explicit and graphic content. Or does it? Recently, Haninger and Thompson (2004b) addressed this very issue—at least in part. In a chart (see Fig. 5.1) that was unpublished (see Haninger & Thompson, 2004b, ¶20) with their *JAMA* article, these scholars showed pictorially that playing and coding games for an hour captures roughly 90% of the content as illustrated by ESRB labels (violence, sex, profanity, substances, gambling). The authors stated, however, that "playing games for only

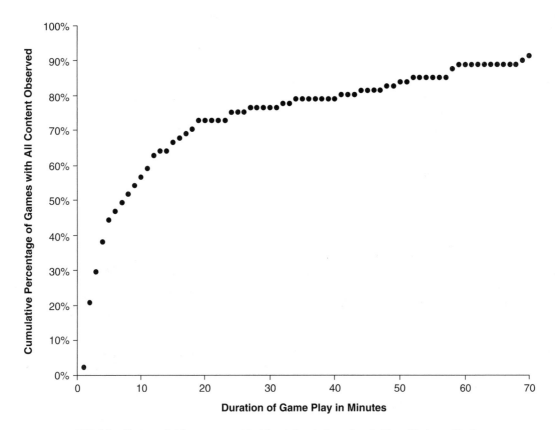

FIG. 5.1.   Content of video games captured in relation to time played. (From Haninger, K., & Thompson, K. M. (2004b). Frequently asked questions about "Content and Ratings of Teen-Rated Video Games," retrieved from http://www.kidsrisk.harvard.edu/faqs4.htm.)

10 minutes would lead to missing 1 or more types of game content over 40% of the time" (Haninger & Thompson, 2004b, ¶20).

In terms of accuracy, Haninger and Thompson's (2004a, 2004b) study suggests that future scholars may want to sample longer time frames in their content investigations. It also will be important to map whether and how violence or any other objectionable content escalates with time. One thing is important to note, however. After controlling for gamers' age and study date, Sherry's (2001) meta-analysis revealed that time spent playing violent video games is *inversely* related to aggressiveness. Two studies were chiefly responsible for this negative effect (Sherry, 2001, p. 425): one with 10 minutes of game play ($r = .90$) and one with 75 minutes ($r = .05$). The largest effect was observed with the shortest amount of time devoted to play.

What this finding cautiously suggests, as Sherry (2001) noted, is that scholars need to think more critically about the theoretical processes (i.e., arousal, general aggression model, catharsis, identification, uses/gratifications) at work when selecting time frames for their experiments. The same rationale should apply to content studies, with a principled theoretical argument crafted for the time period sampled.

Player Skill

Unlike television, the content to be analyzed in a video game is dependent on the individual playing it. Perhaps the greatest challenge to researchers wanting to study content patterns in games is choosing who to play the game and how he or she is told to play it. As illustrated in Table 5.3, some studies (Smith et al., 2003) employ expert gamers with no experience with a particular game (i.e., game-specific novice) whereas others (Haninger & Thompson, 2004a; Haninger et al., 2004) have gamers familiarize themselves with the content for some time (i.e., several hours) before game play is captured (i.e., game-specific experts). Neither approach is right or wrong, but these sampling decisions may yield vastly different estimates of a particular content attribute across studies.

In fact, scholars have indicated that player skill can have an impact on video game content (Christofferson, 2000). Discussing this in terms of novices and experts, Newman (2004, pp. 134–135) argued that preferences and choices of players shift with time and game familiarity. Newman (2004) contended that novices may choose characters or vehicles (see *ATV: Off Road Fury*) with average capabilities across a variety of tasks to learn the inner workings of a game. With repeat play, gamers may forego average capabilities in one area to gain superior skills in others (Newman, 2004), not to mention their increased response times to content features (see Greenfield, deWinstanley, Kilpatrick, & Kaye, 1994) and knowledge of tricks, codes, and cheats that will unlock special competences (i.e., use of special weapons, powers) and/or quests/missions during play. Such sophistication may enable experts to navigate greater levels and types of violence during game play than their novice counterparts.

There are at least two ways to examine the impact of player skill on content patterns, especially in a controversial area such as violence. One approach, as Newman (2004) noted, would be to capture game play of novices longitudinally as they become experts. It may be interesting to also examine how different personality (i.e., aggressiveness, impulsivity, empathy, game efficacy), cognitive (i.e., IQ), or demographic (i.e., age, gender, ethnicity) variables influence this progression or lack thereof over time. Another approach would be to have 10 experts and 10 novices play the same game and look for between and within group variability. Such an investigation would give us valuable information about the static and/or dynamic nature of a content attribute such as aggression and whether it varies tremendously in graphicness and/or explicitness across similar and different types of players.

Type of Character Chosen

In some games, players do not have a choice in terms of character representation (e.g., *The Hulk, Spiderman*). In others, however, they do. Sometimes all of the choices are available at the start of the game. Usually, these types of games allow you to either pick the character you would like to be (e.g., *Def Jam Vendetta*) or construct your character by selecting from a variety of head and body options (e.g., *Syphon Filter, The Omega Strain*). Character choices and exotic outfits sometimes become available with extended play (e.g., *Dynasty Warriors 4, Syphon Filter*). To illustrate, *State of Emergency* gives players five character choices, only two of which can be chosen at the game's beginning (i.e., Libra, McNeil). The remaining three characters (i.e., Spanky, Freak, Bull) are only available after different levels are completed in the Revolution mode. Libra is a thin White female attorney, dressed in a short miniskirt and tight top with a plunging neckline. Spanky is an overweight Hispanic male, with tattoos and a large gold chain around his neck. Clearly, the player's choice of character at the outset or within the duration of the game can have significant ramifications for coding gender, race, and variables related to sexuality.

Not only do character options vary, but so might their skills. Many characters have differing levels of strength, speed, agility, attack/defense, and expertise in weapon use (e.g., *Twisted Metal Black*). Although all of these attributes may affect violence, it is the last feature (character expertise) that may render discrepancies in the types of weapons used and consequences shown across studies. In *Dynasty Warriors 4,* Sun Shang Xiang's weapons of choice are wind and fire wheels and the ability to flip, kick, and make the ground beneath her shake. Huang Gai, on the other hand, uses a staff and grenades to assail his enemies. The game appears to present both characters as equally violent, but engaging in very different *types* of aggression. Thus, a content study featuring 30 minutes of game play with Sun Shang Xiang should yield very different estimates of weapon use than 30 minutes of game play featuring Huang Gai. Hence, character sampling may have important implications for the results of any content study.

The purpose of this section was to bring to the surface three methodological issues confronting scholars executing content analyses of video games. Surely these topics bring up more empirical questions for which we currently have no answers. As video game content evolves and becomes more player specific, researchers approaches to sampling will also have to progress to yield valid estimates of different types of behaviors found in this interactive technology.

RATING SYSTEMS

Several rating systems have been developed to inform consumers about the contents of games. Each system is based on different societal and/or legal standards. As such, the ratings process often directly impacts game content sold and distributed worldwide. As one journalist aptly stated, "Red blood in a game sold in the United States turns green in Australia. A topless character in a European title acquires a bikini in the US. Human enemies in a US game morph into robots in Germany" (Germany: Market tricky for software makers, 2003, p. 1). In this section, four systems will be described. The first two are self-regulatory schemes developed by the interactive industry whereas the last two are legally mandated, controlled by governmental agencies.

Entertainment Software Review Board (ESRB)

Using a framework approved by the video game industry, the *ESRB* has been rating games in the United States since 1994 (ESRB Game Ratings, n.d., ¶1). Rating video games involves two steps, according to the ERSB Web site. First, the publisher completes a survey about the game and provides the ESRB with a videotape of the most typical and graphic portions of the game. Second, the contents are evaluated by three autonomous auditors, who suggest an age-based rating and any accompanying content descriptors. There are five age-based (i.e., EC, E, T, M, AO) categories and over 30 content descriptors a video game can receive, which were reviewed in the previous chapter (see B. Smith, chap. 4, this volume).

Though legislation is pending (see Pereira, 2003), there are no legal restrictions involving the selling of violent games to children in the United States. The ESRB has encouraged retail outlets to implement their "commitment to parents" policy, however. This program states that in the absence of parental approval, retailers will not sell or rent M-rated games to children 16 years of age or younger (ESRB, n.d., ¶14). This policy is rarely followed, however. According to a recent study conducted by the Federal Trade Commission (2000, p. 63), 85% of purchase attempts for M-rated games by unaccompanied 13- to 16-year-olds were successful across 380 retail outlets. In those stores endorsing the "commitment to parents" program, a young consumer could buy an M-rated game 8 times out of 10. A follow-up study employing the same method shows some improvement (FTC, 2003), with 69% of purchase attempts for M-rated games from retailers successful by 13- to 16-year-old consumers.

Pan European Game Information (PEGI)

Beginning in 2003, this new single rating system was developed by the Interactive Software Federation of Europe (ISFE) and is currently being used across 16 countries. The system uses an age-based rating scheme (i.e., 3+, 7+, 12+, 16+, 18+) and up to six content descriptors (i.e., violence, sex, bad language, fear, discrimination, drugs) that are illustrated by logos. According to the Web site for the system, "The intensity of the content is appropriate to the age rating of the game" (Pan European Game Information, n.d., ¶3).

To receive a rating, game publishers have trained auditors fill out an online self-assessment form. The form contains a series of items tapping information about, but not limited to, violence, sex, profanity, drugs, and stereotyping. After submitting the form, the game publisher is given an automatic rating. All game-related information is also reviewed by the Netherlands Institute for the Classification of Audiovisual Media (NICAM). It must be noted that there is some variability in the application of the PEGI rating system. Within the United Kingdom, the Video Standards Council (VSC) administers ratings and content descriptors for PEGI. Some variance in ratings is also observed in Portugal and Finland. Germany has not embraced PEGI and continues to use the *Unterhaltungssoftware SelbstKontrolle* (USK).

Unterhaltungssoftware SelbstKontrolle (USK)

The USK is a nonprofit organization in Germany that has been rating games for the entertainment software industry since 1994. Until recently, the USK ratings functioned as advice about games for interested consumers. On April 1, 2003, a new law went into effect that requires consumers to show proof of age when purchasing video games. To receive a rating (USK, n.d.),[3] publishers submit game-related materials to the USK that go through a three-step process. First, a "tester" at the USK plays the game and writes comments about it. Second, individuals with some expertise in youth work review all game-related information with the tester from the USK and recommend an age-based rating. Third, an elected official from the 16 Bundeslaenders reviews and grants a final rating decision.

According to the USK Web site, games can receive one of five age-based labels: no restrictions (i.e., games with no contents of concern, but may be too difficult for young children), 6+ (i.e., games with fantasy/comedy, complexity may cause fear/unease), 12+ (i.e., games with fighting/competition, aggressive actions would not occur in real life), 16+ (i.e., games with fictional/historical warlike aggression, weapon use against humanlike characters), or 18+ (i.e., game contents may negatively affect development).

Another entity in Germany influences the distribution and sales of video games, however. Authorized by the "Youth Protection Law," the purpose of the Federal Department for Media Harmful to Young Persons (Bundesprüfstelle für Jugendgefährdende Medien or BPjM) is to "protect children and adolescents in Germany from any media which might contain harmful or dangerous contents" (BPjM, n.d., p. 1). The BPjM lists the following video game elements as possibly contributing to harmful effects: Human violence is the sole game objective, severe consequences of violence are shown (blood, gore, viscera, cries of agony), consequences are juxtaposed with humor, and human violence is rewarded (p. 2). Other dangerous attributes include but are not limited to sexual objectification, Nazi glorification, and hatred of different racial groups (BPjM, n.d., p. 2).

BPjM is primarily a reactive agency, evaluating game content only at the request of other institutions. Two thirds of the 16-member board has to agree that a particular game has elements that threaten harm to children. When this occurs, the game is put on the "list of youth endangering media" (BPjM, n.d., p. 1). This list, also known as the "index," legally prohibits games from being sold or rented to children as well as advertised or publicly displayed. Index violations are punishable by law, with one source reporting fines up to 50,000 Euro for every infringement (Germany: Tricky market for software makers, 2003, pp. 1–2). Examples of games that have been indexed include *Wolfenstein 3D*, *Mortal Kombat*, and *Grand Theft Auto* (Kreimeier, 1999).

Office of Film and Literature Classification (OFLC)

Prior to being made available to the public, a video game in Australia must be rated by the Office of Film and Literature Classification (2003). The *Classification Act of 1995* specifies a code and guidelines surrounding the review board's evaluation and rating of video games. This act also specifies that for all age-based ratings of 8 and above, the board must give "consumer advice" about potentially objectionable contents of games (OFLC, 2003, p. 3).

To receive a rating and consumer advice, the OFLC Web site indicates that publishers must submit written information, the game itself, and specification of and accessibility to (i.e., footage of the game) any of the potentially problematic contents in the game. Once in receipt of all relevant materials, a classification board evaluates the information. Taking into account community standards as well as the context of the game, the board assigns one of the five age-based ratings (G, G8+, M15, M15+, or RC).

The last two ratings are controlled by law (see OFLC, Information for Consumers, n.d.). MA 15+ games, or those games "that depict, express or otherwise deal with sex, violence or coarse language in such a manner as to be unsuitable for viewing or playing by persons under 15," cannot be purchased by or rented to children under 15 years of age unless they are accompanied by a parent/legal guardian (OFLC, n.d., p. 4). Games that are refused classification are not to be displayed, purchased, or rented in Australia (OFLC, Information for Consumers, n.d.). These games, per the OFLC (n.d., p. 4) Web site, "(a) depict, express or otherwise deal with matters of sex, drug misuse or addiction, crime, cruelty, violence or revolting or abhorrent phenomena in such a way that they offend against the standards of morality, decency, and propriety generally accepted by reasonable adults to the extent that they should not be classified, or (b) depict in a way that is likely to cause offense to a reasonable adult, a person who is, or who looks like, a child under 16 (whether the person is engaged in sexual activity or not); or (c) promote, incite or instruct in matters of crime or violence; or (d) are unsuitable for a minor to see or play."

CONCLUSION

The aim of this chapter was to explore the amount and types of objectionable content found in video games. Two specific attributes seem to inundate games: violence and sexuality. Violence is in a majority of games, independent of rating or researcher. Sexuality, not sexual content, also seems to have found a niche in video games. Women are often presented in games as ultrasexual, with unrealistic body images and scantily clad.

Both of these attributes, undoubtedly, are designed to appeal to the typical video game user: an adult male. Yet many children and adolescents seek out and play games that are rated for mature audiences (i.e., T or M). As a result, young video game players may be exposed to substantially more violence and sexualized content today than ever before. More research is needed on the effects of negative content in games on children's socioemotional development.

On the positive side, rating systems have been designed and implemented to give parents information about video game content prior to making a purchase. Some research in the United States, however, reveals that parents do not always agree with the suitability of content found in E and T rated games for younger audiences (Walsh & Gentile, 2001). Other studies have shown that content descriptors are not always applied consistently to games with violent content (Thompson & Haninger, 2001), especially those rated E for everyone. These findings suggest that more research is needed on the reliability and validity of the ESRB scheme and other rating systems worldwide.

Finally, the issue of conducting content analytic research on video games was addressed. Much of the content shown in games may be player dependent. As a result, the choice of selecting who to play the games in a content study is critical. This will become even more important as games develop increasingly open-ended formats (i.e., MMOGs), allowing play to be uniquely tailored to the needs, wants, and desires of gamers. Thus, future content analytic research will need to explore how characteristics of gamers influence the dynamic or static nature of messages in this interactive and constantly evolving medium.

NOTES

[1] Partial nudity was defined as anything showing part of a character's chest, stomach, buttocks, or upper thighs.

[2] Sexual behaviors were those acts that communicate a sense of physical or likely intimacy. Such actions could range from an intimate touch to portrayals of intercourse.

[3] The information from the USK Web site had to be interpreted by a translator from German to English for this chapter.

REFERENCES

Advertising Review Council. (2001). *Principals and guidelines: Responsible advertising practices for the interactive entertainment software industry* (2nd ed.). Retrieved May 1, 2004, from http://www.esrb.org/arc.asp.

Anderson, C. A., & Bushman, B. (2001). Effects of violent video games on aggressive behavior, aggressive cognition, aggressive affect, physiological arousal, and prosocial behavior: A meta-analytic review of the scientific literature. *Psychological Science, 12*, 353–359.

Bandura, A. (1965). Influence of models' reinforcement contingencies on the acquisition of imitative responses. *Journal of Personality and Social Psychology, 1*, 589–595.

Bandura, A., Ross, D., & Ross, S. A. (1963). Vicarious reinforcement and imitative learning. *Journal of Abnormal and Social Psychology, 67*, 601–607.

Baron, R. A. (1971a). Magnitude of victim's pain cues and level of prior anger arousal as determinants of adult aggressive behavior. *Journal of Personality and Social Psychology, 17*, 236–243.

Baron, R. (1971b). Aggression as a function of magnitude of victim's pain cues, level of prior anger arousal, and aggressor–victim similarity. *Journal of Personality and Social Psychology, 18,* 48–54.

Baron, R. A. (1978). The influence of hostile and nonhostile humor upon physical aggression. *Personality and Social Psychology Bulletin, 4,* 77–80.

Beasley, B., & Standley, T. C. (2002). Shirts vs. skins: Clothing as an indicator of gender role stereotyping in video games. *Mass Communication & Society, 5,* 279–293.

Berkowitz, L. (1970). Aggressive humor as a stimulus to aggressive responses. *Journal of Personality and Social Psychology, 16,* 710–717.

Berkowitz, L. & LePage, A. (1967). Weapons as aggression-eliciting stimuli. *Journal of Personality and Social Psychology, 7,* 202–207.

Botta, R. A. (1999). Television images and adolescent girls' body image disturbance. *Journal of Communication, 49,* 22–41.

Brand, J., & Knight, S. J. (2002). *Diverse worlds project.* Retrieved June 12, 2004, from http://www.diverseworlds. bond.edu.au/Default.htm.

Braun, C. M. J., & Giroux, J. (1989). Arcade video games: Proxemic, cognitive, and content analyses. *Journal of Leisure Research, 21,* 92–105.

Brownback, S. (2000, March 21). *Brownback examines impact of violent video games on children in commerce committee hearing.* Retrieved August 20, 2004, from http://www.brownback.senate.gov/pressapp/record. cfm?id=175723&.

Bundesprüfstelle für Jugendgefährdende Medien. (BPjM). (n.d.). About the BPjM (Federal Department for Media Harmful to Young Persons). Retrieved June 24, 2004, from http://www.bundespruefstelle.de/Texte/General.htm.

Cassell, J., & Jenkins, H. (Eds.). (1999). *From Barbie to Mortal Kombat: Gender and computer games.* Cambridge, MA: MIT University Press.

Children Now. (2000). *Girls and gaming: A console video game content analysis.* Oakland, CA: Author. Retrieved May 17, 2004, from http://www.childrennow.org/media/video-games-girls.pdf.

Children Now. (2001). *Fair play? Violence, gender, and race in video games.* Oakland, CA: Author.

Christofferson, J. (2000). Classifications of interactive electronic media. In C. von Feilitzen & U. Carlsson (Eds.), *Children and the new media landscape: Games, pornography, perceptions* (pp. 133–137). Nordicom, Swedsen: The UNESCO International Clearinghouse on Children and Violence on the Screen.

Cope-Farrar, K., Krcmar, M., & Nowak, C. (2004). *The general aggression model: Explaining the effects of point of view and blood in violent video games.* Paper presented to the Mass Communication Division at the Annual Conference of the International Communication Association, New Orleans, LA.

Dietz, T. L. (1998). An examination of violence and gender role portrayals in video games: Implications for gender socialization and aggressive behavior. *Sex Roles, 38,* 425–442.

Entertainment Software Review Board (ESRB). (n.d.). *ESRB game ratings.* Retrieved May 1, 2004, from http://www. esrb.org.

Federal Trade Commission. (2000). *Marketing violent entertainment to children: A review of self regulation and industry practices in the motion picture, music recording, and electronic games industries.* Retrieved May 1, 2004, from http://www.ftc.gov/reports/violence/vioreport.pdf.

Federal Trade Commission. (2003). *Results of nationwide undercover survey released.* Retrieved August 20, 2004, from http://www.ftc.gov/opa/2003/10/shopper.htm.

Germany: Tricky market for software makers (2003, June 9). *Los Angeles Times.* Retrieved June 22, 2004, from http://www.ebusinessforum.com/index.asp?doc_id=6530&layout=rich_stoy.

Greenfield, P. M., deWinstanley, P., Kilpatrick, H., & Kaye, D. (1994). Action video games and informal education: Effects on strategies for dividing visual attention. *Journal of Applied Developmental Psychology, 15,* 105–123.

Haninger, K., Ryan, S. M., & Thompson, K. M. (2004). Violence in teen rated video games. *Medscape General Medicine.* Retrieved May 14, 2004, from http://www.medscape.com/viewarticle.

Haninger, K., & Thompson, K. M. (2004a). Content and ratings of teen-rated video games. *Journal of the American Medical Association, 291*(7), 856–865.

Haninger, K., & Thompson, K. M. (2004b). *Frequently asked questions about "content and ratings of teen-rated video games."* Retrieved March 16, 2004, from http://www.kidsrick.harvard.edu/faqs4.htm.

Harrison, K. (1997). Does interpersonal attraction to media personalities promote eating disorders? *Journal of Broadcasting and Electronic Media, 41,* 478–500.

Huesmann, L. R. (1988). An information processing model for the development of aggression. *Aggressive Behavior, 14,* 13–24.

Huesmann, L. R., Moise-Titus, J., Podolski, C. L., & Eron, L. D. (2003). Longitudinal relations between children's exposure to TV violence and their aggressive and violent behavior in young adulthood: 1977–1992. *Developmental Psychology, 39,* 201–221.

Jose, P. E., & Brewer, W. F. (1984). Development of story liking: Character identification, suspense, and outcome resolution. *Developmental Psychology, 20*, 911–924.

Kaiser Family Foundation. (1999). *Kids and media at the new millennium*. Menlo Park, CA: Author.

Kreimeier, B. (1999). Killing games: A look at German videogame legislation. *Gumasutra: The art and science of making games*. Retrieved June 23, 2004, from http://www.gamasutra.com/features/19990827/killing_games_01.htm.

Maccoby, E. E. & Wilson, W. C. (1957). Identification and observational learning from film. *Journal of Abnormal and Social Psychology, 55*, 76–87.

Morris, C. (2004, May 12). Video games get raunchy. *CNN/Money*. Retrieved May 12, 2004, from http://money.cnn.com/2004/05/11/technology/e3_nekkidgames/.

Murphy, C. (2002, April 30). Playing the game: Germany's teenage killer. *BBC News*. Retrieved May 17, 2004, from http://news.bbc.co.uk/1/hi/world/europe/1959632.stm.

Newman, J. (2004). *Videogames*. London: Routledge.

Office of Film and Literature Classification (n.d.). *Information for consumers*. Retrieved August 24, 2004, from http://www.oflc.gov.au/content.html?n=150&p=111.

Office of Film and Literature Classification (n.d.). *National classification code*. Retrieved August 21, 2004, from http://www.oflc.gov.au/resource.html?resource=60&filename=60.pdf.

Office of Film and Literature Classification. (2003). *Guidelines for the classification of films and computer games*. Retrieved May 1, 2004, from http://www/oflc.gov.au.

Paik, H., & Comstock, G. (1994). The effects of television violence on antisocial behavior: A meta-analysis. *Communication Research, 21*, 516–546.

Pan European Game Information (PEGI). (n.d.). *What is Pegi?* Retrieved May 1, 2004, from http://www.pegi.info/pegi.jsp?language=en&content=pegi.

Parvaz, D. (1999, October, 14). Murder, pimping, drugs: Subjects guarantee concern. *Seattle Post-Intelligencer*. Retrieved June 3, 2004, from http://seattlepi.nwsource.com/videogameviolence/game14.shtml.

Pereira, J. (2003, January 10). Just how far does First Amendment protection go? Videogame makers use free speech to thwart proposals to keep violent, adult fare from kids. *Wall Street Journal*, B1.

Perry, D. G., & Perry, L. C. (1976). Identification with film characters, covert aggressive verbalization, and reactions to film violence. *Journal of Research in Personality, 10*, 399–409.

Provenzo, E. F. (1991). *Video kids: Making sense of Nintendo*. Cambridge, MA: Harvard University Press.

Schierbeck, L., & Carstens, B. (2000). Violent elements in computer games: An analysis of games published in Denmark. In C. von Feilitzen & U. Carlsson (Eds.), *Children in the new media landscape: Games, pornography, perceptions* (pp. 127–131). Nordicom, Sweden: The UNESCO International Clearinghouse on Children and Violence on the Screen.

Sherry, J. (2001). The effects of violent video games on aggression: A meta-analysis. *Human Communication Research, 27*, 409–431.

Shibuya, A., & Sakamoto, A. (2005). The quantity and context of video game violence in Japan: Toward creating an ethical standard. In R. Shiratori, K. Arai, & F. Kato (Eds.), *Gaming, simulation and society: Research scope and perspective* (pp. 111–120). Tokyo: Springer-Verlag.

Smith, S. L., & Donnerstein, E. (1998). Harmful effects of exposure to media violence: Learning of aggression, emotional desensitization, and fear. In R. G. Geen. & E. Donnerstein (Eds.), *Human aggression: Theories, research and implications for social policy* (pp. 167–202). San Diego: Academic Press.

Smith, S. L., Lachlan, K., & Tamborini, R. (2003). Popular video games: Quantifying the presentation of violence and its context. *Journal of Broadcasting and Electronic Media, 47*, 58–76.

Smith, S. L., Pieper, K. & Choueiti, M. (2004). [Video game packaging and ad copy: Are gaming publishers in compliance with the ARC?] Unpublished raw data.

Smith, S. L., Wilson, B. J., Kunkel, D., Linz, D., Potter, W. J., Colvin, C., & Donnerstein, E. (1998). Violence in television programming overall: University of California, Santa Barbara. *National television violence study* (vol. 3, pp. 5–220). Newbury Park, CA: Sage.

Thomas, M. H., & Tell, P. M. (1974). Effects of viewing real versus fantasy violence upon interpersonal aggression. *Journal of Research in Personality, 8*, 153–160.

Thompson, K. M., & Haninger, K. (2001). Violence in E-rated video games. *Journal of the American Medical Association, 286*(5), 591–598, 920.

Unterhaltungssoftware SelbstKontrolle. (USK). (n.d.). *Haven-guessed/advised*. Retrieved August 22, 2004, from http://www.usk.de.

Walsh, D. A., & Gentile, D. A. (2001). A validity test of movie, television, and video-game ratings. *Pediatrics, 107*, 1302–1308.

Woodruff, J., & Schneider, B. (1999, April 23). Are guns or society to blame? Lawmakers search for answers. *CNN.com*. Retrieved May 17, 2004, from http://www/cnn.com/ALLPOLITICS/stories/1999/04/23/politics.guns/.

6

Massively Multiplayer Online Games

Elaine Chan
University of Southern California

Peter Vorderer
University of Southern California

In 2004, more than 43% of video gamers played online an hour or more each week, an increase from 31% the previous year (Americans playing, 2004). A recent survey found that, in the United States, women over the age of 40 play the most online games; they spend almost 50% more time each week playing online than men and are more likely to play online daily than men or teens. Among this demographic, word and puzzle games are the most popular, followed by arcade, trivia, and sports games (Greenspan, 2004a). Other studies have found that, in terms of gender, online game players are nearly balanced, with slightly more males (51 to 53%) than females (Greenspan, 2004b).

In terms of variety, the range of games that may be played online is vast and continually expanding. Online games do not comprise a particular genre; rather, traditional genres (e.g., adventure games or sports games) are able to incorporate online play, typically by connecting distant players as either teammates or opponents.

Online play is not a new phenomenon, but as video games have become more complex, so too has the degree to which players have been able to use the Internet to find competitors, teammates, and communities for play. "Online play" has multiple potential meanings. When we use the term "online play" in this chapter, we do not mean local area network (LAN) play, although it also qualifies as being online. We are referring to play that occurs on and over the Internet. Additionally, "online games" refers to computer games of any genre that permit the player to play the games over the Internet.

It is possible to play fairly simple games online as a single player. In this case, the Internet serves as a medium for distribution of computer games. Shockwave or Flash games, which are easily streamed to a user's computer, are one type of single-player stand-alone game, usually played through a Web interface. The technological requirements for these games, both in terms of hardware and bandwidth, are fairly low. The time commitment necessary for the player to learn how to play the game, play it, and reach an appropriate stopping point provided by the game (what we may consider a round of play) is also minimal; in general, the player can expect one round of play to consist of minutes, not hours. Additionally, because no other players are involved, no pressures exist, other than events within the game itself, to constrain duration of play. In other words, it should be easier for the individual to quit the game at any time, or alternatively, to continue playing indefinitely.

Other online games may be slightly more complex, and the online component of computer games can provide opponent and/or teammate matching. Players can connect to and play over some servers for free, although others charge a monthly fee. In this instance, the Internet (server) functions as a bridge for two or more players. This bridge allows players to compete against each other, although competition is often accompanied by cooperative or team aspects of play as well. The number of individuals who may participate in a game simultaneously varies depending on the specific game played, though a range from 2 to 64 players is common. Moreover, the actions of multiple players may occur by turn-taking (as is the case for most card or board games) or simultaneously (as in the case of real-time strategy games). Both direct (head-to-head) and diffuse (based on the individual's score or relative standing) forms of competition may take place. Players of many different kinds of games are able to use the Internet to find competitors—card games and first-person shooters alike may provide the option. Online console gaming has been gradually expanding, with Microsoft's XBox Live and Sony's Playstation 2. Of the 41.1 million video game consoles in U.S. households at the end of 2003, 3.3 million were online. Market researchers expect this figure to increase as the online capabilities of consoles and games improve (Study: Online-game, 2004). Players are often able to maintain their identity over multiple episodes of play (each time they play the game or log onto the server). Thus, in a particular episode, a player may engage in direct competition with another player. The resulting win/loss, however, may be calculated into his or her overall score, which offers a means of assessing a player's overall cumulative success in diffuse or ranked competition. The development of handheld games that will allow competition over wireless LAN connections, and possibly the Internet, is also expected.

Large-scale online games capture the notion of the Internet as a location for virtual communities (Rheingold, 1993). Massively multiplayer online games (MMOGs) connect thousands of players in real-time interaction and communication. The game worlds are continuously accessible online, which allows for the emergence of complex social structures, reputation systems, and economies. Although opportunities for players to engage in direct competition sometimes exist, the majority of MMOG players seem to prefer diffuse competition, if competition is a factor at all. Several games have specific servers or areas for players who are interested in "Pkill" (Player Kill) or "PvP" (Player vs. Player) modes, but the number of players who play "PvP" in comparison to players who do not is very small.

Despite this classification of possibilities for online play, in practice, these categorical boundaries of online play are rather permeable. Communities and complex social dynamics can arise wherever enough interested individuals are available. The *Yahoo! Games* Web site, for example, provides online chat rooms for many of its games, which are played on the site itself. Therefore, playing *Yahoo! Poker* online involves the Internet both as a medium for access to the game and a means for playing with other people. If players choose to maintain their identities over multiple episodes of play, relationships and community may arise.

MMOGs should be of particular interest to communication scholars. These virtual worlds integrate communication and entertainment in a play environment that evolves through user interaction. Players of MMOGs tend to become highly involved, whether during a specific episode of play or over a long period of time. Studies concerning multi-user dungeons (MUDs, their text-based predecessors) have found such games to have great significance to players for identity and community (Turkle, 1995). The high degree of player involvement in MMOGs appears to be quite similar, though as mainstream commercial productions, it occurs on a much larger population scale.

WHAT EXACTLY ARE MASSIVELY MULTIPLAYER ONLINE GAMES?

Massively multiplayer games have acquired a hefty collection of acronyms and labels, which include but are not limited to: MMO (massively multiplayer/multiuser online), MMOG (massively multiplayer online game), MMOPW (massively multiplayer online persistent world), and MMORPG (massively multiplayer online role playing game). Researchers in different disciplines also use terminology with varying emphases—taking cues from the video game industry, scholars have used acronyms with the MMO-prefix most often for the sake of specificity. The somewhat broader terms "persistent world" and "virtual world" are also used by social scientists and humanist scholars, and computer scientists and others researching virtual reality technology have examined "networked virtual environments" (Ellis, 1991; Kent, 2003). For the sake of simplicity, we will use MMOG, that is, massively multiplayer online games (sometimes dropping the "online" part), most of the time. *Massively* refers to the fact that games can support thousands of players at a time. *Multiplayer* obviously means that players are in the world simultaneously. The games are played *online*—unlike other games that may be played online with others, there is no equivalent stand-alone (offline) version of MMOGs. The *game* aspect is less defined, with new genres that involve more open-ended play, but we will include those within our concept of MMOGs as well.

Massively multiplayer games provide a 3-dimensional virtual world for thousands of players to interact with and explore. They evolved out of text-based role-playing games known as multi-user dungeons/domains (or MUDs) and their variants (Kent, 2003). Because of this, MMOGs have also been referred to as graphical MUDs. Yet while some MUDs could host upwards of a hundred players at a time, few actually required that capacity. A MMOG, on the other hand, requires a player base that is significantly larger than would be necessary for a MUD, in order to become and remain profitable as a commercial enterprise.

Multiple genres exist, and the games industry has been experiencing an explosion in both number and variety of massively multiplayer games (Chick, 2003). Still, the majority of what has been written concerning massively multiplayer games, much as it was in the early days of computer games in general, has been in the popular press—in gaming magazines such as *Electronic Gaming Monthly* or *PC Gamer*; in newsmagazines such as *Time* and *Newsweek*, and in newspaper articles. Both socially deviant and socially correct behavior that has emerged within MMOGs has been chronicled (Chmielewski, 2003; Game sparks, 2003; Terdiman, 2004).

Social Science Research on MMOGs

Although there has been little scholarly research, and even less published work, on MMOGs, research has been conducted in many related areas. Schroeder's (2002) edited volume, *The Social Life of Avatars*, for example, contains many essays analyzing the issues of presence (the perceived illusion of nonmediation), particularly social presence, in networked virtual environments (see also Tamborini & Skalski, chap. 16, this volume). In the *Video Game Theory Reader*, Filiciak (2003) examined identity from a postmodern perspective in massively multiplayer games; the notion of postmodern identity and multiple aspects of self on the Internet prior to multiplayer games with graphics was examined by Turkle (1995). Others have also studied issues of gender, race, identity, and deception in several types of online communities, including MUDs (Smith & Kollock, 1999). Research on nonpersistent 3-dimensional games is also applicable to the study of 3-dimensional MMOGs (McMahan, 2003).

Still, research on MMOGs in various disciplines is slowly seeping into refereed journals and gaining wider recognition for its academic impact in areas as diverse as intellectual property law, economics, and psychology (Castronova, 2004; Choi & Kim, 2004; Griffiths et al., 2003; Griffiths et al., 2004).

Playing MMOGs

Traditionally, MMOGs have been played online via computer. Users install game software onto their computers, which then allows them to connect to a server hosting the virtual world. Each server hosts approximately 2,000 players at any one time (Griffiths et al., 2003). Presently the 50 separate worlds that comprise the entire EverQuest universe operate on approximately 1,000 computers, the majority of which are located in San Diego, California. Five of the most recently launched servers are located in Europe (Benson-Lennaman, 2004). Once a player logs onto a server, he or she is able to select his or her character, navigate through virtual space, and interact with the world and other players on-screen.

Recently MMOGs have begun to expand toward console playability. Currently, two MMOGs are playable with Sony's Playstation 2 via its network adapter—Sony's *EverQuest Online Adventures: Frontiers* and *Final Fantasy XI* by Square Enix. LucasArts announced a Playstation 2 version of *Star Wars Galaxies: An Empire Divided*, but subsequently canceled its release. Computer and console play of *EverQuest* occurred on separate servers (also referred to as shards). *Final Fantasy XI*, however, is notable in that it allows both computer players and Playstation 2 players to mingle within the same game space. Previously, whereas games could be accessed through the Internet on both computers and gaming consoles, users from one system did not interact with users from the other. Particular servers were designated for each system. Much in the same manner, PC and Macintosh users were also hosted separately. Finally, it is anticipated that MMOGs will soon spread to mobile devices. In 2004, Sega released the world's first massively multiplayer mobile role-playing game, *Pocket Kingdom: Own The World*, for Nokia's N-Gage, a gaming cell phone that was introduced in 2003. As gaming platforms continue to change and the diversity of player options increases, researchers will face additional challenges of investigating how delivery and context of the medium impact the experience of game play.

The MMOG Industry and the State of Play

Analysts expect revenues from massively multiplayer online games to exceed $1 billion in 2004 and project revenues to grow to more than $4 billion by 2008 (Online games, 2004). Although it is difficult to obtain precise figures concerning the number of players of MMOGs due to variations in how game publishers count their players (e.g., definitions of what counts as an active subscription), attempts to estimate MMOG user counts worldwide have been made. Recent estimates assert that MMOG users comprise a population of nearly 5 million people, and the number is increasing. These users subscribe to a diverse array of MMOGs spanning multiple genres. Until 2005, NCSoft's *Lineage* had the greatest number of active subscribers worldwide with, at its peak, 4 million subscribers. Approximately 2.5 million of these individuals lived in South Korea, 1.5 million were located in Taiwan, and 1 million were in the United States (Kent, 2003). In 2005, Blizzard Entertainment announced that their game *World of Warcraft* had become the largest MMORPG in the world, with more than 3.5 million global customers. Nevertheless, a number of highly anticipated MMOGs continue to enter the market. Woodcock (2005) maintains an analysis of online subscription growth based on data provided by game companies and their employees.

The number of MMOG subscribers is astonishing in part because of the industry's economic model, which is largely based on monthly subscription fees for players. Stand-alone computer games typically cost $30 to $50 for the boxed software—a onetime fee. Maintaining a subscription, however, means paying anywhere from $9.95 per month (for *The Sims Online*) to approximately $15 per month (*Star Wars Galaxies'* month-to-month subscription fee), although some games offer discounted subscriptions for longer periods of time. Sony has also experimented with various subscription rates, offering a premium service for hardcore players of *EverQuest* called *EverQuest Legends*, which offers additional features such as unique game objects, an exclusive server, and player home pages for $39.95 per month. It is plausible to assume that because users are paying a monthly subscription fee, they are subject to the all-you-can-eat meal phenomenon—a license to gorge that could encourage increased play. Economists might refer to a game subscription fee as a sunk cost, that is, a cost already incurred that should not affect future decisions. Therefore, the fact that one has paid $12 to play a given MMOG for 1 month should not influence the amount of time spent playing. Nevertheless, research in behavioral economics indicates that individuals often fall into the sunk-cost trap and will act in such a way as to get their money's worth (Thaler, 1980). One result of this business model is that these games appeal primarily to hardcore players or those who are willing to invest substantial amounts of their time to playing the game. This possibility is supported by a study by Griffiths et al. (2003), which found that on average, a quarter of their sample of *EverQuest* players played for more than 41 hours per week. In comparison, Americans spend an average of 25.8 hours per week working and doing work-related activities, 18 hours per week watching television, 5.5 hours per week socializing and communicating, and 2.1 hours per week participating in sports, exercise and recreation (Time-Use, 2004).

Game developers continue to struggle with the problem of casual play, or more specifically, the difficult task of attracting and retaining individuals who do not want to replace the majority of their real-life activities with an online game. They have yet to figure out how to lure more casual players, who would increase the player market substantially (Kosak, 2003). Naturally, many factors encourage passionate play; in addition to the economic model, structural characteristics of MMOGs encourage a high level of involvement. Other business models, however, have been considered. *There*, a nongenre MMOG, planned to earn revenue not by subscription but through the sale of virtual goods, meaning players pay real-world currency for items within the game (Clark, 2003). Although the sale of virtual goods is not exclusive to *There*—sale of in-game characters and items across MMOGs is rampant on—game publishers have generally discouraged such transactions in the past, citing intellectual property rights. Still, the sale of virtual property can be a lucrative business, with virtual sweatshops of workers harvesting goods in several MMOGs emerging in Asia (Dibbell, 2003; Lee, 2005). In 2005, Sony launched their own marketplace for virtual good transactions allowing players to buy and sell *EverQuest II* items, characters, and currency through auctions.

The MMO Gamers

Surveys of MMOG player demographics indicate a user base that largely consists of young adult males. This is in contrast to the survey data about online game players in general (Greenspan, 2004a), which included individuals who play casual Web-based games. Data collected from online *EverQuest* fan sites found the majority of players (71%) between 10 and 30 years of age and overwhelmingly (approximately 85%) male (Griffiths et al., 2003). Yee's (2004) MMORPG demographic survey findings are similar, compiled from four games—*EverQuest*, *Dark Ages of Camelot*, *Asheron's Call* and *Anarchy Online*—which yielded a combined demographic that was 89% male, 11% female (for *Asheron's Call*, see also the chapter by Axelsson & Regan in this volume). The average respondent age was 26.7, with female players significantly older than male players, spending an average of 24 hours per week playing their respective games. A comparative anomaly in massively multiplayer online gaming is that roughly 60% of *The Sims Online* players are female (Gaither, 2003).

Although MMOGs account for only a small fraction of the international \$32 billion computer games industry, analysts anticipate tremendous increases in revenue as high-speed broadband Internet becomes as ubiquitous in other countries as it is in South Korea (Invaders from, 2003; Kent, 2003).

SPECIFIC CHARACTERISTICS OF MMOGs

Despite several very diverse genres of massively multiplayer online games, a few broad characteristics define the game type as a whole. Massively multiplayer online games share the following characteristics: *persistence*, *physicality*, *social interaction*, *avatar-mediated play*, *vertical game play,* and *perpetuity*. In addition to these key qualities, to be detailed, other technical and design aspects of massively multiplayer games that many individual worlds have in common also exist. Nevertheless, these are the most integral, defining qualities of MMOGs across genres.

Persistence. Persistence is the critically distinguishing design quality that separates MMOGs from other types of games. It occurs on at least two levels—*world persistence* and *avatar persistence*. Most commonly used with the former meaning in mind, "persistence" refers to permanence of the virtual world. In this way persistence means that barring technical difficulties, the world exists (is online and accessible to players) continuously 24 hours a day, 7 days a week. Technologically, this means that the virtual world is housed on one or more computer servers that users access from various geographic locations around the world. Information about player characters are also located on the server and accessed from other locations. The persistent aspect of these games cannot be overemphasized. Thus, some refer to MMOGs as persistent worlds. The element of persistence defines the world; continuance and constancy allow community, economy, reputation, and other social structures to emerge. Persistence allows players to feel that they are not just playing one episode of a trivial game, but entering into and becoming involved with a parallel reality.

World persistence exists in games to varying degrees. We may consider a game's persistence on a continuum that ranges from not at all to fully persistent. The low end of such a construct would include short puzzle and card games that are available on or offline. Such games are typically played in rounds, with each episode of play detached from the previous one. Games that allow players to save their position and later continue from the point at which they left off are moderately persistent. The majority of action and adventure games, for example, involve entry into a particular realm. In this realm, the player's character must accomplish specific tasks and/or move through physical landscapes from one point to another in order to succeed at or beat the game. If the player chooses to stop playing temporarily, he or she can usually save the game to resume from the same point in the future. In this situation, the player experiences both avatar and world persistence. If the status of the character does not change between episodes of play, avatar persistence exists. But persistence in massively multiplayer online games is dynamic; the world continues to change even if the player is not currently playing it; other individuals will continue exploration of the virtual realm. Nevertheless, change in the virtual world is dependent on the presence and action of its players. When a given individual stops playing and exits the virtual realm, the world keeps moving, as long as other players are able to participate online.

Persistence affects game play in several ways. First, there is no pause command for MMOGs. Unlike the majority of stand-alone computer games, virtual worlds are continuously dynamic. In a combat-oriented RPG such as *EverQuest*, for example, a player in a combat situation may be fighting a nonplayer character or a creature of some sort. The player can attempt to withdraw from the battle, but while engaged, he or she cannot stop the rest of the events as they occur in the world. Although the player may sever the Internet connection or shut down her computer, her character is likely to continue in battle, and possibly die.

Play does not break naturally into episodes or rounds. At the same time, there are multiple levels of segmentation, increments by which the player can gauge play. The only natural breaks that occur in the game are leveling. In most MMORPGs, for example, one's character level increases after a certain amount of experience points has been gained by slaying enemies. A player may decide to stop when his or her character gains a level. But if he or she is in a group with other players, it is highly likely that they do not gain a level at the same time. The established norm of not leaving a group mid-combat or in the middle of an excursion may encourage the player to continue the episode of play. Another natural break might be when the group is done fighting. But, the group issue is also problematic because groups are not available all of the time. The player may feel compelled to take any opportunity to group, not knowing when the next opportunity will be.

Herz (n.d.) and Yee (2004) both speculated on the psychological implications of persistence for the MMOG player. The dynamic nature of both the physical (in the virtual sense) and the social world may compel players to engage in that world as much as possible. Players may feel as though they are missing out on game experiences when they are not online. Player-determined events may or may not be related to gaming itself. In-game weddings between two characters or group appointments to hunt or to raid can both be events that the player would not want to miss.

Avatar persistence, however, exists when individuals are able to maintain stable representations of their chosen personality within the online environment. This characterization may represent their actual identity or be a fictional persona. Regardless, it is how each chooses to present him or herself to others. Persistence of identity is a requisite factor for the formation of interpersonal relationships within the game. Without an understanding that the person behind a given character name is the same individual you conversed with as that character last time, it would be very difficult to increase the breadth and depth of communication. It may be the case that the person behind the avatar is not actually one individual. Networked communication creates the potential for individuals to try out different identities. In terms of either character or community avatar, an individual's brother or sister may be sitting at the keyboard rather than the original person himself. Or, it may be the case that someone has sold a particular character and the person playing that role is not the individual who has always played it. Additionally, each individual player may have several identities, each of which is persistent within itself, with a different set of personal characteristics and social contacts. Nevertheless, it is the perception of character or avatar persistence that is required for relationships to develop. This may be within a particular world or game (avatar or character persistence), within the universe (such as a community bulletin board or chat room handle that may or may not be one's character name), or spanning multiple universes and spilling into the real world.

Physicality.   MMOGs are physical by design. Although virtual environments are not tangible, they are digital representations of material things. For the most part, a physical environment is reproduced in a virtual world and the usual notions of time, space, and natural laws all apply, albeit with slight adjustments. Days may pass more quickly in the virtual world than on Earth, space might be traveled through much more quickly, and falling from great heights may be less injurious to the body than one may have suspected. Nevertheless, players must open doors in order to walk through them. They must swim across bodies of water and go around trees.

Whereas there may be the addition of fantasy creatures, supernatural powers, and other extraordinary phenomena in some MMOG genres, for the most part, game designers have not strayed far from reality in their creation of virtual space. Rather, they construct synthetic environments that largely mimic our own. Gravity exists, though it may be overcome by magical spells or supernatural powers. Likewise, distances must be traversed in the absence of other transport systems.

MMOGs are typically 3-dimensional, although they are laid out (or landscaped) on a 2-dimensional grid. The 3-D rendering comes in two flavors—immersive or isometric. Perspective in the immersive environments can usually be manipulated by the player. Preference for first- or third-person perspectives seems to vary according to several factors, including age, gender, and experience (Yee, 2004).

Social Interaction. Early media portrayals of game players often suggested legions of solitary individuals sitting in front of their computer screens, playing games, and becoming increasingly isolated and socially inept. More recently some scholars have argued that playing video games is much more of a social phenomenon than originally perceived. A Pew Internet & American Life study of undergraduate students at U.S. universities found that video and computer games are highly integrated into ordinary social life and are often utilized as an opportunity to interact socially with others (Jones, 2003). Regarding Internet use more generally, an early study of Internet use found minor negative effects—decreases in social involvement and psychological well-being (R. E. Kraut et al., 1998). Nevertheless, these negative effects disappeared between the initial study and the follow-up 3 years later (R. Kraut et al., 2002). Another perspective suggests that individuals are able to use Internet technology to create social capital (Wellman, 2001). A 2-month study of the effect of MMORPG play on measures of social capital supports this latter view (see also Williams, chap. 14, this volume).

Whether or not MMOGs are the first interactive mass medium combining entertainment and communication (online and multiplayer versions of other role-playing, real-time strategy, or first-person shooters also do so), they are one of few to date. MMOGs also afford a much broader range of possibilities in terms of the number of channels players may use to communicate to or among any number of players. Additionally, as Filiciak (2003) argued, interpersonal interaction is not only a clever or useful feature of MMOGs: For the majority of players, it is integral to the experience. The virtual world itself might also be considered a gigantic forum for communication. Similar to other communication media such as telephones or e-mail, both a sender and receiver (a minimum of two individuals) are necessary to gain the benefits of use. Nevertheless, theoretically an individual could turn off all text or voice-based channels of communication and play the game alone, ignoring all other players around him or her. There would be little difference, however, between this sort of game experience and that of many other stand-alone, single-player role-playing games. In addition, MMOGs offer more than an affordance for social interaction.

MMOGs may also contain specific features that allow for interpersonal interaction and communication. The typical MMOG features numerous channels to allow for communication between dyads, conversations among a group, or questions and announcements directed to many individuals at once. More recently, games have also allowed players to add individuals to "buddy lists," an instant-messaging style list which shows which of the player's friends are online at any time. In addition to features, however, specific incentive structures may also be designed into the MMOG algorithm so that the game not only allows for but also actively promotes interpersonal interaction. The process of "skilling" in *The Sims Online* is a very basic example of this. This is the process by which characters develop their abilities by participating in a specific task for some duration of time. There is an upper limit, but, in general, the speed of character learning is directly proportional to the number of characters in the general vicinity learning the same skill. Thus, one character learning how to cook by reading a book on the subject will increase his or her abilities at a rate of 25%, with increases in the rate of learning for each additional participant. Six players in any area learning the same skill maximizes their learning rate at 118%. Of course, although the opportunity for such incentives exists within the game structures, their availability is dependent on the participation of other players. Such incentives, however, may fail to encourage genuine interaction (cf. Steen, Greenfield, Davies, & Tynes, chap 21, this volume). In fantasy games that involve combat, an incentive to interact with other characters and fight dragons, demons, or spiders collectively also exists. A group of players can cooperate to defeat an enemy—possibly by strength in numbers, but even better, by combining a diverse set of skills. For example, the brawny warrior may engage in melee combat while the priest in the party heals and restores members' health, and the mage casts attack spells from a distance.

What is it about social interaction that is so integral to the MMOG experience? One perspective from which to consider this question is the notion of social interaction in games more generally. A very loose definition of social would include any interaction between two human beings. Anecdotal evidence about competition in games suggests that playing against a real live person is inherently more fun than playing against a computer. As expressed in the following quote from a newspaper article about professional gamers, "There is a deeper, more perverse joy in annihilating your opponent if you know that the online avatar represents a real person sitting at a computer terminal, albeit hundreds, perhaps thousands of miles away" (Why virtual, 2004). The importance of social experiences is supported by the idea that humans are inherently social creatures, assigning meaning to objects based on social contexts (Fiske, 1992). Still, sometimes situations occur in which a person would prefer to play alone or with a computer opponent rather than with other individuals. For example, a computer opponent is unlikely to cheat or otherwise participate in unsportsmanlike activities. Or, a human opponent with appropriately matched abilities may not be available. Nevertheless, it seems that all else being equal, the majority of players would choose to play against another person rather than against a computer. Often, social relationships develop out of such competitive and cooperative interactions (cf. Vorderer, Bryant, Pieper, & Weber, chap. 1, this volume).

MMORPGs, however, are particularly notorious for being fertile ground for the development of significant social relationships among their players. These relationships may exist between players in or out of character. For in-character relationships, communication remains within the realm and scope of the game itself, and information exchanged is understood to be from one character to another character, rather than between players. An out-of-character relationship could be described as one in which players communicate as themselves or via their representations of their real-life selves, rather than through their role-played characters from the game. Out-of-character relationships are generally recognized as having greater breadth and depth, and they often grow out of the bounds of the virtual world into additional forms and places for communication, such as e-mail, telephone, or even face-to-face meetings. One survey found that 68% of males and 82% of females surveyed had exchanged e-mail contact information with at least one person they had met playing an MMORPG. Fifty-nine percent of males and 64% of females had given out their instant messenger information, and 28% of males and 52% of females had given their phone number (Yee, 2004). It is important to note the significance of out-of-character relationships in shaping the virtual-world experience and community. As stated in Koster's (n.d.) Laws of Online World Design, "Virtual social bonds evolve from the fictional towards real social bonds. If you have good community ties, they will be out-of-character ties, not in-character ties. In other words, friendships will migrate right out of your world into email, real-life gatherings, etc."

In fact, the most frequently cited breakdown of player types in MMOGs consists of the four player types described by Bartle (2003) in "Hearts, Clubs, Diamonds and Spades: Players Who Suit MUDs." Developers of other types of online games now apply Bartle's types to players of other mainstream online games, including MMOGs. Hearts, one of the four player classifications, are those players whose primary interest in the game is interaction with other players. Clubs, diamonds, and spades describe those who are interested mainly in imposition upon others (often through combat), achievement in the game context (e.g., by accumulating wealth), and exploration of the game, respectively.

Avatar-Mediated Play. The avatars in MMOGs today developed out of text-based avatars in earlier role-playing games. An avatar is the virtual-world, or character, representation of the player in an online game. Text-based avatars in MUD games were usually player-written descriptions of their character's appearance, the length of which could range from one or two lines to several paragraphs. Drawing upon the idea of *avatar* as character description (or depiction) ather than merely an image or figure, we can expand on the technical definition of avatar to a broader notion of the term as it may be used within the context of MMOGs, which are not just virtual spaces, but virtual worlds. Key to what avatars afford is the means by which an individual represents him or herself—a physical and located person/object—to others in the virtual environment and community. Used in this sense, the word *avatar* becomes nearly interchangeable with *character*. The effect of having an avatar to represent one's own or created identity, of interacting and communicating with others through this represented self, is curious and complex.

As is the case with players of character-based non-MMOG games, the extent to which an individual affiliates with his or her on-screen character varies greatly. Even the differences between how players speak of their characters evidence this phenomenon. One person, when referring to (each of) his characters, may use the pronouns "I" or "me" to refer to his avatar. Another player may discuss each of her characters in the third person, mentioning them by name as if they were acquaintances. Some players may view their character as an extension of themselves, whereas others may see their characters as foils through which they can experiment with or, as the advertising campaign for *The Sims Online* suggested, "Be Somebody. Else." Finally, other players may not associate their characters with themselves at all, but see each one as an entity completely separated from themselves or an object to be manipulated. Additionally, persons may be more or less conscious of how active their own role is in constructing their characters. Those who emphasize role-playing, for example, may put more explicit effort into personality creation for their characters than a player whose primary orientation is achievement through killing as many monsters as possible.

Massively multiplayer online games are the most recent development in the long genealogy of avatar games (see Castronova, 2002, for a history). Technically, any figure or image used to represent a person in virtual space is an *avatar*. In the early days of 2-dimensional virtual environments, avatars were 2-D images of a given pixel dimension. Avatars could be used in spaceless chat environments or virtual environments designed with two or three dimensions of space in mind. With technology, avatars gradually developed from static 2-dimensional images to animated 3-dimensional ones of increasing resolution. Avatars in MMOGs have also moved toward increased resolution and realism, with the most recent generation of worlds providing extremely detailed and lifelike depictions, usually with many customizable features. *EverQuest II*, for example, boasts extensive customizability of each character's appearance. Players may choose the avatar's face, age, skin tone, armor, eye, hair color, and body type. The developers of *Star Wars Galaxies* have also made avatar breast and butt size configurable. Avatar selection varies greatly according to genre, ranging from mostly human (nongenre games like *The Sims Online*) to a high proportion of supernatural and anthropomorphized beings (e.g., *World of Warcraft*). In a group or crowd setting, avatars may be used to distinguish one character from another.

Within the social context of the virtual world, one major affordance of avatar-mediated communication is anonymity. In the context of massively multiplayer games, role-playing (fantasy or otherwise) may be acceptable, the norm, or even encouraged. Even in the case of nongenre games, there is an expectation that at least when the topic of conversation remains at the level of the avatar or the character, individuals are speaking as their characters rather than their personal player identity. Thus, one level of anonymity is avatar-level anonymity—the potential for players to maintain completely anonymous identity if they choose to do so, by not disclosing any information about their out-of-character (or real-life) lives. An additional level of anonymity exists in virtual worlds because of the fact that they are online. As is the case for most other online social interactions, the user can mask age, gender, and other identity features. The ability of players to hide their identity may contribute to griefing—disruptively antagonizing behavior in online games (Game sparks, 2003).

Vertical Game Play and Perpetuity. For the most part, a character's progress in an MMOG is assessed by the character's attributes. Games may provide statistics about the number of enemies a character has vanquished or the amount of time a player has spent as that character online. Players may also note the areas they have passed through or explored. But, the main indicator of how much a player has achieved in the game is usually related to his or her character's attributes. In an RPG-style game, this is usually a combination of the character's level (a number), their equipment or gear (quality), and perhaps wealth (gold pieces or platinum accumulated). Nongenre games may or may not have quantitative measures of in-game achievement, though as mirrors to real life, they raise the question of how individuals assess achievement or worth in reality as well. Still, regarding the massively multiplayer experience, it makes little sense for players to say they have beaten the game. As there are no definite markers of completion, play is much less focused on an end goal. Perhaps the closest thing to a mark of finishing an MMOG would be attaining the highest level possible. This could be interpreted as an indication that one has completed the in-game goals. Nevertheless, even after reaching such a pinnacle of play, the player can continue his or her existence in the game in the same way.

In the end, "massively multiplayer online game" is an appropriate label only if we acknowledge that they are more than just games. These virtual worlds are not contained within video game packaging; they are created through social experiences and player interactions. What we can learn from MMOGs is how games and gaming fit into ordinary life more generally, in the moments where the increased use of media as entertainment, the widespread application of networked technologies, and participation in new social experiences intersect.

REFERENCES

Americans playing more games, watching less movies and television. (2004). Retrieved November 22, 2004, from http://www.theesa.com/5_12_2004.html.

Bartle, R. (2003). Hearts, clubs, diamonds, spades: Players who suit MUDs. In J. Mulligan & B. Patrovsky (Eds.), *Developing online games* (pp. 397–435). Boston: New Riders.

Benson-Lennaman, A. (2004). alt.games.everquest EQ FAQ. Retrieved June 2, 2004, from http://webpages.charter.net/lenny13/age.faq.htm.

Castronova, E. (2002). On virtual economies. Retrieved October 29, 2002, from http://ssrn.com/abstract_id=338500.

Castronova, E. (2004). Virtual worlds research widens academic impact. Retrieved May 11, 2004, from http://terranova.blogs.com/terra_nova/2004/01/virtual_worlds_.html.

Chick, T. (2003). MMOs: 2004 and beyond. Retrieved December 20, 2003, from http://www.gamespy.com/amdmmog/week7/.

Chmielewski, D. C. (2003, June 5). Mobs move into "Sims Online" power vacuum. Retrieved July 11, 2003, from http://www.siliconvalley.com/mld/siliconvalley/6019958.htm.

Choi, D., & Kim, J. (2004). Why people continue to play online games: In search of critical design factors to increase customer loyalty to online contents. *CyberPsychology & Behavior, 7*, 11–23.

Clark, D. (2003, Jan. 8). Online game hopes to convert virtual cash into real revenue. *The Wall Street Journal*, p. B.1.

Dibbell, J. (2003). Serfing the Web. Retrieved November 24, 2004, from http://www.juliandibbell.com/texts/blacksnow.html.

Ellis, S. (1991). Nature and origins of virtual environments: A bibliographical essay. *Computer Systems in Engineering, 2*, 321–347.

Filiciak, M. (2003). Hyperidentities: Postmodern identity patterns in massively multiplayer online role-playing games. In M. J. P. Wolf & B. Perron (Eds.), *Video game theory reader* (pp. 87–102). New York: Routledge.

Fiske, A. P. (1992). The four elementary forms of sociality: Framework for a unified theory of social relations. *Psychological Review, 99*, 689–723.

Gaither, C. (2003). Battle for the sexes. Retrieved January 13, 2003, from http://www.boston.com/globe/search/stories/reprints/battlefor011303.htm.

Game sparks bad behavior. (2003, July 6, 2003). Retrieved July 6, 2003, from http://www.wired.com/news/culture/0,1284,59539,00.html.

Greenspan, R. (2004a). Girl gamers grow up. Retrieved December 2, 2004, from http://www.clickz.com/stats/sectors/demographics/article.php/3312301.

Greenspan, R. (2004b). Online gaming revenue to quadruple. Retrieved December 2, 2004, from http://www.clickz.com/stats/sectors/software/article.php/3403931.

Griffiths, M. D., Davies, M. N. O., & Chappell, D. (2003). Breaking the stereotype: The case of online gaming. *CyberPsychology & Behavior, 6*, 81–91.

Griffiths, M. D., Davies, M. N. O., & Chappell, D. (2004). Online computer gaming: A comparison of adolescent and adult gamers. *Journal of Adolescence, 27*, 87–96.

Herz, J. C. (n.d.). Gaming the system: What higher education can learn from multiplayer online worlds. Retrieved May 11, 2004, from http://www.educause.edu/ir/library/pdf/ffpiu019.pdf.

Invaders from the land of broadband. (2003, December 13). *The Economist,* 57–58.

Jones, S. (2003). Let the games begin: gaming technology and entertainment among college students. Retrieved July 10, 2003, from http://www.pewinternet.org/pdfs/PIP_College_Gaming_Reporta.pdf.

Kent, S. L. (2003). Alternate reality: The history of massively multiplayer online games. Retrieved November 5, 2003, from http://www.gamespy.com/amdmmog/week1/.

Kosak, D. (2003). The future of massively multiplayer gaming. Retrieved December 30, 2003, from http://www.gamespy.com/amdmmog/week8/.

Koster, R. (n.d.). The laws of online world design. Retrieved June 22, 2004, from http://www.legendmud.org/raph/gaming/laws.html.

Kraut, R., Kiesler, S., Boneva, B., Cummings, J., Helgeson, V., & Crawford, A. (2002). Internet paradox revisited. *Journal of Social Issues, 58*, 49–74.

Kraut, R. E., Patterson, M., Lundmark, V., Kiesler, S., Mukhopadhyay, T., & Scherlis, W. (1998). Internet paradox: A social technology that reduces social inolvement and psychological well-being? *American Psychologist, 53*, 1017–1032.

Lee, J. (2005, July 5). From sweatships to stateside corporations, some people are profiting off of MMO gold. Retrived July 5, 2005, from http://www.Iup.com/do/feature?cID=3141815.

McMahan, A. (2003). Immersion, engagement, and presence: A method for analyzing 3-d video games. In M. J. P. Wolf & B. Perron (Eds.), *Video game theory reader* (pp. 67–86). New York: Routledge.

Online games make serious money. (2004, January 19). Retrieved November 19, 2004, from http://news.bbc.co.uk/1/hi/technology/3403605.stm.

Rheingold, H. (1993). *The virtual community: Homesteading on the electronic frontier.* Reading, MA: Addison-Wesley.

Schroeder, R. (Ed.). (2002). *The social life of avatars: Presence and interaction in shared virtual environments.* London: Springer.

Smith, M. A., & Kollock, P. (1999). *Communities in cyberspace.* New York: Routledge.

Study: Online-game revenue to skyrocket. (2004). Retrieved December 2, 2004, from http://news.com.com/2100-1043-5266062.html.

Terdiman, D. (2004). Mr-President bids for re-election. From http://www.wired.com/news/games/0,2101,62635,00.html.

Thaler, R. (1980). Toward a positive theory of consumer choice. *Journal of Economic Behavior and Organization, 1*, 39–60.

Time-use survey—first results announced by BLS. (2004). Retrieved September 20, 2004, from http://www.bls.gov/news.release/pdf/atus.pdf.

Turkle, S. (1995). *Life on the Screen: Identity in the Age of the Internet.* New York: Simon & Schuster.

Wellman, B. (2001). Does the Internet increase, decrease, or supplement social capital? Social networks, participation, and community commitment. *American Behavioral Scientist, 45*, 436–455.

Why virtual entertainment means real money. (2004). Retrieved January 4, 2004, from http://www1.chinadaily.com.cn/en/doc/2004-01/02/content_295214.htm.

Woodcock, B. S. (2005). An analysis of MMOG subscription growth. Retrieved July 28, 2005, from http://www.mmogchart.com/.

Yee, N. (2004). The Daedalus Project. Available at http://www.nickyee.com/daedalus/.

MOTIVATION AND SELECTION

7

Why People Play Games: An Industry Perspective

G. Christopher Klug and Jesse Schell
Carnegie Mellon University

We have heard it said that no one really knows why people play games. In fact, many times in design meetings deep within the bowels of game development companies, we hear those exact words uttered. Professional game designers often take it for granted that people just "want" to play our games. We rarely examine the psychology of our gaming audience. Instead, we examine opinion polls, where we discover that, say, 43% of our audience wants a game with violent combat or that 17% want romantic interludes (both numbers totally invented off the top of our heads for the sake of this chapter). But, rarely does the game industry examine what truly motivates our players.

Often, what we do is look at what evidence the marketplace has given us (in other words, what games by other publishers have been hits) and then try to emulate that success, adding in a little bit of a twist. This chapter is our attempt to summarize over 30 years of game design experience, gathered through firsthand successes and failures, seat of the pants guesses, late-night arguments in hotel bars, and lots and lots of rolling dice and pushing joystick handles.

THESIS I: WHAT MOTIVATES PLAYERS TO PLAY GAMES

Play theorists have identified a number of types of players, each with a different need that gets met by the type of game they play. Currently the game industry delivers product that meets some of these needs, but not yet all. Some more prominent types are identified as: *The Competitor, The Explorer, The Collector, The Achiever, The Joker, The Director, The Storyteller, The Performer*, and *The Craftsman*.

The Competitor plays to be better than other players.

The Explorer plays to experience the boundaries of the play world. He plays to discover first what others do not know yet.

The Collector plays to acquire the most stuff through the game.

The Achiever plays to not only be better now, but also be better in rankings over time. He plays to attain the most championships over time.

The Joker plays for the fun alone and enjoys the social aspects.

The Director plays for the thrill of being in charge. He wants to orchestrate the event.

The Storyteller plays to create or live in an alternate world and build narrative out of that world.

The Performer plays for the show he can put on.

The Craftsman plays to build, solve puzzles, and engineer constructs.

Most players are a combination of two or more types, the motivations meshing together in various combinations, often changing emphasis depending on what game they are playing. A particularly passionate game player we know can gain equal enjoyment from a vicious head-to-head game of *Unreal Tournament* (*The Competitor*) as well as clearing a level of *Tetris* (*The Craftsman*). These play types influence in varying ways a particular player's desire to play any game. Players play games to meet their needs as defined by their type in the following ways:

THESIS II: PLAYERS PLAY GAMES TO CONTROL THEIR ENVIRONMENT

In our experience, we find many people play games in part to escape from their real world, like any form of popular entertainment. But rather than simply escape, as they do when they read a novel or when they watch a movie, games allow players to become actively involved in the world they escape into. This is part of what makes 3-D virtual worlds so compelling, as the players become agents in those worlds. Storytellers make up little stories about the characters in the games they play. These little stories make the alternate game-world reality more real. This lends importance to the decisions players make during the game. The more real the world, the more important the players' influence over that world becomes and the more important the decisions made by the players become. What are players trying to do as they make decisions during game play? They are trying to gain some amount of "control" over their gaming world, resulting in an increased sense of agency.

The "control" the players seek is a curious phenomenon. Control is not what they really want, although it is the word they will use if asked. They want the illusion of control, the feeling that their actions make a difference. Now in most games the player is not in total control—after all, there are opponents (both human and artificial intelligence, or AI) and obstacles whose purpose is to defeat the player's plans. But, the player does want influence over the game system in a predictable way, mixed in with some degree of unpredictability (randomness). An example would be the chess player's desire to have, for example, a rook *always* move in a straight line parallel with the grain of the chessboard and not, say, move diagonally at random times. This allows both the player and his opponent not only to observe the current position of every rook on the board and predict all potential rook moves both on their part as well as the part of the opposing player, but to plan future moves armed with that knowledge. The player feels in control because he can predict what his opponent is going to do on the next turn, insofar as the moves of the rook are concerned. However, as he projects all rook moves further and further into the future, and while all the moves *are* predictable, the human player's ability to store and analyze all the potential combinations of those moves grows progressively more difficult. Compounding that, we have the fact that the player cannot predict in any fashion *which* of the potential moves available to the opposing player will, in fact, be chosen. Thus, we have the delicious combination of predictability, control, and randomness that is the essence of exciting strategy games.

The rook's behavior is predictable and therefore somewhat "controllable." Players would become very frustrated indeed if they had planned a certain move for that rook but when they took their hand away from the piece after finishing their move, it shifted one square at random from the player's chosen destination. The player would feel as if his strategic planning no longer meant anything. The player's frustration would vary by how often it performed this random move, how predictable this random pattern would be, and whether the piece could accidentally capture that way (because then the player could capture an enemy piece that otherwise he could not have reached—if, say, the rook could move randomly diagonally). But, if the player could *use* the random move to his advantage, because it was semipredictable, then some "controllable" excitement might be brought into play. It would become a different game, for sure, but perhaps experimenting in that way might lead to an interesting *new* type of game.

Many players want to exert control through the choice of the kind of game they purchased in the first place. What we mean by this is that players have a love–hate relationship with innovative features. They essentially want the game to give them what they "bought into." Example: Some MMORPG (Massively Multiplayer Online Role-Playing Game) players play only to accumulate items (Collector) and better their character's abilities (Achiever). In that vein, they want the game to be predictable (controllable), because to them the most important thing when they play is to maximize the efficiency of their online time.

Theses players can boast about achieving the most levels per hour, the quickest time to level 50, and so on. They always want, for instance, to go to a specific hunting location in the universe and find the object of their hunt right there, ready to be plucked. To have the prey not be there or (worse) having left never to return is to them a tragedy, because that will lessen their chances of leveling up that night, as well as lessening their chance of setting the record time to level 50. Because MMORPG players tend to be Achiever/Collectors, they exhibit desire for this kind of control to a finely honed degree. These people tend to view MMORPG games as a way to gain control in an alternate universe that is "sort of" like the one they actually live in, but is much more predictable. This forms an alternative to the world they live in, which feels (to them) random, heartless, and insensitive to their needs. In the experience of one of the authors, while listening to feedback from this audience about his MMORPG designs, he concluded that many of these people in their real life seem like control freaks. He believes that they express a need to control in games because they feel the more they control life, the safer they are. To some extent, all people play games to exist and compete in a world where they can be safe, but the degree to which MMORPG games appeal to this type of player is extreme, in this author's experience.

One last thought about Control players: Often they are willing to make any kind of sacrifice to ensure that their experience is controlled. In our experience, we have discovered that to them, the game they are playing can be quite illogical so long as it is controllable by them. If the game has a story, for instance, fictional inconsistencies are absolutely allowable to this group of players if they get their control in return for that illogic. They will often rationalize an illogical world or simply not care about the internal logic of the world if that world is delivering their controllable experience. They would like their roller coaster to be as repeatable as possible: The drops are the same every time down the ride; The incline on the way up is identical. No surprises for them, even if that might make the world seem more real or the story more interesting. They do not want it.

THESIS III: PEOPLE PLAY GAMES TO VICARIOUSLY EXPERIENCE SOMETHING THEY KNOW OF BUT OTHERWISE ONLY AS AN OBSERVER

Virtual gaming worlds allow participants to experience a universe they may have only imagined. Fantasy and history often fit into this category. This phenomenon is why many people play war game simulations. They have read all about Stonewall Jackson's famous flanking maneuver at Chancellorsville, and they want to "see it for themselves." This can be done by controlling Stonewall's forces in that battle in a virtual Chancellorsville battlefield against a computer AI. One of the authors has designed pencil-and-paper war games, the kind where you push around cardboard pieces on paper maps and combat is resolved by rolling dice. Most of those players were indeed students of history, but they also were great fans of the question "What if?" What if in the battle of Gettysburg, Longstreet had not waited so late to attack on the 2nd day? What if Stonewall had not been killed 2 months earlier and Lee had him at his side at that famous battle? These people are Storytellers, for the most part. They want to rewrite history to see what it might have been or to prove a point they have long held dear. They love to put themselves in the place of Lee or Napoleon and make decisions for them. In the old days, war game designers would call these gamers "armchair generals."

Another genre that attracts these types of players is sports games. These customers are not content to watch sports on TV—they want to get right in there with the stars they follow and feel as if they are on the same field with them. Often, they are people who really did want to play a professional sport but were not blessed with the physical talent. In the last few years, these games also have added "management" modules. These allow the players to not only be the player but also the manager or the general manager of their sports teams. These modules often allow rapid replays of whole seasons one after the other, so the player, in the role of the team's general manager, can, for example, trade the team's star player, play out the season in a few minutes, see how the team did after making the deal, see if the player's conclusions about whether the team would be better without the star were correct, and then undo the trade if they wish and go back before the deal. In other words, create the fictional alternatives and see how they play out. They also are Storytellers.

THESIS IV: PEOPLE PLAY TO VICARIOUSLY LIVE ELSEWHERE AND ELSEWHEN

Similar to Storytellers, these players have a lot in common with those who enjoy traditional media such books and movies. They often play games to escape into an alternate reality, to see and explore and interact with every nook and cranny of that reality. They are Explorers, Collectors, Performers, and Craftsmen. To these players, having control over that environment is not as important to them as it is to have the environment seem as real and be as fully fleshed out as possible. This group can stomach (and in many cases expect) and even want some random events, because that kind of randomness brings the virtual world closer to the reality of the world they know. These players want surprises; they want storms to buffet the ship. They want these things precisely because they understand that the unexpected is part of any experience, any story. These people want this alternate world to be as "real" and "consistent" as possible, because those qualities make it more immersive.

They are less likely to put up with fictional inconsistencies, because those are the things they would reject if it were a movie or a novel ("the character wouldn't say that"). The randomness in their virtual world can not be illogical or come out of the blue. If the virtual world is going to have some sort of volcanic eruption, they have to be told far in advance in as clear a manner as possible that (a) the world has volcanoes and (b) that the volcanoes can erupt at any moment. Then, lastly, they have to be told that an eruption is imminent. What that then becomes, to these players, is an invitation to "come and watch the fireworks." These players do not necessarily want to control the eruption—to some extent they do not mind being killed by the eruption—they just get excited at the possibility that there will *be* an eruption and it could happen at any minute.

Then, if the game designer gives these players a chance to role-play during the eruption (Performers) or get stuff that falls from the sky during/after the eruption (Collectors) or rebuild the devastation after the eruption (Craftsmen), the designer has started to push more buttons and the reality of the virtual world starts to take on an even greater texture, which these players enjoy even more.

Perhaps it might be said that these people are more interested in the journey and not so much in the destination. They are the most likely players to enjoy the Dream of the Holodeck (see below).

THESIS V: PEOPLE PLAY TO COMPETE (BUT IN A STRUCTURED WAY, WITH RULES)

Game designers know who these people are very well; they are the people that most outsiders tend to think of when they think of "gamers." The stereotype is hardcore, frag-minded, trash-talking, head-to-head gamers playing *Doom* or *Quake* on the Internet and bragging about their conquests afterwards. They fall into the categories of Competitor, Achiever, and, to a lesser extent, Director and Performer. One interesting note to consider here is that in computer and console gaming, for many years these people did not have much to talk about because they could only play against the computer, which was not as rewarding (computer egos are harder to insult). They could and did compete against the computer AI, but given the state of computer opponent AI, that was not much to write home about.

Competitive games give people a way to express their combative, aggressive tendencies in a safe, socially acceptable way. One could argue that violence expressed in a game is not that much different from violence expressed in the real world. These people play games so they can feel good about themselves because they have won something. This proves in their mind that they are better than someone else, and often for these people, it is the only way they *can* feel good about themselves. Following the stereotype even further, these gamers are sometimes the ones who are extremely maladjusted socially. They are also for the most part engineers, who perhaps lack certain social graces. Their success in games many times is a substitute for social acceptance and success in the real world. Let us be clear that we are talking about the extremes in these cases.

Many times, these people are similar in makeup to those who play competitive real-life sports because of, in part, the adrenaline rush of competition and the need to establish dominance in some arena. Many professional athletes are also big fans of video games, especially video sports games. For these people, it is totally about the game as a vehicle for establishing a pecking order, regardless of how friendly or unofficial the competition is. Achievers need the environment to be organized, because that lets them derive standings, ladders, and rankings, all of which are important for bragging rights. We have observed this dynamic time and time again among the players in the computer and video gaming field that are rarely observed among the players in the war game or role-playing field. The authors have designed games in all the genres, and the players who played one of the authors' sports games were absolutely more concerned about their success or lack thereof than either of the other two genres. In many ways, Competitors play games solely for the establishment of that pecking order.

The extremists in this category tend to hide behind, to a greater or lesser extent, the illusion of "it's all just a mental exercise" or "we're just testing our hand/eye coordination" or some such innocuous description of why they play. But, we have seen many times that the individual player's choice of which kind of game they want to play has everything to do with which one they feel they can excel in as opposed to which one they enjoy the most (although that line is very fuzzy; do they enjoy it because they can excel at it?). Example: *Magic the Gathering*. We know many people who enjoy it for the arcane nature of its rules. This arcana supports a devoted following who enjoy obtaining mastery over it so in turn those masters can pound into submission opponents who have not yet achieved mastery. The exchange of cards at game's end plays into this dynamic perfectly, because then the game appeals to both the Collectors (described above) and the Competitors. For many of the other types of gamers, *Magic* has almost no appeal because they do not need to achieve mastery over the game so they can beat others, and so the thinly veiled fantasy motif holds no fascination for them because it is not consistent and engaging enough. The Storytellers or those who enjoy enveloping themselves in an alternate world do not get enough satisfaction from the world that *Magic* delivers.

THESIS VI: PEOPLE PLAY TO EXPLORE FANTASY RELATIONSHIPS SAFELY

Satisfying the desire of many people to escape into a true fantasy world is, in our opinion, the big gateway into the mass market for games. We are appealing here to the following categories: Explorer, Joker, Director, Storyteller, and Performer. Big pieces of evidence pointing to that conclusion are the successes of industries that play to that tendency in both men and women: adult entertainment and romance novels, which many have said are two sides of the same coin. Men seek to engage their fantasies in a safe manner and their fantasies and daydreams take the form of physical couplings with women. Women, on the other hand, while seeking to engage their fantasies in an equally safe manner, are looking for couplings of the emotional kind. In our country, it is really our puritanical societal roots that have categorized one form of entertainment as unacceptable and the other as acceptable.

Games have done very little in this area to date. Role-playing games (RPGs) have come the closest, and if you talk to fans of that genre, in those games it is all about the story. And then if you talk to the women who play role-playing games (and this is one genre in the game industry where women do indeed form a sizeable group), they often discuss how the romantic possibilities within the role-playing universe are the very reason they play. Even if the game designer does not explicitly deal with romance in the story, the female gamers will invent it in their own head. Male fans of role-playing games long for this as well. One of the authors has designed role-playing games for Nintendo (perhaps the least overtly sexual gaming platform on the market) and had the publisher insist that the story include a girlfriend/potential love interest in the story for the hero because "all great role-playing games have a love story." And this is in addition to the scantily clad women that tend to adorn the box covers of RPGs.

Women especially can be very frank in their desire that RPGs allow them to role-play and indulge in their fantasies. In the world of pencil-and-paper RPGs, one female player, socially well-adjusted and physically attractive, told me flat-out that she created characters in these virtual worlds that were facets of her own personality that she could never feel safe expressing in the real world. One of those facets was that of the promiscuous female who coupled with multiple partners. And that was the type of character she played in the world, much to the delight of the male players in the world, who enjoyed role-playing with that version of the woman, engaging in conversation with her that they might have fantasized about in real life but never had the nerve or opportunity to engage in.

Is adult entertainment a part of the game industry then? It has long been known that while online games have a hit-and-miss record of profitability, interactive Web sites that deal in adult entertainment are extremely profitable. If we compare the video game industry to the adult entertainment industry: In fiscal 2002, EA (Electronic Arts, the biggest and most consistent game publisher) made $1.7 billion. The video game business grossed approximately $7 billion in 2002. Now, a leading producer of adult entertainment, Vivid, grossed (according to industry estimates) $1 billion in that same year and the adult industry grossed $10 billion in that same year. And that income, in both the video game and adult entertainment industries, is derived mostly from men. So, the adult entertainment industry grossed more than twice what the game industry did in that year, and the leader in the adult entertainment industry approximately grossed half as much business as the video game leader, and much more than any other company in the game business that year.

What does that mean? It *may* mean that if the game business can figure out a way to provide greater exploration of the kinds of fantasy relationships that men and women both want, in ways that do not raise the anger and resentment of society too much, that kind of content could really grow the industry exponentially. And the game industry has longed to enter the mass market and especially get women to play games for years, but really has not made any kind of progress. I am not saying that adult entertainment is the killer app; what I am saying is that in avoiding the kind of entertainment that plays to our collective fantasies, we are avoiding what may indeed be some kind of new mass market. We should starting thinking about creating the kind of entertainment that adults clearly would be willing to pay for.

In summary, people play games for a variety of reasons, in the main related to their desire to escape from the mundane world and enter into some kind of special world. Alternatively, they play to achieve something in that game world that earmarks them as special or allows them to establish dominance over others in this world. So, how do we begin to design games to take this knowledge into account?

HOW GAMES ARE DESIGNED

It is one thing to talk about why people enjoy games and quite another to use that information to design a game. To understand how designers use what they know about human psychology in the design of games, it helps to first understand the game design process. For those who have never done it, there is a lot of mystery and confusion surrounding the question of how computer games are actually designed. Many people are under the misguided impression that a game designer sits down and writes a design document of several hundred pages, thus completing the design, which can then be given to a development team of programmers and artists who will produce the game to that specification. If only it were that simple! Because there are so many constraints on making a successful computer game, what really happens during design is much more complicated.

Where to Start

When designing a game, where is the right place to begin? To quote Reiner Knizia, the famous board game designer, "If you always start in the same corner, you always end up in the same corner" (Tinsman, 2002). Different games require different approaches. Most game designs, however, do have one thing in common: They begin with a single idea, which is at the core of the game design process and which the rest of the design process will crystallize around. Interestingly, as the design develops, it may be the case that the original seed idea is no longer at the core of the design—the design may have evolved a new core. However, without that original seed idea, the design process would never have begun.

Generally, there are three types of seed ideas that game designs crystallize around:

Technology ideas. New technologies make new types of game play possible. These technologies are often new software or hardware systems. For example, new algorithms for the simulation of realistic physics, new rendering technique, or new types of handheld input devices might all be central technology ideas that a game crystallizes around. When Sony's Playstation2 was released, it featured the ability to draw realistic fireworks effects in real time. This became the seed idea for *Fantavision*, a puzzle game all about launching fireworks. For games like this, the seed idea is an answer to the question, "What can I do with this new technology that I couldn't do before?"

Theme or story ideas. Sometimes the seed idea is a setting, or some characters, or a plotline. The designer envisions a game world that would be interesting to visit or explore, characters that would be interesting to meet, or a story that would be fun to experience. Because stories, worlds, and characters are often the most visible parts of many games, many novices to game design assume that all games begin with these stories. And some of them certainly do. Once the designer has this kind of seed idea, he or she next has to answer the question, "What kind of game would be fun in this world?"

Game mechanic ideas. This last one is often the hardest for nongame designers to comprehend. "Game mechanic" is a term that refers to the structure of a game, independent of all of the story, theming, and other content that makes the game easier to comprehend and play. For example, the game mechanics of chess have everything to do with the layout of the board and the way the pieces move, and nothing to do with the labels and images put on each piece. From a game mechanics standpoint, the king could just as easily be "piece type 1," the knight "piece type 2," and so on. The game mechanic of what moves are possible and how the pieces interact would be exactly the same. Often, a new mechanic is a combination or modification of existing mechanics. Experienced game designers are skilled at separating the game mechanics from the thematic elements in their minds and thinking about them independently, as structures unto themselves. Novice game designers have a harder time with this, and are forever tangling the ideas of mechanics, story, and technology in their minds and in their designs. When a designer comes up with a new game mechanic, questions arise such as: What other mechanics would work well with this one? What technologies would best facilitate this mechanics? and In what stories and themes would this mechanic work well?

Growing the Seed

Once the seed idea has been established, the designer then starts to grow the game through the application of what we term "filters." Each filter is a perspective on the game. Each one encourages the game to grow in certain directions, and each one imposes limits on the game. The application of these filters is often a team process and happens gradually over time. They certainly are not applied in order, and they are applied multiple times, as the application of each filter can change the design substantially. When the design is complete, the game must withstand scrutiny from the perspective of each of these filters. Most games are not fully designed until partway through the development process, or perhaps not until the very end. The fact that these different filters provide so many different conflicting constraints is what makes game design such a challenging task. The primary filters that most designers apply to their design are:

1. *Artistic Impulse.* Basically, designers consider what they would like to see this game become, based on their natural impulses and personal preference. Most game designers put a lot of stock into "what feels right" to them. Of course, the game is not for them, and sometimes a designer's natural instincts about how to grow an idea are wrong, but that is what the other filters are for.

2. *Demographic Considerations*. Designers have to ask, Who is this game for and what would those people like? Based on the age, gender, and interests of the intended audience, the design can change dramatically. If the game is meant to please multiple demographics, this filter becomes all the more challenging. Designers' ability to successfully use this filter is contingent on how well they truly understand the psychology of the intended audience. When we consider these demographics, sometimes we express them as the types described above and sometimes it is simply age, gender, and so on.

3. *Experience Design*. Game designers have many rules of thumb about what a successful experience should be like, several of which were explained earlier. These are principles about the types of decisions a player should be asked to make, the pacing and intensity of the experience, the level of difficulty that is appropriate, and many more. Based on these principles, which differ from designer to designer and are usually based on opinions about human psychology, modifications are made to a game in an attempt to make the game optimally enjoyable.

4. *Innovation*. The world of game design is a world of innovation. The pleasure of novelty is strong, and interactive entertainment that is enabled by new technologies is fertile soil for creating novel experiences. Designers will ask the question, "What is truly new in this game?" and try to make the most of that, often discarding features that are clichéd in favor of features that better enhance the parts of the game that are innovative and new.

5. *Business and Marketing*. The games business is a business, and designers who want their games to sell must consider the realities of this and integrate them into their game's design. This involves many questions. Are the theme and story going to be appealing to consumers? Is the game so easily explainable that one can understand what it is about just by looking at the box? What are the expectations consumers are going to have about this game based on the genre? How do the features of this game compare to other similar games in the marketplace? Will the cost of producing this game be so high as to make it unprofitable? The answers to these and many other questions are going to have an impact on the shape of the design. Ironically, the innovative idea that drove the initial design may prove to be completely untenable when viewed through this filter.

6. *Engineering*. Until you have built it, a game idea is just an idea, and ideas are not necessarily bound by the constraints of what is possible or practical. For this filter, the designer has to ask, "How are we going to build this?" The answer may be that the limits of technology do not permit the idea as originally envisioned to be constructed. Novice designers often grow frustrated with the limits that engineering imposes on their designs. However, the engineering filter can just as often grow a game in new directions, because in the process of applying this filter, the designer may realize that engineering makes possible features for the game that did not initially occur to the designer. The ideas that appear during the application of this filter can be particularly valuable, as the designer can be certain that they are practical.

7. *Social/Community*. Many considerations arise during the application of this filter. If the game is inherently multiplayer, either because players play together in the same room or via computer networking, a host of issues arise involving the psychology of human interaction and communication. The designer will consider many things, including the balance between collaboration and competition and the type of social interaction best facilitated by this type of game. Even if the game is inherently single-player, community issues still apply, as the player may well be part of a community of players who discuss the game. The designer may change the game to increase the amount of community discussion about the game, or by giving players ways to help one another by sharing

information about the game and feel rewarded by doing so. The designer will consider not just pleasures within the game, but how the game can be designed so that it facilitates pleasures outside the game as well.

8. *Playtesting.* Once the game has been developed to the point that it is playable, the playtesting filter, which is arguably the most important of all the filters, is applied. It is one thing to imagine what playing a game will be like and quite another to actually play it. Most designers want to get to a playable stage as soon as possible, because they know that when they actually see their game in action, important changes that must be made will become obvious. Designers learn a lot by playing their own games and even more by watching others play their games. While surveys of playtesters will reveal some important data, designers usually get most of their useful information by watching people play and occasionally asking them questions, all in an attempt to see if attempts to motivate and entertain players have been successful. In addition to modifying the game itself, the application of this filter often changes and fine-tunes the other filters, as the designer starts to learn more about the psychology of the intended audience.

Do not mistake the above for a simple eight-step process. Instead, the process is an iterative one, where the design is constantly in flux, being changed, considered, and tested, until it is deemed good enough via each of the eight filters (and perhaps others)—or until time runs out. There is an old saying that "A work of art is never finished, only abandoned," and this is perhaps more true for video games than any other medium.

THE DREAM OF THE HOLODECK

She spread the elastic headband and settled the trodes against her temples—one of the world's characteristic human gestures, but one she seldom performed. She tapped the Ono-Sendai's battery-test stud. Green for go. She touched the power stud and the bedroom vanished behind a colorless wall of sensory static. Her head filled with a torrent of white sound.

Her fingers found a random second stud and she was catapulted through the static wall, into cluttered vastness, the notational void of cyberspace, the bright grid of the matrix ranged around her like an infinite cage. (Gibson, 1988, p. 49)

William Gibson's apocalyptic vision of cyberspace, written years before the Internet became a household word, has, for the most part, remained tantalizingly close yet frustratingly far away. However, the direction games have gone in the 17 years hence has unerringly been toward the realization of Gibson's dreams. If indeed people desire to live in alternate realities, if they want to shift away from the cold harsh here and now and slip into the warm fantasy of a virtual world, it seems as if the game industry will lead the way there.

Will we ever get to interact with the "Holodeck" in the way Gibson envisioned or Captain Picard enjoyed in Star Trek? It is hard to say for sure, but we are on that road and the wind is at our back.

REFERENCES

Gibson, William. (1988). *Mona Lisa overdrive*. New York: Bantam.
Tinsman, Brian. (2002). *Game inventors guidebook*. Iola, WI: Krause Publications.

8

Why Play? An Evolutionary Perspective

Peter Ohler
University of Technology Chemnitz

Gerhild Nieding
University of Würzburg

Play researchers of various disciplines agree that an unambiguous definition of play has not been offered in the history of play research. Sutton-Smith (1997) even assumed that its ambiguity is the peculiar quality of play. Nevertheless, researchers have built theories for the underlying psychological processes of play: for example, the classical pre-exercise theory of Groos (1899, 1901), the physical exercise theory of Smith (1982), the cognitive-developmental theory of Piaget (1962), and the arousal modulation theories of Hutt (1979), Shultz (1979) and Ellis (1973), to mention a few.

More recent psychological approaches of play differ from more classical ones with respect to their breadth. Compared with speculations and theories of the 19th century (e.g., Groos, 1899, 1901; Gulick, 1898; Lazarus, 1883; Spencer, 1864, 1873), which tried to explain the entire phenomenon of play, psychological approaches of the 20th and 21st centuries only focus on specific aspects of play. The 20th-century play researchers criticize those earlier approaches for their one-sidedness. The early theories tried to explain the entire phenomenon of play but mostly with a narrow focus on one psychological function. On the other hand, the early approaches show a closer match to what we would now call a unified theory of play in animals and humans.

But now the theoretical prerequisites are available to try once again to construct a unified theory of play in animals and humans, an attempt that neglects neither the variability of play forms and play behaviors nor the variability of play functions. In an evolutionary perspective, we are able to answer the question, "Why play?"

In this chapter, we briefly introduce the principles of evolutionary biology. This is followed by an outline of two versions of evolutionary psychology that differ with regards to culture as an evolutionary mechanism. After that, we introduce an evolutionary theory of play in animals and humans and make some inferences of that theory for various play forms. The framework of this theory also allows predictions for the field of playing computer games. We will therefore take a closer look at two inferences of the theory by briefly reporting two experiments we conducted on computer games. One experiment examines whether playing induces a broader range of strategies compared to framing as a problem-solving task. The other experiment is concerned with the connection of being in a playful mode and the probability of showing aggressive behavioral acts while playing first-person shooter games.

AN EVOLUTIONARY APPROACH TO PLAY IN ANIMALS AND HUMANS

The starting point of evolutionary considerations about play is the phenomenon that not only (young) individuals in the species *Homo sapiens sapiens* show play behavior but also other species (Bekoff & Byers, 1998; Burghardt, 1998a). All mammals, some birds, and perhaps even nonavion reptiles (turtles [*trionyx triunguis*]; Burghardt, 1998b) play. This means that the behavior system "play" is much older than modern humans, it is even much older than the hominids (all members of the family *hominidae*), and it had already existed for millions of years when the first australopithecines appeared. For a deeper understanding of play behavior and its adaptive function, it is therefore necessary to study this phenomenon within an evolutionary perspective. Play as a psychological phenomenon has to be studied within the framework of evolutionary psychology (Buss, 1999; Cosmides & Tooby, 1992; Tooby & Cosmides, 1992).

Evolutionary psychology (EP) is a synthesis of modern evolutionary biology with modern (cognitive) psychology. It focuses on the influence of natural selection on constructive features of the psychological mechanisms that control behavior. In a narrow cognitive version (Cosmides & Tooby, 1997), EP can be defined as the description of information-processing mechanisms designed by natural selection whose interplay form the human mind or its cognitive architecture, respectively. EP, sometimes also called Darwinian psychology (Plotkin, 1994), should therefore be congruent with valid principles of classical and modern evolutionary biology.

The Foundation of Evolutionary Psychology in Principles of Evolutionary Biology

The classical assumption of natural selection by Charles Darwin (1859/1995) is still a main basis of evolutionary psychology. Darwin's argument included the following steps: (a) the number of individuals of a species grows faster than the necessary resources that are available (principle of Malthus, 1826); (b) individuals of a species vary in structural and behavioral features; (c) the variations can be inherited; and (d) if individuals of a species acquire a competitive advantage in the acquisition of resources caused by variation, this will increase their chance to reproduce. It follows that the variation will manifest itself gradually in the population or a subpopulation of that species. In modern evolutionary biology, the reproductive success of an individual, including the ability to survive until reproduction, is called fitness. Darwin's (1859/1995) principle of natural selection is equivalent to reproductive fitness, sometimes also called Darwinian fitness.

An important modification of the classical approach to fitness (which had certain shortcomings) was developed by Hamilton (1964a, 1964b) with his theory of inclusive fitness. Inclusive fitness is the sum of direct (reproductive) fitness and so-called indirect fitness, which is the reproductive success that can be acquired through the reproduction of genetic relatives. It follows that not only the reproductive success of the individual itself, but also its actions and consequences that influence the reproductive success of its kin, are relevant for the survival of the individual's genes.

The parental investment theory of Trivers (1971, 1972) explains the conditions of sexual selection for both sexes: short-term and long-term mating strategies of males and females. This theory, which is an elaboration of Darwin's (1871) theory of sexual selection, predicts that the sex that invests more resources in the rearing of offsprings should be more choosy in mate selection (e.g., human females), whereas the other sex should be more competitive with members of their own sex (e.g., human males) in accessing members of the opposite sex. It should be mentioned that the preference for more aggressive playing in males—for example, a higher frequency of play-fighting in boys and the preference for first-person shooter games in male adolescents—can be attributed to such inherited sex differences.

Kimura (1983) showed that gene frequencies can also change randomly over time and that, therefore, specific features can manifest themselves in populations even if they are not adaptive. Wilson's (1975) book on sociobiology was the first synthesis that integrated humans and cultural tools produced by humans into the range of phenomena ultimately to be explained by evolutionary principles.

These principles of evolutionary biology suffice to construct an evolutionary psychology of the human cognitive architecture and the emotional and motivational makeup of our species.

Principles of Evolutionary Psychology

Cosmides and Tooby (1997) argued that the human brain works like a biocomputer and its neural circuits enable the production of behavior, which is adapted to a specific environment. The neural circuits were designed by processes of natural selection (they are solutions of adaptive problems). Adaptive problems are problems that occur very often in the evolutionary history of a species—they are evolutionary recurrent situations (Cosmides & Tooby, 1994). Only the results of some (integrated) high-level neural circuits can be processed consciously; the majority of processes operate automatically. Different neural circuits are specialized to solve different adaptive problems. Solutions can only be achieved by very specific functionally distinct mechanisms (Cosmides & Tooby, 1994; Tooby & Cosmides, 1992). These specialized modules increase the inclusive fitness, but something that fits one domain does not necessarily fit another domain. To solve that problem, many distinctive domain-specific modules evolved.

On such a basis, we can distinguish two principal types of evolutionary psychology. The classical form of EP assumes that the cognitive architecture and socioemotional makeup of *Homo sapiens sapiens* was finally formed in an *environment of evolutionary adaptedness* (EEA)[1] that existed between 100,000 and 200,000 years ago in the savannahs of Africa (Cann, Stoneking, & Wilson, 1987), when the species lived as hunters and gatherers (e.g., Barkow, Cosmides, & Tooby, 1992). This approach takes for granted that the later cultural improvements of the species (e.g., parietal art and portable art; cf. Conkey, 1999) circa 40,000 years ago are entirely the result of the cognitive architecture of modern humans. The cognitive architecture, which secured the survival of hunters and gatherers in the Pleistocene, is still at work today. There was not enough time in evolutionary dimensions (variation qua mutation, inheritance, and selection) to change the cognitive architecture of *Homo sapiens sapiens* fundamentally. Therefore, culture does not influence the way we think.

The other type of EP favors a coevolutionary approach, assuming a close linkage between brain and culture that accelerated human evolution in the last 40,000/50,000 years (Donald, 1991, 2002). The highly social mind of modern humans is able to use external representations (e.g., parietal art, development of writing) as a cultural strategy that improves remembering and problem solving. Human anatomy and behavior seem to have evolved slowly before 50,000 years ago. Afterwards, anatomy remained unchanged but behavior, including cultural traditions and use of some first media, accelerated dramatically. Perhaps this change was caused by a mutation (Klein & Blake, 2002).

THE BEHAVIOR-DIVERSIFICATION PROTO-COGNITION THEORY OF PLAY IN ANIMALS AND HUMANS

These introductions on evolutionary psychology now allow us to explain the core structure of our behavior-diversification proto-cognition theory (BD-PC theory) of play in animals and humans (Ohler, 2000; Ohler & Nieding, 2001). We will begin with the behavior-diversification part of the theory. The basic assumption is that the behavior system "play" was selected in evolution because of its potential to generate behavior variants. Once upon a time, a vertebrate species existed and in this species a mutation appeared, which caused a very curious behavior that probably was not observed in any other species before. The curious behavior was to produce random sequences of acts, which were originally part of other behavior systems (e.g., fight, flight, hunting, nutrition, reproduction). The behavior acts are sequenced in a combinatory fashion, combining acts across the borders of behavior systems (Bekoff, 1995). Individuals of that species, which possess this behavioral feature, are able to retrieve a repertoire of behaviors more effectively than other conspecies. If this species lives in an EEA (much older than the environments that shaped the hominids) with variable changing niches—but with changes that can be reacted to with responses on a behavioral level—these individuals have an advantage in fitness (an adaptive advantage). Individuals who possess the feature "play" show higher reproductive success. Over many generations the feature will manifest itself in the genetic pool of that species, and all members of the population or isolated subpopulation will play. This is the first argument of BD-PC theory.

This part of the theory is not new. The central assumption of *behavior diversification* in BD-PC theory is close to ideas formulated by Sutton-Smith (1978, 1997) using labels like *adaptive variability* (Sutton-Smith, 1997) or *adaptive potentiation* (Sutton-Smith, 1978) or ideas formulated by Fagen (1981) like *functional flexibility*. Both authors want to express that playing species adapt to their environments with the aid of flexible behavior patterns and may even change their environments in this process (Fagen, 1981). One major difference seems to be that Fagen (1981) and Sutton-Smith (1978, 1997) believed that play behavior *uses* different behavioral elements whereas the BD-PC theory assumes that play behavior originally *is* identical with the combination of behavioral elements of different behavioral systems. The other approaches look at diversification as a play function, whereas in BD-PC theory diversification is the defining element of play when it first appeared in evolution.

Fagen's (1981) approach is restricted to animal play; he did not try to integrate animal and human play forms. For Sutton-Smith (1997) and the BD-PC theory, animal play is only the starting point for assumptions integrating animal and human play. Both approaches agree that the evolutionary function of play is not practice (Smith, 1982) or pre-exercise (Gross, 1899) but behavioral adaptation. However, for BD-PC theory it is not an enlargement of adaptive behavioral potencies (Sutton-Smith, 1997); it is a behavior system that increases inclusive fitness. Therefore, we think Sutton-Smith's (1997) main mistake is to take the evolutionary function of play too metaphorically.[2] Our conceptualization of adaptation via behavioral diversification is to be understood in a strictly literal sense.

We can now continue with the proto-cognition argument of the BD-PC theory. The individuals of the first species with a fixed play allele in its genome differ from the individuals of that species before this mutation occurred. The neural circuits and control mechanisms, which generate play behavior, are now part of the psychological architecture of the playing species, that is, the individuals of the species now possess a play module. The module is distinct from other neural circuits that were selected in the course of solving other adaptive problems (Cosmides & Tooby, 1994). It is always triggered when specific environmental cues occur, resembling the adaptive problems in the EEA. In correspondence with the adaptive problems that were in function during the establishment of the play module, it will always be triggered when *nonreducible novelty* appears in the stimulus field of an individual of the playing species. If environmental cues are associated with nonreducible novelty,[3] individuals of that species will respond with the activation of the play module. As a result, those individuals will execute the typical combinatory behavior sequences transcending the borders of behavior systems.

If a specific feature is fixed in the genome of a species, it will only be extinguished again if it causes detrimental effects on the individuals of the species under the conditions of a change in the biotic and/or a-biotic environment. This was not the case with the play module of our hypothetical first players. Had it been extinguished, humans and other animal species would not play nowadays.

We will make a jump in the evolutionary period of time. The evolved play mechanism is preserved even in species with more highly developed cognitions. We have now approximately reached the complexity of cognitive systems that can be observed in mammals like the canines (Bekoff, 1995, 1998). The general principle of the play module still is: If it is active, it modifies the units of those other systems systematically that are activated at the same time. But the module is now—in contrast to the situation of the original first players—not restricted to circuits controlling behavior but can also operate on cognitive modules.

Then a species occurred—probably either a common ancestor of the families *hominidae* and *pongidae* (of humans and great apes) or an ancestor in the hominid line—that possessed a cognitive architecture with a capacity allowing a qualitatively new type of mental representation. Up to this point in evolution, all species were only capable of operating cognitively on the basis of primary representations (Leslie, 1987; Perner, 1991). The contents of primary mental representations—also called cued representations (Gärdenfors, 1995)—are always related to entities that are available in the present situation of the representing organism. Maybe the entity is really present in the stimulus field of the representing organism or at least some cues in the environment exist, which refer to the represented entity (e.g., a cat sitting in front of a mouse hole, where a mouse has disappeared). Primary or cued representations are always tied directly or indirectly (via cues still present) to the perception of the representing organism. Primary representations may be selective—that is, they may not represent every aspect of the environment—but they do not represent aspects that are not at least indirectly present. At this point in the evolution of representational systems, no individual of any species was capable of representing entities that are only imagined.

In this situation, the play module would enable a quantum leap in the functioning of the representational system. Our premise is that a species has evolved with at least a few individuals who possess a cognitive architecture capable of producing a new type of representation. When those individuals have activated primary representations that trigger the play module—genuine novelty or inconsistencies within the represented content may be the causes—the play module will work in its well-established fashion. It will force every unit to combine in the activated part of the cognitive system with every other unit. This systematical combination of units in primary representations will lead to the emergence of a semiotic function that was not realized before. A new relation between elements is established: A mental element is able to represent another mental element.

With this rudimentary novel semiotic function, the first secondary representations (Povinelli, 1998a; Perner, 1991) are established in the course of evolution. Secondary representations— also called detached representations (Gärdenfors, 1995) or decoupled representations (Leslie, 1987)—refer to entities that are *not* available in the present situation of the representing organism and *not* present in its stimulus field. The elements of secondary representations are *not* triggered by cues in the environment. The content of secondary representations consists of mental elements referring to mental elements. Now entities can be represented that are only imagined. Secondary representations are the prerequisite for all cognitive operations that allow hypothetical and/or counterfactual thinking (Mitchell & Riggs, 2000).

This means that the play module is responsible for the quantum leap to the first secondary mental representations in phylogenesis, of which some authors think that they are fundamental for the difference between humans and all other primate species (Gärdenfors, 1995; Povinelli, 1998a). The capacities of the new representation system in its original state were surely very small. Only a rudimentary anticipatory planning was made possible. However, for the individuals possessing such a system the possibility nevertheless suddenly emerged to play cognitively with behavior alternatives and not just to pick alternatives at random and to act blindly based on trial and error. If this allowed an individual to avoid at least one tragedy from birth to reproduction, the direct fitness of that individual was increased enormously. Again, over many generations the feature will manifest itself in the genetic pool of that species, and all members of the population will have secondary mental representations at their disposal.

We can now take a closer look at the consequences that should be expected from this phylogenetic pattern for the ontogenesis of human children. Ontogenesis is not a recapitulation of phylogenesis as assumed by some authors of the 19th century (Haeckel, 1874; for recapitulation theory of play: Gulick, 1898; Hall, 1920), but they possess some analogue elements in their developmental logics (Lock & Peters, 1999; McKinney, 1998). This would mean that early pretense play, which appears at around 12 to 13 months in humans, should be the ontogenetically first psychological domain, in which secondary mental representations are realized. Even in ontogenetically later play forms—for example, constructive play, social play, and games with rules—the combinatory diversified depth structure of activities triggered by the play module remain in function but with different surface features in the behaviors executed in the different play forms. These are the central assumptions of BD-PC theory of play, which tries to explain animal and human play within a unified evolutionary framework.

Some Inferences From Behavior-Diversification Proto-Cognition Theory of Play for Other Play Forms

We will infer some propositions from BD-PC theory to show that the theory allows the deduction of some psychologically very interesting and also empirically testable hypotheses. A main consequence of BD-PC theory concerns the underlying neural substratum of the evolved play module. A distinct neural topography realized as a distributed neural network should exist, which is always active when an organism is playing. This should be found in the central nervous systems of all playing species and should also be identifiable across the different play forms of humans when they are playing. To assume a distinct neural topography associated with play does not exclude a partial overlap between those brain areas active in play and those involved in other behaviors. The combination of different behavior acts is the original defining feature of play. Therefore, neural circuits associated with those other behaviors will also be activated during play. Nevertheless, if a play module exists, there should also be a unique neural fingerprint of that module.

The human brain with a mammalian cerebral cortex (but a much more developed neocortex compared to all other species) should show different connections of the neural circuits, which are active during play, to areas of the association cortex. The unique neural fingerprint of the play module should show enriched connections to the association cortex in humans. This would match the differences in the behavior of humans toward other species: Some play forms are executed only by humans. On the other hand, our argument emphasizes that in evolutionary terms, the play module is much older than the appearance of the first primates. Therefore, it should not be located dominantly in regions of the brain that are unique to humans and perhaps not even in regions that are unique to the mammalian brain. The play substratum should show connections to evolutionary older regions of central nervous systems.

Some empirical evidence already exists that this inference from the BD-PC theory is not speculative. Siviy (1998) researched the neurobiological substrates of a social play form in mammals: the rough and tumble play of juvenile rats of different strains (Siviy, Baliko, & Bowers, 1996). The parafascicular area of the thalamus (PFA) seems to be an area of the brain that may be a candidate for the unique fingerprint of the play module.

> PFA might be a major interface for integrating somatosensory information encountered during play and relaying this information in a manner that facilitates the motor patterns used for playing. (Siviy, 1998, p. 234)

Analyses of immediate-early genes that indicate brain areas that were active in behavioral sequences enacted before, showed that rats in the experimental group (30 minutes of rough-and-tumble play) had increased values in PFA but also in areas of the cortex and of the hippocampus. Siviy (1998, p. 234) stated that "future studies that use this technique could attempt to isolate different components of play in order to reduce the background noise." This effect and other neurobiochemical studies of the Siviy group (e.g., Siviy & Baliko, 2000; Siviy et al., 1996) to find the neurobiological substrates of mammalian playfulness indicate that the play module stated by BD-PC theory can be proven on a neurophysiological level.

Another inference from BD-PC theory touches the question whether great apes should show early pretense play or not. Early pretense play in humans starts at about 12 to 13 months of age with play acts like the following: A child crawls and articulates "bow-wow" pretending to be a dog, although there is no dog present in the stimulus field of this child. Another example is one female subject in our laboratory who at the age of 12 months exhibited a very complete early pretense script of feeding a doll with a pretense meal. The BD-PC theory holds that early pretense in humans should be the ontogenetically first psychological domain, in which secondary mental representations are realized.

But, whether great apes should also exhibit early pretense not only depends on the BD-PC theory but also on the state of the art in the animal "theory of mind" discussion. Some researchers believe that great apes are able to attribute mental states in other members of their group (Fouts, 1997; Whiten, 1997, 1998), which means that the apes consider other apes to be intentional agents that possess the same mental states as themselves. Other researchers report that great apes are only "clever behaviorists" (Povinelli, 1996, 1998b; Reaux, Theall, & Povinelli, 1999; Tomasello, 1999). Despite the insightful and complex problem-solving strategies great apes show in their physical interaction within the environment (Povinelli, 2000), they are not able to capture their social world on the basis of intentional concepts. The researchers believe that the deception activities even of great apes living in the wild are only the result of complex reinforcement schedules (Heyes, 1998) and that their ability to recognize themselves in mirrors (pass the mark-test; Gallup, 1982, 1983) are evidence only of a kinesthetic self and not of a mental self (Povinelli, 1998b).

If this last picture is true, it is left open by the BD-PC theory whether great apes should show early pretense play or not. May be there is no early pretense, and therefore great apes do not acquire a theory of mind later on. Maybe they show early pretense but are not developed enough to ever reach the stage of an intuitive psychology of about 4-year-old children. But if the first picture is true and research finds strong evidence that great apes possess a theory of mind, BD-PC theory will make a clear inference: Great apes should necessarily show early pretense play. But whether great apes really show early pretense play is not definitely clear in the literature (Ohler, 2000; Tomasello & Call, 1994) to date. The data of Jensvold and Fouts (1993) on pretend play of chimpanzees do not seem very convincing to us (Ohler, 2000).

The reason why the BD-PC theory makes the strong inference that great apes will show early pretense play if they possess a theory of mind, is that the theory favors what we call a "pole-position-approach" of early pretense play, phylo- and ontogenetically. Therefore, we assume that early pretense play is the first domain, in which decoupled representations will emerge in the cognitive development of human children. Contrary to that position, most authors assume that early pretense play is only one of at least a few functions that can be executed on the basis of detached mental representations as soon as they emerge in ontogeny (Leslie, 1987; Perner, 1991). Our approach, on the other hand, is compatible with play as a zone of proximal development (Vygotzki, 1973, 1978). Play should also allow the training of theory of mind tasks, another inference that can be drawn from the BD-PC theory and that was confirmed in our research (Nieding & Ohler, 2002; Ohler, 2000).

Some Inferences From Behavior-Diversification Proto-Cognition Theory of Play for Computer Games

A Study on Playing and Problem Solving

The BD-PC theory assumes that the adaptation resulting in playful behavior is still functional and present during the development of different play forms in the ontogeny of human children. That means that the combinatory diversified depth structure of activities triggered by the play module remain in function in ontogenetically later play forms, for example, constructive play, social play, and games with rules. The surface features in the behaviors executed in the different play forms will change but the general diversification mechanism will remain. Regarding the field of computer games, the principle holds true for strategy games that playing strategy games is associated with a higher diversity of strategies in comparison with the same scenario presented as a problem-solving task (Ohler & Nieding, 2000).

Playing a computer game shows much more interactive dynamics than solving a problem that is presented (e.g., via a booklet), and this factor of confoundation must be experimentally controlled. We therefore constructed a so-called bottleneck scenario of the real-time strategy game *WARCRAFT II*. The bottleneck scenario reduces the variability of possible states within the game, especially at the bottlenecks.

Design and Method.   Sixteen students were randomly assigned to one of the two levels of the factor "playing vs. problem solving." The play group played the bottleneck scenario of the computer game *WARCRAFT II*. Each participant of the problem-solving group received the screenshots of a matched pair member in the play group and other materials and had to solve the problems that were presented at the various bottlenecks. Both groups were instructed to think aloud about their goals, strategies, and tactics they used to solve the various bottleneck problems. The verbal comments of both groups and the change of the user interfaces while playing were recorded on video. The problem-solving group was allowed to use paper and pencil to make notes.

Result.   The main result of this study was that the play group interindiviually used two to three times more strategies and tactics than the problem-solving group in each of the eight bottlenecks. This result could be replicated with turn-for-turn strategy computer games (e.g., with *CIVILIZATION II*). Human computer interaction is more rapid in real-time games compared with turn-for-turn games. This indicates that an average of a 2.5 higher strategy diversity is not caused by the rapidity of the interaction. The playful mode of thinking causes higher diversification. We conclude that the principle of playful diversification also holds true in the area of games with rules, in our case computer strategy games. We therefore conclude that the principle holds true in a play form that emerges relatively late in ontogeny. The evolutionary play mechanism postulated by BD-PC theory is still present in modern media-based play forms.

A Study on Playing First-Person Shooter Games and Real-Time Strategy Games

A second inference from BD-PC theory for computer games concerns the issue of violent content in computer games. If users are really motivated to play (are cognitively and motivationally in playful mode), the violent content of computer games—for example, of first-person shooter games—should not influence the amount of aggression during play. The hypothesis assumes that the number of aggressive acts performed while playing a computer game is less influenced by the amount of violence in the game and more by the motivation to win and the elaborateness of strategies used to achieve that goal.

Design and Method. Twenty adolescent males ranging between 13 and 17 years played a total of seven computer game scenarios. Four scenarios were taken from the real-time strategy game *AGE OF EMPIRES II* (AoE II) and three first-person shooter game scenarios were taken from *UNREAL TOURNAMENT* (UT). The four scenarios of AoE II were constructed by the experimentors. These two games embody two levels of the independent variable 1 "violence factor" (IV 1; *within subjects*). In the game with stronger violent content, UT, the player experiences a 3-D graphic of the scenario in a subjective perspective and operates various types of weapons. The main action of the game revolves around shooting, and resource management is not necessary. Compared to the first-person shooter game, AoE II contains a lower violence factor. Here fighting is initiated but not enacted, iconic figures and larger parts of the scenario are seen from a bird's eye view (distant nadir or opaque perspective), and management of resources takes priority over fighting.

The participants' play sequences and their think-aloud commentaries were recorded on video. The data were converted into a questionnaire by three experts. The questionnaire surveyed specific items relating to each scenario. It measured how strategically the players acted and the amount of aggression the players displayed. Only aggressive acts that were not necessary to reach the goal of winning the scenario were labeled as aggressive. This distinction was necessary because players were urged to use symbolic aggression in order to win, but due to the levels of IV 1, each game type required differing amounts of necessary, symbolic aggression.

Results. Only 25% of the participants displayed no aggressive acts unnecessary to reach the goal when playing the first-person shooter games; 75% of the participants showed at least sometimes and in a few dimensions (two out of the nine defined) aggressive acts that were not necessary within the limits of the first-person shooter game scenarios. Only 3 of the 20 players showed aggressive acts in more than five dimensions.

Table 8.1 shows the correlations between the calculated strategy coefficients in the real-time strategy games, the strategy coefficients in the first-person shooter games, and the aggressive acts in the first-person shooter games across the 20 participants. It can be seen that the measures of strategy orientation in real-time strategy games and first-person shooter games are significantly and highly significantly negatively correlated with the measure of aggressive acts in first-person shooter games. This pattern reveals that a player with a high strategy orientation in real-time strategy games and first-person shooter games will commit just a few aggressive acts,

TABLE 8.1

Correlations of Aggressive Acts While Playing First-Person Shooter Games with Measures in the Strategy
Orientation in Real-Time Strategy Computer Games and First-Person Shooter Games (Pearson Correlations)

	Aggressive Acts: First-Person Shooter Game	Strategy Coefficient: Real-Time Game	Strategy Coefficient: First-Person Shooter Game
Aggressive Acts: First-person shooter game	—	−.46*	−.80**
Strategy Coefficient: Real-Time Game		—	.50*
Strategy Coefficient: First-person shooter game			—

Note. *p <.05, **p < .01.

whereas a player with low strategy orientation will also commit a higher number of aggressive acts.

Strategy orientation in first-person shooter games very clearly predicts the absence of aggressive acts with $R = .82$ (with negative Beta) in a multiple regression (correlation: $r = −.80$ in Table 8.1), but there is also a medium connection between strategy orientation in real-time strategy games and aggressive acts in first-person shooter games with $R = .57$ (with negative Beta) in a multiple regression (correlation: $r = −.46$ in Table 8.1). This means that a player with a high strategy orientation in another play type will show fewer aggressive acts in first-person shooter games. The medium correlation between the strategy coeffients in real-time games and first-person shooter games ($r = .50$ in Table 8.1) reflects that some participants prefer one play type more than the other and vice versa, which is also attributed to the different skills possessed in the preferred and unpreferred play type.

We inferred from BD-PC theory that users who are in a playful mode are not expected to show unnecessary aggressive acts during play even when playing computer games with a great deal of violent content. The hypothesis predicts that the number of aggressive acts performed while playing a computer game is less influenced by the amount of violence of a game and more by the strength of the intention to win and the elaborateness of the strategies used by the player. This pattern was confirmed in the data of the reported experiment.

CONCLUSION

The BD-PC theory of play in animals and humans can give play research a new direction. It contributes to one of the most important psychological riddles still to be solved: How does the process of anthropogenesis work, and how do we become the cognitive animals we are? The theory postulates a play module that enacts variability at different behavioral and cognitive levels. Variability is the invariant key component of the play module that remains stable despite new ontogenetically emerging play forms. Therefore, the BD-PC theory allows for the integration of play research with the new emerging field of developmental evolutionary psychology (Bjorklund & Pellegrini, 2000). BD-PC theory also allows us to make many inferences that are contrary to conventional play research and enables predictions to be made for the ontogenetically late play form "games with rules," for example, for the playing of computer games. In a playful mode, users of computer games employ a broader range of strategies. Consequently, despite playing a violent first-person shooter game, players are not expected to display many unnecessary aggressive acts when highly motivated to play and their cognitive system is shifted in a playful mode initiated by the play module, which is very old in evolutionary terms.

Perhaps not only the playing of computer games can be explained in the perspective of an evolutionary play psychology. Stephenson (1988) argued that the reception of mass media should be modeled as a subjective play of the recipients. Vorderer (2001) suggested that entertainment experiences share many aspects with experiences during play and that therefore entertainment could be reconstructed as a form of play. Steen and Owens (2001) proposed that media entertainment is a culturally more elaborate form of pretense that phylogenetically evolved as a cognitive adaptation for pretense play. We believe that the creative explosion of modern humans that took place within dispersed populations some 50,000 years ago could be modeled as an interplay of secondary internal representational systems and external representational systems (e.g., parietal art). If this picture is true, play really is a phylogenetic cornerstone for modern media entertainment.

NOTES

[1] The EEA "is not a specific time or place. It is the statistical composite of selection pressures that caused the design of an adaptation" (Cosmides & Tooby, 1997). The constant terrestrial illumination conditions that formed the design of the verebrate eye are part of the EEA of *Homo sapiens sapiens* but also the selection pressures that caused the mental adaptations in the hominid line in the last 2 million years.

The EEA that caused the fine-tuning of the modern human mind 100,000 years ago consisted of such tasks as: wandering around in the savannahs of Africa in small bands, being confronted with the adaptive problems to find food, to find mates, to detect and avoid predators, to cooperate with the other members of the group, and to detect cheaters, etc.

[2] This can be shown in various formulations such as, "[play is P.O. & G.N.] . . . some kind of reinforcement of realistic adaptive variability" (Sutton-Smith, 1997, p. 224). Sutton-Smith (1997) undermined the evolutionary appeal of his appraoch to play by assuming that play reinforces other prior and "real" adaptive functions of variability.

[3] In assuming that epistemic novelty causes the play mechanism, the BD-PC theory strictly contradicts all variants of an activation theory of play (e.g., Ellis, 1973; Hutt, 1979; Shultz, 1979). At least in its evolutionary origin, the play mechanism was not triggered to increase the activation level of a central nervous system to an optimal level. The play mechanism was only triggered when the activation level was rather high, for it is always a stressful situation for the organism to cope with epistemic novelty.

REFERENCES

Barkow, J. H., Cosmides, L., & Tooby, J. (Eds.). (1992). *The adapted mind. Evolutionary psychology and the generation of culture*. New York: Oxford University Press.

Bekoff, M. (1995). Play signals as punctuation: The nature of social play in canids. *Behaviour, 132*, 419–429.

Bekoff, M. (1998). Playing with play: What can we learn about cognition, negotiation, and evolution? In D. D. Cummins & C. Allen (Eds.), *The evolution of mind* (pp. 162–182). Oxford, England: Oxford University Press.

Bekoff, M., & Byers, J. A. (Eds.). (1998). *Animal play: Evolutionary, comparative, and ecological perspectives*. Cambridge, England: Cambridge University Press.

Bjorklund, D. F., & Pellegrini, A. D. (2000). Child development and evolutionary psychology. *Child Development, 71*, 1687–1708.

Burghardt, G. M. (1998a). Play. In G. Greenberg & M. M. Haraway (Eds.), *Comparative psychology. A handbook* (pp. 725–735). New York: Garland.

Burghardt, G. M. (1998b). The evolutionary origins of play revisited: Lessons from turtles. In M. Bekoff & J. A. Byers (Eds.), *Animal play. Evolutionary, comparative, and ecological perspectives* (pp. 1–26). Cambridge, England: Cambridge University Press.

Buss, D. M. (1999). *Evolutionary psychology. The new science of the mind*. Boston: Allyn & Bacon.

Cann, R. L., Stoneking, M., & Wilson, A. C. (1987). Mitochondrial DNA and human evolution. *Nature, 325*, 31–36.

Conkey, M. W. (1999). A history of the interpretation of European "palaeolythic art": Magic, mythogram, and metaphors for modernity. In A. Lock & C. R. Peters (Eds.), *Handbook of human symbolic evolution* (pp. 288–349). Oxford, England: Blackwell.

Cosmides, L., & Tooby, J. (1992). Cognitive adaptation for social exchange. In J. H. Barkow, L. Cosmides, & J. Tooby (Eds.), *The adapted mind. Evolutionary psychology and the generation of culture* (pp. 163–228). New York: Oxford University Press.

Cosmides, L., & Tooby, J. (1994). Origins of domain specifity: The evolution of functional organization. In L. A. Hirschfeld & S. A. Gelman (Eds.), *Mapping the mind: Domain specifity in cognition and culture* (pp. 85–116). New York: Cambridge University Press.

Cosmides, L., & Tooby, J. (1997). *Evolutionary psychology: A primer*. Retrieved July 21, 2001, from http://www.psych.ucsb.edu/research/cep/primer.html.

Darwin, C. (1871). *The descent of man, and selection in relation to sex. 2 vols*. London: John Murray.

Darwin, C. (1995). *On the origins of species. A facsimile of the first edition* (14th ed.). Cambridge, MA.: Harvard University Press (original work published 1859).

Donald, M. (1991). *Origins of the modern mind. Three stages in the evolution of culture and cognition*. Cambridge, MA: Harvard University Press.

Donald, M. (2002). *A mind so rare: The evolution of human consciousness*. New York: Norton.

Ellis, M. J. (1973). *Why people play*. Englewood Cliffs, NJ: Prentice-Hall.

Fagen, R. M. (1981). *Animal play behavior*. New York: Oxford University Press.

Fouts, R. (1997). *Next of kin. What chimpanzees have taught me about who we are*. New York: Morrow.

Gallup, G. G., Jr. (1982). Self-awareness and the emergence of mind in primates. *American Journal of Primatology, 2*, 237–248.

Gallup, G. G., Jr. (1983). Toward a comparative psychology of mind. In R. L. Mellgren (Ed.), *Animal cognition and behavior* (pp. 473–510). Amsterdam: North Holland Publishing.

Gärdenfors, P. (1995). Cued and detached representations in animal cognition. *Lund University Cognitive Studies (LUCS), 38*.

Groos, K. (1899). *The play of animals*. New York: Appleton.

Groos, K. (1901). *The play of man*. New York: Appleton.

Gulick, L. H. (1898). *A philosophy of play*. New York: Scribner's.

Haeckel, E. (1874). *Anthropogenie: Keimes- und Stammesgeschichte des Menschen*. [The evolution of man]. Leipzig, Germany: Engelmann.

Hall, G. S. (1920). *Youth*. New York: Appleton.

Hamilton, W. D. (1964a). The genetical evolution of social behaviour. I. *Journal of Theoretical Biology, 7*, 1–16.

Hamilton, W. D. (1964b). The genetical evolution of social behaviour. II. *Journal of Theoretical Biology, 7*, 17–52.

Heyes, C. M. (1998). Theory of mind in nonhuman primates. *Behavioral and Brain Sciences, 21*, 101–148.

Hutt, C. (1979). Exploration and play. In B. Sutton-Smith (Ed.), *Play and learning* (pp. 175–194). New York: Gardner.

Jensvold, M. L. A., & Fouts, R. S. (1993). Imaginary play in chimpanzees (*Pan troglodytes*). *Human Evolution, 8*, 217–227.

Kimura, M. (1983). *The neutral theory of molecular evolution*. Cambridge, MA: Cambridge University Press.

Klein, R. G., & Blake, E. (2002). *The dawn of human culture. A bold new theory on what sparked the "big bang" of human consciousness*. New York: Wiley.

Lazarus, M. (1883). *Über die Reize des Spiels*. [About the attractiveness of play] Berlin, Germany: Ferdinand Dümmlers Verlagsbuchhandlung.

Leslie, A. M. (1987). Pretense and representation: The origins of "theory of mind." *Psychological Review, 94*, 412–426.

Lock, A., & Peters, C. R. (1999). Editorial introduction to part III: Symbolic development and symbolic evolution. In A. Lock & C. R. Peters (Eds.), *Handbook of human symbolic evolution* (pp. 371–399). Oxford, England: Blackwell.

Malthus, T. R. (1826). *An essay on the principle of population; or, a view of its past and present effects on human happiness, with an inquiry into our prospects respecting the future removal or mitigation of the evils which it occasions. 2 vols*. (6th ed.). London: John Murray (original work published 1798).

McKinney, M. L. (1998). Cognitive evolution by extending brain development: On recapitulation, progress, and other heresies. In J. Langer & M. Killen (Eds.), *Piaget, evolution, and development* (pp. 9–31). Mahwah, NJ: Lawrence Erlbaum Associates.

Mitchell, P., & Riggs, K. J. (Eds.). (2000). *Children's reasoning and the mind*. Hove, England: Psychology Press.

Nieding, G., & Ohler P. (2002, March). *Effekte unterschiedlicher Rahmungen von False-Belief-Aufgaben* [Effects of different framing of false-belief-tasks]. Paper presented at the 44th conference of Experimentally Working Psychologists, Chemnitz, Germany.

Ohler, P. (2000). *Spiel, Evolution, Kognition. Von den Ursprüngen des Spiels bis zu den Computerspielen* [Play, evolution, cognition. From the origins of play to the computer games]. Habilitationsschrift an der Technischen Universität Berlin, Germany.

Ohler, P., & Nieding, G. (2000). Was lässt sich beim Computerspielen lernen? Kognitions- und spielpsychologische Überlegungen [What can be learned when playing computer games. Considerations of cognitive psychology and play psychology]. In R. Kammerl (Ed.), *Computerunterstütztes Lernen* (pp. 188–215). Munich, Germany: Oldenbourg Verlag.

Ohler, P., & Nieding, G. (2001). The behavior-diversification proto-cognition theory of play in animals and humans. In University of Erfurt (Ed.), *Play and toys today. Conference Proceedings of the 22nd World Play Conference (CD: 115KB)*. Erfurt, Germany: TIAW-Verlag.

Perner, J. (1991). *Understanding the representational mind*. Cambridge, MA: MIT Press.

Piaget, J. (1962). *Play, dreams, and imitation in childhood*. New York: Norton (original work published 1946).

Plotkin, H. (1994). *Darwin machines and the nature of knowledge*. Cambridge, MA: Harvard University Press.

Povinelli, D. J. (1996). Chimpanzee theory of mind: The long road to strong inference. In P. Carruthers & P. K. Smith (Eds.), *Theories of theories of mind* (pp. 243–329). Cambridge, England: Cambridge University Press.

Povinelli, D. J. (1998a, September). Chimpanzees, children and the evolution of explanation: Theory of mind and beyond. *Main lecture at the 41st congress of the German Society for Psychology*, Dresden, Germany.

Povinelli, D. J. (1998b). Can animals empathize? Maybe not [Electronic version]. *Scientific American, 9*, 67–75.

Povinelli, D. J. (2000). *Folk physics for apes: The chimpanzee's theory of how the world works*. New York: Oxford University Press.

Reaux, J. E., Theall, L. A., & Povinelli, D. J. (1999). A longitudinal investigation of chimpanzees' understanding of visual perception. *Child Development, 70*, 275–290.

Shultz, T. R. (1979). Play as arousal modulation. In B. Sutton-Smith (Ed.), *Play and learning* (pp. 7–22). New York: Gardner.

Siviy, S. M. (1998). Neurobiological substrates of play behavior: Glimpses into the structure and function of mammalian playfulness. In M. Bekoff & J. Byers (Eds.), *Animal play: evolutionary, comparative, and ecological perspectives* (pp. 221–242). Cambridge, England: Cambridge University Press.

Siviy, S. M., & Baliko, C. N. (2000). A further characterization of alpha-2 adrenoceptor involvement in the rough-and-tumble play of juvenile rats. *Developmental Psychobiology, 37*, 25–34.

Siviy, S. M., Baliko, C. N., & Bowers, K. S. (1996). Rough-and-tumble play behavior in Fisher-344 and Buffalo rats: Effects of social isolation. *Physiology and Behavior, 61*, 597–602.

Siviy, S. M., Fleischhauer, A. E., Kerrigan, L. A., & Kuhlman, S. J. (1996). D_2 dopamine receptor involvement in rough-and-tumble play behavior of juvenile rats. *Behavioral Neuroscience, 110*, 1168–1176.

Smith, P. K. (1982). Does play matter? Functional and evolutionary aspects of animal and human play. *The Behavioral and Brain Sciences, 5*, 139–184.

Spencer, H. (1864). *The principles of biology. Vol. 1*. London: Williams & Norgate.

Spencer, H. (1873). *Principles of psychology. Vol. 2*. (3rd ed.). New York: Appleton.

Steen, F. F., & Owens, S. (2001). Evolution's pedagogy: An adaptationist model of pretense and entertainment. *Journal of Cognition and Culture, 1*, 289–321.

Stephenson, W. (1988). *The play theory of mass communication*. New Brunswick, NJ: Transaction Publishers.

Sutton-Smith, B. (1978). *Die Dialektik des Spiels. Eine Theorie des Spielens, der Spiele und des Sports*. [The dialectics of play. A theory of playing, play and sports]. Schorndorf, Germany: Verlag Karl Hofmann.

Sutton-Smith, B. (1997). *The ambiguity of play*. Cambridge, MA: Harvard University Press.

Tomasello, M. (1999). *The cultural origins of human cognition*. Cambridge, MA: Harvard University Press.

Tomasello, M., & Call, J. (1994). *Primate cognition*. New York: Oxford University Press.

Tooby, J., & Cosmides, L. (1992). The psychological foundations of culture. In J. H. Barkow, L. Cosmides, & J. Tooby (Eds.), *The adapted mind. Evolutionary psychology and the generation of culture* (pp. 19–136). New York: Oxford University Press.

Trivers, R. L. (1971). The evolution of reciprocal altruism. *Quaterly Review of Biology, 46*, 35–57.

Trivers, R. L. (1972). Parental investment and sexual selection. In B. Campbell (Ed.), *Sexual selection and the descent of man: 1871–1971* (pp. 136–179). Chicago: Aldine.

Vorderer, P. (2001). It's all entertainment—sure. But what exactly is entertainment? Communication research, media psychology, and the explanation of entertainment experiences. *Poetics, 29*, 247–261.

Vygotski, L. S. (1973). Das Spiel und seine Rolle für die psychische Entwicklung des Kindes. [Play and its function in the psychological development of the child] *Ästhetik und Kommunikation, 11*, 16–37.

Vygotski, L. S. (1978). *Mind in society*. Cambridge, MA: Harvard University Press.

Whiten, A. (1997). The Machiavellian mindreader. In A. Whiten & R. W. Byrne (Eds.), *Machiavellian intelligence II: Extensions and evaluations* (pp. 144–173). Cambridge, England: Cambridge University Press.

Whiten, A. (1998). Evolutionary and developmental origins of the mindreading system. In J. Langer & M. Killen (Eds.), *Piaget, evolution, and development* (pp. 73–99). Mahwah, NJ: Lawrence Erlbaum Associates.

Wilson, E. O. (1975). *Sociobiology: The new synthesis*. Cambridge, MA: Harvard University Press.

The Influence of Personality Factors on Computer Game Choice

Tilo Hartmann and Christoph Klimmt

Hanover University of Music and Drama

Selection of computer games certainly depends on many factors like situational factors (e.g., time resources), technological factors (e.g. system infrastructure), and individual factors (e.g., motivation; see Bryant & Davies, chap. 13, this volume). This chapter focuses on the influence of enduring personality factors on computer game choice. The chapter begins by explicating the central terms "media choice" and "personality factors." Next is a review of related empirical studies that have been conducted so far, in which we consider the relationships between personality factors and choice of computer games. The chapter concludes with a discussion about how the empirical findings might be structured into a more coherent theoretical framework.

CLARIFYING THE TERMS: MEDIA CHOICE AND PERSONALITY FACTORS

Media Choice as a Part of Selective Exposure to Media

Selective exposure to media is a kind of intentional human action (cf. Zillmann & Bryant, 1985a). Thus, the basic structure that underlies all human action also applies to selective media exposure: Based on situational determinants, a specific medium or content (cf. Donsbach, 1989) is chosen, which leads to specific results and consequences. Selective media exposure combines a chain of processes—(a) the media choice, which includes formation of a concrete *preference*; (b) volition as a higher-order cognitive process; and (c) an observable action that initializes media exposure (e.g., launching a computer game by double clicking on a desktop item), which might be described as an "*actual media choice*" (cf. Webster & Wakshlag, 1983). In addition, selective exposure implies that the medium is the "primary perceptual activity" (Zillmann & Bryant, 1985a, p. 5) for a considerable period of time, which might be labeled as "persistent usage." Thus, the term "selective exposure" addresses a variety of distinguishable actions. Thus, in order to minimize the semantic overlap between "selective exposure" and terms that more specifically refer to reception processes like "usage," it is necessary to note that "selective exposure" primarily denotes the process of *actual media choice*. As the quality and length of exposure indicate its intensity, both reveal something about the relevance or meaningfulness of given selections. For example, an individual who frequently selects computer games, but always plays rather mindlessly for only very short periods of time likely exhibits an equally meaningful pattern of selective exposure as an individual who only selects computer games rarely, but with much longer durations of exposure.

Personality and Media Choice

Media choice develops as the result of the situation-specific interplay of external aspects and individual dispositions (Zillmann & Bryant, 1985a). At the individual level, selective exposure to media is influenced by situation-specific constructs, like affective and cognitive states that reflect salient needs or motivations; a set of situational preferences; and domain-specific attitudes, values, and norms (cf. Donsbach, 1989; Webster & Wakshlag, 1983; Zillmann, 2000). However, on a general level and across a variety of situations, individuals often exhibit consistencies in the type or content of media they prefer. In other words, individual patterns of selective exposure are likely to endure across different situations. Therefore, it is plausible to assume a stable cause behind such cross-situational patterns of exposure. If selective exposure remains consistent across *different* situations, such a stable cause likely lies in an enduring *individual* structure, the *personality* (cf. Buss, 1987; Magnusson & Endler, 1977).

What is personality? A wide range of definitions exist, according to the heterogeneous structure of the psychology of personality and individual differences. It goes far beyond the scope of this chapter to structure the diverse field of personality research (cf. for an overview Pervin & Oliver, 2001; Ryckman, 1997). However, to evaluate research on personality and computer games better, several clarifications are necessary.

Trait theorists assume that a person's personality is a set of global enduring dispositions, partly biological or inherited, and formed partly through socialization (cf. for an review on trait approaches Matthes & Deary, 1998). According to trait theorists, individual dispositions have a stable influence on behavior across different situations, in different domains (cross-situational consistency of behavior) as well as at different times (temporal consistency of behavior). From this perspective, personality can simply be considered "that which tells what [a person] will do when placed in a given situation" (Cattell, 1965, pp. 117–118). In turn, personality factors are regarded as biologically rooted, general and enduring constellations of *action tendencies* (cf. Amelang & Bartussek, 1997). Computer game choice is primarily a function of enduring individual dispositions. For example, an individual with a strong need for competition would be expected to show competitive behavior in a range of situations that might also include computer games.

Trait theorists have been criticized by strict behaviorists who view similarity of *situation*, rather than enduring personality factors, as the cause of comparable behavior across individuals. However, with the advent of neobehaviorist cognitive theories of learning, particularly social–cognitive theories (cf. Bandura, 1986; Krampen, 2000; Rotter, Chance, & Phares, 1972), the distinct divisions between trait theories and situation-oriented theories of individual behavior began to blur. Modern learning theory acknowledges the role of individual cognitive structures, arguing that these can be stabilized across different situations as an effect of learning and reinforcement. Such "memory traces," in turn, can become salient in similar future situations and thus moderate behavior similarly to enduring individual factors. Not surprisingly, researchers have attempted to integrate the strengths of both trait and neobehaviorist perspectives into one coherent framework (Mischel & Shoda, 1999). Most researchers now agree that personality factors exist as enduring individual dispositions that influence behavior across a range of *functionally equivalent situations* (cf. Schmitt, 2004). Using the previously mentioned example, competitive individuals could indeed display rivaling behavior in several situations, as long as these have a functionally equivalent meaning for them. Thus, they may play sports to compete, but ignore computer games because they do not associate them with competition cognitively.

The idea of personality as a *hierarchy* of enduring individual dispositions has become quite popular. The hierarchy clamps on the globality of dispositions, that is, the range of their influence on behavior across different situations, with the broadest dispositions on the top and the most specific dispositions on the bottom. Both trait theorists (Bartussek, 1996; Matthes & Deary, 1998) and social–cognitive approaches (e.g., "generalized vs. domain-specific expectations," Rotter & Hochreich, 1975) have presented comparable assumptions. However, although the hierarchical dispositions paradigms appear to be similar at a first glance, the many individual constructs addressed within each do not add up to a unified personality framework (cf. Schmitt, 2004).

Empirical research about personality influences on computer game choice can be conducted following different perspectives, each of which is accompanied by different *methodological implications*. For example, empirical data might suggest stronger relationships between a *specific* enduring personality construct (e.g., a computer game-specific motivation to compete) and a *specific* behavior (e.g., choice of computer games). Likewise, the probability of finding strong relationships between *global* dispositions and *specific* behaviors like computer game choice is much lower. Also, the method applied to assess the relationship between personality and computer game exposure needs to be discussed. For example, experimental settings frequently target the behavior expressed by an individual in a single (and often artificial) situation. As situational factors can impede the effect of enduring personality factors, valid results are more likely if an individual's behavior is observed and aggregated across similar situations, for example, through questionnaires about stable *habits* of media choice. Also, the specificity of functional equivalent situations implies that a single personality factor is probably insufficient to account for a given behavior. For example, an enduring and relatively high tendency toward aggressive action does not necessarily mean that an individual will act aggressively in computer games as well. Or, concerning computer game choice, it is not necessarily the case that he or she prefers to select violent games; other personality factors may impede such an exposure. Therefore, empirical studies building on multivariate theoretical models, which hypothesize the impact of intertwined personality factors, should be more useful for explaining a highly domain-specific behavior like computer game choice. These considerations provide a basis for a review of the empirical research conducted so far on the relationship of personality factors and selective exposure to computer games.

PERSONALITY FACTORS AND COMPUTER GAME CHOICE: CURRENT RESEARCH

Most existing literature on personal factors in computer game use is dedicated to global personal factors, namely gender (e.g., Cassell & Jenkins, (1998) and age. These variables are not of interest in this chapter, however, as we focus on psychological–personological variables (which are frequently associated with gender and/or age, but are conceptually distinct and more informative than those global variables). Other chapters (e.g., Raney, Smith & Baker, chap. 12, this volume; Bryant & Davies, chap. 13, this volume) provide comprehensive information on gender and age issues.

The relevant literature offers a variety of empirical studies addressing the link between personality factors and computer game choice. We organize our review as follows: We begin by examining results on global action tendencies or temperaments and continue with the role of general aggressive tendencies, competitive tendencies, challenge and achievement motivations, and fantasy and escape preferences. Subsequently, studies on the role of frustration tolerance in computer game use are discussed, and finally, the importance of specific skills and capabilities is mentioned. An evaluative summary of all findings concludes the review.

Global Action Tendencies/Temperaments

In general, significant relationships between global dispositions and computer game choice as a specific behavior can hardly be expected. Not surprisingly, Nelson and Carlson (1985, experiment b) found no significant relationship between the extent of an individual's extroversion, neuroticism, and psychoticism and his or her preference for antisocial or prosocial driving games. Similarly, Kestenbaum and Weinstein (1984) did not observe significant differences between high video and low video users in neuroticism and extroversion. Also, Gibb, Bailey, Lambirth, and Wilson (1983) reported only weak (if any) ties between global personality factors and computer game choice. No connection between the global factors "conscientiousness" and "introversion" and video game play occurred in the study of Barnett et al. (1997). However, introversion was positively associated with a desire to play video games to fantasize and to escape from daily concerns. Overall, these results justify the assumption that global personality factors are more related to the choice of *specific* game types than to a preference for computer games in general. However, it remains to future studies to shed more light on this topic.

General Aggressive Tendencies

Many computer games demand aggressive, fast-paced acting. Therefore, numerous empirical studies have assessed the relationship between computer game exposure and aggressiveness (Buckley & Anderson, chap. 23, this volume; Lee & Peng, chap. 22, this volume; Weber, Ritterfeld, & Kostygina, chap. 24, this volume); however, only a few of them make explicit statements about a *general preference to act aggressively* (trait aggression) and its influence on game choice (cf. Slater, Henry, Swaim, & Anderson, 2003). Most of the studies conducted are concerned with the effects of computer game playing on aggressiveness (cf. Anderson, 2004). Some of them report correlations, which can also be interpreted in the reverse direction. That is, aggression may function as a determinant of computer game exposure. Both literature reviews and meta-analyses identify a positive relationship between aggression and amount of computer game play (Anderson, 2004; Anderson & Bushman, 2001; Sherry, 2001a; Dill & Dill, 1998). These findings, however, are of limited use for explaining the relationship between aggressiveness as a stable disposition and computer game choice. The studies have applied inconsistent measures of aggressiveness, and information distinguishing measures of trait aggression and aggressive states is often unreported in meta-analyses and reviews. Furthermore, most of the existing meta-analyses in particular mix correlational and experimental data. Thus, it remains unclear whether the reported effects can really be interpreted in that aggressiveness fosters game play (see Anderson, 2004, for an exception).

A number of studies particularly addressing aggression or closely related constructs (e.g., trait hostility) and game exposure confirm positive links between aggressiveness and computer gaming. For example, Colwell and Payne (2000) reported significant relationships between an individual's aggressiveness and game exposure, both in relation to frequency as well as duration or years of play (for similar results, see also Fling et al., 1992; Lin & Lepper, 1987). In the only study we know of that explicitly addresses the impact of trait aggression on computer game exposure and vice versa, Slater et al. (2003) found both effects of aggressiveness on violent media use and lagged effects of violent media use on aggressiveness. Other studies suggest that individuals with a highly aggressive disposition in particular are more exposed to games with aggressive *content* (Colwell & Payne, 2000; Wiegman & van Schie, 1998; see also S. L. Smith, chap. 5, this volume; B. P. Smith, chap. 4, this volume). Regarding *gender*, analyses show that boys choose more aggressive games than girls (Colwell & Payne, 2000). This finding is closely related to the well-demonstrated finding that in several boys have stronger aggressive personality factors than girls (Maccoby & Jacklin, 1980). Other studies, however, did not reveal significant relationships between aggressiveness or related trait constructs and computer game choice (Nelson & Carlson, 1985; Gibb et al., 1983). Slater (2003) observed effects of trait aggressiveness on exposure to violent media in general, but not on exposure to violent computer games. Therefore, although the results might be interpreted to indicate a positive link from enduring aggressive dispositions to computer game choice, specifically to games with aggressive content, future research is necessary to clarify this relationship.

General Competitive Tendencies

Many computer games involve some sort of (violent or nonviolent) competition against a social entity (a computer opponent or a real human player) and/or against natural hindrances. It could be argued that an individual's affinity for competitive situations may predict general computer game choice. Vorderer, Hartmann, and Klimmt (2003) found that individuals with a competitive disposition barely showed a stronger computer game-specific motivation to challenge and surpass others than individuals with a less competitive disposition, but *if* the user had a computer game-specific motivation to compete and surpass others, he or she was more likely to use the computer game (see also Kestenbaum & Weinstein, 1985). Moreover, the user's general motivation to compete in computer games was closely linked to the use of competitive game genres such as "real-time strategy" and "ego shooters." In addition, gender specific analyses demonstrated that the motivation to compete was especially high among male players and that exposure to competitive computer games was significantly higher in males (Hartmann, 2003). In sum, an individual's general preference for social competition does not necessarily imply that he or she will conduct such behavior in computer games. Rather, strong links are more likely between game exposure and domain-specific individual tendencies (here: computer game-specific motivation to compete and surpass others).

General Risk, Challenge, and Achievement Tendencies

A number of researchers have proposed that choice of computer games may also be related to sensation seeking (Zuckerman, 1979, in press)—a general tendency to enter risky, challenging, or performance-oriented situations (Slater, 2003; Gibb, Bailey, Lambirth, & Wilson, 1983). Apparently, exposure to computer games is an indicator for one specific facet of sensation seeking, namely the need to seek for thrill without taking vital risks, but not for sensation seeking in general (cf. Gniech, Oetting, & Brohl, 1993; Andresen, 1990). Accordingly, Slater (2003) found no significant impact of an individual's sensation-seeking tendency (measured by two items that refer to situations with vital risk) and exposure to violent computer games.

Some weak empirical evidence also suggests that an individual's general *achievement* and *competence motivation* (cf. Heckhausen, Schmalt, & Schneider, 1985) fosters preference for computer game play. Gibb et al. (1983) found a significant relationship between females' achievement motivation and computer game experience. They concluded that "females high in achievement motivation may have found the games attractive because they could strive for

progressively higher levels of achievement and receive immediate feedback on their performance level" (p. 165). Similarly, Kestenbaum and Weinstein (1984) found that accomplishing
good results in computer games is more important for frequent players than for others who
play less frequently. Fritz, Hönemann, Misek-Schneider, and Ohnemüller (1997) reported that
children who prefer computer games over other play alternatives have a significantly higher
fear of failure. Computer games may function as "secure" platforms for players to exhibit
and assert their own skills, and they are especially attractive if demonstrating skills in other
domains fails or is, at least, insecure (see also Sakamoto, 1994). In typical frequent players, a
general fear of failure and achievement anxiety could thus be accompanied by a high domain-
specific achievement motivation and the desire to affirm one's own skills (Nelson & Carlson,
1985). Roe and Muijs (1998) supported the idea that motivation, rather than a general domain-
specific achievement, facilitates computer game preference. They reported that heavy players
score significantly lower than other players on all indicators of academic achievement, which
is as an important facet of general achievement motivation. In sum, an individual's tendency to
expose skills, show competence and strive for achievements in the domain of computer games,
accompanied by a general fear of failure, seems to endure as a motivational factor for computer
game exposure.

As most computer games offer experiences of success that elevate mood and self-esteem (cf.
Klimmt, 2003), it might be argued that individuals who specifically have a stable and salient
need to *maintain their self-esteem* are most likely to play (see also Bryant & Davies, chap.
13, this volume). A number of empirical studies are based on the rationale that individuals
with low self-esteem seek opportunities to maintain their self-evaluation, including computer
games. However, it may also be argued that individuals with a strong self-esteem choose
computer games in order to strive for additional ways of coping (Grodal, 2000). In fact,
the empirical results on the relationship between self-esteem and computer game choice are
mixed. Fritz et al. (1997) found that intense computer game players report a more positive self-
concept than individuals who play continuously, but only for short periods of time. In contrast,
Barnett et al. (1997) observed that the general extent of video game exposure was unrelated
to self-esteem, except that "individuals with low self-esteem tended to perceive video games
as companions to a greater extent" (p. 1332). Similarly, Colwell and Payne (2000) found no
meaningful association between self-esteem and computer game exposure, except for a weak
negative relationship between boys' self-esteem and frequency of play. Funk and Buchman
(1996) found a significant negative relationship between time devoted to computer games and
self-esteem for girls. In an earlier study, they identified a strong preference for violent video
games to be associated with lower self-concept scores in boys (Funk & Buchman, 1994). Roe
and Muijs (1998) found that frequent computer game players scored significantly lower than
infrequent players in academic self-concept and overall self-esteem. In sum, the studies revealed
some evidence for a link between an individual's self-esteem and computer game choice,
although there are some methodological comparability problems (e.g., the subpopulations, the
game types, and the criteria of computer game choice that were applied; for similar findings
see Colwell, Grady, & Rhaiti, 1995; Fling et al., 1992; Gibb et al., 1983; Dominick, 1984).

General Fantasy and Escape Tendencies

Some studies report on personality factors that are associated with an escapistic use of computer games. A survey by Sherry, Lucas, Rechtsteiner, Brooks, and Wilson (2001) revealed that
the motive "fantasy" was positively associated with games of the category "imagination" (e.g.,
role-playing games). Barnett et al. (1997) assessed the tendency to imagine oneself experiencing the feelings of fictitious others. Although they found no significant connection to the
general extent of video game play, they reported that "relatively high scores on the trait fantasy

empathy measure were associated with a preference for videogames that stimulate the imagination [. . .] and enable the player to fantasize helping other individuals" (p. 1328). Again, the finding suggests the need to differentiate between game types. Thus, shortcomings in the differentiation of the assessed exposure might be the reason why Kestenbaum and Weinstein (1984) found no differences between high and low video game users in tendency toward fantasy escapism. However, involvement with interesting and exotic activities appears to be an important factor of computer game enjoyment (Klimmt, 2003), so it is reasonable to assume that individuals with high fantasy tendencies (trait absorption; see Wild, Kuiken, & Schopflocher, 1995) will have a greater affinity toward computer games, despite mixed findings in the studies mentioned.

Frustration (In)Tolerance

Individuals differ with regard to strategies for avoiding unpleasant situations like moments of boredom or frustration (Salovey, Mayer, Goldman, Turvey, & Palfai, 1995; Wegner & Erber, 1993). As computer gaming is an effective way to regulate an individual's mood (cf. Mehrabian & Wixen, 1986; Grodal, 2000), it is plausible that some individuals with a particularly low tolerance of unpleasant situations and a high tendency to regulate moods tend to select computer games more frequently than others (cf. Swinkels & Guiliano, 1995; Zillmann, 2000). Computer games might be a good vehicle for hedonistic mood-contrasting activities, as absorption in a task can be an effective way to dissolve negative moods (Erber & Tesser, 1992; see also Bryant & Davies, chap. 13, this volume). Some investigators report evidence that frequent players might indeed revert to this procedure by getting involved in gaming. Fritz et al. (1997) observed that those who play video games frequently tend to minimize frustrative situations and to play down the impact of frustrative events more often than those who play infrequently. This might be due to a higher intolerance of frustrating incidents. Kestenbaum and Weinstein (1984) observed that "highvideo subjects seemed to have more difficulty with delay of gratification and frustration tolerance than lowvideo subjects" (p. 331). In sum, although evidence is scarce, computer game players might be sensitive-mood managers, susceptible to annoying or unpleasant situations and strongly inclined to alter unwanted moods. Thus, they frequently choose computer games as distracting activities.

Skills and Efficacy Beliefs

Motives or general action tendencies are not the only engine of human behavior, and this holds also true for computer game choice. Rather, as different theories of action suggest (cf. Rotter, Chance, & Phares, 1972; Krampen, 2000), an individual must also be able to enact a motive in appropriate behavior. In the context of selective exposure to computer games, for example, competitive individuals might be attracted by multiplayer computer games, but inhibit their media choice because they believe themselves unable to face the expected challenges sufficiently. Also, while they are actually playing, users may learn that they do not have the skills needed to perform well, and they may reduce or even abandon future exposure.

In general, video games require concentration, memory, hand–eye coordination, and fine motor skills, as well as quick reactions (cf. Goldstein et al., 1997) and spatial–perceptual skills (cf. McClurg & Chaille, 1987; Griffith, Voloschin, Gibb, & Bailey, 1983). Primarily successful performances generate reinforcing experiences and increase the likelihood of future computer game exposure, while losses and defeats increase frustrations that repel individuals from exposure to games. Thus, it seems plausible that computer game choice occurs more frequently among individuals that endue skills relevant for playing. However, we know of no study to have addressed the impact of specific individual skills on computer game choice.

Sakomoto (1994) investigated the relationship between sociocognitive *abilities* and exposure to computer games. Based on two surveys and a panel study, he found that boys with lower sociocognitive abilities in particular prefer to play computer games. Lower scores of empathy, cognitive complexity, and cognitive abstractness were associated with a higher frequency of exposure. The panel study demonstrated the causal effect of sociocognitive abilities on game choice. Interestingly, the effect only occurred in boys. As sociocognitive abilities are important aspects of an individual's overall capability to adjust socially, Sakomoto (1994) concluded that children who do not interact well with others might tend to prefer activities that do not require social interactions, such as playing computer games. However, additional factors that affect media choice, like social norms, could lead girls with low sociocognitive abilities to select alternative leisure activities.

Even more important than "objective" skills are the individual's perceived capabilities, that is, one's own belief about one's ability to master a given situation. The research on efficacy convictions (e.g., Bandura, 1977) and computer game play is discussed in detail in another chapter (Klimmt & Hartmann, chap. 10, this volume).

Evaluative Summary

The empirical research on the relationship between personality factors and computer game choice provides valuable insights. Findings offer some evidence that general action tendencies like aggressive or competitive dispositions as well as individual skills and efficacy beliefs can cause increased exposure to computer games. According to the literature, the "typical computer game player" has a comparatively high tendency to act aggressively and a strong computer game-specific motivation to challenge and surpass others. He or she displays a need to demonstrate skill and to perform well in computer games, which might be fueled by his or her general anxiety of failing in other domains of life. In addition, the typical player is quite intolerant of frustrating situations and might regard computer games as an attractive technique to dissolve bad moods. Furthermore, he or she has rather low sociocognitive abilities but strong skills like concentration, coordination, and perception that aid in game-playing performance. Also, the typical player believes in his or her competencies to perform well in computer games and thus is unafraid of the (encountered) challenges.

The conceptual integration of these findings is difficult, as most of the studies did not explicitly investigate users' computer game choice, but were concerned with outcomes of exposure. In some cases, correlational data can be reinterpreted as illuminating exposure-related processes. The main problem, however, is that almost none of the studies was embedded in a coherent theoretical framework. Given the complexity of entertainment through computer games (Klimmt, 2003; see also Vorderer, Bryant, Pieper, & Weber, chap. 1, this volume), the existing studies may have neglected the role of some important personality factors. The lack of theoretical foundations also makes it difficult to connect the findings to existing frameworks of communication research (cf. Sherry et al., 2001b). Consequently, the findings remain sporadic and eclectic. Herein computer game research mirrors well-known problems of studies on personality in other domains, for example, exposure to television (Weaver, 2000): The reviewed investigations differ intensely with regard to (a) the specificity or hierarchical level of the regarded individual constructs, (b) the range of situations computer game choice was observed in, and (c) the extent to which the underlying approach regarded multivariate effects of different personality factors. In addition, the studies differ with respect to (d) the measures applied to assess both individual constructs and computer game choice and (e) whether exposure was related to computer games in general or to specific types of games.

However, in spite of the heterogeneous approaches, two basic individual factors of computer game choice seem to emerge from the discussed studies. One factor could be labeled as "general action tendencies" and covers needs, motives, general preferences, positive attitudes, and other emotional dispositions. The other basic factor includes "capabilities, competencies, and skills." It seems plausible that both individual action tendencies and capabilities determine computer game exposure in a combined way. However, most of the existing studies focused on singular influences instead.

Finally, the review also shows the importance of distinguishing between game types: The landscape of computer games appears to be too diverse to hypothesize about the role of personality factors in gaming in general (see B. P. Smith, chap. 4, this volume). Consequently, the majority of studies related to computer games per se had difficulties identifying significant relationships, whereas the studies regarding a specific game type were more likely to report noteworthy findings.

In sum, more theoretically derived empirical studies could help to advance the current research, if they increase the level of detail of both involved personality constructs as well as preferred and selected computer games. As a conclusion, we therefore introduce some considerations toward a theoretical framework that is in line with the present empirical findings, addresses the interplay of personality factors and computer game choice in line with the present empirical findings, and may guide future studies on personality and computer game choice.

PROSPECT: TOWARD A SYSTEMATIC APPROACH ADDRESSING RELATIONSHIPS BETWEEN PERSONALITY FACTORS AND COMPUTER GAME CHOICE

A coherent theoretical framework of personality and computer game exposure could be based on traditional models of uses-and-gratifications. The outline of the framework to be proposed begins with a brief review of those models and suggests expansions and modifications that help to better integrate personality factors in the modeled process of computer game choice. For this second step, we refer to psychological models of action (Krampen, 2000). Finally, we discuss how the framework might advance future research on the role of personality in computer game (and in general: media) choice.

Traditional Models of Media Choice

Selective exposure can be regarded as a special kind of reasoned action. Therefore, it is plausible that it is directly influenced by an individual's *expectancies or beliefs* about anticipated actions, outcomes, and their affective evaluations (*values or valences*). These are the key variables in models of communication research that try to predict media choice. According to uses-and-gratifications research, selective exposure is a rational choice that depends primarily on *general beliefs* or *expectations* about media attributes and outcomes of media use (which, in turn, are regarded as parts of an individual's media image or naïve media theory; c.f. Lichtenstein & Rosenfeld, 1983), and the *affective evaluation* of these anticipated effects (Palmgreen, Wenner, & Rosengren, 1985; van Leuven, 1981; for computer games: Schlütz, 2002). Positively evaluated anticipated effects ("gratifications sought"; cf. Palmgreen & Rayburn, 1985) increase the probability of media selection.

Most of the uses-and-gratification-based explications suggest expectancies and values as enduring individual tendencies that have a stable impact on cross-situational patterns of exposure (e.g., Palmgreen & Rayburn, 1985). Also, generalized expectancies about media experiences are apparently viewed as enduring individual constructs (although it is argued that they can

FIG. 9.1. Enduring individual constructs involved in media choice as suggested by uses-and-gratifications-based expectancy-value approaches.

be altered by media use; cf. Palmgreen & Rayburn, 1985). Other uses-and-gratifications the-orists argue that personality factors are not the same as, but are strongly associated with the evaluation of media outcomes, as needs generate expectations of the mass media or other sources (cf. Sherry et al., chap. 15, this volume). Many researchers agree that expected media gratifications are at least partially affected by enduring individual needs and other differen-tial combinations of individual characteristics (cf. Rosengren, 1974; McGuire, 1974; McLeod & Becker, 1981; Sherry, 2001b; Weaver, Brosius, & Mundorf, 1993; Weaver, 2000). Conse-quently, some researchers try to elaborate the role of individual differences in media choice by integrating established models of the psychology of personality (Sherry, 2001b; Weaver, 2000).

In sum, following the expectancy-value approaches of uses-and-gratifications research, computer game exposure can be regarded as a rational choice that builds on an individual's general beliefs about media experiences as well as on needs and other promotive enduring action tendencies (e.g., motives, temperaments) that may lead to a positive evaluation of these anticipated experiences (cf. Fig. 9.1).

Expansions and Modifications

Although we basically agree with the notion that computer game choice is influenced by individual anticipations of the gaming experience and their evaluation, we believe that the current uses-and-gratifaction-based models lack some important aspects, specifically when they are adapted to selective exposure to computer games.

First, as recent psychological models of action suggest (Krampen, 2000), affective evalua-tions can refer to quite *different elements of an action*, like the action itself ("intrinsic valence"; cf. Heckhausen, 1989, p. 186), its outcomes, or subsequent consequences ("extrinsic valence"; cf. Heckhausen, 1989, p. 186). However, most uses-and-gratifications-based conceptualiza-tions do not clearly distinguish between positive valenciations of anticipated actions, outcomes, or consequences (which are all covered by the umbrella term "gratifications sought"). Accord-ingly, these conceptualizations do not cover different influences of general action tendencies on distinct aspects of action. However, it could be assumed, for example, that individuals differ in their general preference for either extrinsic or intrinsic gratifications. As computer games offer primarily intrinsic forms of entertainment (Klimmt, 2003), goal- or outcome-oriented in-dividuals might not be attracted by this leisure activity, although their remaining motivational structures would perfectly match the experiences inherent to game play.

Second, we think that the uses-and-gratifications-based models disregard the role of individ-ual capabilities. For example, Palmgreen and Rayburn (1985) did not address individual skills or capabilities throughout their explications. Probably, this aspect remained underspecified as the models traditionally refer to conventional easy-to-use media like television. However, in contrast to noninteractive media, computer games urge the users to become active in a sophisticated way (Klimmt, 2003; Vorderer, 2000; Lee, Park, & Jin, chap. 18, this volume). Because of the performance and achievement aspects of computer games, it is necessary to consider skills and competencies as well as the individual's needs when trying to explain computer game choice (cf. Klimmt, 2003). More specifically, we think it is important to take individual action-related *self-beliefs* into account to advance the concept of subjective expectancies beyond mere anticipations of possible media experiences or "obtainable gratifications" (cf. Palmgreen & Rayburn, 1985; see Klimmt & Hartmann, chap. 10, this volume).

Psychological Models of Action and Personality

Due to the shortcomings of existing expectancy-value models of media choice, we refer to elaborated psychological conceptualizations of action in order to elaborate the modeling of selective exposure to computer games. Detailed theoretical conceptualizations of action distinguish four integral elements: (a) the situation in which an action occurs, (b) the action in its narrow sense, (c) outcomes of the action, and (d) consequences that are determined or at least influenced by the outcome (cf. Krampen, 2000). Individuals cognitively represent all four elements before and while they conduct an action. In terms of selective media exposure, the cognitive realization would trigger "neutral" beliefs about requisitions and experiences the individual thinks the use of a medium generally involves. In the context of the choice of computer games, "situation" would refer to the anticipated circumstances of exposure (e.g., sitting in front of a computer monitor), the "action" element in its narrow sense would refer to imaginations of (inter)active engagement in the game (e.g., directing virtual soldiers), the "outcome" would be the expected result of the exposure (victory or a defeat), whereas the "consequences" would refer to events assumed to follow on the anticipated outcome (e.g., mood enhancement or feelings of power). Such anticipations closely resemble what uses-and-gratifications researchers understand as expectancies of media attributes. We think that media beliefs, as an enduring individual factor, are central in determining media exposure. Individuals who lack or have an unclear perception about the experiences involved in computer game play do not have a "cognitive" foundation for a subsequent affective evaluation and preference formation, which are crucial for selective exposure.

As action-oriented models further suggest (cf. Krampen, 2000), individuals also have generalized expectations about their own impact on the different elements of action (which resemble the status of personality factors; we address them as *self-beliefs* in contrast to media beliefs). More specifically, it is argued that:

- Individuals hold general beliefs about how many different ways they will be capable of *acting* when placed in a given situation (cf. "self-efficacy," Bandura, 1977; see also Krampen, 2000; Klimmt & Hartmann, chap. 10, this volume).
- Individuals hold general beliefs about their personal impact to achieve a given *outcome* if they conduct an action (cf. "locus of control," Rotter, Chance, & Phares, 1972; "generalized action–outcome expectancies," Krampen, 2000).
- Individuals hold general beliefs about *consequences* that follow on a given outcome of an action (cf. "level of conceptualization," Krampen, 2000). In contrast to the two self-beliefs addressed above, this construct is apparently quite similar to general "neutral" expectations of media exposure, which do not regard the effect of an individual's own capabilities. Therefore, we would argue that expectations about consequences are rather a part of the subjective media beliefs than a self-belief.

Generalized self-beliefs affect the choice of computer games by determining the estimated likelihood that one will experience positively evaluated actions or to receive anticipated positive outcomes or consequences of exposure. If the individual does not expect him- or herself to be capable to achieve the expected rewards, he or she may impede selection. Individuals with a low self-efficacy in the domain of computer game input devices, for example, might think that they are unable to make use of the action possibilities offered. Even if they evaluate the variety of possible actions positively, exposure is probably inhibited. Individuals with a low internal locus of control might think that they are not able to control desired outcomes (e.g., success) by their performance and thus will not expect to receive positively evaluated outcome-related experiences (e.g., feelings of power).

In sum, we think that selective exposure to computer games is strongly affected by different enduring personality factors (see Fig. 9.2). First, individuals make assumptions about the experiences computer games are able to deliver. These include media beliefs about the situation computer game exposure is typically placed in as well as general expectations about the actions, (experiential) outcomes, and consequences evoked by computer games. Second, individual general action tendencies (needs, values, etc.) impact the affective evaluation of the aspects cognitively anticipated before computer game exposure. For example, individuals with a strong affinity to engage in absorbing experiences (Wild et al., 1995) may particularly favor anticipated actions inherent to computer game play; individuals with a strong achievement motivation (Heckhausen, Schmalt, & Schneider, 1985) should evaluate anticipated successful performances of computer game exposure very positively; and individuals with a strong disposition to perform better than others could appreciate social–competitive outcomes, which are typical for multiplayer computer games. Third, exposure should be affected by an individual's self-beliefs, which moderate the expected likelihood to achieve positively evaluated outcomes and consequences.

FRAMEWORK SUMMARY AND PERSPECTIVES

The framework addresses the role of personality factors involved in the anticipation and evaluation preceding selective exposure to computer games as a rational choice. In general, it reflects the assumptions of expectancy-value approaches, but it further differentiates both constructs by taking the range of different elements of action into account. In addition, it includes self-beliefs as an important class of general expectations, which has been neglected by past conceptualizations of selective media exposure. In sum, the framework is consistent to expectancy-value models of uses-and-gratifications research, but advances these approaches by considering explications of action-oriented models of psychology of personality (Krampen, 2000) and by identifying specific conceptual points where different personality factors connect to the process of computer game selection.

We think the framework allows for more precise hypotheses and explanations concerning the relationship between personality factors and computer game choice. For example, the low average age of computer gamers that is often reported (cf. McClure & Mears, 1984 or *http://www.theesa.com/* for recent data) might be due to (a) poorly conceived media beliefs about computer games, (b) a lack of salient promotive action tendencies, and/or (c) weak general self-beliefs about the own capability to achieve appreciated outcomes and consequences in computer games among the elderly. Clearly, the framework suggests that the involved personality factors have a combined effect on computer game exposure. Thus, for example, individuals with a "perfect" structure of promotive needs and values might nevertheless refuse to play computer games if they do not believe themselves capable of performing well. In contrast to conventional expectancy value approaches of uses-and-gratifications research, the proposed framework is capable to cover this role of (general) self-beliefs.

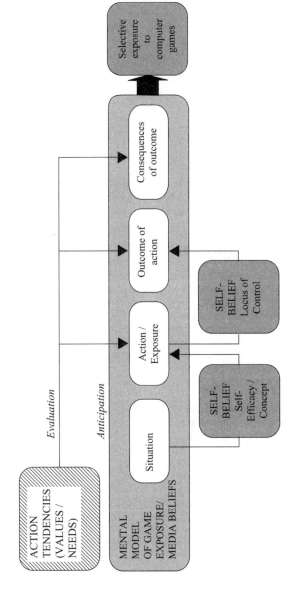

FIG. 9.2. Influence of personality factors (general action tendencies, media beliefs, and self-beliefs) on selective computer game choice/exposure.

This chapter has reviewed the existing literature on personality and computer game choice. Given the broad diversity of concepts dealt with under the notion of personality and the limited number of available empirical investigations, the general insight of this review is that there is a strong demand for more theoretical coherence and conceptual integration. Although there is evidence for a prominent role of personality factors in computer game choice, the mixed and eclectic findings need a solid theoretical foundation to advance this domain of computer game research in the future. Such efforts would of course also inform the discussion about personality factors and media choice in general and contribute to the progress of communication research in a much broader sense. The outlined theoretical framework may function as a starting point for this venture and might help to solve the conceptual problems that the literature review has disclosed.

REFERENCES

Amelang, M., & Bartussek, D. (1997). *Differentielle Psychologie und Persönlichkeitsforschung [The psychology of individual differences]*. Stuttgart: Kohlhammer Verlag.

Anderson, B. (1990). Reizsuche- und Erlebnissuche II: Sekundärfaktoranalysen, Invarianzprüfung und Ableitung eines MISAP-Strukturmodells. [Seeking sensations and excitement: A factor analytic approach]. *Zeitschrift für Differentielle und Diagnostische Psychologie, 11*(2), 65–92.

Anderson, C. A. (2004). An update on the effects of playing violent video games. *Journal of Adolescence, 27,* 113–122.

Anderson, C. A., & Bushman, B. J. (2001). Effects of violent video games on aggressive behavior, aggressive cognition, aggressive affect, physiological arousal, and prosocial behavior: A meta-analytic review of the scientific literature. *Psychological Science, 12,* 353–359.

Bandura, A. (1977). Self-efficacy: Toward a unifying theory of behavioral change. *Psychological Review, 84,* 191–215.

Bandura, A. (1986). *Social foundations of thought and action: A social cognitive theory*. Englewood Cliffs, NJ: Prentice-Hall.

Barnett, M. A., Vitaglione, C. D., Harper, K. K., Ouackenbush, S. W., Steadman, L. A., & Valdez, B. S. (1997). Late adolescents' experiences with and attitudes toward videogames. *Journal of Applied Social psychology, 27,* 1316–1334.

Bartussek, D. (1996). Faktorenanalytische Gesamtsysteme der Persönlichkeit [Factor-analytic full systems of personality]. In M. Amelang (Ed.), *Enzyklopädie der Psychologie, Differentielle Psychologie und Persönlichkeitsforschung [Encyclopedia of psychology: Individual differences and personality research]* (pp. 51–105). Göttingen: Hogrefe.

Buss, D. (1987). Selection, evocation and manipulation. *Journal of Personality and Social Psychology, 53,* 1214–1221.

Caspi, A. (2000). The child is father of the man: Personality continuities from childhood to adulthood. *Journal of Personality and Social Psychology, 78,* 158–172.

Cassell, J., & Henkins, H. (Eds.). (1998). From Barbie to mortal combat: Gender and computer games. Cambridge, MA: MIT Press.

Cattell, R. B. (1965). The Scientific analysis of personality. Harmond worth: Penguin.

Colwell, J., Grady, C., & Rhaiti, S. (1995). Computer games, self-esteem and gratification of needs in adolescents. *Journal of Community and Applied Social Psychology, 5*(3), 195–206.

Colwell, J., & Payne, J. (2000). Negative correlates of computer game play in adolescents. *British Journal of Psychology, 91,* 295–310.

Dill, K. E., & Dill, J. C. (1998). Video game violence: A review of the empirical literature. *Aggression & Violent Behavior, 3,* 407–428.

Dominick, J. R. (1984). Videogames, television violence, and aggression in teenagers. *Journal of Communication, 34*(2), 136–147.

Donsbach, W. (1989). Selektive Zuwendung zu Medieninhalten. Einflußfaktoren auf die Auswahlentscheidung der Rezipienten [Selective exposure to media content: Determinants of recipients' selection decision]. In M. Kaase & W. Schulz (Eds.), *Massenkommunikation. Theorien, Methoden, Befunde [Mass communication: Theories, methods, findings]* (pp. 392–405). Opladen: Westdeutscher Verlag.

Erber, R., & Tesser, A. (1992). Task effort and mood regulation: The absorption hypothesis. *Journal of Experimental Social Psychology, 28,* 339–359.

Fling, S., Smith, L., Rodgriguez, T., Thornton, D., Atkins, E., & Nixon, K. (1992). Video games, aggression, and self-esteem: A survey. *Social Behavior and Personality, 20,* 39–46.

Fritz, J., Hönemann, M., Misek-Schneider, K., & Ohnemüller, B. (1997). Vielspieler am Computer [Heavy computer game players]. In J. Fritz & W. Fehr (Eds.), *Handbuch Medien: Computerspiele [Handbook media: Computer games]* (pp. 197–205). Bonn: Bundeszentrale für politische Bildung.

Funk, J. B., & Buchman, D. D. (1994, October). *Video games and children: Are there "high risk" players?* Paper presented at the International Conference on Violence in the Media, New York.

Funk, J. B., & Buchman, D. D. (1996). Playing violent video and computer games and adolescent self-concept. *Journal of Communication, 46*(2), 19–32.

Gibb, G., Bailey, J., Lambirth, T., & Wilson, W. (1983). Personality differences between high and low electronic video game users. *Journal of Psychology, 114,* 159–165.

Gniech, G., Oetting, T., & Brohl, M. (1993). *Untersuchungen zur Messung von "Sensation Seeking" [Studies on the measurement of "sensation seeking"]*. Bremen: Department of Psychology, Bremen University.

Goldstein, J., Cajko, L., Oosterbroek, M., Michielsen, M., van Houten, O., & Salverda, F. (1997). Video games and the elderly. *Social Behavior and Personality, 25,* 345–352.

Griffith, J. L., Voloschin, P., Gibb, G. D., & Bailey, J. R. (1983). Differences in eye–hand motor coordination of videogame users and non-users. *Perceptual and Motor Skills, 57,* 155–158.

Grodal, T. (2000). Video games and the pleasures of control. In D. Zillmann & P. Vorderer (Eds.), *Media entertainment. The psychology of its appeal* (pp. 197–212). Mahwah, NJ: Lawrence Erlbaum Associates.

Hartmann, T. (2003, November). Gender differences in the use of computer-games as competitive leisure activities. Poster presentation at "Level Up—The first conference of the Digital Games Research Association," Utrecht, The Netherlands.

Heckhausen, H. (1989). *Motivation und Handeln [Motivation and action]*. Berlin: Springer.

Heckhausen, H., Schmalt, H. D., & Schneider, K. (1985). *Achievement motivation in perspective*. New York, London: Academic Press.

Kestenbaum, G., & Weinstein, L. (1985). Personality, psychopathology and development issues in male adolescent video game use. *Journal of American Academy of Child Psychiatry, 24,* 329–333.

Klimmt, C. (2003). Dimensions and determinants of the enjoyment of playing digital games: A three-level model. In M. Copier & J. Raessens (Eds.), *Level Up: Digital Games Research Conference* (pp. 246–257). Utrecht: Faculty of Arts, Utrecht University.

Krampen, G. (2000). *Handlungstheoretische Persönlichkeitspsychologie [Action-theoretical psychology of personality]*. Göttingen: Hogrefe.

Leuven, J. van (1981). Expectancy theory in media and message selection. *Communication Research, 8,* 425–434.

Lichtenstein, A., & Rosenfeld, L. (1983). Uses and misuses of gratifications research: An explication of media functions. *Communication Research, 11,* 393–413.

Lin, S., & Lepper, M. R. (1987). Correlates of children's usage of video games and computers. *Journal of Applied Social Psychology, 17,* 72–93.

Maccoby, E. E., & Jacklin, C. N. (1980). Sex differences in aggression: A rejoinder and reprise. *Child Development, 51,* 964–980.

Magnusson, D., & Endler, N. S. (1977). *Personality at the crossroads: Current issues in interactional psychology.* Hillsdale, NJ: Lawrence Erlbaum Associates.

Matthes, G., & Deary, I. J. (1998). *Personality traits.* New York: Cambridge University Press.

McClure, R. F., & Mears, F. G. (1984). Video game players: Personality characteristics and demographic variables. *Psychological Reports, 55,* 271–276.

McClurg, P. A., & Chaille, C. (1987). Computer games: Environments for developing spatial cognition? *Educational Computing Research, 3*(11), 95–111.

McGuire, W. J. (1974). Psychological motives and communication gratification. In J. G. Blumler & E. Katz (Eds.), *The uses of mass communications. Current perspectives on gratifications research* (pp. 167–196).Beverly Hills, CA: Sage.

McLeod, J. M., & Becker, L. B. (1981). The uses and gratifications approach. In D. D. Nimmo & K. R. Sanders (Eds.), *Handbook of political communication* (pp. 67–99). London: Sage.

Mehrabian, A., & Wixen, W. J. (1986). Preferences for individual video games as a function of their emotional effects on players. *Journal of Applied Social Psychology, 16,* 3–15.

Mischel, W., & Shoda, Y. (1999). Integrating dispositions and processing dynamics within a unified theory of personality: The cognitive affective personality system (CAPS). In L. A. Pervin & O. John (Eds.), *Handbook of personality: Theory and research* (pp. 197–218). New York: Guilford.

Nelson, T. M., & Carlson, D. R. (1985). Determining factors in choice of arcade games and their consequences upon young male players. *Journal of Applied Social Psychology, 15,* 124–139.

Palmgreen, P., & Rayburn, J. D. (1985). An expectancy-value approach to media gratifications. In K. E. Rosengren, L. A. Wenner, & P. Palmgreen (Eds.), *Media gratifications research: Current perspectives* (pp. 61–72). Beverly Hills, CA: Sage.

Palmgreen, P., Wenner, L. A., & Rosengren, K. E. (1985). Uses and gratifications research: The past ten years. In K. E. Rosengren, L. A. Wenner, & P. Palmgreen (Eds.), *Media gratifications research: Current perspectives* (pp. 11–37). Beverly Hills, CA: Sage.

Pervin, L. A., & Oliver, P. J. (Eds.). (2001). *Handbook of personality: Theory and research* (2nd ed.). New York: Guilford.

Roe, K., & Muijs, D. (1998). Children and computer games: A profile of the heavy user. *European Journal of Communication, 13*, 181–200.

Rosengren, K. E. (1974). Uses and gratifications: A paradigm outlined. In J. G. Blumler & E. Katz (Eds.), *The uses of mass communications: Current perspectives of gratifications research* (pp. 269–286). Beverly Hills, CA: Sage.

Rotter, J. B., Chance, J. E., & Phares, E. J. (Eds.). (1972). *Social learning theory of personality.* New York: Holt, Rinehart & Winston.

Rotter, J. B., & Hochreich, D. J. (1975). *Personality.* Glenview, IL: Scott, Foresman.

Ryckman, R. M. (1997). *Theories of personality.* Pacific Grove, CA: Brooks/Cole.

Sakamoto, A. (1994). Video game use and the development of sociocognitive abilities of children: Three surveys of elementary school students. *Journal of Applied Social Psychology, 24*, 21–42.

Salovey, P., Mayer, J. D., Goldman, S., Turvey, C., & Palfai, T. P. (1995). Emotional attention, clarity, and repair. In J. W. Pennebaker (Ed.), *Emotion, disclosure, and health* (pp. 125–154). Washington, DC: American Psychological Association.

Schlütz, D. (2002). *Bildschirmspiele und ihre Faszination: Zuwendungsmotive, Gratifikationen und Erleben interaktiver Medienangebote [Computer games and their fascination: Motives of exposure, gratifications and experience of interactive media].* München: R. Fischer.

Schmitt, M. (2004). Persönlichkeitspsychologische Grundlagen [Foundations from the psychology of personality]. In R. Mangold, P. Vorderer, & G. Bente (Eds.), *Lehrbuch der Medienpsychologie [The handbook of media psychology]* (pp. 151–173). Göttingen: Hogrefe.

Sherry, J. L. (2001a). The effects of violent video games on aggression. A meta-analysis. *Human Communication Research, 27*, 409–431.

Sherry, J. L. (2001b). Toward an etiology of media use motivations: The role of temperament in media use. *Communication Monographs, 68*, 274–288.

Sherry, J. L., Lucas, K., Rechtsteiner, S., Brooks, C., & Wilson, B. (2001, May). *Video game use and gratifications as predictors of use and game preference.* Paper presented at the ICA convention, Video Game Research Agenda Theme Session Panel.

Slater, M. (2003). Alienation, aggression, and sensation seeking as predictors of adolescent use of violent film, computer, and website content. *Journal of Communication, 53*(1), 105–121.

Slater, M., Henry, K. L., Swaim, R. C., & Anderson, L. L. (2003). Violent media content and aggressiveness in adolescents: A downward spiral model. *Communication Research, 30*, 713–736.

Swinkels, A., & Guiliano, T. A. (1995). The measurement and conceptualization of mood awareness: Attention directed towards one's mood states. *Personality and Social Psychology Bulletin, 21*, 934–949.

Vorderer, P. (2000). Interactive entertainment and beyond. In D. Zillmann & P. Vorderer (Eds.), *Media entertainment: The psychology of its appeal* (pp. 21–36). Mahwah, NJ: Lawrence Erlbaum Associates.

Vorderer, P., Hartmann, T., & Klimmt, C. (2003). Explaining the enjoyment of playing video games: The role of competition. In D. Marinelli (Ed.), *Proceedings of the 2nd International Conference on Entertainment Computing (ICEC 2003), Pittsburgh* (pp. 1–8). New York: ACM.

Weaver, J. B. (2000). Personality and entertainment preferences. In D. Zillmann & P. Vorderer (Eds.), *Media entertainment: The psychology of its appeal* (pp. 235–248). Mahwah, NJ: Lawrence Erlbaum Associates.

Weaver, J. B., Brosius, H. B., & Mundorf, N. (1993). Personality and movie preferences: A comparison of American and German audiences. *Personality and Individual Differences, 14*, 307–316.

Webster, J., & Wakshlag, J. J. (1983). A theory of television program choice. *Communication Research, 10*, 430–446.

Wegner, D. M., & Erber, R. (1993). Social foundations of mental control. In D. M. Wener & J. W. Pennebaker (Eds.), *Handbook of mental control* (pp. 36–56). Englewood Cliffs, NJ: Prentice-Hall.

Wiegman, O., & van Schie, E. G. M. (1998). Video game playing and its relations with aggressive and prosocial behavior. *British Journal of Social Psychology, 37*, 367–378.

Wild, T. C., Kuiken, D., & Schopflocher, D. (1995). The role of absorption in experiential involvement. *Journal of Personality and Social Psychology, 69*, 569–579.

Zillmann, D. (2000). Mood management in the context of selective exposure theory. In M. E. Roloff (Eds.), *Communication Yearbook* (pp. 123–145). Thousand Oaks, CA: Sage.

Zillmann, D., & Bryant, J. (1985a). Selective-exposure phenomena. In D. Zillmann & J. Bryant (Eds.), *Selective exposure to communication* (pp. 1–10). Hillsdale, NJ: Lawrence Erlbaum Associates.

Zillmann, D., & Bryant, J. (1985b). Affect, mood, and emotion as determinants of selective exposure. In D. Zillmann & J. Bryant (Eds.), *Selective exposure to communication* (pp. 157–190).Hillsdale, NJ: Lawrence Erlbaum Associates.

Zuckerman, M. (1979). *Sensation-seeking: Beyond the optimal level of arousal*. Hillsdale, NJ: Erlbaum Associates.

Zuckerman, M. (in press). Sensation seeking. In J. Bryant & P. Vorderer (Eds.), *The psychology of entertainment*. Mahwah, NJ: Lawrence Erlbaum Associates.

10

Effectance, Self-Efficacy, and the Motivation to Play Video Games

Christoph Klimmt and Tilo Hartmann
Hannover University of Music and Drama

When communication researchers consider computer games (and video games) as a new form of media entertainment, they typically highlight those games' interactivity because it is the main feature that distinguishes them from most other kinds of entertainment (Grodal, 2000; Vorderer, 2000). Not only do game players actively process information provided by the medium (as do viewers, readers, and users of other noninteractive media), but they also contribute substantially to the quality and progress of the media product itself. Their decisions and actions determine how a game looks, how it develops, and how it ends. Consequently, most theoretical work on the enjoyment of playing computer games has focused on the issue of interactivity and player action during game play (Klimmt, 2003, 2005). In order to handle modern entertainment software successfully, users must stay alert for most of the playing time and be able to respond quickly and appropriately to incoming new information. Some, if not many, of these responses may be automatized (Bargh, 1997) for efficient execution and conservation of cognitive processing resources. But, in general, the use of computer games should be modeled as a complex and multifaceted kind of action. Therefore, research on the psychology of action (e.g., Gollwitzer & Bargh, 1996; Heckhausen, 1977) offers theories and empirical findings that may help explain why people play and what they do during playing. Adopting the perspective of the psychology of action means searching for motifs (and motivations) of playing.

The same holds true for the domain of selective exposure (Bryant & Davies, chap. 13, this volume; Hartmann & Klimmt, chap. 9, this volume; Raney, Smith, & Baker, chap. 12, this volume; Sherry et al., chap.15, this volume). Compared to other kinds of media entertainment, the technical requirements and the costs of playing computer games are significant. Computer games are expensive, not only because they require high-end hardware, but also because the costs for the software itself are remarkable. Even illegally copied games are expensive to acquire because of the costs of special hardware and an Internet connection. In other words, more effort is required to supply a household with the technical equipment to use a computer game than to provide the platform for using TV or reading an enjoyable book. Similarly, the actual consumption process of computer games is more demanding. Installing a game or simply running it takes more time and effort than activating a TV set, for example. Even more important is the effort required to learn to play a computer game. Watching a new TV program is easy compared to learning how to use a new game: Novice players must memorize the functions of different keys and input devices as well as understand the causal connections and regulatory mechanisms in the game world. Thus, before exposure to a computer game can be fun, players have to invest much more than they would have to for an enjoyable evening in front of a television set or with a book. For this reason, we should consider the process of choosing to play a computer game a well-considered and intentional action. Because of numerous obstacles to playing a game that must be overcome (i.e., costs, technical difficulties, learning requirements), players need a strong motivation to achieve an enjoyable playing session. It is reasonable to expect that people who are only weakly motivated to play a computer game will most often terminate the selection or preparation process because they are not prepared to devote the energy or resources required to begin game play or make it fun. As people may so thoroughly consider the costs and benefits of computer gaming, it is useful to apply the psychological concept of action to both the selection of games and the process of using them. Relevant psychological models incorporate such multidimensional anticipations and evaluations as important determinants of the intention formulation, action selection, and action maintenance (Heckhausen & Kuhl, 1985).

Adopting the perspective of intended actions allows us to investigate theories and concepts from this domain of psychology that may be useful for computer game research. This chapter discusses the importance of two well-established action-related constructs that promise to explain selective exposure to computer games and user activity during game play. The first is White's (1959) notion of effectance motivation, and the second is Bandura's (1997) concept of self-efficacy. Before these theories are connected to playing computer games, the links between the selection of computer games and the usage experience are elaborated, because this relationship is the foundation on which the function of both effectance and self-efficacy within a theory of computer game use can be explicated.

THE LINK BETWEEN ANTICIPATED PLAYING EXPERIENCE AND SELECTION INTENTIONS

The psychological perspective on play is dominated by the idea that play is a special kind of action (Oerter, 1999; Sutton-Smith, 1997). According to Heckhausen and Kuhl (1985), goals that people intend to achieve through a given action may be situated outside of the actual action (*consequence* orientation), on the level of the action's *result,* or within the process of acting (*action* orientation). Most often, human actions are devoted to multiple goals that are situated at different levels (process, result, and/or consequence), even if one goal (level) dominates the original behavioral intention.

From this perspective, one defining element of playful actions (and media entertainment alike; e.g., Klimmt, 2005; Vorderer, 2001; Vorderer, Klimmt, & Ritterfeld, 2004) is the absence of any consequence orientation. Play is performed in order to achieve goals on the level of the acting process (e.g., juggling) or on the level of results (e.g., winning a competitive game; Vorderer, Hartmann, & Klimmt, 2003). Consequences such as earning money or social status outside of the game situation are not intended by players. The absence of consequence-related action motivations has been labeled intrinsic motivation (Deci & Ryan, 1975). It distinguishes playful actions and entertainment consumption from other types of action, such as labor, which typically involve (and are dominated by) consequence-related action goals (Klimmt, 2005).

If playful actions and the consumption of media entertainment are directed toward result-related or action-related goals, a process-based view of computer game play is most useful. Enjoyment is generated by action processes and/or action results. Therefore, it is most promising to model the complex cognitive and emotional responses and interactive operations related to enjoyment in basic psychological categories of action (Klimmt, 2005): In each stage of one action episode, specific processes influence the player's experience. For example, most episodes in a computer game begin with the presentation of a task (e.g., an approaching enemy must be repelled). During the first step of the action process, players analyze the current situation. They identify the task and formulate a desired action goal (e.g., to defeat the enemy). Different cognitive and emotional reactions to both the situation analysis and the derived goal can be assumed, for example, fright reactions because of the enemy's strength or curiosity about the episode's outcome (Grodal, 2000). Similarly, a wide variety of experiential processes may arise during the subsequent stages of the action episode (action and result; e.g., Klimmt, 2003, 2005). Such processes of cognitive and emotional experience form the computer game-specific kind of enjoyment. We assume that computer game players experience multiple processes throughout game play, that the processes are involved in the formation of enjoyment, and that the anticipation of those processes is a key factor in psychological actions of media choice.

THE PSYCHOLOGY OF INTENDED ACTIONS

The psychology of action is based on a fundamental assumption that human beings are capable of reflecting on their own psychological status, for example, their mood (Mayer & Gaschke, 1988) or their behavioral intentions (Heckhausen, 1977). People can make well-informed decisions to alter their own condition and use media entertainment in order to improve their well-being. Zillmann's (1988) mood management theory proposes that such decisions are based on reinforcement and conditional learning; that is, memory traces from past (media) experiences guide future behavior by influencing the choice of stimuli in a given situation. This perspective is rejected by action psychologists, who argue that such decisions are (or at least can be) made because of a much broader knowledge of what happens to oneself during exposure to a given stimulus arrangement (e.g., Salovey, Hsee, & Mayer, 1993).

Researchers in cognitive psychology have advanced the concept of mental models (e.g., Johnson-Laird, 1983) — cognitive representations of situations or objects based on perception (bottom-up components) and prior experience or knowledge (top-down components). They can also refer to processes and action sequences, as demonstrated in the psychology of labor (Hacker, 1996). Therefore, a mental model functions as an internal representation of what one's own future (emotional, cognitive) condition will be if a particular action (for example, using media entertainment) is selected and carried out. Over time, such mental representations evolve through repeated exposure to the corresponding objects or processes. After each session of play, players' ideas of what they experience during a certain type of action (computer game play) are expanded, modified, and completed. Therefore, those who reflect on a long personal playing history should have valid mental representations of their own psychological condition during game play, even if their mental models do not include every detail of each experience. They can use this knowledge for future decisions and activity choice. In actions of selective exposure, they can evaluate their anticipated experience against their current personal preferences. If the expected experience matches the desired one, the individual selects and performs the action (Klimmt, 2005). This idea of a rationally calculated media choice displays some similarities with the uses and gratifications perspective (Sherry et al., chap. 15, this volume), but differs from this approach as it also refers to psychological categories of mental representation and mental control (Wegner & Pennebaker, 1993).

In spite of the complexity and the amount of personal cognitions involved, such decision processes do not need to be very demanding or time-consuming, in part because they become automatic over time. Using media entertainment, like playing games, is a frequently repeated activity (Vorderer, 2001). This allows individuals to learn and routinize even complex cognitive and behavioral processes over time (Bargh, 1997). Rapid and intuitive decisions for receiving a certain type of media entertainment are not considered effects of an activated memory trace, as proposed by Zillmann (1988), but as complex, intended actions that have been automatized through frequent performance. Consequently, the motivation to play computer games is the result of self-reflection; individuals compare expectations of what their own cognitive and emotional conditions will be during game play to what they would like their cognitive and emotional condition to be. As computer gamers know to a certain extent what will happen to them during game play, the strength of their motivation to begin a gaming session depends on both their current status and on personal evaluations of what they expect to occur during game play. The role of effectance and self-efficacy as factors in this decision process can be explicated within this framework. We consider how effectance influences the individual's decisions regarding events and experiences that he or she expects during game play, whereas self-efficacy is a component of the self-evaluations connected to the game play process that also contributes to the overall playing motivation.

EFFECTANCE: THE ENJOYMENT OF CAUSING CHANGE IN THE ENVIRONMENT, ACHIEVED THROUGH GAME PLAY

The Concept of Effectance Motivation

The concept of effectance was introduced by White (1959), who formulated a general motivational theory to overcome traditional explanations of human behavior that were based on drive concepts or anxiety reduction. White (1959) portrayed people's motivational system as being energized by an urge toward competence, that is, making progress in the knowledge and abilities that support the individual's struggle for survival. In evolutionary terms, the acquisition of competence is adaptive, and according to White (1959), the human motivational system is laid out to secure adaptive behavior, which means to gain new competences. However, people are not aware of this superordinated function of their activities. Rather, they perform actions that lead to competence gain because of more immediate and situation-based reasons.

White explained the difference between the evolutionary function of human actions and the subjective motivation to perform those actions by referring to the example of sexual behavior. Clearly, the key function of sexual behavior is reproduction, which holds an adaptive benefit for the individual. However, reproduction is usually not the reason why people engage in sexual activities; the enjoyment derived from performing such actions is their primary motivation. White (1959) argued that the general function of adaptive actions and the subjective motivation to perform those actions are linked because the human motivational system declares adaptive actions as rewarding. Having sex is fun because this way, the individual is sufficiently motivated to engage in adaptive behaviors that lead to reproduction. Similarly, the motivation to perform other kinds of action is explained by White's (1959) theory. Especially in childhood, exploratory behavior and manipulation of objects in the environment are crucial to acquire competence and progress in cognitive, emotional, and motor development (Ohler

& Nieding, chap. 8, this volume; Steen et al., chap. 21, this volume). Children often display a very strong motivation to engage in their environment, and frequently explore and rearrange their immediate surroundings. Adolescents and adults exhibit similar kinds of interest in their environment but prefer more complex objects and activities, for example, machine mechanics. In all mentioned cases, engagement in the named behavior is obviously rewarding for the agent and often leads to new insights and acquisition of competences.

According to White (1959), the subjective reward of such activities that people perceive during performance is *the satisfaction of having imposed an effect on the environment*. It is inherently rewarding to bring about an event, whether it is turning a light on and off again repeatedly or repairing an old car. White (1959) labeled these kinds of experiences *efficacy*. Achieving the gratifying experience of efficacy provides the immediate motivation to deal with one's environment. Because this experience is rewarding, people develop a stable motivation to enter this condition, and this motivational disposition ensures that the individual continues the active exchange with his or her environment to gain new competence. White (1959) called it *effectance motivation*. Harter (1978, p. 35) has elaborated the concept further by deconstructing White's general notion of effectance into "a) the organism's desire to produce an *effect* on the environment, b) the added goal of dealing *effectively* or competently with the environment, c) the resulting feelings of *efficacy*." She formulated a developmental model that can explain the emergence of effectance motivation over multiple episodes of actions and results. The model incorporates much more complexity than the original idea of effectance as the rewarding experience of imposing an effect on the environment. Harter (1978) elaborated, among other aspects, the notions of task, challenge, success, and failure, which have been employed in many other motivational theories as well. While all these concepts obviously apply to computer game play, which contains tasks, challenges, success and failure in virtually every game play session (Klimmt, 2003), we believe that White's (1959) original and rather simple proposition of effectance motivation is especially useful in computer game research.

The Experience of Interactivity in Computer Games

The explanatory value of White's (1959) theory is related to the key characteristic of computer game usage that has been labeled "interactivity" (Steuer, 1992; Vorderer, 2000, and introduction of this chapter). It is difficult to formulate a comprehensive definition of interactivity (Vorderer, 2000). Nevertheless, we can explicate its typical components, especially in the realm of computer game usage. Interactivity is the possibility of a continuous exchange between players and the game software. Each allowable action or input from the players is received and processed by the game software and contributes to change in the game system's condition. Thus, the program automatically generates a certain response to the players' input, which, in turn, the players can consider as they plan and perform their next input. Over time, a sequence of action–reaction loops between players and the computer game emerges (so called input–output-loops or I-/O-loops; cf. Klimmt, 2003). Imagine this sample I-/O-loop: In a first-person shooter game, a new I-/O-loop is activated by the program, which "sends out" an autonomous agent programmed to attack the player's character. As a result, the computer screen shows players an approaching enemy. This situation already has implications for the player's entertainment experience (see above), but for the progress of the situation, the game's interactivity is the most important factor. This is because the players are able (and, in fact, are forced) to respond to the current threat. They may decide to defend themselves (or their character, which does not make a difference in experience; cf. McDonald & Kim, 2001), which means that they aim at the enemy and begin to fire.

The computer game's interactivity permits these reactions. The software transforms the player's inputs into comprehensive actions, for example, body movements of the player's character or firing a weapon. The display of those actions represents the second stage of the I-/O-loop, that is, the game's output, which is based on the players' input and at the same time presented to the players, allowing them to incorporate the new information (e.g., "weapon has been fired") in their further decision making. A typical computer game session contains a very large number of such simple I-/O-loops. For example, the conflict and resolution with one single enemy mentioned above may comprise dozens or even hundreds of single I-/O-loops, as each input (e.g., each small movement of the computer mouse) that produces an output (e.g., shift of the player's crosshair by a few millimeters) is considered one loop. Because playing a combat game means fighting hordes of enemies, the overall number of I-/O-loops during the course of one session of game play is substantial. From a process perspective, the interactivity of computer games causes a continuous sequence of player actions and game software responses, each related to the other (Klimmt, 2003).

Implications of Game Interactivity for Efficacy Experiences

White's (1959) concept of effectance allows a theoretical connection of this important characteristic of computer game usage to the players' entertainment experiences. As computer games automatically respond to every single input, the perception of creating an effect within the game environment is especially salient. Players experience themselves as causal agents within the game environment — a state that is sustained by the regularity of the game's responsiveness to incoming inputs. Thus, computer game play addresses the player's effectance motivation; it produces many individual experiences of efficacy and causal agency. The temporal stability of the game's responsiveness allows the player to continue the rewarding experience for long periods of time.

With respect to the quality of the efficacy experience, two characteristics of computer games are important. The first is the *immediacy* of the game's response to the player's input. If there is no delay between a player's action and the game's response, the temporal congruency between action and result provides a very strong sense of effectance, as the outcomes of one's own actions are transparent and easy to comprehend. The importance of temporal congruency between action and effect has been stressed very early as factor in the perception of causality (e.g., Hume, 1739/2003). The immediacy of response removes ambiguity from the perception of causal agency, making the experience of effectance intuitive and requiring little cognitive effort.

The second relevant game characteristic that facilitates a forceful effectance experience is the *ratio between input and output*. Most games allow players to modify the game world substantially through only a few inputs. For example, in a combat game, players often need only a few mouseclicks to fire a powerful weapon and cause spectacular destruction. The ability to cause such significant change in the game environment supports the perception of effectance, as players can regard themselves as the most important (if not the only) causal agent in the environment. Computer games utilize their interactivity to offer a continous and "high-quality" experience of effectance through these properties (immediacy of response and an attractive input–output ratio). They allow players to perceive clearly and unambiguously the effects they impose on their environment. As White (1959) argued, the resulting feelings of efficacy are very pleasurable and rewarding. Therefore, effectance is valuable for explaining the enjoyment of playing computer games (Klimmt, 2003, 2005; Landauer, 1995).

More complex aspects of the gaming experience, such as the pride of success or the identification with a certain role (Sherry et al., chap. 15, this volume; Raney et al., chap. 12, this volume; Salisch, Oppl, & Kristen, chap. 11, this volume), build on the basic experience of effectance. For example, the belief that one causes observable changes in the game environment allows the player to attribute positive events (e.g., death of an antagonist) to him or herself, which leads to pleasurable emotions like joy and pride (Weiner, 1985). Moreover, feeling one's direct influence on the objects and characters of a game adds to the perceived completeness, realism, and pervasiveness of the narrative world depicted on the screen. Effectance therefore facilitates player involvement with the global structures of a game (such as the underlying plot) and stimulates the experience of being part of the game world, which is called presence (Tamborini & Skalski, chap. 16, this volume). In sum, effectance is both rewarding in itself and an important foundation for other parts of the entertainment experience (for more elaborate connections between effectance and other dimensions of game enjoyment: Klimmt, 2005).

Due to the very basic nature of efficacy experiences in computer game play, players may not be aware that the enjoyment they feel is partly derived from effectance. In fact, it is reasonable to expect that experienced players understand effectance as a matter of course because it occurs in every computer game, at each input–output loop. Examining the link between a person's anticipation (or mental representation) of a future media experience and his or her motivation to use that medium, we find that people do not appear to consider effectance intentionally in their decision making. It is conceivable, however, that certain circumstances could highlight the importance of effectance for enjoyment, namely, if effectance seems absent.

Imagine running a brand-new computer game on an old computer. From time to time, the computer's processing capacities are overwhelmed by the tremendous amount of data that the game software produces. Consequently, the game's audio and video are choppy. In addition, the game is unable to respond immediately to player input, and the playing experience begins to "splutter" as the continuity of events is broken up. In such cases, players will lose the perception of causality between their actions and the results produced in the game environment, because of the unexpected delay between input and output. This is striking, because there are never delays in normal game play. Nonsystematic observations suggest that players react negatively to such incidents, which extinguish the entertainment experience. In this way, computer problems may eliminate a player's sense of effectance within the game and destroy the fun of playing. Thus, even if the effectance dimension of enjoyment is not among the player's intentional considerations that precede selection decisions, the playing experience that lacks effectance is so strikingly unenjoyable that players will remember it and keep it in mind when selecting between computer games in the future.

Efficacy Experiences and the Motivation to Play Computer Games

Even if a person's anticipatory mental representation of the enjoyment derived from game play does not include the basic principle of effectance, it may contain more complex forms of experience that rely on effectance (see above). If someone is asked why computer games are fun, a typical response might be, "because I can do something, move around, and try things out" (e.g., Mallon & Webb, 2000). Most players seem to reflect the perception of their own activities and the opportunity to manipulate the game world, and this awareness is often combined with notions of control, power, or dominance (e.g., Jantzen & Jensen, 1993; Grodal, 2000). Schlütz (2002) has used the term "agency" to describe this kind of experience, one that most players are obviously aware of.

If the decision for or against playing a computer game is modeled as a psychological action that takes the anticipated experience during game play into account (see above), effectance may play a crucial role when the final motivational disposition is "computed." When positive and negative facets of experience are anticipated or mentally "simulated" (Taylor & Pham,

1996), effectance will certainly be an important factor if its absence in game play is likely (e.g., if the computer to be used is too old for the game under consideration). Effectance (or related, more complex phenomena such as agency, control, etc.) should also appear in the motivational calculation if its absence is improbable. The individual may, for example, decide that a given computer game does not offer "enough things to do," which would mean that the assumed input-to-output ratio is not sufficiently attractive. In this case, considering effectance could cause an individual to select a modern 3-D combat game with stunning visual, auditory, and narrative outputs over an old-fashioned game like Tetris, which is unimpressive in terms of player-produced output. So there are good reasons to assume that effectance is an important factor for enjoyment during game play and a crucial element in computer game selection. Consequently, effectance should be considered an important factor in explaining the motivation to play computer games.

SELF-EFFICACY: THE EXPECTATION OF MASTERY, CONTINUOUSLY CHALLENGED BY GAME OPPONENTS

Bandura's Concept of Self-Efficacy

Bandura (1977a, p. 193) defined self-efficacy as "the conviction that one can successfully execute the behavior required to produce the outcomes" that are expected to be caused by that behavior. The concept was originally formulated in the context of psychotherapy and behavioral change and was intended to guide the development of more effective interventions. An integral part of Bandura's social–cognitive framework (Bandura, 1977b, 1986, 2001), it has also been applied in nontherapeutic contexts, for example, the modeling and improvement of student performance (Schwarzer, 1992). Bandura's central proposition is that the more individuals believe they are able to deal with given tasks or situations successfully, the stronger their motivations to engage in those situations and the more effort they will invest to resolve them. Individuals with low self-efficacy, that is, those who believe they cannot handle their current situation, will try to escape the situation or, if they engage in the task at hand, will not display strong or sustained effort. Many empirical studies have demonstrated the explanatory power of self-efficacy in numerous domains of human behavior (Bandura, 1997). These investigations indicate that self-efficacy not only varies in strength, but also in generalizability. With respect to different domains of life (e.g., intellectual versus sports performance), people may hold very different self-efficacy expectations. For example, the effective use of computer technology depends heavily on users' computer-related self-efficacy (e.g., Wang & Newlin, 2002). Nevertheless, people with high levels of computer-related self-efficacy do not necessarily hold high efficacy beliefs in other domains of life (cf. Hartmann & Klimmt, chap. 9, this volume).

When an individual is confronted with a task, his or her self-efficacy convictions affect the decision-making process. Compared to other forms of media entertainment, computer games include many tasks and stressful events, and they demand various skills to be used effectively for entertainment. If players do not overcome at least a few opposing forces in a game, game play is not enjoyable; frustration and other negative emotions dominate the experience (Klimmt, 2005). Therefore it is reasonable to assume that people evaluate their own chances of success in a given computer game when they decide whether they want to use it (Vorderer et al., 2003). Applying the notion of computer-related self-efficacy as a domain-specific efficacy expectation, one could presume the existence of still more specific kinds of self-efficacy, that is, computer game-related self-efficacy or even efficacy convictions limited to a particular game product. The motivation to play computer games or one single game under consideration would then be influenced by the individual's deliberations on self-efficacy.

The same is true for engagement in playing once the decision for selecting a computer game has been realized. People who believe in their competences will invest more effort to overcome opposing forces in a game and to master the game, whereas players who doubt their skills will be more reluctant to devote as much energy to playing. As a consequence, the motivation to sustain the activity of computer game play will be higher in people with stronger game-related efficacy beliefs, as they would not withdraw from the activity when difficulties (e.g., powerful opponents) occur. Self-efficacy is therefore an important motivational factor in both selection processes and gaming activity itself.

Bandura (1977a, 1997) has posited that self-efficacy is shaped by numerous factors, the most important one being the experience of mastery. Successful coping in a situation increases the conviction that one will be able to deal effectively with similar situations in the future. In contrast, failures will decrease efficacy expectations. Observing another person successfully master a challenge may also contribute to one's self-efficacy, but Bandura (1977a) demonstrated empirically that personal mastery experiences are the most effective way to affect efficacy beliefs.

Self-Efficacy and Computer Game Selection

If self-efficacy determines much of the motivation to play computer games, the playing activity must offer sufficient opportunities to sustain and enhance efficacy beliefs (Lee, 2000), or at least avoid substantial decreases in self-efficacy. Otherwise, people would have a lower motivation to play, and alternative pursuits would become more attractive. Hence, a reasonable number of mastery experiences is a precondition for game enjoyment (see above) and the efficacy-based playing motivation. On the other hand, it has been demonstrated that the enjoyment of playing games partly depends on emotional reactions to challenges (Klimmt, 2003; Malone, 1981; Vorderer, Hartmann, & Klimmt, 2003). Playing task-based or competitive games is fun only as long as the resolution of the given tasks remains uncertain. If the game is too easy and mastery of game situations is predictable, important components of the entertainment experience, such as suspense and curiosity, will not be present (Friedman, 1995). To be enjoyable, computer games offer a carefully designed balance between challenge and mastery. Players should be able to achieve goals and perceive themselves as successful, but at the same time feel suspense and curiosity that is generated from the uncertainty of whether they will be able to cope with challenges as they arise (Klimmt, 2003). As a player's skills increase during game play (through automatization of motor activities and learning processes), the difficulty level of the game increases continuously in order to preserve the balance between mastery and challenge (Greenfield, 1984).

The design of typical computer games is therefore able to address the self-efficacy factor in players' motivation processes: They offer mastery experiences even to novice players and thus support the development of game-specific efficacy expectations, which in turn increase the players' motivation to sustain the activity even when they face opposition and obstacles, and to return to the game later when the current session is terminated. At the same time, players are most often left with uncertainty about whether they will be able to cope with the challenges ahead, and must deliver their optimum performance to achieve their goals within the game. Thus, computer game play establishes a cyclic relationship between self-efficacy and mastery: Initial mastery experiences facilitate the increase of game-specific self-efficacy, while this efficacy conviction is a motivational precondition of the maximum performance that is required for new and enjoyable mastery experiences.

These considerations demonstrate the importance of self-efficacy for explaining the motivation to play computer games. The interactivity of game use implies exposure to numerous tasks of variable difficulties. In contrast to the audiences of noninteractive entertainment, such as crime drama on TV, computer game players' *own* abilities to deal with the tasks presented are crucial if the desired entertainment experience is to occur. Mastery is a key component of both game players' experience and motivation, and it is reasonable to predict that people will select only those computer games for enjoyment that they believe they can master to a satisfying degree. Someone may expect a game to be too difficult, for example, because a game review indicates a mismatch between the game's average task complexity and his or her perceived capabilities. If this is the case, he or she is less likely to select the game, because his or her self-efficacy does not recommend the product under consideration. On the other hand, players who hold a very strong efficacy belief, those convinced that they can easily master the game at hand, may refuse to play because they anticipate mastery experiences that are not preceded by real challenges. Such success is not considered as enjoyable (Klimmt, 2003, 2005). From this perspective, computer games continuously call players' efficacy beliefs into question and, at the same time, offer opportunities for players to demonstrate that their level of self-efficacy is appropriate. Thus, the enjoyment of performance-based interactive entertainment such as computer games appears to be a complex process in which changes of efficacy beliefs are closely linked to the motivational determinants of selecting a computer game and sustaining the engagement in a selected game.

SUMMARY AND OUTLOOK: COMPUTER GAMES AS TESTBED OF PERFORMANCE ORIENTATION

This chapter has illuminated two concepts of the psychology of action that can explain the motivation to play computer games. Both effectance (White, 1959) and self-efficacy (Bandura, 1977a) are relevant to a player's experience and enjoyment. Because anticipated enjoyment and activity selection are closely linked, effectance and self-efficacy contribute to the formation of action intentions concerning selective exposure to computer games. Although a more complete formula of all the variables that people may consider when choosing whether to play a computer game or engage in another activity would have to include much more information than deliberations on effectance (or related experiences like agency, control, power, or dominance) and self-efficacy (see the various related chapters in this volume), effectance and self-efficacy would be very prominent entries in the list of those factors. Evidence for this assumption stems from the often repeated insight that girls and women display less interest in computer games, devote less time to playing, and on average achieve lower levels of mastery in computer games (Brown, Hall, Holtzer, Brown, & Brown, 1997; Cassell & Jenkins, 1998). Gender differences in computer game playing have often been explained through variables of narrative content (e.g., Dietz, 1998; Jansz & Martis, 2003), but perhaps the very nature of the gaming experience itself causes a gender difference in playing motivation. One may argue that girls are (for various reasons, including parental gender stereotypes) less likely to be familiar with computer use and thus establish lower average self-efficacy beliefs with respect to computer games than boys (Dickhäuser & Stiensmeier-Pelster, 2003; Hartmann, 2003). The perception that one cannot cope with the challenges of a particular game would then weaken the motivation to play the game and to sustain engagement in order to deliver optimal performance, which could explain the findings reported by Brown et al. (1997). Players who never enter the cyclic process of mastery, increase of efficacy beliefs, performance gain, and new mastery experiences will not display a strong general disposition that favors engagement in the given activity. For this reason, new sorts of games have recently been published that explicitly invite girls to get familiar with gaming. This way, girls may raise their low levels of computer-game-related efficacy convictions step by step, and initial experiences of success may increase their motivation to play computer games.

An observation of gender differences also provides evidence for the importance of effectance motivation in explaining the preference for playing computer games. Men exhibit more interest in other activities that provide effectance experiences through immediate and/or spectacular feedback, such as driving cars at high speed (Krahé & Fenske, 2002) or using firearms (e.g., Simon, Crosby, & Dahlberg, 1999), than women do. This does not necessarily mean that there is a general gender difference in effectance motivation, but there may be specific characteristics of effectance experiences (for example, types of experiences related to immediacy of feedback or perceptions of being in control) that are more appealing to males than to females. Playing computer games might provide effectance qualities that are of differential preference value (that is, they are more appealing to males than to females). This speculation remains to be tested empirically, of course, but should be considered a plausible argument for the importance of effectance in explaining the motivation to play (and the enjoyment of playing) computer games.

The general conclusion of this chapter is that both effectance motivation and self-efficacy are presumably variables that explain a recognizable portion of variance in the motivation to play computer games. Effectance as a very basic experience occurs during the use of any computer game, as it is directly generated through interactive media consumption. It facilitates more complex dimensions of game enjoyment because it serves as foundation for attribution, immersion, and comprehension processes that make players feel triumphant as part of the game-world and as heroes of great stories (Klimmt, 2003, 2005). The concept of self-efficacy appears to be in effect at various levels of analysis as well, because there are efficacy convictions of varying generalizability. A person's belief about her/his ability to deal with computers in general will affect her/his general tendency to play any computer game, but even a very self-confident computer user may feel insecure about her/his capabilities to handle a specific game. A general computer-related (and maybe computer-game-related) self-efficacy is an important prerequisite for the overall motivation to play computer games, whereas more specific efficacy convictions (e.g., those beliefs related to a game genre or one concrete product) explain preferences within the landscape of available computer games. So theoretical and empirical assessments of the motivation to play computer games should refer to both effectance and self-efficacy but also carefully consider which specific mechanisms are in operation at the chosen level of analysis.

Given the portrayed properties of the process of selecting and playing computer games, namely the perspective of such games as complex environments that deal with performance, victory and defeat, the broader perspective on future games research should consider interactive entertainment a test environment for skill-related and action-related facets of the user's self-concept. Games in general function as coping activities that enable subjects to deal with their real life (Ohler & Nieding, chap. 8, this volume; Vorderer, Steen, & Chan, in press). Computer game playing specifically targets performance-related issues of life — sustained engagement, excellence orientation, motivational self-control, and so on. As the social environment of Western societies imposes more and more performance-related pressures onto adolescents' development (e.g., in terms of school performance, social acceptance through peer groups), the need for coping with motivational and performance-related issues may have become more important. This would explain the enormous and still growing motivation (especially among young people) to play computer games (Raney et al., chap. 12, this volume): Playing allows for active coping through actualization and individualization of tasks, challenges, and skills (e.g., personalization of computer games' difficulty level). Moreover, computer games provide a complex environment in which the performance-related parameters can be defined in such a way that engagement is rewarding and functions as compensation for frustrating performances in real life: For people who suffer from failure or performance-based social rejection, gaming offers escape into simulated worlds with challenges, a substantial number of victories and controllable number of defeats. It produces competence gains, elevates (domain-specific) self-efficacy, and restores subjects' feelings of being able to influence the environment (effectance). Although one must not neglect the narrative facets of game play and its implications for playing motivation (see the according chapters of this volume), effectance and self-efficacy appear to be dominating determinants of the motivation to play computer games that fit well into the broader personal and social contexts that breed motivational dispositions and tendencies.

REFERENCES

Bandura, A. (1977a). Self-efficacy: Toward a unifying theory of behavioral change. *Psychological Review, 84*, 191–215.

Bandura, A. (1977b). *Social learning theory*. Englewood Cliffs, NJ: Prentice-Hall.

Bandura, A. (1986). *Social foundations of thought and action: A social cognitive theory*. Englewood Cliffs, NJ: Prentice–Hall.

Bandura, A. (1997). *Self-efficacy: The exercise of control*. New York: Freeman.

Bandura, A. (2001). Social cognitive theory of mass communication. *Media Psychology, 3*, 265–299.

Bargh, J. (1997). The automaticity of everyday life. In R. S. Wyer (Ed.), *The automaticity of everyday life: Advances in social cognition* (pp. 1–61). Mahwah, NJ: Lawrence Erlbaum Associates.

Brown, R. M., Hall, L. R., Holtzer, R., Brown, S. L., & Brown, N. L. (1997). Gender and video game performance. *Sex Roles, 36*, 793–812.

Cassell, J., & Jenkins, H. (Eds.). (1998). *From Barbie to mortal combat: Gender and computer games*. Cambridge, MA: MIT Press.

Deci, E. L., & Ryan, R. M. (1975). *Intrinsic motivation and self-determination in human behavior*. New York: Plenum.

Dickhäuser, O., & Stiensmeier-Pelster, J. (2003). Gender differences in the choice of computer courses: Applying an expectancy-value model. *Social Psychology of Education, 6*, 173–189.

Dietz, T. L. (1998). An examination of violence and gender role portrayals in video games: Implications for gender socialization and aggressive behavior. *Sex Roles, 38*, 425–442.

Friedman, T. (1995). Making sense of software: Computer games and interactive textuality. In S. G. Jones (Ed.), *CyberSociety: Computer-mediated communication and community* (pp. 73–89). Thousand Oaks, CA: Sage.

Gollwitzer, P. M., & Bargh, J. A. (Eds.). (1996). *The psychology of action: Linking cognition and motivation to behavior*. New York: Guilford.

Greenfield, P. (1984). *Mind and media: The effects of television, video games, and computers*. Cambridge, MA: Harvard University Press.

Grodal, T. (2000). Video games and the pleasures of control. In D. Zillmann & P. Vorderer (Eds.), *Media entertainment. The psychology of its appeal* (pp. 197–212). Mahwah, NJ: Lawrence Erlbaum Associates.

Hacker, W. (1996). Handlungsleitende psychische Abbilder ("Mentale Modelle") [Action-guiding psychological representations ("mental models")]. In J. Kuhl & H. Heckhausen (Eds.), *Enzyklopädie der Psychologie Band C 4/4 [Encyclopedy of Psychology, Volume C 4/4]* (pp. 769–794). Göttingen: Hogrefe.

Harter, S. (1978). Effectance motivation reconsidered: Toward a developmental model. *Human Development, 21*, 34–64.

Hartmann, T. (2003, November). *Gender differences in the use of computer-games as competitive leisure activities*. Poster Presentation at "Level Up," the 1st Conference on Digital Games, Utrecht, The Netherlands, November 4–6, 2003.

Heckhausen, H. (1977). Achievement motivation and ist constructs: A cognitive model. *Motivation and Emotion, 1*, 283–329.

Heckhauen, H., & Kuhl, J. (1985). From wishes to action: The dead ends and short cuts on the long way to action. In M. Frese & J. Sabini (Eds.), *Goal directed behavior: The concept of action in psychology* (pp. 134–161). Hillsdale, NJ: Lawrence Erlbaum Associates.

Hume, D. (1739/2003). *An inquiry in human understanding*. Oxford, England: Clarendon.

Jansz, J., & Martis, R. (2003). The representation of gender and ethnicity in digital interactive games. In. M. Copier & J. Raessens (Eds.), *Level up: Digital games research conference* (pp. 260–269). Utrecht: Utrecht University.

Jantzen, G., & Jensen, J. F. (1993). Powerplay—power, violence and gender in video games. *AI & Society, 7*, 368–385.

Johnson-Laird, P. N. (1983). *Mental models. Towards a cognitive science of language, inference, and consciousness*. Cambridge, England: Cambridge University Press.

Klimmt, C. (2003). Dimensions and determinants of the enjoyment of playing digital games: A three-level model. In M. Copier & J. Raessens (Eds.), *Level Up: Digital Games Research Conference* (pp. 246–257). Utrecht: Faculty of Arts, Utrecht University.

Klimmt, C. (2005). *Computerspielen als Handlung: Dimensionen und Determinanten des Erlebens interaktiver Unterhaltung. [Computer game play as action: Dimensions and determinants of the experience of interactive entertainment]*. Koeln: von Halem.

Klimmt, C., & Vorderer, P. (2003). Media psychology "is not yet there": Introducing theories on media entertainment to the Presence debate. *Presence: Teleoperators and Virtual Environments, 12*, 346–359.

Krahé, B., & Fenske, I. (2002). Predicting aggressive driving behavior: The role of nacho opersonality, age, and power of car. *Aggressive Behavior, 28*, 21–29.

Landauer, T. K. (1995). *The trouble with computers. Usefulness, usability, and productivity*. Cambridge, MA: MIT Press.

Lee, K. M. (2000). MUDs and self-efficacy. *Educational Media International, 37*, 177–183.

Mallon, B., & Webb, B. (2000). Structure, causality, visibility and interaction: Propositions for evaluating engagement in narrative multimedia. *International Journal of Human-Computer Studies, 53*, 269–287.

Malone, T. W. (1981). Toward a theory of intrinsically motivating instruction. *Cognitive Science, 4*, 333–369.

Mayer, J. D., & Gaschke, Y. N. (1988). The experience and meta-experience of mood. *Journal of Personality and Social Psychology, 55*, 102–111.

McDonald, D. G., & Kim, H. (2001). When I die, I feel small: Electronic game characters and the social self. *Journal of Broadcasting and Electronic Media, 45*, 241–258.

Oerter, R. (1999). *Psychologie des Spiels. Ein handlungstheoretischer Ansatz [The psychology of play: An action-theoretical approach]*. Weinheim: Beltz.

Salovey, P., Hsee, C. K., & Mayer, J. D. (1993). Emotional intelligence and the self-regulation of affect. In D. M. Wegner & W. Pennebaker (Eds.), *Handbook of mental control* (pp. 258–277). Englewood Cliffs, NJ: Prentice-Hall.

Schlütz, D. (2002). *Bildschirmspiele und ihre Faszination. Zuwendungsmotive, Gratifikationen und Erleben interaktiver Medienangebote [Computer games and their appeal: Motifs of exposure, gratifications and experience of interactive media]*. München: Reinhard Fischer.

Schwarzer, R. (Ed.). (1992). *Self-efficacy: Thought control of action*. Washington, DC: Hemisphere.

Simon, T. R., Crosby, A. E., & Dahlberg, L. L. (1999). Students who carry weapons to high school: Comparison with other weapon-carriers. *Journal of Adolescent Health, 24*, 340–348.

Steuer, J. (1992). Defining virtual reality: Dimensions determining telepresence. *Journal of Communication, 42*, 73–93.

Sutton-Smith, B. (1997). *The ambiguity of play*. Cambridge, MA: Harvard University Press.

Taylor, S. E., & Pham, L. B. (1996). Mental simulation, motivation, and action. In P. M. Gollwitzer & J. A. Bargh (Eds.), *The psychology of action: Linking cognition and motivation to behavior* (pp. 219–235). New York: Guilford.

Vorderer, P. (2000). Interactive entertainment and beyond. In D. Zillmann & P. Vorderer (Eds.), *Media entertainment. The psychology of its appeal* (pp. 21–36). Mahwah, NJ: Lawrence Erlbaum Associates.

Vorderer, P. (2001). It's all entertainment, sure. But what exactly is entertainment? Communication research, media psychology, and the explanation of entertainment experiences. *Poetics, 29*, 247–261.

Vorderer, P. (2003). Entertainment theory. In J. Bryant, D. R. Roskos-Ewoldsen, & J. Cantor, (Eds.), *Communication and emotion: Essays in honor of Dolf Zillmann* (pp. 131–154). Mahwah, NJ: Lawrence Erlbaum Associates.

Vorderer, P., Hartmann, T., & Klimmt, C. (2003). Explaining the enjoyment of playing video games: The role of competition. In D. Marinelli (Ed.), *Proceedings of the 2nd International Conference on Entertainment Computing (ICEC 2003), Pittsburgh* (pp. 1–8). New York: ACM.

Vorderer, P., Klimmt, C., & Ritterfeld, U. (in press). Enjoyment: At the heart of media entertainment. *Communication Theory, 14*, 388–408.

Vorderer, P., Steen, F., & Chan, E. (in press). Motivation. In J. Bryant & P. Vorderer (Eds.), *The psychology of entertainment*. Mahwah, NJ: Lawrence Erlbaum Associates.

Wang, A. Y., & Newlin, M. H. (2002). Predictors of Web students' performance: The role of self-efficacy and reasons for taking an on-line class. *Computers in Human Behavior, 18*, 151–163.

Wegner, D. M., & Pennebaker, W. (Eds.). (1993). *Handbook of mental control*. Englewood Cliffs, NJ: Prentice-Hall.

Weiner, B. (1985). An attribution theory of achievement motivation and emotion. *Psychological Review, 92*, 548–573.

White, R. W. (1959). Motivation reconsidered: The concept of competence. *Psychological Review, 66*, 297–333.

Zillmann, D. (1988). Mood management: Using entertainment to full advantage. In L. Donohew, H. E. Sypher, & E. T. Higgins (Eds.), *Communication, social cognition, and affect* (pp. 147–171). Hillsdale, NJ: Lawrence Erlbaum Associates.

11

What Attracts Children?

Maria von Salisch
Universität Lüneburg (Germany)

Caroline Oppl
Astrid Kristen
Freie Universität Berlin (Germany)

WHICH CHILDREN PLAY ELECTRONIC GAMES?

Even a cursory glance at the droves of youngsters playing at the computers for sale at large department stores indicates that children and adolescents are attracted to computer games and other electronic media products. Surveys confirm this impression and add that the use of computers and computer games has rapidly increased over the last few years. We have evidence for the unprecedented historical trend that computer ownership—in a representative sample of German households with school-age children—doubled between 1990 and 1999 (Feierabend & Klingler, 1999) and increased in the 2 years between 2000 and 2002 by another 10% . By now 81% of the households with 12- and 13-year-olds own a computer in Germany, and 23% of these youngsters are able to work and play on a computer of their own (Feierabend & Klingler 2003). Figures from a representative sample of 8-to-13-year-olds from the United States are about the same and are related to community income (Roberts, Foehr, Rideout, & Brodie 1999; Subrahmanyam, Kraut, Greenfield, & Gross 2001). Computers are mostly used for playing computer games, in Germany as well as in the United States.[1] By middle childhood, playing computer and video games has become one of the favorite leisure-time activities for boys and (less so) for girls in Western industrialized countries (Feierabend & Klingler 2003; Subrahmanyam et al., 2001).

Around the time children enter primary school, they begin playing computer games on a regular basis. That most children begin using computers somewhere between their 6th and 12th birthdays is evidenced by the proportion of children not using computers that declined from 61% among first graders to 18% among sixth graders (Feierabend & Klingler, 2003; Subrahmanyam et al., 2001). Whether children below school age will adopt playing electronic games in the future, as they have watching TV in the past, remains to be seen. So far it seems not to be the case (Roberts et al., 1999).

German children own on average 7.1 computer games. Most of these 6-to-13-year-olds (85%) received electronic games as gifts from parents and other adults, but some (20%) indicated that these were given by or shared with friends. About a third of the children reported buying these games themselves, more so with increasing age. Among young teenagers, 44% determined on their own which games they wanted to take home to play (Feierabend & Klingler, 2003). Even though some parents and school authorities restrict the use of computer games, many parents do not know or obstruct their children's interest in these games (Roberts et al., 1999). Instead, friends are the best source of news about electronic games (Feierabend & Klingler, 2003).

These findings underscore the active role that children play when it comes to playing electronic games. The main thesis of this chapter is that children select electronic games and other media contents in accordance with the uses and gratifications they expect (Sherry et al., chap. 15, this volume) in order to get ahead with their developmental tasks and other difficulties they encounter while growing up (even though when asked directly they may not be able to say that they are using media to this end). This assumption of an active child fits well with a contextualist or transactional model of development that posits that children and adolescents are not passive recipients of their environment (which includes media messages) but active in constructing their environment (Lerner, 2002), which, of course, includes their media experiences. Although only a few children are actively involved in producing media, all are active in selecting to which media (content) they attend, in attributing meaning to media messages, in interpreting and evaluating these messages, and in including them in their action plans and conceptions of reality (Bonfadelli, 1981). This rather general statement also applies to the use of electronic games. Few children are busy programming new interactive games (Kafai, 1996; Feierabend & Klingler, 2003). Nevertheless, children in general determine which computer or video games they want to open, when and where they want to work on them, and with which other persons they want to go about playing these games. In addition, children show considerable mental activity when playing electronic games, because they understand media and their messages depending on their level of (social-) cognitive and emotional development (Huston & Wright, 1998; Bonfadelli, 1981; Valkenburg & Cantor, 2000). In sum, children are active users (although not always rational ones) who are attracted to electronic games that

1. address their developmental tasks,
2. offer possibilities for escapism and possibly mood management, and
3. match their level of development.

These three points will be elaborated in the rest of this chapter.

CHILDREN LIKE ELECTRONIC GAMES THAT ADDRESS THEIR DEVELOPMENTAL TASKS

One of the most famous explanations of what "people do with media" is the *uses and gratifications approach*, that is, very simply: People seek media (content) in order to reach goals or to fulfill other needs or interests (Rubin, 1994). Many goals and needs change with age. A good way to describe the age-graded changes in long-term goals and needs is afforded by the concept of developmental tasks (Havighurst, 1953) because all members of a given age group face these tasks (similar to Erikson's [1963] crises). Developmental tasks are formulated for the whole life span. In a given period, such as middle childhood, they are interdependent insofar as the advancement in one task furthers the advancement in another. They are overlapping because individuals can work at more than one task at a time. Over time they interlock because the on-time solution of these tasks helps in coping with the tasks of the next developmental period(s). Developmental tasks tend to change over historical time (and tend to vary between cultures) because they include expectations of societies (or subcultures) of what a person should accomplish at a given age (Flammer, 1996). Further sources of developmental tasks are the level of a person's physical maturation and his or her individual goals. Because individuals differ in their aspirations and values, they also differ in which goals they want to reach, in which order they want to work on them, and how they want to accomplish them (Flammer, 1996). Because individuals are viewed both as influenced by their social–cultural context and as active in pursuing their own goals, the concept of developmental tasks can be subsumed under the transactional model of development outlined above (Lerner, 2002).

When these theoretical formulations are translated into the relationship between children and media products, children's interests in media topics, plots, formats, characters, or graphics are influenced by the goals they have set for themselves in the context of the developmental tasks lying ahead of them (Bonfadelli, 1981). Children's attraction to particular computer games is thus a function of the developmental tasks they are working on (more or less consciously). To put it succinctly: Children choose specific electronic games as a leisure-time activity—over other activities—because playing these games may help them in getting ahead with some developmental tasks. In particular, these games allow them to gather information, to coordinate social perspectives, and to explore new roles in the safe context of a game, without the risk of being held accountable for the consequences of their actions in the real world (Oerter, 1999). Table 11.1 includes the nine tasks Havighurst (1953) postulated for middle childhood, the age span between roughly 6 and 12 years of age. While playing computer games in order to develop or reinforce moral standards, attitudes, and values (Rubin, 1994) may be achieved by almost every electronic game, particular games are best suited for working on the five developmental tasks outlined below.

Developing Basic Skills in Reading, Writing, and Arithmetic

A typical developmental task in middle childhood (Havighurst, 1953) is learning how to read and write. This is normally accomplished in school, but strategy games that require reading and typing words (with correct spelling!) into the computer may also further children's skills in

TABLE 11.1
Developmental Tasks of Middle Childhood

Developmental Tasks

---

1. Learning the physical skills that are necessary for playing children's games.

2. Developing a positive attitude to oneself as a growing person.

3. Learning to get along with peers.

4. Learning appropriate male or female gender role behavior.

5. Developing basic skills in reading, writing, and arithmetic.

6. Acquiring concepts and schemata that are necessary for everyday life.

7. Developing conscience, moral standards, and values.

8. Achieving personal autonomy.

9. Developing attitudes toward social groups and institutions.

---

this area. Educational games make use of the appeal of animated characters and the interactive mode to motivate children to improve their knowledge in nearly all subjects ranging from basic skills in reading, writing, and spelling in the child's native language and arithmetic to advanced topics in foreign languages, history, geography, and the natural sciences. More sophisticated educational software challenges youngsters' skills with the help of simulations, tutorials, and microworlds (Schaumburg, 2002) that vary in their form of presentation. Forty-six percent of the representative sample of German 6-to-13-year-olds indicated that they worked with educational software at least once a week (Feierabend & Klingler, 2003). In the Kaiser Millenium study from the United States, 25% of the 8-to-13-year-olds noted in their media diary that they had played this kind of game the previous day (Roberts et al., 1999). Nevertheless, academic software is usually not found among children's favorite games. Only 14% of the German children mentioned these type of games when they were asked for their "top three" games (Feierabend & Klingler, 2003). A substantial minority (25%) of the children liked playing these games *less* or *not at all*, with many of them attending lower tracks in secondary school (Hauptschule; Feierabend & Klingler, 2001).

Acquiring Concepts and Schemata
That Are Necessary for Everyday Life

Computer games are similar to television and other media insofar as they help children to look beyond the little world of their own family and gather information about the wider world. This corresponds not only to children's needs to master the knowledge and the technology of their own culture (Erikson, 1963), but also with developmental tasks having to do with "acquiring concepts and schemata in thinking that are necessary for everyday life" (Havighurst, 1953), because many computer games challenge children's skills in problem solving. Adventure games, for example, confront children with problems that they have to solve by their own means, using their intelligence, their creativity, their knowledge of the rules of the game, or the physical strength of their play character (which can be changed in some adventure stories). Different levels of difficulty present adequate challenges for novices and experts alike. Finding evildoers, helping the helpless, or unraveling hidden secrets are strong motivations for school-age children who have been shown to prefer mysteries and detective novels (Hurrelmann, 2002).

Sociologists tend to point out that we live in an age-segregated society (Steinberg, 2002) where youngsters can no longer find the real-life challenges they had once encountered when they had easier access to adults and their work out of the home. Thus, it is no surprise that in the representative German sample, 33% of the children counted strategy games and 27% simulation games among their three favorite games. In the Kaiser Millenium study from the United States, 37% of the 8-to-13-year-olds indicated that they had played adventure games on their video console and 15% on their computer the previous day. Simulations and strategic games were played by 9% (as a video game) and 12% (as a computer game; Roberts et al., 1999). When faced with a limited range of possible experiences at home, at school, or in the neighborhood, media dependency tends to increase (Rubin, 1994). Exploring exciting places, finding adventures, and mastering challenges on the screen seems to be the better alternative, especially if this can be carried out safely in the context of a game and in the familiar surroundings of the home.

Learning the Physical Skills Necessary
for Playing Children's Games

A third developmental task is "to develop the physical skills necessary for playing children's games" (Havighurst, 1953). A large number of electronic games require some coordination between players' eyes who watch a moving object and their hands that manipulate it in some way. Coordinating vision and movement is required in jump-and-run-games like *Mario Brothers* just as much as in action games that require shooting at a moving target. It is also an integral part of coordination games like *Tetris*. In sports or racing competitions other motor skills may be called upon, such as steering a car or moving a tennis ball, but all of these games require some kind of sensorimotor skills. As every computer player has experienced, hand–eye coordination quickly improves by exercise, and improved skills are needed in order to move on to more difficult levels. Honing their fine motor skills to be an expert in these games may be a motive for some school-age children when they spend endless hours playing these somewhat repetitive (video) games, especially for boys. More than 20% of the "tween" boys in the Kaiser Millenium study noted in their media diary that they had played a video game for over an hour the previous day (girls: 5%; Roberts et al., 1999). In the representative German sample, 20% of the children liked jump-and-run games best, and 14% counted sports games among their favorites. High in the favor of German boys were electronic games that imitated playing soccer or coaching a soccer team (Feierabend & Klingler, 2003). Indeed, playing interactive games improved North American boys' and girls' nonverbal intelligence, in particular their ability to anticipate spatial paths (Subrahmanyam & Greenfield, 1996) and to mentally rotate objects in space (Okagaki & Frensch, 1996). In addition, it helped U.S. participants keep track of events at multiple locations, thus furthering their ability to divide their attention (Greenfield, deWinstanty, Kilpatrick, & Kaye, 1996). Studies on long-term or cumulative effects of playing electronic games on these skills are, however, still missing.

Learning Appropriate Male or Female Gender Role Behavior

Although children's need for sex typing becomes less urgent in primary school as compared to the preschool period, and more deviations from the stereotypes of male and female attitudes and behavior are now tolerated (Serbin, Powlishta, & Gulko, 1993), boys and girls in middle childhood are nevertheless faced with the developmental task "to learn appropriate male and female gender role behavior" (Havighurst, 1953). Media protagonists are believed to offer some guidance in this regard (Huston & Wright, 1998). Media stories and formats that help boys and girls formulate their gender role are still attractive for this age group, when characters and situations are more realistic than in productions designed for preschoolers (Valkenburg & Cantor, 2000). *Barbie* games, for example, are fashioned in pink like the original toy and require children to look out for the young, create fancy outfits, or solve simple problems. No wonder that this game, with its focus on nurturance and arts and crafts, appeals only to girls: 5% of the girls and 0% of the boys included it among their three favorite electronic games (Feierabend & Klingler, 2003).

Most adventures, simulations, jump-and-run, or action games, however, feature male characters as main protagonists who play leading and attractive roles (see S. Smith, chap. 5, this volume); in a sample of 33 of the most popular video games in 1995, 41% of the games with human or humanlike characters contained no female characters (Dietz, 1998). Content analyses of 44 German computer and video games demonstrate that this bias is not restricted to the relatively simple jump-and-run or action games, but can also be found in more elaborate adventure and strategy games. In fact, there were only a handful of games on the market in which female heroes played leading roles as clever and assertive girls or amazon fighters (Fromme & Gecius, 1997). If females were present at all, they were cast in supporting roles, like the proverbial "damsel in distress" who has to be liberated by the male hero of the game (21% of the games in Dietz' sample). But also in simulation games, such as *Pizza Connection*, no female character possesses the competencies necessary for running a successful business. In this game women characters have special skills—you guessed it—only as cooks in the kitchen. In most computer or video games male heroes perform some sort of mission that sometimes requires cleverness and creativity, but often also some sort of physical strength (see S. Smith in this volume). In 79% of the games included in Dietz' (1998) sample of video games, some type of aggressive behavior was required in order to reach the next level, but in some cases (27%) the aggressive behavior was socially acceptable because it was embedded in sports competitions, like football, hockey, or wrestling. More alarming may be those 50% of the games in which it was necessary to direct aggressive behavior at other (human) characters in order to get ahead with the game. Thus, most video games portray males and females in stereotypical ways that may affect a further developmental task: children's attitudes toward social groups. What happens when electronic games disorient children because they do not adequately meet their developmental needs is an open question for further research (and a question of much public debate).

Considering that most entertaining electronic games contain attractive roles only or mostly for male characters, it comes as no surprise that boys tend to play more often per week and that they tend to spend longer time in front of the console or computer screen than girls (Subrahmanyam et al., 2001; Wright et al., 2002). More heavy players are boys. In the German sample about twice the number of boys spent more than 1 hour playing a particular game in one sitting—21%—compared to 11% of the girls (Feierabend & Klingler, 2003).

The Development of Girls' and Boys' Preferences for Electronic Games in the KUHL Study

Nearly all of the figures presented so far came from cross-sectional data sets that present a snapshot of children's preferences for electronic games at the particular time of measurement. Because these studies cannot capture stability or change of these preferences, we decided to conduct a longitudinal study that focused on children, computers, hobbies, and learning (Kinder, Computer, Hobby, Lernen, in short: KUHL) (Kristen, 2004; Oppl, 2004; von Salisch, Kristen, & Oppl, 2004[2]. Measurement took place in the fall of 2002 (time 1) and again exactly 12 months later in the fall of 2003 (time 2), so that seasonal effects could be ruled out. A total of 324 children in the third and fourth (later fourth and fifth) grades participated in primary schools in Berlin (Germany). In order to control for influences of socioeconomic status, the schools were located in neighborhoods of high or low unemployment that belonged to the Eastern or the Western part of the formerly divided city of Berlin (see Table 11.2).

The 155 boys and 169 girls included in the KUHL study (von Salisch, Kristen, & Oppl, 2004) came from different ethnic backgrounds. Although the majority was born in Germany (87%), 13% were born abroad, mostly in Turkey, the Near Eastern countries, or Russia. The mean age of the children at time 1 was 8.9 years; at time 2 it was 9.9 years. In order to examine different aspects of development as well as different views on a person, a triangulated system of measurement was used. Self-report questionnaires were supplemented with a peer rating, a teacher rating, and a structured diary on leisure-time activities. The self-report questionnaires included an instrument on the use of interactive electronic games that asked the children to write down a maximum of six of their favorite computer and video games. The children named about 350 different game titles at time 1 and about 240 additional titles at time 2. With the help of an internet databank provided by a rating board of experts, the *Unterhaltungssoftware Selbstkontrolle* (*www.zavatar.de*), all nominated titles were classified according to their genre. Figures 11.1 and 11.2 present the results on gender differences in children's preferences for

TABLE 11.2
The Longitudinal Sample of the KUHL Study (Berlin, Germany)

| | | Proportional Unemployment Rate | | | | |
| City Area | High | | Low | | Total | |
	n	(P)	n	(P)	N	(P)
East Berlin	81	(25%)	84	(26%)	165	(51%)
West Berlin	99	(31%)	60	(18%)	159	(49%)
Total	180	(56%)	144	(44%)	324	(100%)

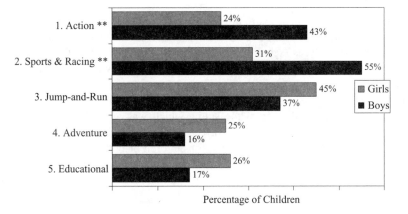

Note. Boys n = 139. Girls n = 141. *p < .05. **p < .01.

FIG. 11.1.   Genre preferences of boys and girls at time 1(KUHL Study).

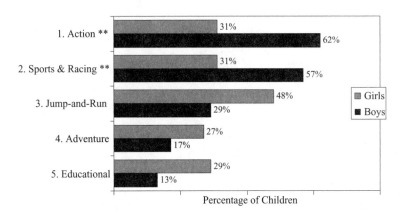

Note. Boys n = 143. Girls n = 151. *p < .05. **p < .01.

FIG. 11.2.   Genre preferences of boys and girls at time 2 (KUHL Study).

electronic games at time 1 and time 2. Significant differences between groups of boys and girls are marked with asterisks.

Figure 11.1 shows that boys attending third- and fourth-grade classrooms in Berlin favored action, sports, and racing games (similar: Feierabend & Klingler, 2003), just like their age mates in two studies from the United States (Roberts et al., 1999; Wright et al., 2002). Primary school girls preferred jump-and-run, adventure, and educational games at time 1, but the gender difference was not significant (just like in the Kaiser study from the United States; Roberts et al., 1999). One year later, gender differences in genre preferences became more pronounced and reached statistical significance. More boys continued to like action, sports, and racing games while a higher number of girls continued to prefer jump-and-run, adventure, and educational titles (Fig. 11.2). Thus, gender differences in genre preferences on a group level seem to be stable in the KUHL study, even though the longitudinal measurement shows that there were some children who had started playing electronic games in the meantime. Neither was the stability of these gender differences affected by the fact that the favored titles differed between time 1 and time 2 (von Salisch, Kristen, & Oppl, 2004).

The longitudinal stability of genre preferences within the groups of boys and girls in the KUHL study (von Salisch, Kristen, & Oppl, 2004) was examined with two repeated measures MANOVAs. These analyses (see Table 11.3) show that boys' nominations for action titles increased over the 1-year time span while jump-and-run titles dwindled in their favor. At the same time, sports and racing games declined in girls' preference.

Table 11.4 presents the stability of the genre preferences among individual children in the KUHL study over the 1 year of the study, which was again broken down by gender groups (von Salisch, Kristen, & Oppl, 2004). The most stable individual genre preference was for educational software. This was the same for children of both sexes. Individual girls and boys who liked to play educational games (whether by choice or by necessity) were likely

TABLE 11.3
Longitudinal Changes in Genre Preferences (KUHL study)

| | Multivariate Analysis of Variance (Repeated measurement) | | | | | | | | | | | |
| | Boys | | | | | | Girls | | | | | |
Source	T1 M	T2 M	df	error df	F	p	T1 M	T2 M	df	error df	F	p
	Within Subjects						Within Subjects					
Overall			5	123	4.16	.00			5	126	9.11	.48
1. Action	.18	.24	1	127	4.06	.05	.10	.11	1	130	.02	.89
2. Sports & Racing	.28	.24	1	127	1.83	.18	.16	.11	1	130	3.77	.05
3. Jump-and-Run	.15	.08	1	127	9.97	.00	.22	.21	1	130	.07	.80
4. Adventure	.06	.04	1	127	2.06	.15	.11	.11	1	130	.00	97
5. Educational	.07	.04	1	127	3.46	.07	.11	.13	1	130	.60	.44

Note: T = Time of measurement. *p < .05. **p < .01.

TABLE 11.4
Stability of Genre Preferences Among Individual Children
(KUHL study)

	Boys	Girls
	Correlation Coefficients Indicating Stability Over a 1-year Period	
Genre	r	r
1. Action	.23**	ns
2. Sports & Racing	.26**	.32**
3. Jump-and-Run	.15	.13
4. Adventure	ns	.17*
5. Educational	.50**	.49**

Note: Boys n = 128. Girls n = 131. * p < .05. ** p < .01.

to continue to do so 1 year later. Preferences for jump-and-run titles were not so stable, nei-ther among individual boys nor among individual girls, because this genre may be growing out of all children's favor toward the end of childhood (see means in Table 11.3). A liking for action games was stable only among the boys, possibly because the protagonists in these games tend to be brave, dominant, successful, and independent characters who may serve as male role models (Valkenburg & Cantor, 2000). Girls who had nominated an action game among their favorite games at time 1 were not more (or less) likely to do so again at time 2, although more girls played action games (see Fig. 11.2), and although the overall proportion of action games named by girls as a group did not change much over this time (see mean in Table 11.3). This pattern of results suggests that individual girls tended to try out or "taste" action games but did not include them in their regular "diet" of electronic games, possibly because they did not like the fighting or because they could not manipulate attractive protag-onists of their own gender. Why was liking adventure game titles (moderately) stable among female individuals? Adventure games offer more variety in terms of possible actions. Fight-ing, if necessary, is typically combined with problem solving. Usually the action is framed in a story that involves other (human) characters. The more social character of adventure games may add to their appeal to girls (Valkenburg & Cantor, 2000). But these are specula-tions. Asking boys and girls about the appeal of these types of electronic games may reveal still other reasons for stability and change in genre preferences among individual boys and girls.

Learning to Get Along With Peers

One of the developmental tasks as well as social goals among school-age children and ado-lescents (see Raney et al., chap. 12, this volume) is to gain the acceptance of their age mates and to increase their autonomy vis-à-vis their parents. Playing electronic games is embedded in children's relationships to their peers in many ways. Children like to play these games in the company of their friends. Forty-three percent of the girls and 59% of the boys in Germany reported that they played computer games with children at least once a week (Feierabend & Klingler, 2003). Children's growing orientation toward peers is also evident when looking at where children regularly spend time with computers and video game consoles. While computer use in the company of friends in the home or in a friend's home increased with age (especially among the boys), use with parents or siblings declined toward adolescence. Nevertheless, parents were still the most important source of information when children encountered problems they could not solve by themselves; only 30% turned to their peers and 15% to their siblings for advice (Feierabend & Klingler, 2001). Peers, however, were key figures when children wanted to learn about new games: 72% of them got their information from their age mates (Feierabend & Klingler, 2003). Talking about computer games becomes a major topic of conversation with peers (Feierabend & Klingler, 2001), especially among preadolescents, who also tend to use common preferences for media formats or contents as a feature for distinguishing between different peer groups. Expertise in electronic games can also serve to break the ice, that is, new students who are knowledgeable about an attractive game stand a good chance to be admitted into an existing peer group. Social status in a peer group may be maintained by the ability to give advice to others—or at least to play at a certain level of difficulty. Therefore, playing electronic games may fulfill children's affiliative needs just as much as their developmental task to gain acceptance in a group of peers. This works, however, only if children play the game titles that are appreciated by their local peer group. Toward the end of childhood, talking about, playing together, copying, borrowing, and exchanging electronic games becomes part of some peer cultures. Friends are sought out, when it comes to exchanging computer and video games which always involves some trust because sometitles may not be legally acquired (Schwab & Stegmann, 1999).

Playing interactive electronic games may further children's acceptance by their peers in another, less obvious, way: through their effects on children's emotions and emotion regulation. Many action and adventure games challenge children to overcome frustrating obstacles or to cope with surprise or anxiety in the face of monsters appearing in dark tunnels. "Don't panic, act" is the message of many of these games. With attention focused on the task, practice in sensorimotor skills, resourceful thinking, and knowledge of the game, these challenges can be mastered over time. In fact, the sequence of frustration and anger that leads to increased effort and to eventual mastery with accompanying feelings of pride and control may be a major emotional motive for playing these electronic games (Fritz & Misek-Schneider, 1995; Grodal, 2000). When playing electronic games over a longer period of time, children may thus not only acquire the cognitive and motor skills needed to succeed in the game, but also some emotional competencies, such as tolerance for frustration and fright arising from surprise attacks (Valkenburg & Cantor, 2000).

Children in primary school tend to value children who are (or appear to be) emotionally unperturbed in many situations. Excessive emotional behavior is a reason for rejection in the peer group; school-aged children often call this type of behavior *babyish*. Children showing emotional behavior like crying are criticized in their peers' gossip, which also serves to reinforce the children's rather strict norms of "nonexpression" of emotions (Gottman & Mettetal, 1986). Primary school children tended to expect more negative reactions from their peers (than from their parents) when they expressed anger, sadness, or pain in front of them (Zeman & Garber, 1996), and only about one third of the boys shared their fears about a monster story with their peers and friends (Rimé, Dozier, Vandenplas, & Declercq, 1996; Saarni, 1988). Between 9 and 13 years of age, children increasingly reported that they would regulate their anger at a friend by distancing themselves, that is, by ignoring the friend on purpose or by shifting their attention to more pleasurable activities. Thus, peers seem to challenge children to downregulate their emotions and to act in cold blood, even when faced with surprising, frightening, or anger-provoking situations (von Salisch, 2001). Playing electronic games may be a way children (especially boys) train themselves to gain or to maintain peer acceptance by their ability to control their emotional expressions (and experiences), that is, to stay "cool" when emotionally aroused. There is some evidence that German male adolescents attending vocational tracks in high school (Hauptschule) tended to "dare" each other to show no emotional reaction when watching horror movies (Gleich, 2004), and that peer influences also apply to watching scary programs on U.S. television (Valkenburg & Cantor, 2000), but further investigation of this possibility in the context of electronic games is definitely needed.

ESCAPISM AND POSSIBLY MOOD MANAGEMENT AS MOTIVES FOR PLAYING ELECTRONIC GAMES

In playing electronic games children not only pursue developmental tasks but are also motivated by their fluctuating emotions and moods (Finn & Gorr, 1988); which may have to do with accomplishing or not accomplishing developmental tasks but which may also be independent of these long-range goals. Children's play is intrinsically motivated by the wish to manipulate reality, to escape into fantasy worlds, and to explore possible solutions for the problems encountered while growing up (Oerter, 1999). Children, especially young children, often feel confused, frustrated, or powerless when they are confronted with complex situations or (adult) decisions they cannot understand. Older children tend to experience helplessness when they are faced with (excessive) demands of parents or school or when they are ostracized by their peers (Gottman & Mettetal, 1986). Electronic games may help children in disentangling themselves from these (and other) predicaments, because they offer a fantasy world in which figures act in somewhat predictable ways (Grodal, 2000). In this regard, interactive electronic games are similar to TV (and other media), whose association with family conflicts has long been known (e.g., Rosenblatt & Cunningham, 1976). When asked to select situations in which they would turn to particular media, 10% of the representative sample of German school-age children mentioned that they would turn to the computer when they wanted to forget something (Feierabend & Klingler, 2001). For older children and adolescents, conflicts with their friends are an important source of stress (Seiffge-Krenke, 1995). Those children (especially boys) who tended to distance themselves emotionally when they were angry at a friend were also the ones who demonstrated more interest in electronic games (von Salisch & Bretz, 2003). Although computer games (10%) were endorsed by fewer children when asked where they would fulfill these escapist needs than television (24%; Feierabend & Klingler, 2001), distraction is still one of the functions many electronic games satisfy, especially those whose fast speed calls for *total immersion.*

But these interactive games have an added advantage over TV. In electronic games, children can be active and master difficulties themselves (Grodal, 2000). In fact, there is a high pressure to act, especially in jump-and-run, action, and adventure games. Acting is not difficult, because most scenarios of these games picture a simple world of good guys and bad guys (Fromme & Gecius, 1997). Moving their protagonists on the computer or video screen may thus help children in (re-) gaining a sense of power or control that they often feel to be lacking in their life in general and in frustrating situations in particular. Mastery is furthermore predictable because it can be achieved through practice. Developing a potential for mastering challenging situations is an important motive for playing electronic games, and is supported by empirical evidence. A majority (61%) of male adolescents in a large Swiss study endorsed this as their most important motive for playing these games (Süss, 2000). Both distraction and mastery tend to go along with the pleasurable sensations of fun (16%), excitement (14%), or good mood (9%) that German school children noted when asked which needs were gratified by certain media (Feierabend & Klingler, 2001). Because these survey data did not ask about participants' feelings before turning to the media, mood management in the sense of seeking arousal when bored and calmness when overstressed may not be differentiated from simple escapist needs.

In this same representative German study, 7% of the schoolchildren reported that they would switch on the computer when they were frustrated or angry (TV: 16%). More boys (10%) were among these "frustration players" than girls (5%; Feierabend & Klingler, 2001). They were not asked which games they played on the computer, but we know that aggressively disposed adolescents tend to play for longer periods of time and tend to prefer electronic games with violent content (Gentile et al., 2004). In their study, Gentile and colleagues asked ($N = 607$) eighth- and ninth-grade students from the United States to complete a survey. The more time these adolescents spent playing and the more violence-laden their games were, the more likely they were to get into arguments with their teachers and into physical fights with their schoolmates. A path analysis showed a mediating effect of hostility, that is, hostility mediated the effects of both amount of play and violence exposure on grades and aggressive behaviors (Gentile et al., 2004). This cross-sectional evidence on electronic games confirms what we know from many years of research on television. Violent content tends to appeal to male individuals who are more aggressive and extroverted than other persons, who are high in sensation seeking and curious about unusual situations. Some of them are more socially isolated or emotionally troubled, but all need to be able to distance themselves from violent media content (Gleich 2004; see also Hartmann & Klimmt, chap. 9, this volume). When and how this emotional distancing ability develops in childhood is an interesting question for further research.

THE DEVELOPMENT OF PREFERENCES FOR VIOLENT ELECTRONIC GAMES IN THE KUHL STUDY

In the KUHL study we also examined associations between children's aggressive behavior and their preference for violent media content (von Salisch, Kristen, & Oppl, 2004). All game titles that the children wrote down as their favorite computer and video games were rated by experts (e.g., owners of game stores, male and female adolescent hobby players) on whether plot and presentation of the game was "brutal and bloody" and on whether the game was a first-person shooter game. Because the ratings of the experts were highly correlated ($r = .71$), mean ratings on the brutality and bloodiness of each game were calculated and Z-*standardized* (that is, 0 is the mean value). These values were summed up over the games that each child had nominated and divided by the number of games mentioned. Thus, every child received an average rating on how brutal and bloody their favorite games were. The repeated measures MANOVA depicted in Table 11.5 shows that the preference for egoshooters did not change from time 1 to time 2, among either boys or girls. Electronic games with a brutal and bloody plot became significantly more attractive for boys and significantly less repellent for girls as they grew older. Nevertheless, as the negative values show, playing brutal electronic games seems to be a favorite pastime for only a few children. Most participants of the KUHL study (especially girls) tended to be attracted to other kinds of electronic games (von Salisch, Kristen, & Oppl, 2004).

Because subtypes of aggressive behavior need to be differentiated we took measures in the KUHL study to obtain ratings on children's openly and relationally aggressive behavior both from their teachers and their peers. These ratings on aggressive behavior were positively correlated with the amount of time boys and girls spent playing electronic games per week. As shown in Table 11.6, the more time boys and girls spent playing electronic games, the more physically and verbally aggressive they were rated by their peers and teachers alike. Boys who tended to play electronic games for longer hours were also rated as more

TABLE 11.5

Development of Childrens' Preferences for Violent Electronic Games (KUHL study)

| | Univariate Analysis of Variance (Repeated measurement) | | | | | | | | | | | |
| | Boys | | | | | | Girls | | | | | |
Source	T1 M	T2 M	df	error df	F	p	T1 M	T2 M	df	error df	F	p
	Within Subjects						Within Subjects					
1. Egoshooter	.07	.08	1	127	.43	.52	.03	.05	1	132	1.22	.28
2. Brutal & Bloody Plot of the Game	-.13	.04	1	128	6.55 *	.01	-.44	-.39	1	133	4.16 *	.04

Note: $^*p < .05.$ $^{**}p < .01.$

TABLE 11.6

Correlations Between Time Spent Playing Electronic Games
and Aggressive Behavior (KUHL study)

| | | Time Spent Playing Electronic Games | |
| | | Boys | Girls |
Rated by	Type of Aggression	r	r
Peers	1. relational	.24**	.12
	2. physical & verbal	.17*	.18**
Teachers	3. relational	.25*	.07
	4. physical & verbal	.19*	.14*

Note: Boys $n = 105 - 146$. Girls $n = 118 - 165$. $^*p < .05.$
$^{**}p < .01.$

relationally aggressive by their teachers and peers (Kristen, 2004) that is, they were more inclined to hostile socially manipulative behavior, such as spreading lies or excluding other children.

Adolescents and young adults who tend to play violent electronic games for longer periods of time are also at risk for aggressive behavior, which has been documented for college student samples (e.g., Anderson & Dill 2000). Our study is the first to extend this association downward to the age group of primary school students. Nine-to-eleven-year-olds who were rated as more aggressive by their teachers and peers tended to spend more time playing electronic games in the KUHL study (von Salisch, Kristen, & Oppl 2004). Whether they played these games in order to escape from frustrating experiences or whether to smoothe their emotional turbulences, cannot be determined so far (if at all).

Electronic Games and Level of Development

Which games children like not only depends on their developmental tasks and on their efforts in managing their moods, but also on their level of (social-) cognitive and emotional development (Bonfadelli, 1981). In the following section we focus on two aspects of cognitive development, attention and memory, that are integral parts of information processing. We extrapolate from a substantial body of knowledge generated when studying how children understand television (e.g., Huston & Wright 1998; Valkenburg & Cantor 2000). Investigations that confirm these findings or explore the differences due to the interactive nature of electronic games (Grodal 2000) are definitely needed.

Attention governs the way in which information from the electronic game is taken up and interpreted. Which information is selected depends on the salience and the affective valence of the figures on the screen in the sense that attractive protagonists are given more attention. Which media characters are attractive for children? Attractive figures are generally similar to the child because they are of the same sex and about the same age group as the child as well as figures who possess valued characteristics. What is valued, however, tends to change with age. School-age children tend to prefer a more realistic depiction of screen characters and give attention to detail. Insufficient graphics will be criticized by them (Valkenburg & Cantor 2000). Because speed of information processing increases and memory expands greatly over the primary school years (Schneider & Pressley 1997), action can be faster paced and more complex actions can be understood and performed on the screen. How game information is transformed and structured depends on children's level of cognitive organization, specifically their ability to decode symbolic (visual) codes and to store them in memory. Next to general knowledge about the world, which develops rapidly in childhood, prior knowledge may aid in playing a particular game. Youngsters learn to navigate complicated spatial arrangements and social scenarios in electronic games because they play the same game again and again and slowly build up a store of knowledge on characters, weapons, scripts, or spatial maps to use in future sessions. Prior experience (with the same or similar games) also helps them find out which cues are important in order to win the game. Experienced players are usually able to recognize even the subtle cues novice players would have overlooked. Individual differences in cognitive skills, such as memory, spatial navigation, and logical thinking, are also likely to influence which game a child finds appealing.

Selective attention is also relevant to playing computer games, because children need to filter out all information that is irrelevant or distracting from the task at hand. Blumberg (1998) investigated developmental differences in boys' and girls' ($N = 46$ second and $N = 58$ fifth graders) performance on a popular (nonviolent) video game. After playing the game, the children were asked what they would tell someone else who has never played the game before. Results indicated that younger children's comments reflected a greater emphasis on personal feelings about the game, whereas those of older children reflected a greater emphasis on specific goals or personal standards for mastering the game. Blumberg (1998) therefore concluded that the better performance of the fifth graders combined with their greater reference to game strategies provides evidence for their ability to focus their attention on the relevant strategies in order to improve their performance.

Children's attraction to an electronic game is also influenced by their arousal. Attraction is likely to be low when the child's autonomic arousal is too high, making results poor, especially in fast-paced games that depend on good hand–eye coordination. Repeated playing tends to reduce arousal and to exercise motor skills (Grodal 2000). All in all, children are most likely to be attracted to those games that match their level of sensorimotor skills and their general and game-specific (cognitive) sophistication. Because fine-motor skills and game-specific knowledge tends to improve quickly, advanced levels of the same game offer new challenges. The ability to distance themselves emotionally is needed when children play violent electronic games. How emotional distancing develops in concert with children's growing distinction between fantasy and reality (Huston & Wright 1998) is an interesting question for future research.

CONCLUDING THOUGHTS

Which electronic games are attractive for children is based on the complex interrelationship between their goals resulting from their developmental tasks, their fluctuating emotions and moods, and their cognitive and emotional understanding of these games. Because developmental tasks change and information processing capabilities greatly improve over the primary school years, patterns of what is attractive tend to vary between different age groups. Whether escapist and habitual motives for playing electronic games generally decline toward adolescence as they do for watching TV (Bonfadelli 1981), is an open question. Perhaps computer games will substitute television as the medium sought out for escape in the future because they can be played in the solitude of the bedroom.

Not included in this review were all the other factors that are likely to influence children's attraction to and use of electronic media, such as their social class, their ethnic group, or the patterns of use in their family or subculture. How all of these external factors facilitate or constrain children's use of computer games, directly or indirectly (Roberts et al., 1999; Subrahmanyam et al., 2001), is another question for future research.

Which games appeal to children determines how much time they spend playing them. Both length of exposure, and social–cognitive understanding of electronic games is likely to influence which "effects" playing them will have on children, because in social learning theory, modeling depends on exposure to potential models, attention to their actions and cognitive representation of them, as well as practice and some form of reinforcement for them (Bandura 1994). The "effects" of electronic media products thus have to be analyzed in the context of their use (Bonfadelli, 1981; Rubin, 1994). In addition, we need to consider personality influences. Aggressive children, for example, tend to prefer violent electronic games (Gentile et al., 2004). But, in which direction is the effect to be interpreted? In other words: Does violent computer game play increase children's aggressive behavior over time? Or, do aggressive children select more violent computer games over time? These questions that have long been asked in research on the effects of television (Huesmann, Moise-Titus, Podolsky, & Eron 2003) need to be addressed in a longitudinal field study on electronic games that examines the direction of effects between preferences for violent games and aggressive behavior over time (von Salisch, Kristen, & Oppl 2005). Only such a study can give us answers on possible self-selection processes by aggressive youngsters (which in turn may increase their exposure to violent media messages and intensify their "effects").

Fortunately, playing electronic games is only one leisure-time activity among many others. Developmental tasks and emotional needs may be more directly served by playing with other children, by exercising in sports, or by working on arts and crafts projects, to name just a few activities school-age children tend to enjoy. Interacting with real instead of virtual age mates and engaging in firsthand (instead of secondhand) experiences in the outdoors—that is, playing with friends and spending time outside the house—are indeed more attractive alternatives for most children (Feierabend & Klingler 2003). Thus, electronic games always have to compete with the appeal of these activities (and those of other media) for their share of children's attention.

NOTES

[1] In the following, data from the representative sample of German 6-to-13-year-olds will be used to illustrate age trends and gender differences because this study includes first and second graders and because results are broken down by smaller age groups. When age groups are comparable, patterns seem to be rather similar in the United States and Germany. Diverging national trends will be noted in the text.

[2] We would like to thank the German Ministry of Family, Senior Citizens, Women and Youth for the financial support of the data collection of the KUHL study as well as all children, teachers, and school administrators involved.

REFERENCES

Anderson, C. A., & Dill, K. E. (2000). Video games and aggressive thoughts, feelings, and behavior in the laboratory and in life. *Journal of Personality and Social Psychology, 78,* 772–790.

Bandura, A. (1994). Social cognitive theory of mass communication. In J. Bryant & D. Zillman (Eds.), *Media effects: Advances in theory and research* (pp. 61–90). Hillsdale, NJ: Lawrence Erlbaum Associates.

Blumberg, F. C. (1998). Developmental differences at play: Children's selective attention and performance in video games. *Journal of Applied Developmental Psychology, 19,* 615–624.

Bonfadelli, H. (1981). *Die Sozialisationsperspektive in der Massenkommunikationsforschung.* Berlin: Verlag Volker Spiess.

Dietz, T. L. (1998). An examination of violence and gener role portrayals in video games: Implications for gender socialization and aggressive behavior. *Sex Roles, 38,* 425–442.

Erikson, E. (1963). *Childhood and society.* New York: Knopf.

Feierabend, S., & Klingler, W. (1999). Kinder und Medien 1999. Ergebnisse der Studie KIM 99 zur Mediennutzung von Kindern. *Media Perspektiven, 12,* 610–625.

Feierabend, S., & Klingler, W. (2001). Kinder und Medien 2000: PC/Internet gewinnen an Bedeutung. *Media Perspektiven, 14,* 345–357.

Feierabend, S., & Klingler, W. (2003). Kinder und Medien 2002. *Media Perspektiven, 16,* 278–300.

Finn, S., & Gorr, M. B. (1988). Social isolation and social support as correlates of television viewing motivations. *Communication Research, 15,* 135–158.

Flammer, A. (1996). *Entwicklungstheorien.* Bern: Huber.

Fritz, J., & Misek-Schneider, K. (1995). Computerspiele aus der Perspektive von Kindern und Jugendlichen. In J. Fritz (Ed.), *Warum Computerspiele faszinieren* (pp. 86–125). Weinheim: Juventa.

Fromme, J., & Gecius, M. (1997). Geschlechtsrollen in Video- und Computerspielen. In J. Fritz & W. Fehr (Eds.), *Handbuch Medien: Computerspiele* (pp. 121–135). Bonn: Bundeszentrale für politische Bildung.

Gentile, D. A., Lynch, P. J., Linder, J. R., & Walsh, D. A. (2004). The effects of violent video game habits on adolescent hostility, aggressive behaviors, and school performance. *Journal of Adolescence, 27,* 5–22.

Gleich, U. (2004). Medien und Gewalt. In R. Mangold, P. Vorderer, & G. Bente (Eds.), *Lehrbuch der Medienpsychologie* (pp. 587–618). Göttingen: Hogrefe.

Gottman, J. M., & Mettetal, G. (1986). Speculations about social and affective development: Friendship and acquaintanceship through adolescence. In J. Gottman & J. Parker (Eds.), *Conversations of friends. Speculations on affective development* (pp. 91–113). Cambridge, England: Cambridge University Press.

Greenfield, P. M., deWinstantley, P., Kilpatrick, H., & Kaye, D. (1996). Action video games and informal education: Effects on strategies for dividing visual attention. In P. M. Greenfield & R. R. Cocking (Eds.), *Interacting with video* (pp. 187–206). Norwood, NJ: Ablex.

Grodal, T. (2000). Video games and the pleasures of control. In D. Zillmann & P. Vorderer (Eds), *Media entertainment. The psychology of its appeal* (pp. 197–214). Mahwah, NJ: Lawrence Erlbaum Associates.

Havighurst, R. J. (1953). *Developmental tasks and education.* New York: Longman.

Huesman, L. R., Moise-Titus, J., Podolsky, C.-L., & Eron, L. D. (2003). Longitudinal relations between childrens exposure to TV violence and their aggressive and violent behavior in young adulthood: 1977–1992. *Developmental Psychology, 39,* 210–221.

Hurrelmann, B. (2002). Sozialhistorische Rahmenbedingungen von Lesekompetenz sowie soziale und personale Einflussfaktoren. In N. Groeben & B. Hurrelmann (Eds.), *Lesekompetenz* (pp. 123–149). Weinheim: Juventa.

Huston, A. C., & Wright, J. C. (1998). Mass media and children's development. In I. E. Sigel & K. A. Renninger (Eds.), *Handbook of child psychology: Vol. 4. Child psychology in practice* (pp. 999–1058). New York: Wiley.

Kafai, Y. B. (1996). Gender differences in children's construction of video games. In P. M. Greenfield & R. R. Cocking (Eds.), *Interacting with video* (pp. 39–66). Norwood, NJ: Ablex.

Kristen, A. (2004). *Realität und Virtualität: Eine Längsschnittstudie zum Zusammenhang zwischen aggressivem Verhalten und gewalthaltigen Bildschirmspielen bei Jungen.* Unpublished Dissertation, Freie Universität Berlin.

Lerner, R. M. (2002). *Concepts and theories of human development* (3rd ed.). Mahwah, NJ: Lawrence Erlbaum Associates.

Oerter, R. (1999). *Psychologie des Spiels. Ein handlungstheoretischer Ansatz.* Weinheim: Beltz Verlag.

Okagaki, K., & Frensch, P. A. (1996). Effects of video game play on measures of spatial performance: Gender effects in late adolescence. In P. M. Greenfield & R. R. Cocking (Eds.), *Interacting with video* (pp. 115–140). Norwood, NJ: Ablex.

Oppl, C. (2004.). *Lara Crofts Töchter? Eine Längsschnittstudie zu (gewalthaltigen) Bildschirmspielen und aggressiven Verhaltensweisen bei Mädchen.* Unpublished Dissertation, Freie Universität Berlin.

Rimé, B., Dozier, S., Vandenplas, C., & Declercq, M. (1996). Social sharing of emotion in children. In N. Frijda (Ed.), *Proceedings of the 9th Conference of the International Society for Research on Emotions, Toronto, Canada* (pp. 161–163). Storrs, CT: ISRE Publications.

Roberts, D. F., Foehr, U., Rideout, V., & Brodie, M. (1999, November). *Kids & media @ the new millenium. A comprehensive national analysis of children's media use.* Menlo Park, CA: Kaiser Family Foundation.

Rosenblatt, P. C., & Cunningham, M. R. (1976). Television watching and family tensions. *Journal of Marriage and the Family, 38,* 105–110.

Rubin, A. M. (1994). Media uses and effects: A uses-and-gratifications perspective. In J. Bryant & D. Zillman (Eds.), *Media effects: Advances in theory and research* (pp. 417–436). Hillsdale, NJ: Lawrence Erlbaum Associates.

Saarni, C. (1988). Children's understanding of the interpersonal consequences of dissemblance of nonverbal emotional–expressive behavior. *Journal of Nonverbal Behavior, 12,* 275–294.

Salisch, M. von. (2001). Children's emotional development: Challenges in their relationships to parents, peers, and friends. *International Journal of Behavioral Development, 25,* 310–319.

Salisch, M. von, & Bretz, H. J. (2003). Ärgerregulierung und die Nutzung von (gewalthaltigen) Bildschirmspielen bei Schulkindern. *Zeitschrift für Medienpsychologie, 15,* 122–130.

Salisch, M. v., Kristen, A., & Oppl, C. (2004). Aggressives Verhalten und (neue) Medien. In I. Seiffge-Krenke (Ed.), *Aggressionsentwicklung zwischen Normalität und Pathologie* (pp. 198–237). Göttingen: Vandenhoek & Ruprecht.

Salisch, M. V., Kristen, A., & Oppl, C. (2005). Playing violent electronic games and aggressive behavior among children: A Longitudinal study on what influences what. Ms. Under review.

Schaumburg, H. (2002). Besseres Lernen durch Computer in der Schule? Nutzungsbeispiele und Einsatzbedingungen. In L. Issing & P. Klimsa (Eds.), *Information und Lernen mit Multimedia und Internet* (pp. 335–344). Weinheim: Beltz Verlag.

Schneider; W., & Pressley, M. (1997). *Memory development between two and twenty* (2nd ed.). Mahwah, NJ: Lawrence Erlbaum Associates.

Schwab, J., & Stegmann, M. (1999). *Die Windows-Generation. Profile, Chancen und Grenzen jugendlicher Computeraneignung.* München: KoPäd-Verlag.

Seiffge-Krenke, I. (1995). *Stress, coping, and relationships in adolescence.* Mahwah, NJ: Lawrence Erlbaum Associates.

Serbin, L. A., Powlishta, K. K., & Gulko, J. (1993). The development of sex-typing in middle childhood. *Monographs of the Society for Research in Child Development,* Serial No. 232, Vol. 58. Chicago: University of Chicago Press.

Steinberg, L. (2002). *Adolescence.* New York: Knopf.

Subrahmanyam, K., & Greenfield, P. M. (1996). Effects of video game practice on spatial skills in girls and boys. In P. M. Greenfield & R. R. Cocking (Eds.), *Interacting with video* (pp. 95–114). Norwood, NJ: Ablex.

Subrahmanyam, L., Kraut, R., Greenfield, P., & Gross, E. (2001). New forms of electronic media: The impact of interactive games and the internet on cognition, socialization, and behavior. In D. G. Singer & J. L. Singer (Eds.), *Handbook of children and the media* (pp. 73–100). Thousand Oaks, CA: Sage.

Süss, D. (2000). *Kinder und Jugendliche im sich wandelnden Medienumfeld. Eine repräsentative Befragung von 6- bis 16-Jährigen und ihren Eltern in der Schweiz.* Institut für Publizistikwissenschaft und Medienforschung der Universität Zürich.

Valkenburg, P., & Cantor, J. (2000). Children's likes and dislikes of entertainment programs. In D. Zillmann & P. Vorderer (Eds), *Media entertainment. The psychology of its appeal* (pp 135–152). Mahwah, NJ: Lawrence Erlbaum Associates.

Wright, J. C., Huston, A., Vandewater, E., Bickham, D., Scantlin, R., Kotler, J. A., Caplovitz, A. G., et al. (2002). American children's use of electronic media in 1997: A national survey. In S. Calvert, A. Jordan, & R. Cocking (Eds). *Children in the digital age: Influences of electronic media on development* (pp. 35–54). Westport, CT: Praeger.

Zeman, J., & Garber, J. (1996). Display rules for anger, sadness, and pain: It depends on who is watching. *Child Development, 67,* 957–973.

12

Adolescents and the Appeal of Video Games

Arthur A. Raney, Jason K. Smith, and Kaysee Baker
Florida State University

Teens and video games.[1] If ever a match was made in media heaven, it was between these two. Several chapters in this volume have already detailed the various connections between the youth and video game markets. Our goal is to help explain why those connections are so strong. To do so, we look at a variety of factors that lead to the attraction between adolescents and video games: Some are psychological in nature, others are sociological; some are content based, others individually based; some are stable over time, others are more situationally bound. Suffice it to say, the attraction and appeal of video games is a complex phenomenon. We hope that the following pages provide some clarity.

It is difficult to discuss what features of video games attract adolescent use without an introduction to the larger body of scholarship related to media enjoyment, in particular the research on selective exposure to media and media uses and gratification. Fortunately, excellent overviews of these perspectives have been provided in chapters 13 and 16, respectively; we commend them to you.

In short, selective exposure refers to the tendency for individuals to opt for media content that is presumed to be (for the most part) consistent with their existing attitudes, beliefs, and thoughts. Support for selective exposure is widespread (e.g., Sweeny & Gruber, 1984; Vidmar & Rokeach, 1974). For entertainment theorists, selective exposure naturally extends to include media content that not only is congruous with prevailing attitudes, but that furthermore presumably brings pleasure to the viewer. Thus, it follows that fans of certain genres of programming or media experiences seek out those contents and experiences because of the presumed benefits of doing so. Video game players will prove to be no different: Those who find pleasure in playing video games—whatever the reason—seek them out in heavy doses.

The uses and gratifications approach to media effects also informs the current project. Blumler and Katz (1974) outlined the basics of the approach: Individuals have social and psychological needs, which they presume can be met through exposure to certain media content. Consequently, individuals seek out different media contents at different times depending on those needs, with the expectation that their needs will be gratified through media consumption. Use of and support for the uses and gratification approach with a variety of media content and media channels is extensive (e.g., Johnson, 1995; Perse, 1986; Vincent & Basil, 1997).

The approach thus assumes that individuals are motivated to expose themselves to (and to avoid) particular media content and experiences at different times. Of course, video games are one of those contents and game playing is one of those experiences. Generally speaking, uses and gratifications researchers contend that media use is motivated by many needs including diversion and escape, emotional release, companionship, socialization and social utility, self-exploration and awareness, value reinforcement, and surveillance (cf. Herzog, 1940; Katz, Blumler, & Gurevitch, 1974; McQuail, Blumler, & Brown, 1972). As we will see, the needs associated with and gratifications sought through video game playing are similar to those with other media.

To summarize, we assume that individuals experience psychological and sociological needs in their daily lives that drive them to certain mediated experiences with the expectations that those experiences will address and satisfy their needs. The gratifications sought through media use vary between individuals of different ages, gender, and stages in life, among other factors, as well as within individuals given situational factors such as mood, time of day, and stress. With this in mind, it can be expected that individuals might turn to video game playing to meet the various needs they experience.

One note at the outset: As a part of a larger research project, a series of focus groups were conducted with male and female video gamers ranging in age from 10 to 21 during January 2003 (Maxwell et al., 2003). Appropriate responses from the groups are included below to add depth and richness to the descriptions of the various appeals. Finally, as mentioned above, adolescents have consistently reported playing video games for various psychological and social reasons. Both of these general categories include a variety of specific motivations and appeals; however, these categories are not necessarily mutually exclusive. Our intention in using the more abstract category headings is not to create a hard-and-fast typology of appeals, but rather to provide some structure to the discussion.

PSYCHOLOGICAL APPEAL OF VIDEO GAMES

Pleasure and Enjoyment

Simply stated, teens are drawn to video games because they enjoy playing them. Exactly what makes the games enjoyable (and thus appealing) is discussed below, but we must note that enjoyment is a motivation in itself. Though the field of entertainment theory has gained momentum over the past decade, the key concept of *enjoyment* has yet to be fully explicated (Raney, 2003). While we might suggest one definition for our purposes here—the pleasure experienced from playing video games—this does little to aid our understanding of what leads to this pleasure, how and why it differs between individuals, and how it differs between different game types, systems, and playing conditions.

But, like Justice Stewart said of hard-core pornography, we know enjoyment when we see it or, better yet, when we feel it. That is, though enjoyment surely involves a great deal of cognitive activity, most argue that enjoyment is experienced as an affective response to (in our case) a media stimulus. Overwhelmingly, avid gamers report that they play video games because of the emotional rewards that they receive from doing so. For instance, Phillips and colleagues noted that nearly three fourths (72.8%) of the 11–16-year-olds they surveyed played video games for the sake of enjoyment (Phillips, Rolls, Rouse, & Griffiths, 1995). In fact, more than 6 in 10 said they enjoyed playing so much that they typically spend more time playing than they originally intend to.

Though the resulting emotional reactions are not always positive in nature—for instance, frustration can occur when challenges within a game cannot be met—positive outcomes are surely hoped for and thus can potentially lead to habitual playing. One might consider, in fact, that the emotional highs and lows that are experienced throughout the course of a game are actually part of the appeal of video gaming; scholars have noted this emotional rollercoaster as an appeal of sports spectating (e.g., Gan, Tuggle, Mitrook, Coussement, & Zillmann, 1997). In general terms, teens play video games because they expect to experience positive emotions while doing so. In fact, teens consistently report experiencing feelings of delight or joy while playing these games (Calvert & Tan, 1994; Fleming & Rickwood, 2001; Greenfield, 1984; Mehrabian & Wixen; 1986; Morlock, Yando, & Nigolean, 1985). One focus group participant noted, "If it gets you into it, like, you don't want to do anything else" (male, age 11; Maxwell et al., 2003, p. 88). These feelings of joy and delight are doubtlessly caused or enhanced by a variety of factors, some of which are player-dependent while others are content-dependent. We will continue our discussion with the former, while addressing the latter shortly.

Excitation and Arousal

Previous research indicates that video game playing can lead to physiological changes associated with increased arousal. For instance, Segal and Dietz (1991) demonstrated pronounced effects with 16- to 25-year-old males and females playing *Ms. Pac Man*, including increases in heart rate, systolic and diastolic blood pressures, and oxygen consumption. Furthermore, the energy expenditure of the player also increased, leading the researchers to conclude that effects of video game playing on metabolic rate and cardiovascular stimulation can be similar to mild-intensity exercise.

Fleming and Rickwood (2001) reported a significant increase in both heart rate and self-reported arousal among pre- and early adolescents after playing a violent video game. Calvert and Tan (1994) reported similar findings with a slightly older population: Pulse rate and self-reported arousal increased among college students (mean age 20.5 years) who played a virtual-reality game. Furthermore, Mehrabian and Wixen (1986) found that playing arcade-style video games was associated with higher levels of self-reported arousal and that the games that led to more arousal were preferred by adolescent players. Kubey and Larson (1990) found the same among adolescents males ages 9 to 15 for video game and arcade game playing, but not for females.

Similarly, video game players in previous research consistently self describe the experience as arousing, amazing, enjoyable, exciting, fun, "like a powder keg ready to explode," "off the hook," and thrilling (Calvert & Tan, 1994; Fleming & Rickwood, 2001; Maxwell et al., 2003; Mehrabian & Wixen; 1986; Morlock, Yando, & Nigolean, 1985). At times, as one focus group participant expressed, the experience can be indescribable.

The first time I played *Dragon Ball*, I don't know what, I was sitting there pushing these buttons . . . I was doing special moves . . . I'm, like, I didn't know I could do that . . . I don't know where it came from . . . the adrenaline. (male, age 15; Maxwell et al., 2003, p. 150)

Without a doubt, video games impact levels of excitation and arousal.

Mood Enhancement and Management

Research consistently demonstrates that various media content—including video games—can alter or enhance a viewer's mood. It is typically assumed that viewers are familiar with the mood-enhancing power of media (whether they completely understand it or not). Mood-management theory posits that viewers utilize specific media content to minimize the life and intensity of bad moods and maximize the life and intensity of good moods (see Oliver, 2003, for a recent summary). The principles of mood-management theory are generally supported in the literature (e.g., Biswas, Riffe, & Zillmann, 1994; Knobloch & Zillmann, 2002).

While only a handful of studies have investigated the relationship between video game playing and mood, the connection between the two seems obvious. In one of the only investigations with adolescents, Fleming and Rickwood (2001) found that the general mood of pre- and early-adolescent players became more positive after playing a violent video game, as compared to a paper-and-pencil game. This finding was consistent across genders. However, neither males nor females reported being in a more aggressive mood after the violent game as compared to the nonviolent (and paper-and-pencil) game. Also, adolescent males have reported feeling more and feeling better after playing video games, as compared to other media experiences (Kubey & Larson, 1990).

Despite the dearth of mood-related studies with adolescent video game players, it is reasonable—based on our previous discussion of arousal—to hypothesize that many teens use video games for the express purpose of excitation. Working under the assumption that excitation is primarily associated with a positive mood state, then we can expect that the games can serve a mood-management function. Therefore, one might expect that adolescents are drawn to video games at times for that purpose. However, given the lack of evidence in this area, researchers should more closely examine the relationship between video game playing and mood in adolescents.

Mastering the Challenge

Adolescents are also drawn to video games because of the challenges—and primarily the cognitive ones—the games themselves present (Greenfield, 1984; Grodal, 2000; Morlock et al., 1985). In truth, at times these challenges are more physical than others; for example, a player might be initially limited in her motor ability to produce the sequence of button and arrow punches required to accomplish a move. However, when teens report being motivated by their desire to master or "beat" the game, it is obvious that they are typically referring to the mental challenges offered by the games. One focus group participant said, "I don't like E-games, [i.e., games rated "E for Everybody"]; they're too easy to beat. T-games [i.e., games rated "T for Teens"] are harder" (male, age 12; Maxwell et al., 2003, p. 130). Another participant, commenting on the wide availability of shortcut and "cheat" information on the Internet, boasted, "I don't like cheats. I like to pick around at stuff myself" (female, age 11; Maxwell et al., 2003, p. 66).

Even with the earliest of video games, Greenfield (1984) found that a major attraction for adolescents was the presence of a challenging goal. Furthermore, Morlock et al. (1985) found that frequent video game players are motivated by a desire to beat their previous score, to continue improving, and to feel like they have mastered the game. Similarly, frequent game players preferred games that required them "to try harder to win" (p. 249). It is reasonable to expect that meeting the challenges presented in the game generates many of the emotional reactions previously discussed, but it also seems that the mental challenges themselves keep players coming back for more. As one teen noted: "I like the games where you're trying to achieve an ultimate goal ... [The best ones] keep you playing until you can beat it" (female, age 15; Maxwell et al., 2003, p. 149). Vorderer, Hartmann, and Klimmt (2003) lent further support to this claim, suggesting that video gamers expect higher enjoyment for games that involve more competitive elements (e.g., the extent to which players are required to act on sudden challenges).

Interestingly, some researchers have noted that this appeal may be experienced differently by males and females. While ultimate mastery over a game (and fellow players) may be paramount for males, females tend to be willing to play without regard to their score (e.g., Morlock et al., 1985). Many males actually reported being upset by poor scores. A variety of reasons could be offered to explain this finding, but it might simply be that many video games do not engage a desire in females to master a game for its own sake.

The thrill that comes from mastering play is surely experienced on a variety of levels (cf. Grodal, 2000; Vorderer, 2000). As mentioned, players at times simply want to beat their previous high score. They desire to defeat other players. They want to unlock the mystery of a certain section of the game. They "want to see what will if happen if I do this." They are motivated to find the bells and whistles, tricks, and shortcuts often hidden by programmers. And ultimately, many want to reach the last level, to overcome the final challenge, to defeat the enemy, to save civilization from destruction, to conquer the universe, to beat the game. It is reasonable to expect that these various levels of mastery keep players interested in playing the same game over and over. It is also reasonable, then, to expect that certain games prove too challenging. Others may not be challenging enough to motivate frequent play; one respondent reported, "I played *Spyro the Dragon* with PlayStation. I beat that game in three hours. Then I took up playing *Spyro Year of the Dragon*. I beat down that one within an hour and a half . . . so I was just sitting there" (male, age 16; Maxwell et al., 2003, p. 152). So, for many teens it seems the video games must be sufficiently but also realistically challenging to be appealing.

In terms of function, the mastery of video games may serve a similar psychological and sociological role as the mastery of threat, which has been identified with adolescent viewing of horror films. Zillmann and his colleagues have argued that horror films provide teen males with the opportunity to demonstrate (especially to their female counterparts) their conquering of fear (Zillmann, Weaver, Mundorf, & Aust, 1986; see also chap. 14 in this volume). By acting courageous, bored, or even amused in the face of danger, adolescent males prove their bravery to peers and to themselves. In this way, horror films serve as a rite of passage requiring little cost, risk, or skill. Perhaps displaying mastery over the challenges and tests in video games serves a similar psychological and sociological purpose: By mastering the various situations, regardless of how realistic, the game-playing teen displays and hones skills that mirror those necessary to master similar problems in reality. This seems quite plausible when one considers the appeal of games—especially many first-person shooter, action/adventure, and role-playing titles—in which the player assumes an adult role, in an adult world, dealing with adult situations. When a player experiences and eventually masters these situations in a video world, it seems reasonable to expect that confidence to be able to do so in reality someday would be a natural consequence.

In fact, scholars for decades have noted that play serves to prepare adolescents for adulthood by "reflect[ing] the activities and roles found in the larger adult society" (Goldstein, 1998b, p. 62; see also Parker, 1984; Roberts & Sutton-Smith, 1962). Play provides children with a safe setting in which to learn and practice skills necessary for survival in later life. This perspective has been labeled the "practice hypothesis." Some researchers contend that the function of play, which is often rough-and-tumble in nature, is as a replacement for hunting and warring skills (e.g., Humphreys & Smith, 1984). Whether this evolutionary perspective is useful is of little consequence for our discussion: The place of video game playing within the practice hypothesis is clear. In fact, video games, especially those that closely mimic reality, might actually enhance the ability of play to serve this function. Granted, many critics might contend that the life skills learned by teens playing certain mature games (e.g., kidnapping, robbing, and murdering a prostitute) are not the ones that will necessarily benefit society in the long run. Extreme examples aside, mastering the challenges and solving the intricate problems in many video games can help adolescents develop into adulthood.

IMPORTANCE OF GAME CONTENT
TO PSYCHOLOGICAL APPEALS

A natural question to ask at this point is, "What about the video game experience generates all of the reactions noted above?" Without a doubt, the content of the actual video games is a primary source. Below we discuss the ways in which various gaming features generate the psychological reactions that add to the appeal of games.

Interactive Features

Regardless of the genre, video game playing requires some combination of cognitive and motor skills. Players must quickly process new information and stimuli that the game generates in response to previous moves or decisions, while formulating and translating a response that takes the form of a keystroke, a mouse click, or a series of pressed buttons and arrow keys, with each reaction influencing all subsequent actions and outcomes. The demands placed on the players can be taxing indeed. But for many teens, the demands are welcomed because they are the byproduct of what makes the games appealing in the first place: the opportunity to interact with the content (see chap. 10 and 19 in this volume).

The terms *interactive* and *interactivity* have been applied in numerous ways to new media technologies. For our purposes, Goertz (1995, as cited in Vorderer, 2000) identified several features of new media content that promote interaction and interactivity: selectivity, malleability or modification, linearity, and senses-activating content. Goertz argued that a high number or amount of these features is associated with a high degree of interactivity. Based on this proposal, video games clearly offer almost limitless interactive features to users. For example, in describing a game one youngster giddily noted, "The aliens call at you!" (female, age 10; Maxwell et al., 2003, p. 48). Previous research reports that children and adolescents are attracted to video games because of these features. But why? We contend that the reasons lie within the characteristics identified by Goertz; to further this claim, we rely heavily on a similar discussion provided by Vorderer (2000).

The video game experience offers a tremendous amount of selectivity. Not only do players get to select what game to play from the myriad titles available, but they also select when to play, in what setting, for how long, and with whom (if anyone). Furthermore, as Vorderer noted, "the user is [also] able to select the level of difficulty, the presentation, and the outcome of the game" (2000, p. 26). Similarly, many video games offer a high degree of modification, as users are given the opportunity to create characters, change backgrounds, and select music and other audio effects. With an increasing frequency, users can personalize and tailor games to their specific intentions and interests. As previously noted, the availability of video game titles (and genres for that matter) is seemingly endless; thus, the quantity of different content is remarkable. Furthermore, though the cost of many games and systems is prohibitive to many, the availability of used, reduced-price, rentable, and downloadable games makes them one of the more available media contents for teens today. Plus, as will be noted below, youth tend to be quite open to sharing games with their friends. In fact, many of the teens in the focus groups said they regularly share and trade games with their classmates (Maxwell et al., 2003).

While most games do have a certain beginning and end, many also provide countless paths between the two. Therefore, the degree of linearity provided by current video games allows youth to chose what, how, and when the outcome will be revealed. Finally, video games contain sights and sounds that delight, arouse, and wow teen players. Many console systems offer "rumble pack" or vibrating controls that help the user actually feel the action. Still others—for instance, many of the popular dance games—use sensory pads that measure a player's steps, punches, jumps, or kicks, inputting the data directly into the game. More and more, users are offered the opportunity to utilize an increasing number (and degree) of senses while playing video games.

In fact, as Vorderer (2000) and others have noted, a game player's involvement at times seems to be so strong that the experience has been referred to as immersive (see also chap. 17 in this volume). According to Biocca and Delaney (1995), the term *immersive* refers to the degree to which a "virtual environment submerges the perceptual system of the user in computer-generated stimuli. The more the system captivates the senses and blocks out stimuli from the physical world, the more the system is considered immersive" (p. 57). Several of the focus group participants described the allure of interactivity and immersion in their own terms. One teenager disturbingly related: "It makes me feel, like when I am playing, like, *Nightfire*, that I'm actually shooting a gun. If you have, like, a scope gun, it makes it look . . . it looks like you look through it and shoot somebody" (male, age 11; Maxwell et al., 2003, p. 108).

Suspenseful Content

No doubt a primary source of the excitement and arousal noted above is the suspenseful nature of games themselves. In discussing suspenseful media in general, researchers (Vorderer, Wulff, & Friedrichsen, 1996; Zillmann, 1991, 1994) have theorized that both conflict and the level of uncertainty about the conflict resolution fuels suspense, resulting in excitatory anticipation on the part of the viewer/user. Several scholars have suggested that sports enjoyment is similarly derived (Bryant, Rockwell, & Owens, 1994; Gan et al., 1997). It is reasonable to expect that the same is the case with video games. In fact, Malone (1981) reported that males are attracted to video games to the extent to which they contain an element of uncertainty.

Most games contain a variety of suspenseful elements. For instance, the play itself—as dependent upon quick actions and reactions—is a source of uncertainty and suspense. Each command, keystroke, or joystick movement alters the game. For the first-time player, the way the game might respond to each move is a mystery. With more experience, players come to anticipate the effects that each move will have on the game environment. While suspense per-haps decreases for some players as they become more familiar with games, the final outcome is typically in question, especially in games that pit players in competition with one another. Even for seasoned players, pressing an incorrect button or delivering the correct command at the wrong time increases the uncertainty (and thus suspense) of the game. So, the suspense gen-erated by uncertainty in video games can be tremendously high, making the games extremely exciting to players.

While conflict and uncertainty—including the likelihood and fear of undesirable outcomes—play large parts in suspense, other content elements contribute as well. For in-stance, music and other sound effects play a similar role in adding to suspense as they do in motion pictures. The audio track warns the player of impending danger or conflict; the music echoes the relative successes and failures of the player. The digitized crowd cheers, the computerized gunshot victim screams, the ever-increasing pace of the music tells of imminent doom. Certain visual cues operate in a similar manner, with flashing lights, glowing weapons, simulated gauges, and indicators all enhancing the emotional reactions of players.

Finally, many sports games contain simulated commentary like that found on televised games. Previous investigations into televised sports repeatedly have shown that the accompa-nying commentary can greatly impact perceptions of suspense among viewers (Bryant, Brown, Comisky, & Zillmann; 1982; Comisky, Bryant, & Zillmann, 1977). It is quite possible that the commentary that accompanies some video games serves a similar purpose, making those games even more attractive to players.

Violent Content

Another source of the excitement experienced by teens when playing video games is the perceived violence contained in the action. In fact, the following pattern seems to emerge when talking with teens about video games: The more violent the game is, the more I like it! A cursory glance at the most popular titles (see Table 12.1) among our focus group participants will seemingly confirm this.[2] Moreover, the confessions of the focus group participants are even more telling:

> I like *007* because I like killing people. I like to act vicious. (male, age 10; Maxwell et al., 2003, p. 108)

> You can ride over people! (male, age 11; Maxwell et al., 2003, p. 118)

> My mother's friend's husband has a game. He's got two kinds of guns, and he'll hook it up to his PlayStation with his TV and play a little shooting game. And I'll be shooting people that be in a casino. (male, age 10; Maxwell et al., 2003, p. 109)

> My friend . . . had this game, and you could get like bazookas and stuff. And you could blow them up and see the liver, the brain, and then you'd see all these broken bones. (male, age 11; Maxwell et al., 2003, p. 132)

> [I like] beating up innocent people. (female, age 17; Maxwell et al., 2003, p. 149)

Adolescent (especially adolescent male) attraction to violent television and movies is well documented (see Goldstein, 1998a, for a complete summary). It seems reasonable to assume that video game violence is similarly appealing. As previously noted, teens often play video games to be aroused; research repeatedly has shown that violent media content can increase both physiological and self-reported levels of arousal. Therefore, it is not surprising that violent content would be appealing for this reason. Many teen players report feeling a "rush" (presumably in adrenaline) when they are playing violent games. A desire to experience this emotional arousal surely leads many teens to seek out violent video games.

Furthermore, some have proposed that violent content allows teens to vicariously play the role of the heroic aggressor who renders justice and brings order to a chaotic world. One focus group participant said, "*Mario Sunshine* is really neat because you collect this thing where you can fly. It hovers you, and you have a rocket and stuff, and you're trying to beat . . . like the Bad Mario or something" (male, age 10; Maxwell et al., 2003, p. 109). First-person shooter games no doubt encourage vicarious play by giving the player the visual perspective of the (often heroic) game character. While little empirical support exists for this hypothesis, it seems reasonable given the widespread popularity of games and other types of teen media that are based on a heroic or superheroic protagonist. This seems even more probable given the practice hypothesis discussed above.

Conversely, though, some research indicates that teens might be just as likely drawn to the villain as to the hero. Cantor (1998) indicated that enjoyment of violent television programming and horror films among adolescents is often associated with empathy with the aggressor or "killer" rather than with the hero or victim; perhaps video game players are motivated in a similar manner. In fact, the opportunity to empathize with and vicariously experience an aggressive perpetrator or outlaw seems to be one that many teen gamers cherish; several of the quotes above reflect this perspective. In fact, because this seemingly contradicts disposition theories of media enjoyment (cf. Raney, 2003), the phenomenon of aggressor-empathic reactions to video games might prove a fertile ground for video game research.

TABLE 12.1
Favorite Video Games of Focus Group Participants (Maxwell et al., 2003)

Title	Genre	Rating	Content Information
007 (series)	Fighting	T	Suggestive themes, violence
Basketball (various series)	Sports	E	
Baseball (various series)	Sports	E	
Big Game Hunter	Sports	T	Teen
Bounty Hunter	Action Adventure	T	Violence
DBZ	Fighting/Action Adventure	NR-T	Animated blood, violence
Dead or Alive 2	Action Adventure	T	Animated violence, suggestive themes
Dead to Rights	Fighting	M	Blood, mature sexual themes, violence
Defender	Fighting	T	Violence
Desert Storm	Action Adventure	T	Blood, violence
Donkey Kong	Arcade	E	
Dragon's Lair 3D	Action Adventure	T	Mild violence
Football (various series)	Sports	E	
Ghost Recon	Action Adventure	M	Blood and gore, violence
Gran Turismo (series)	Driving	E	
Grand Theft Auto (series)	Driving, Fighting	M	Violence, strong language, blood and gore, strong sexual content
Halo	Fighting	M	Blood and gore, violence
Harry Potter (series)	Action Adventure	E	
Hit Man (series)	Action	M	Mature
Hunter the Reckoning	Action Adventure	M	Blood and gore, violence
Just Bring It (WWE Wrestling)	Sports	T	Animated violence, mature sexual themes, mild language
Kingdom Hearts	Role-Playing	E	Violence
Kirby (series)	Action	E	
Lord of the Rings (series)	Action Adventure	T	Blood, violence
Mario (series)	Action Adventure/Sports	E	
Matt Hoffman Pro BMX 2	Sports	T	Blood, mild lyrics, violence
Max Payne	Action Adventure	M	Blood, violence
Medal of Honor (series)	Fighting	T	Violence
Mission Impossible	Fighting	T	Animated violence
Mortal Kombat (series)	Fighting	M	Blood, gore, and violence
Nascar (series)	Sports	E	
Odd World	Action	T	Comic mischief, violence
Pac Man (series)	Puzzle/Aracade	E	
Rachet & Clank	Action	T	Mild violence
Resident Evil 3	Adventure	M	Blood and gore, violence
Rocky	Sports	T	Violence
Shut Your Mouth (WWE Wrestling)	Action/Fighting	T	Mature sexual themes, strong lyrics, violence
Simpson's Road Rage	Driving	T	Mild language, suggestive themes, violence
Spiderman: The Movie	Action	T	Violence
Splinter Cell	Action	T	Blood, drug reference, violence
Spyro the Dragon (series)	Action	E	
Star Fox	Action Adventure	T	Animated blood, mild violence
Star Wars (series)	Action	T	Violence
Tekken (series)	Fighting	T	Violence
Tony Hawk (series)	Sports	E-T	Blood, mild lyrics, suggestive themes
U.S. Socom Navy Seals	Action Adventure	M	Blood, violence
Zelda	Role Playing	E	

Yet another appeal of some video games might be explained by the so-called forbidden fruit effect (Bushman & Stack, 1996; Cantor, 1998; Gunter, 2000). Simply stated, teens might be motivated to play some violent video games because their parents or guardians try to restrict their access to them. Support for the effect is mixed across the media spectrum. However, it seems reasonable to expect that the video games rating system used by the industry might provide young teens with an incentive to play games that are not age-appropriate. A quick glance at the number of games in Table 12.1 that are rated M for mature shows that teens are in fact drawn to these games.

However, if our focus group participants are typical, then the "fruit" may be a lot less "forbidden" than you might expect. One youth reported, "I *look* for mature video games" (male, age 10; Maxwell et al., 2003, p. 48). Another added, "It's like how there's kids that are allowed to go to see rated-R movies at the movie theater which you go to, but you have to be 18 or older. It does not matter on games because you can just go buy yourself two of these even if you're 8-years-old. Just go buy a game" (male, age 11; Maxwell et al., 2003, p. 90). Others reported how older siblings or friends get them whatever games they want. Of course, many reported that their parents used the ratings and content warnings to limit access to some games. One youngster further explained, "The only games I'm not allowed to get is that like you blow up people, and you can actually see all the blood, and you can see their intestines and stuff like that" (female, age 12; Maxwell et al., 2003, p. 132). Regardless of the effectiveness of the ratings and content descriptions in preventing young gamers from coming in contact with age-inappropriate materials, it seems reasonable to conclude that the forbidden pleasures (and gore) that await players in some games add to their appeal.

Other Content Factors

To be accurate, violence may not be the only allure (cf. chap. 4 and 5 in this volume). Several other types of antisocial content seem to attract teens. For instance, sexual content seems to offer a similar appeal, though teens are much less open about discussing it. One youth said, "I like the stealing and stuff. You get more money. . . . I use the cheats to get the money" (male, age 12; Maxwell et al., 2003, p. 48, 52). Another teen noted, "On *GTA* [*Grand Theft Auto*], you can just keep stealing cars, and do whatever you want to do . . . pick up prostitutes, or whatever" (male, age 15; Maxwell et al., 2003, p. 214). And, as you might imagine, a combination of antisocial content is even more enjoyable; this is from a 17-year-old female: "OK, [*Grand Theft Auto*] *Vice City* is funny. It's hysterical. I mean, the prostitutes in there, and then you see the car rocking, and then you get out and shoot them and you get your money back. That is hysterical" (female, age 17; Maxwell et al., 2003, p. 222-223).

Many teens also play certain video games and buy particular gaming systems because of the graphics. One teen noted, "If you played a football game on a PlayStation, you couldn't, like, see their [uniform] numbers. But then if you played it on an X-Box, you can see the people blink. That's cool" (male, age 11; Maxwell et al., 2003, p. 99). In fact, when given the option to play an interesting game with poor graphics or a dull game with great graphics, a great majority of the focus group participants said they would choose the latter. What is clear from the previous literature and our focus group discussions is that adolescent players prefer graphics that are realistic. Apparently, realism in this case refers both to visuals that appear lifelike and to characters that display behaviors and movements like those found in reality.

Not only are teens drawn to a game's graphics because of their realism but also because of their novelty. Greenfield (1984) identified this tendency in one of the earliest studies of video games, noting that attention-attracting power of novel visual imagery. Many critics were outraged in the mid-1990s by the release of the video game *Mortal Kombat*, which featured seven free-bleeding characters who could be made to rip the head off their defeated foes. No doubt the violent content alone led many teens to play the game; but, it is also reasonable to think that the novelty of seeing the digitized blood was also a big draw. In fact, many teens seem to be drawn to some video games by the "what will they think of next" allure.

Finally, adolescents also seem to be drawn to video games that feature the familiar characters from other media content, such as motion pictures, television shows, sports, and music. Specifically, one young player stated, "I like the cartoon games, like TV based" (male, age 10; Maxwell et al., 2003, p. 109). This tendency to be drawn to video games that feature favorite characters and athletes is to be expected given the strong affective dispositions that media consumers form toward those character and athletes. The various disposition-based theories of enjoyment (see Raney, 2003 for a recent summary) would reasonably foreshadow this appeal and could help explain why great enjoyment is derived from games containing favored characters.

SOCIAL APPEAL OF VIDEO GAMES

Video gaming is quite often a social experience. This is not to say that teens do not play video games by themselves; surely, they do. As one player noted, "I play by myself because you get a longer time" (male, age 12; Maxwell et al., 2003, p. 85). Like few other media experiences today, though, video game playing invites social interaction. In some games, like those based on sports, head-to-head competition drives the interaction. In others—for example, military-based games—the need for cooperation encourages a social environment. As a result, games specifically designed for multiple players are often the favorites of teens; many teen boys reported enjoying video game playing more when they play with friends (Kubey & Larson, 1990). In fact, several focus group participants noted that they only seek out games that offer multiplayer capabilities (Maxwell et al., 2003). They spend hard-earned dollars and birthday wishes on securing additional peripherals and plug-ins that allow for or enhance multiplayer action. At times, though, even single-player games promote social interaction as teens huddle on a crowded sofa vying to top one another's high score or extreme moves. Regardless of the specific scenario, it is obvious that video gaming can be, and quite often is, a communal experience (Vorderer & Ritterfeld, 2003).

As one might imagine, these social interactions involve siblings and/or friends gathering together in someone's home after school or on the weekends for hours of playing. In fact, one 11-year-old reported: "I'll go to my friend's house and play his or play with friends [at my house] or something like that as long I want, all day [on Saturday]" (male, age 11; Maxwell et al., 2003, p. 55). Others related a similar Friday night routine: Go to the local movie-rental store, rent as many games as the store will allow, and then stay up playing as late as possible.

Playing video games in a group setting is appealing for a variety of reasons. First, the companionship offered is something that most teens—in fact, humans of any age—seek out. The game-playing environment often engenders a teamlike camaraderie similar to sports. More generally speaking, game-playing ability and knowledge seems to operate as a criterion for group membership on a variety of levels. In some respects, to *be a teen today* is to play or at least have some working knowledge of popular games, similar to music, movies, television, and fashion. The teens in the focus group said that they usually buy the games that are most talked about. One stated, "If I hear a bunch of people talking about it and they say it's good, then I'll just go get it" (male, age 15; Maxwell et al., 2003, p. 219).

Therefore, video game knowledge and ability serve as a form of social capital for many teens. Being the first in the school to have a certain game or system or conquering a game faster than anyone else increases your value among certain peers. Another major source of social capital is knowledge of the latest cheats and where (particularly online) to get them. Not unexpectedly, video gaming and Internet experience are often highly correlated, especially among adolescent males (cf. Cassell & Jenkins, 1998). As a result, gaming masters (à la Lazarsfeld's opinion leaders) are easily identifiable and are given their due respect. Game ownership, knowledge, and mastery become points of contact for peers; they create the ties that bind some individuals together in friendship. Furthermore, hard-core gamers—for instance, loyal players of massively multiplayer online games (MMOGs; see chap. 6 in this volume) such as *Everquest* and *Star Wars Galaxies*—constitute a subgroup in the adolescent society, similar to the Dungeons-and-Dragons loyalists of the 1980s. Games, in these situations, become a primary purpose for social relations. Gaming itself defines the subgroup; the games are the *raison d'etre* for the friendships.

Second, playing games together allows many teens—in some cases, those who lack exceptional athletic or academic prowess—to compete against and defeat peers. While it is beyond the scope of this chapter to explore the psychological necessity for such domination, teens consistently note the joy derived from conquering a friend or the game itself. Mastery of a game can serve as a source of pride and self-esteem for young players. As with many adolescent activities, the players (often the male ones) do not waste the opportunity to lord their victory over their friends: "With the multiplayer [games], I mean, once you finished a game, you got to play again just to talk trash to each other" (male, age 16; Maxwell et al., 2003, p. 153). The psychological and sociological role of this experience, while difficult to explain, cannot be denied. Video gaming offers these experiences to players of various heights, strengths, body types, IQ levels, and income brackets. On some level, these social experiences surely serve as an appealing byproduct of game playing.

Finally, many individuals enjoy playing video games in groups because they simply like watching others play. Part of this appeal is purely functional: A great deal of hands-on education happens as players share tips, moves, shortcuts, and cheats with spectators as they play. Others, though, derive enjoyment from watching video game playing in much the way that they do viewing television programs or motion pictures. The games are suspenseful, with the outcomes always in doubt. Sitting with others as they play seems to also make the experience (at least appear) more interactive. For example, the rooting for or attempts at warning a player of impending doom appear to be much more efficacious (because perhaps they actually are) than doing so while watching a suspenseful program or movie. The video game spectator can actually rejoice with the player in victory or wince with them in defeat; they are truly a part of the entertainment experience.

Some females reported that watching boyfriends or love interests play games has become a regular part of their teen relationships. However, as one 17-year-old noted, this practice may not be mutually satisfying: "I got [my boyfriend the game] *Halo*. Every Friday night, [he and his friends] speak like they're just gods. They hook all the X-Boxes up, and they play *Halo* for hours. It's crazy" (female, age 17; Maxwell et al., 2003, p. 148). Indeed, spectatorship, whether for educational, entertainment, or romantic purposes, adds to the appeal of video games.

While the social appeal of video games is present for adolescents, a recent study suggests that the appeal may increase over time. In a study of the online game *Everquest*, Griffiths and colleagues (2004) reported that adult players tended to favor the social aspects of the game more than their teen counterparts. Of course, these findings may simply apply to the game in question or to online (as opposed to console) games. Whatever the case, it is apparent that many adolescents enjoy and are motivated by the social aspects of video gaming.

CONCLUDING THOUGHTS

In this chapter, we have attempted to outline and explain the appeal of video games to youth today. Indeed, the relationship between the two is highly complex; many important factors must me considered. Simply stated: Video games satisfy psychological and sociological needs in adolescents. We all utilize media for specific purposes in our lives. Teens are no different; they use music, television, films, magazines, video games, and more. Video games appeal to teens on a variety of levels—emotional, physiological, cognitive, behavioral, and social—at different times and with different results. In general, however, video games play a tremendous role in the socialization and lives of many teens today. From our perspective, researchers must continue to acknowledge and confront head on the complexities inherent in the relationship between teens and video games as the union between the two promises to only grow stronger in time.

NOTES

[1] As this volume attests, video games are quite varied from arcade style to at-home consoles to desktop computer software. While teens no doubt play each of these types, the focus of this chapter is at-home gaming systems such as Sony's PlayStation series, Microsoft's X-Box, and Nintendo's GameCube. However, we contend that the attractions discussed herein surely apply to all types of video games.

[2] We also want to note that even though violent video games garner much interest by parents, educators, and government official, many, many popular video games contain no violence. To assume that the entire (and perhaps even a sizable portion of the) gamer population is constantly bombarded with violent images would be incorrect.

REFERENCES

Biocca, F., & Delaney, B. (1995). Immersive virtual reality technology. In F. Biocca & M. R. Levy (Eds.), *Communication in the age of virtual reality* (pp. 57–124). Hillsdale, NJ: Lawrence Erlbaum Associates.

Biswas, R., Riffe, D., & Zillmann, D. (1994). Mood influence on the appeal of bad news. *Journalism Quarterly, 71*, 689–696.

Blumler, J. G., & Katz, E. (Eds.). (1974). *The uses of mass communication: Current perspectives on gratifications research*. Beverley Hills, CA: Sage.

Bryant, J., Brown, D., Comisky, P., & Zillmann, D. (1982). Sports and spectators: Commentary and appreciation. *Journal of Communication, 32*(1), 109–119.

Bryant, J., Rockwell, S. C., & Owens, J. W. (1994). "Buzzer beaters" and "barn burners": The effects on enjoyment of watching the game go "down to the wire." *Journal of Sport & Social Issues, 18*, 326–339.

Bushman, B. J., & Stack, A. D. (1996). Forbidden fruit versus tainted fruit: Effects of warning labels on attraction to television violence. *Journal of Experimental Psychology: Applied, 2*(3), 207–226.

Calvert, S. L., & Tan, S. (1994). Impact of virtual reality on young adults' physiological arousal and aggressive thoughts: Interaction vs. observation. *Journal of Applied Developmental Psychology, 15*, 125–139.

Cantor, J. (1998). Children's attraction to violent television programming. In J. H. Goldstein (Ed.), Why we watch: The attractions of violent entertainment (pp. 88–115). New York: Oxford University Press.

Cassell, J., & Jenkins, H. (1998). *From Barbie to Mortal Kombat: Gender and computer games*. Cambridge, MA: MIT Press.

Comisky, P., Bryant, J., & Zillmann, D. (1977). Commentary as a substitute for action. *Journal of Communication, 27*(3), 150–153.

Fleming, M. J., & Rickwood, D. J. (2001). Effects of violent versus nonviolent video games on children's arousal, aggressive mood, and positive mood. *Journal of Applied Social Psychology, 31*, 2047—2071.

Gan, S-L., Tuggle, C. A., Mitrook, M. A., Coussement, S. H., & Zillmann, D. (1997). The thrill of a close game: Who enjoys it and who doesn't? *Journal of Sport & Social Issues, 21*, 53–64.

Goertz, L. (1995). Wie interaktiv sind neue Medien? Auf dem Weg zu einer Definition von Interaktivität. *Rundfunk und Fernsehen, 43*, 477–493.

Goldstein, J. H. (1998a). *Why we watch: The attractions of violent entertainment*. New York: Oxford University Press.

Goldstein, J. H. (1998b). Immortal kombat: War toys and violent video games. In J. H. Goldstein (Ed.), *Why we watch: The attractions of violent entertainment* (pp. 53–68). New York: Oxford University Press.

Greenfield, P. M. (1984). *Mind and media: The effects of television, video games, and computers.* Cambridge, MA: Harvard University Press.

Griffiths, M. D., Davies, M. N. O., & Chappell, D. (2004). Online computer gaming: A comparison of adolescent and adult gamers. *Journal of Adolescence, 27*, 87–96.

Grodal, T. (2000). Video games and the pleasures of control. In D. Zillmann & P. Vorderer (Eds.), *Media entertainment: The psychology of its appeal* (pp. 197–214). Mahwah, NJ: Lawrence Erlbaum Associates.

Gunter, B. (2000). Avoiding unsavoury television. *Psychologist, 13*(4), 194–199.

Herzog, H. (1940). Professor quiz: A gratification study. In P. Lazarsfeld (Ed.), *Radio and the printed page* (pp. 64–93). New York: Duell, Sloan, & Pearce.

Humphreys, A. P., & Smith, P. K. (1984). Rough-and-tumble in preschool and playground. In P. K. Smith (Ed.), *Play in animals and humans* (pp. 241–266). Oxford, England: Blackwell.

Johnson, D. D. (1995). Adolescents' motivations for viewing graphic horror. *Human Communication Research, 21*, 522–552.

Katz, E., Blumler, J. G., & Gurevitch, M. (1974). Utilization of mass communication by the individual. In J. G. Blumler & E. Katz (Eds.), *The uses of mass communication: Current perspectives on gratifications research* (pp. 19–32). Beverley Hills, CA: Sage.

Knobloch, S., & Zillmann, D. (2002). Mood management via the digital jukebox. *Journal of Communication, 52*, 351–366.

Kubey, R., & Larson, R. (1990). The use and experience of the new video media among children and young adolescents. *Communication Research, 17*, 107–130.

Malone, T. W. (1981). Toward a theory of intrinsically motivating instruction. *Cognitive Science, 5,* 333–369.

Maxwell, M., Raney, A. A., Rhodes, N., Dinu, L., Fosu, I., Harriss, C., Zhu, H., & Bryant, J. (2003, February). *Movie Gallery focus groups: Birmingham, AL. Report to Movie Gallery.* Tuscaloosa, AL: Institute for Communication Research.

McQuail, D., Blumler, J. G., & Brown, J. R. (1972). The television audience: A revised perspective. In D. McQuail (Ed.), *Sociology of mass communications* (pp. 135–165). Middlesex, England: Penguin.

Mehrabian, A., & Wixen, W. J. (1986). Preferences for individual video games as a function of their emotional effects on players. *Journal of Applied Social Psychology, 16,* 3–15.

Morlock, H., Yando, T., & Nigolean, K. (1985). Motivation of video game players. *Psychological Reports, 57,* 247–250.

Oliver, M. B. (2003). Mood management and selective exposure. In J. Bryant, J. Cantor, & D. Roskos-Ewoldsen (Eds.), *Communication and emotion: Essays in honor of Dolf Zillmann* (pp. 85–106). Mahwah, NJ: Lawrence Erlbaum Associates.

Parker, S. T. (1984). Playing for keeps: An evolutionary perspective on human games. In P. K. Smith (Ed.), *Play in animals and humans* (pp. 271–293). Oxford, England: Blackwell.

Perse, E. M. (1986). Soap opera viewing patterns of college students and cultivation. *Journal of Broadcasting & Electronic Media, 30*, 175–193.

Phillips, C. A., Rolls, S., Rouse, A., & Griffiths, M. D. (1995). Home video game playing in schoolchildren: A study of incidence and patterns of play. *Journal of Adolescence, 18*, 687–691.

Raney, A. A. (2003). The enjoyment of sports spectatorship. In J. Bryant, J. Cantor, & D. Roskos-Ewoldsen (Eds.), *Communication and emotion: Essays in honor of Dolf Zillmann* (pp. 397–416). Mahwah, NJ: Lawrence Erlbaum Associates.

Roberts, J. M., & Sutton-Smith, B. (1962). Child training and game involvement. *Ethnology, 1*, 166–185.

Segal, K. R., & Dietz, W. H. (1991). Physiologic response to playing a video game. *American Journal of Diseases of Children, 145*, 1034–1036.

Sweeny, P. D., & Gruber, K. L. (1984). Selective exposure: Voter information preferences and the Watergate affair. *Journal of Personality and Social Psychology, 46*, 1208–1221.

Vidmar, N., & Rokeach, M. (1974). Archie Bunker's bigotry: A study in selective perception and exposure. *Journal of Communication, 24*, 36–47.

Vincent, R. C., & Basil, M. D. (1997). College students' news gratification, media use and current events knowledge. *Journal of Broadcasting & Electronic Media, 41*, 380–392.

Vorderer, P. (2000). Interactive entertainment and beyond. In D. Zillmann & P. Vorderer (Eds.), *Media entertainment: The psychology of its appeal* (pp. 21–36). Mahwah, NJ: Lawrence Erlbaum Associates.

Vorderer, P., Hartmann, T., & Klimmt, C. (2003). Explaining the enjoyment of playing video games: The role of competition. Proceedings of the 2nd International Conference in Entertainment Computing (ICEC). Retrieved August 19, 2004, from http://delivery.acm.org/10.1145/960000/958735/p1-vorderer.pdf?key1=958735&key2 =8618543901&coll=GUIDE&dl=GUIDE&CFID=26256417&CFTOKEN=30782271.

Vorderer, P., & Ritterfeld, U. (2003). Children's future programming and media use between entertainment and education. In E. L. Palmer & B. Young (Eds.), *The faces of televisual media: Teaching, violence, selling to children* (2nd ed., pp. 241–262). Mahwah, NJ: Lawrence Erlbaum Associates.

Vorderer, P., Wulff, H. J., & Friedrichsen, M. (Eds.). (1996). *Suspense: Conceptualizations, theoretical analyses, and empirical explorations.* Mahwah, NJ: Lawrence Erlbaum Associates.

Zillmann, D. (1991). Television viewing and physiological arousal. In J. Bryant & D. Zillmann (Eds.), *Responding to the screen: Reception and reaction processes* (pp. 103–133). Hillsdale, NJ: Lawrence Erlbaum Associates.

Zillmann, D. (1994). Mechanisms of emotional involvement with drama. *Poetics, 23,* 33–51.

Zillmann, D., Weaver, J. B., Mundorf, N., & Aust, C. F. (1986). Effects of an opposite-gender companion's affect to horror on distress, delight, and attraction. *Journal of Personality & Social Psychology, 51,* 586–594.

13

Selective Exposure to Video Games

Jennings Bryant
University of Alabama

John Davies
University of North Florida

In retrospect, the first video games to make their way into homes were as audiovisually bland and cognitively undemanding as most other pioneering forms of electronic media have been. *Pong*, an electronic form of table tennis, was the first popular video game. It was little more than two rectangular bars batting a square "ball" back and forth across the television screen. Even so, players apparently were fascinated with the simple game, and so with little fanfare began what was to become a worldwide adoption of increasingly sophisticated video games, which in their current iteration are fast becoming an almost universally popular pastime. According to the Entertainment Software Association (2004), sales of computer and video games reached $7 billion in 2003. In 1997, boys 12 years and younger in the United States spent slightly greater than 6 hours per week on average playing video games, and 12-year-old girls averaged nearly $5\frac{1}{2}$ hours per week playing video games. A half decade later, American eighth- and ninth-grade boys averaged 13 hours of game play per week, and girls averaged nearly 6 hours per week (Gentile, Lynch, Linder, & Walsh, 2004). A British survey found that one in five adolescents aged 12 to 16 could be classified as *dependent* on video games, and another 6.8% were *addicted* to such games, even employing the relatively stringent DSM–III criteria for dependency and addiction (Griffiths & Hunt, 1998).

Electronic games are not just the domain of children and adolescents either. The Entertainment Software Association (2004) reported that the so-called average gamer is 29 years old, and 17% of all gamers are 50 years or older. Clearly, video games are a popular way for adults to spend time. In fact, nearly half (46.6%) of the adult players of a massively multiplayer online role-playing game (MMORPG) reported that they sacrificed sleep, work, education, or socializing with others to play their games (Griffiths, Davies, & Chappell, 2004; see Chan & Vorderer, chap. 6, this volume). These normative numbers were even higher among adolescent players.

Given the popularity of video games coupled with a growing body of research detailing negative effects on some gamers of playing some types of games (e.g., Anderson, 2002;

Anderson & Bushman, 2001; Gentile et al., 2004), understanding how and why people choose to play video games becomes a crucial question. Unfortunately, the empirical literature to date offers few answers to this question when it is narrowly defined. Research on the uses and enjoyment of video games in general is relatively limited, and research on selective exposure to video games is virtually nonexistent. Fortunately, however, research on selective exposure to other media has been abundant of late. In fact, a recent content analysis revealed that selective exposure was the 10th most frequently employed theory in mainstream mass communication journals between 2001 and 2004 (Bryant & Miron, 2004). Moreover, selective exposure theory and much of its attendant research on media other than electronic games would appear to have direct application to electronic gaming, despite the unique features that differentiate video games from other media.

In this chapter, we examine the selective exposure literature to show how this research has application for video games as well as situations in which selective exposure theory seems ill-equipped to predict or explain game playing. We identify some of the unique challenges posed by video games for selective exposure researchers, and finally we outline what needs to be done before we are able to reach solid conclusions based on robust empirical evidence about the nature of selective exposure to video games.

WHY PLAY VIDEO GAMES?

For the uninitiated who have somehow managed not to play electronic games (yet!), a basic question about video games is why do people play? Parents may wonder what their child finds so appealing about navigating an electronic character through virtual space. They may be concerned that their child is attracted to the violence in many games or may even seem addicted to game play. Or, those in intimate relationships may be concerned that their partner seems to express more interest in playing games than in him or her.

For avid game players, the answer is clear. Common reasons for playing include enjoyment (Phillips, Rolls, Rouse, & Griffiths, 1995), fun and challenge (Griffiths & Hunt, 1995; McClure & Mears, 1984), and social aspects or opportunities (Colwell & Payne, 2000; Griffiths, Davies, & Chappell, 2004; Kim, Park, Dong, Hak, & Ho, 2002; Selnow, 1984). Furthermore, players and parents of players have more favorable attitudes toward video games than do nonplayers or their parents (Sneed & Runco, 1992). Gamers obviously find the experience of playing enjoyable, but that is only a surface explanation that is unlikely to satisfy a concerned public, and it certainly does not sate the epistemic curiosity of theoretically oriented scholars.

One compelling explanation for the appeal of video games is based on the effects they have on players' emotions. Grodal (2000) argued that the interactive nature and unique features of video games allows users to achieve optimal levels of arousal. In essence, games are tools with which to regulate emotion.

SELECTIVE EXPOSURE THEORY: AFFECT-DEPENDENT THEORY OF STIMULUS ARRANGEMENT AND MOOD MANAGEMENT THEORY

A large body of literature supports Grodal's (2000) contention that emotional regulation is a basic motive of media use. Much of what we know about mood management through media is derived from research on Zillmann and Bryant's (1985) affect-dependent theory of stimulus arrangement (see also Oliver, 2003; Zillmann, 1988, 2000). The theory is based on two basic premises. First, the theory assumes that people are motivated to minimize exposure to negative, noxious, or aversive stimuli on the one hand, and second, are motivated to maximize exposure to positive, pleasurable stimuli on the other. The theory further maintains that individuals will try to arrange external stimuli to maximize chances of achieving these goals. The ease with which media can be used to attain the hedonic objective means that individuals often prefer media as a means to this end. Video games, according to Grodal's (2000) speculations, are particularly potent mood managers and should therefore be an especially attractive medium.

A brief examination of the four key elements of using media as a mood manager should make this argument clear. According to the affect-dependent theory of stimulus arrangement, these four elements are excitatory homeostasis, intervention potential of a message, message-behavioral affinity, and hedonic valence.

Excitatory Homeostasis

Excitatory homeostasis refers to the tendency of individuals to choose entertainment to achieve an optimal level of arousal (i.e., if the prevailing affective state is considered aversive, understimulated persons tend to choose arousing material; overstimulated persons to choose calming fare). Bryant and Zillmann (1984) demonstrated that bored individuals selected arousing media messages and avoided relaxing fare when given the opportunity to do so. Individuals who experienced stress chose to view relaxing programs more than did bored individuals. Mastro, Eastin, and Tamborini (2002) extended Bryant and Zillmann's (1984) findings to the Internet. In a pretest, they observed that rapidly surfing a Web site was more likely to be associated with stress than was slowly surfing a Web site. In contrast, participants found the experience of slowly surfing a Web site to be more boring than rapidly surfing the Web site. In the experiment proper, they found the Web-surfing behaviors of stressed individuals were associated with a slow-surfing pattern (which produced boredom in the pretest), and bored individuals tended to surf the Web more rapidly (which was associated more closely with stress than slow surfing). In another study, researchers used a series of surveys and studied selective exposure to television in the context of stressful life events (Anderson, Collins, Schmitt, & Jacobvitz, 1996). Overall, the studies found support for mood management by television but revealed some gender differences. As predicted, stressed men paid more attention to television than did men who were under less stress, but stressed women and less-stressed women paid equal attention to television. Stressful life events were associated with TV addiction for women, but not for men. Stressed men and stressed women viewed more comedy than did their bored peers, as was expected, but stressed men also watched more violent/action/horror programs than did men under less stress.

Although to our knowledge no study has explicitly investigated behaviorally assessed preferences for playing video games as a function of an initial mood state, the second most frequently occurring self-reported reason for playing a favorite computer game in one study was that the game is exciting (Griffiths & Hunt, 1995). Mehrabian and Wixen (1986) examined emotional reactions to video games on three dimensions: arousal, valence, and dominance (Study 2). They reported that arousal was a significant predictor of preferences for games among male and female patrons of an arcade. They also asked participants to mentally play a game that they had played before and think about how the game made them feel (Study 1). Games that made participants feel more dominant and more aroused were preferred. That is, preference for games was predicted by feelings of dominance and arousal. These studies suggest that the stimulating quality of video games is an important component in their selection.

Further support for the notion that players use video games as a means of managing mood was found in a study of computer game addiction, which utilized a scale for pathological gambling adapted from DSM–3–R criteria (Griffiths & Hunt, 1998). Computer gamers who were classified as *dependent* on the medium were more likely than others to report either being in a bad mood or being highly excited before, after, and during play. They also were more likely to report being in a good mood prior to and following play. It is unclear from these findings whether initial mood state fostered dependency on video games, or if the emotional effects of the games was the cause of dependency (or if there was even a causal connection between mood and dependency). It is also curious that dependent players associated stronger good and bad moods with playing. One possibility is that video games are more important to people whose emotional experiences are subjectively stronger than others, because game playing represents a potentially powerful means to alter negative moods and achieve positive feelings. Affect intensity is a characteristic associated with stronger perceptions of and reactions to emotional stimuli (Larsen & Diener, 1987). Individuals who score higher on measures of affect intensity typically feel emotions more strongly, independent of whether the affect is bad or good. Perhaps the emotions of more dependent players are particularly salient; therefore, high emoters experience a greater need to manage those feelings, and then they experience more intense mood states as a result of playing.

Griffiths and Dancaster (1995) predicted and supported a similar hypothesis. They found that individuals with Type A personality, which is associated with greater reactivity to psychologically demanding situations than to physically demanding ones, experienced greater physiological arousal when playing a computer game than did individuals with Type B personalities. They also noted that twice as many Type A individuals as Type B participants reported being addicted to video games, although this difference was not statistically significant. Griffiths and Hunt (1998) also found that dependent players were also more likely than others to say that they started playing games because there was nothing else to do. This finding is perfectly in line with earlier research in mood management. For instance, Bryant and Zillmann (1984) observed that individuals in a state of boredom selected arousing stimuli more than they chose calming materials, when given a constellation of choices.

Taken as a whole, the research on video games suggests that computer gamers play their games because of the arousal-inducing properties of the games. Research has shown that the content of many games is likely to elicit arousal. For example, Dietz's (1998) content analysis of video games found that almost 80% of the sampled games contained violent content. Of these, 21% featured violence toward women. In addition, the structural features of video games are likely to elicit arousal. Moving images have been shown to increase self-reported and measured physiological arousal relative to still images (Detenber & Simons, 1998) and video games routinely feature animation of some sort, even in electronic versions of popular board games such as *Monopoly*.

Intervention Potential

The cognitive intervention potential of a message refers to the ability of a message to engage or absorb an aroused individual's attention or information-processing resources. Clearly, the more absorbing and engaging a media message, the greater the attraction it holds. Research has demonstrated that engaging messages elicit selective exposure behavior. For example, readers choose to read news articles that are accompanied by threatening images more frequently and for longer periods of time than articles accompanied by innocuous images (Knobloch, Hastall, Zillmann, & Callison, 2003; Zillmann, Knobloch, & Yu, 2001). Furthermore, highly engaging messages supposedly disrupt cognition rehearsal related to a particular affective state and tend to diminish the state's perceived intensity. Bryant and Zillmann (1977) reported that the intervention potential of a message moderated research participants' retaliation against a person who annoyed them. That is, highly absorbing messages reduced retaliation, whereas minimally absorbing messages failed to reduce retaliation.

Video games would be expected to have a high intervention potential because they are highly interactive, thereby demanding tactile as well as cognitive engagement (e.g., Vorderer, 2000). Intuitively, we would expect the cognitive intervention potential of a game to have more substantial effects on someone who is actually interacting with the game than someone who is merely observing the action. Calvert and Tan (1994) observed such a phenomenon in undergraduate students who either participated in an aggressive virtual reality game or viewed the game on video monitors. Participants who actually played the game had significantly higher heart rates than did their peers who either watched the game or a control group who engaged in exercises that incorporated the motions used by game players. Players also had more aggressive thoughts than did observers or participants in the control group. Interacting with the game apparently increased the likelihood that participants would not only feel excited, but also that they would be thinking about constructs associated with aggression.

The ability to interrupt cognitive rehearsal is a crucial element of intervention potential. Media with relatively greater intervention potential should be more attractive to individuals seeking to escape negative thoughts and moods associated with, or even partially derived from, those thoughts. One study found that consuming media messages with a high intervention potential not only impeded ability to recall a prior negative experience, but also influenced postconsumption mood (Davies, 2004). Video games are programmed to demand input from the user (see Klimmt & Hartmann, chap. 10, this volume) and therefore should have a high intervention potential relative to less demanding media, although no study seems to have specifically tested this proposition.

The interactive nature of video games essentially demands the user's attention if the gamer is to perform well. Survival (or the attainment of some goal) in most games, especially action games, depends on the player's response to various stimuli. Failure to attend to these stimuli usually results in some negative consequence. Thus, players must exercise constant vigilance or risk negative outcomes—a loss of points, failure to solve a puzzle, or even computer-simulated death. Attention is typically (or eventually, after repeated playing) rewarded with an avoidance of these outcomes and is therefore negatively reinforced by the structural features of the game. This continuous demand on attention comes at a cost, however. Eventually, players experience fatigue, especially if the pace of the game is frenetic. Indeed, the participants in Nelson and Carlson's (1985) experiment reported that extended playing of arcade games resulted in fatigue and discomfort.

Grodal (2000) explained that most action video games are designed so that players can navigate through either player-generated zones, where users can rest from the demands on attention, or computer-generated zones, where players cannot prevent the onset of stimuli and their consequences. In addition, he noted that players can further control the schedule of reinforcing stimuli through the use of pause features, saving intermediate results, and playing at different levels of difficulty. Thus, players can customize their gaming experience and control the degree of intervention potential offered by the game.

Message-Behavioral Affinity

Message-behavioral affinity refers to the degree of similarity between communication content and affective state. Generally, messages that have a high degree of similarity with a person's affective state have less potential to alter or diminish moods. This explanation accounts for the finding by Zillmann, Hezel, and Medoff (1980) that participants who were insulted in an experiment tended to avoid hostile comedy when given the opportunity to view television. Medoff's (1982) results provided strong support for the prediction that annoyed people would be motivated to avoid comedy programs that remind them of their emotional state. Bryant and Zillmann's (1977) investigation had initially demonstrated this phenomenon in an investigation that focused on the cognitive intervention potential of media messages. Wakshlag, Vial, and Tamborini (1983) manipulated apprehension about crime by showing film viewers either a graphic crime documentary film or a control film. When given the choice to view another film, apprehensive viewers chose films that strongly portrayed the fulfillment of justice.

Experimental studies on message-behavioral affinity and video games are nonexistent, to our knowledge. However, several studies have identified negative correlations between self-esteem and computer game use in adolescent samples (Colwell, Grady, & Rhaiti, 1995; Colwell & Payne, 2000; Dominick, 1984; Funk & Buchman, 1996; Roe & Muijs, 1998). One study failed to find a link between self-esteem and game playing in adolescents, but the authors speculated that the teachers who rated the students on self-esteem might have misunderstood the researchers' operationalization of self-esteem (Fling et al., 1992). Slater (2003) also found that alienation from peers predicted use of violent video games among eighth graders. Although that study did not directly examine self-esteem, it is probable that isolation from peer groups would result in a substantial drop in self-esteem, given the importance of peer friendships in that age group. Unfortunately, we cannot make causal inferences as to the nature of self-esteem and computer game use from these investigations, because all were correlational in nature. It may be that video games cause reductions in self-esteem, or teens low in self-esteem may prefer video games, or the effect may be bi-directional.

Nevertheless, the negative correlation between self-esteem and playing video games is consistent with the research on message-behavioral affinity. Video games typically do not present a narrative in the same way that we are used to experiencing it in other media, such as television or movies. Rather, a narrative in a computer game may be used to set up a scenario in which the player then dictates the outcome depending on his or her skill, reactions, and choices. In a computer game, the unintended and implicit story line is "my actions result in rewarding effects." Klimmt and Hartmann (this volume) point out that a computer game is typically structured so that outcomes depend to a large degree on user inputs. Generally, in virtual time and space the relationship between an input and its outcome is overstated, as compared to inputs and outcomes in the real world, and users may experience an exaggerated sense of efficacy. Thus, there is an element of control in video games that is not found in the stories in films, books, or television programs—and certainly games offer players a degree of control that is rare in real life. It is possible that this element of control is one of the implicit messages that players learn from the narrative of video games. It is this same sense of control that may be lacking in the lives of heavy users with low self-esteem.

Of course, this link between low self-esteem and preference for video games is speculative, but it is consistent with theory. Zillmann (1988) pointed out that bad moods can be diminished by consuming media that are both highly involving and have low behavioral affinity with an initial bad-mood state. For persons with low self-esteem, video games apparently meet both criteria. Indeed, one of the motivations for playing video games in one survey was to cheer oneself up (Fling et al., 1992). Another survey found that frequent players were more likely to play to escape pressures than were gamers who played less frequently (McClure & Mears, 1984). Perhaps both surveys revealed different aspects of the motivation to play. That is, it seems reasonable that players turn to video games to not only *forget* bad moods, but also to *alter* them as well.

Another finding from the computer game literature that is consistent with the notion of a link with self-esteem and selective exposure to video games is the effects games have on moods. Fling et al. (1992) observed that adolescents reported that their postgame feelings depended on success or failure in a game. Although this finding should not be too surprising—after all, it is in line with findings from the enjoyment of participating in or watching sporting evens, which are also games (e.g., Bryant, Raney, & Zillmann, 2002)—it does suggest that players are motivated by a desire to achieve a positive outcome, rather than merely deriving enjoyment from the experience of playing irrespective of the conclusion. Therefore, for players with low self-esteem the outcome of a game must be successful to be positive; failure results in a bad mood. Of course, without experimental testing, these propositions remain speculative.

Very little research has examined the positive aspects of video games. In contrast, a great deal of research focuses on the negative aspects of computer game messages, particularly in terms of violent content (see Buckley & Anderson, chap. 23, this volume). This research has consistently found connections, correlational and experimental, between violent content and aggressive thoughts, feelings, and behaviors (e.g., Anderson & Dill, 2000; Gentile et al., 2004; Krahé & Möller, 2004). Much of this research supports the general affective aggression model proposed by Anderson and his colleagues (Anderson, Deuser, & DeNeve, 1995) and does not directly address selective exposure to video games. However, the effects found in these studies are consistent with theorizing on message-behavioral affinity and intervention potential. Bryant and Zillmann (1977) observed that the intervention potential of a media message successfully diminished aggressive reactions, unless the message contained a high degree of message-behavioral affinity. In other words, watching an exciting, nonviolent game show diminished aggressive reactions in provoked participants, but viewing an exciting hockey game, which featured players brawling, exacerbated the aggressive responses of viewers. Likewise, the literature consistently finds differences in nonviolent and violent content on aggression.

Hedonic Valence

Presumably, all communication messages could be classified on a continuum of positive to negative. Positive messages might be described as uplifting, reassuring, amusing, happy, and so on. The list of adjectives that describe negative messages includes threatening, noxious, distressing, sad, etc. In the media context, hedonic valence refers to the positive or negative nature of a message. Affect dependent theory of stimulus arrangement posits that people in a bad mood will prefer media (or experiences) with a positive valence. Consistent with this prediction, Knobloch and Zillmann (2002) demonstrated that when students in a negative mood were given the opportunity to listen to music, they opted to listen to joyful music for longer periods than did students who were in a good mood.

Stated in the most general terms, in cases where the hedonic valence of a message is opposite to the valence of an individual's affective state, the message will effectively diminish the intensity of the affective state. However, this relationship has proven to be reliable only under certain conditions. Message valence in combination with the excitatory potential of a message yields different effects. For example, a meta-analysis of the effects of pornography on aggression revealed differently valenced correlations between aggression and specific types of erotic stimuli (Allen, D'Alessio, & Brezgel, 1995). In that study, exposure to nudity was associated with diminished levels of aggression, whereas exposure to nonviolent or violent pornography yielded positive associations with aggression. Another study produced analogous findings—messages with a positive valence diminished the effects of a negative mood if the message had low excitatory potential, but exacerbated aggressive reactions if the excitatory potential was high (Zillmann, Bryant, Comisky, & Medoff, 1981).

Particularly strong support for the notion that people manage moods through media has been demonstrated in studies that have explored the link between women's menstrual cycle and mood management. The literature indicates that women are more likely to view comedies when their cycle makes them most likely to be depressed (Helregel & Weaver, 1989; Meadowcroft & Zillmann, 1987), and they are more likely to choose programs with higher sexual and romantic content when their cycle makes them more likely to have an increased libido (Weaver & Baird, 1995).

Although Mehrabian and Wixen (1986) did not consider the initial mood state of participants in their study, they found that the emotional component of pleasure predicted preferences for arcade games among college students. This finding is consistent with research on hedonic valence and preferences for other media.

The vast majority of mood management studies has focused on the effects of aversive affective states on selectivity. In survey research, the focus has been on correlations between life events or situations and television viewing. In experimental studies, the researcher has induced a "bad" mood in participants and then given them an opportunity to view television programs. In general, the literature has found support for mood management through media. People have been observed to use media to alleviate boredom (Bryant & Zillmann, 1984), stress (Anderson, Collins, Schmitt, & Jacobvitz, 1996), apprehension (Wakshlag, Vial, & Tamborini, 1983; Wakshlag, Bart et al., 1983), annoyance (Medoff, 1982), and depression (Dittmar, 1994; Helregel & Weaver, 1989; Meadowcroft & Zillmann, 1987; Potts & Sanchez, 1994; Weaver & Baird, 1995). Other research has shown selective avoidance of television in response to annoyance (Christ & Medoff, 1984).

In contrast, very few studies have investigated the effects of positive affect on selectivity. When given the choice, first and second graders were more likely to select programs with fast-paced humorous inserts (Wakshlag, Day, & Zillmann, 1981). Similar results were found with the same age group in a study of the effects of tempo of background music (Wakshalg, Reitz, & Zillmann, 1982). Perse's (1998) investigation of channel changing revealed that positive affect did not deter people from switching programs, but negative affect increased the likelihood a viewer would change the channel. However, these studies only considered emotional reactions to television programs; none of them considered initial mood state.

Zillmann (1988) has argued that good moods can be maintained or enhanced through consumption of media message that are: (a) minimally involving, (b) have a high behavioral affinity to an initial positive mood, or (c) are highly pleasant. Given the limited research, it is difficult to reach firm conclusions about the effects of positive mood on selectivity. In terms of video games, it seems doubtful that players could maintain an initially positive mood through playing the sorts of games that have been studied by most researchers. This is because most video games in these studies have tended to be highly involving and/or have had a negative valence. Highly involving games would be more likely to disrupt cognitive rehearsal of thoughts related to positive mood. This pattern of effects was demonstrated in Nelson and Carlson's (1985) research. They found that the experience of playing a racing game resulted in diminished positive moods and increased negative moods, regardless of whether participants played a game with violent or with nonviolent content. Games with negative valence (i.e., violent content) would be likely to exacerbate negative moods. Evidence for this pattern of effects is found in studies linking exposure to violent content and aggressive behavior (e.g., Gentile et al., 2004).

THE CHALLENGES OF CONDUCTING SELECTIVE EXPOSURE RESEARCH ON VIDEO GAMES

One obvious challenge video game research poses to selective exposure theory is the countervailing assumption that gamers may not really be interested in managing their moods via game playing. Instead, they simply may want to compete for "the thrill of victory." If sensation seeking or other personality or situational factors drives gamers to seek challenges independent of mood management, they may even be willing to accept "the agony of defeat" in order to acquire the possibility of triumph, however remote. With electronic gaming and other interactive entertainment, the possibility of a satisfactory outcome is rarely as certain as it is in formulaic mass entertainment (e.g., studio movies, best-sellers, television programs; see Vorderer, 2000, 2001; Vorderer, Klimmt, & Ritterfeld, 2004), unless, of course, game players are willing to stick with familiar, previously mastered games or at a level of difficulty that clearly has already been mastered. Unfortunately, we are not aware of any research that has been conducted on such telic hedonism. Until evidence to support such contentions becomes available, it seems more reasonable to stick with the well-established findings from selective exposure research that the ultimate goal is maximizing pleasure, recognizing that some gamers may be willing to sacrifice immediate gratification for the greater enjoyment that awaits them as soon as they master that next level.

To this point, we have examined the selective exposure and computer game literature as four distinct streams of research—excitatory homeostasis, intervention potential, message-behavioral affinity, and hedonic valence. In reality, a great deal of overlap exists between theoretical components, and it is the combination of some constellation of these elements (plus others, undoubtedly) that will be most likely to determine patterns of selective exposure. Experimentally manipulating these elements is no easy task, however. For example, it is relatively simple to test the effects of cognitive intervention potential on selective exposure by finding a computer game that is highly involving and one that is less involving, but finding games that vary on multiple dimensions is easier said than done. This is because content, for instance, that is highly negative or positive typically is inherently involving; it may be impossible to find content that is highly negative *and* minimally involving. These are issues for selective exposure theory in general, however, that go beyond the purview of this chapter (see Zillmann, 1988, for further discussion of this issue).

In terms of video games, several patterns of selectivity are worth examining in further detail. First, the literature demonstrates that age is an important predictor of exposure (Bickham, et al., 2003; Buchman & Funk, 1996; Funk, Buchman, & Germann, 2000; Wright et al., 2002; also see Durkin, chap. 27, this volume; Raney, Smith, & Baker, chap. 12, this volume; and von Salisch, Oppl, & Kristen, chap. 11, this volume). Game playing apparently fluctuates with age, with the youngest children and older adolescents playing the least and younger adolescents playing the most. No single study has comprehensively documented game playing throughout the developmental course, however, so it is difficult to reach a firm conclusion by comparing data across studies. Second, nearly all studies show distinct gender differences with males playing more frequently and for longer periods than females, although such gender differences appear to be diminishing over time. Bickham et al. (2003) found that age and gender were the most important predictors of video game use; more so than other sociological variables, such as ethnicity, family conflict, or socioeconomic status.

A third important pattern found in the selectivity of video games is repetition. Although no study has explicitly investigated repeated playing of a single game, players must continually practice in order to develop the skill sets necessary to successfully survive or thrive in a game. In studies of television viewing, repeated viewing of a program is associated with increased learning gains in preschoolers (Crawley, Anderson, Wilder, Williams, & Santomero, 1999), suggesting that the content may be in some sense novel and unique each time a preschooler watches the same videotape or program. Among adults, Furno-Lamude and Anderson (1992) noted that watching a rerun was mainly associated with nostalgia and recall. They speculated that watching a program a second time would require less mental effort and therefore would be more appealing. Alternatively, from a mood management perspective, recall could be associated with the opportunity to rehearse pleasurable cognitions rather than relief from mental exertion. For young computer gamers, repeated playing is likely associated with the novelty of learning new information. This makes sense, because younger children are more likely than older children to play educational games (Wright et al., 2002). For older players, the opportunity to mentally rehearse a triumphant battle or successfully navigate a complex maze may be a key motivator. Psychologists commonly use mental rehearsal as a means of inducing mood, and it may be pleasurable for players to ponder their own victories as a means of self-inducing positive mood.

This explanation may seem reasonable to any parent who has overheard their child or children discussing computer game triumphs, but further evidence is found in the fact that many games now also include an option to archive a game-playing episode for future observation. Some Web sites even feature archived games available for downloading (e.g., www.gamespot.com, www.fifa2004.com). Thus, players can relive the thrill of scoring a golden goal to win the World Cup or experience the rush of single-handedly warding off an onslaught of mutant soldiers.

These three factors—age, gender, and repetitious playing—have important implications for selective exposure to video games and raise several important questions for computer game research: (a) How does repeated playing contribute to selective exposure to video games? (b) How do we account for the origins of selectivity in young children? (c) What role does gender play in the selective exposure process? In the next section, we briefly address these three issues.

A key factor of affect-dependent theory of stimulus arrangement (Zillmann & Bryant, 1985) is past exposure to stimuli. Individuals learn over time that a particular arrangement of stimuli will result in a particular emotional effect. Thus, the theory assumes that any exposure to video games today is partly because players have learned in the past that they are likely to achieve a certain emotional state through game play. Because video games have variable levels of difficulty, the probability of absolutely achieving a desired emotional state is less than certain (recall that players' postgame feelings depend on their success or failure in the game; Fling et al., 1992). Thus, achieving a positive mood through playing is likely only achieved intermittently, which, in learning-theoretic terms, only serves to increase motivation to play.

If players learn from past experience what video games result in positive emotional outcomes, then it is imperative that we account for initial exposure to games. This is likely to become increasingly difficult as the industry targets younger and younger children, and exposure to computers and video games occurs at an increasingly younger age (Kaiser Family Foundation, 2004). Nevertheless, we know that preschool-age children are more likely to play educational games than are school-age children, and the decision to play, and what to play, is more likely to occur under the guidance of parents or caregivers with younger children. As children grow older, they typically take more control over their media-use decisions, especially in the case of electronic games. Funk, Hagan, and Schimming (1999) observed that the majority of parents failed to correctly identify the favorite game of their third- to fifth-grade children. Furthermore, children of these parents described their favorite game as violent in 70% of the cases where a parent failed to correctly identify the favorite game. We also know that younger players (i.e., adolescents) are more likely to report that they like the violent aspect of games than are adult players (e.g., Griffiths et al., 2004; von Salisch et al., chap. 11, this volume). Furthermore, a high prevalence of violence is found among the favorite games of children (e.g., Funk & Buchman, 1996; Krahé & Möller, 2004). Thus, it appears that children are attracted to and play violent video games, especially prior to the onset of more sophisticated levels of social and moral judgment. A great deal of correlational evidence shows a link between violent video games and frequency of play (e.g., Anderson & Dill, 2000; Gentile et al., 2004; Funk et al., 2000). Even though some of these studies have demonstrated experimentally that exposure causes increased aggressive predispositions (e.g., aggressive intent), this fact does not preclude the possibility that aggression increases the appeal of violent games. Therefore, we must account for the appeal of violence in video games.

One possible explanation for the appeal of violent video games is excitation-transfer theory (e.g., Zillmann, 1971). To illustrate, consider the following example: A young child plays an educational game, is excited to learn new things, and is captivated by the interesting graphics and appealing sounds of the game. The child begins to associate pleasure with exposure to games. As the child grows older, more opportunities to play a variety of games arise as parents become less involved with media decisions. Consequently, the child plays a video game with violent imagery and experiences arousal. At the same time, the child experiences a heightened sense of self-efficacy as his or her inputs in the game are rewarded with higher scores, exciting graphics and sound effects, and the elimination of negative consequences. Ultimately, the child becomes proficient at the game and wins. The child searches for an explanation for the heightened feelings of arousal and misattributes those feelings to the experience of efficacy, and ultimately to winning the game. Thus, the child learns to associate pleasurable outcomes from violent content. For some children (e.g., Type A individuals, aggressively predisposed individuals, etc.) the physiological effects are even more potent, and the likelihood that these children turn to violent content in the future would be expected to increase.

Finally, how do we explain that females historically have been less attracted to video games than have males? The literature suggests several possibilities, including that females apparently prefer different types of violent content (i.e., fantasy violence as opposed to human violence; Funk & Buchman, 1996), that males are more likely than females to be socialized to have an interest in computers (Lin & Lepper, 1987), and that the industry has failed to produce games that appeal to female interests (Wright et al., 2002). Another possibility is that females are more likely to perceive heightened arousal as negative, whereas males may tend to perceive enhanced arousal as positive. In studies of physiological reactions to emotional stimuli, arousal was correlated with positive affect in men and negative affect in women (Bradley & Lang, 2000). In any case, all of these factors may contribute to gender differences in preferences for video games.

THE FUTURE OF SELECTIVE EXPOSURE AND VIDEO GAMES

Ultimately, we hope to be able to account for selective exposure to video games. Before we are able to do this, we must first confirm or refute several assumptions and speculations made in this chapter. First and foremost, it seems prudent to confirm and extend the findings of earlier research in selective exposure to other media to video games. Do people use video games as a means to manage moods? Theoretically, it appears they do, but experimentally this has yet to be determined.

In terms of the intervention potential of video games, we assumed that interactivity was equated with the ability of games to disrupt cognitive rehearsal and therefore alter moods. If this assumption proves valid, we would expect that games that require more interaction would have a greater potential to change moods. Furthermore, less interactive games should be less effective in their ability to alter moods. Therefore, for people who are originally in negatively valenced mood states, the game with the greatest interactivity should be the most appealing, because it has the most potential to counteract negative moods (cf. Vorderer, 2000). Again, this is a notion that is theoretically plausible, but it has not yet been confirmed.

We also examined the literature on self-esteem and video games. If a relationship exists between low self-esteem and preference for games, then it seems reasonable to expect that individuals who are placed in a bad mood would be more likely than those high in self-esteem to prefer video games as a means to alter their mood.

We also need to assess preferences for video games in terms of their attraction relative to other media, in addition to variables of content within the medium (i.e., violent and sexual content), which might be expected to affect selective exposure. Cross-medium comparisons in selective exposure would seem to have tremendous ecological validity in today's pluralistic media environment.

We also identified two important rudimentary aspects of selective exposure research that need further investigation—age and gender—before we even begin to build more psychologically complex selective exposure models. The world of the information age's "sovereign consumer" is so complex as to challenge the best of our cognitive and behavioral models; therefore, we would be wise to build on the sturdy foundation of well-established building blocks, such as organismic and demographic variables. Once this firm foundation is laid for research on selective exposure to video games, constructing and testing more complex, even hierarchical, aspects of our exposure models with video games should prove to be one of the true challenges of the 21st century.

REFERENCES

Allen, M., D'Alessio, D., & Brezgel, K. (1995). A meta–analysis summarizing the effects of pornography II: Aggression after exposure. *Human Communication Research, 22,* 258–283.

Anderson, C. A. (2002). Violent video games and aggressive thoughts, feelings, and behaviors. In S. L. Calvert, A. B Jordan, & R. R. Cocking (Eds.), *Children in the digital age: Influences of electronic media on development* (pp. 101–116). London: Praeger.

Anderson, C. A., & Bushman, B. J. (2001). Effects of violent video games on aggressive behavior, aggressive cognition, aggressive affect, physiological arousal, and prosocial behavior: A meta–analytic review of the scientific literature. *Psychological Science, 12,* 353–359.

Anderson, C. A., & Dill, K. E. (2000). Video games and aggressive thoughts, feelings, and behavior in the laboratory and in life. *Journal of Personality and Social Psychology, 78,* 772–790.

Anderson, C. A., Deuser, W. E., & DeNeve, K. M. (1995). Hot temperatures, hostile affect, hostile cognition, and arousal: Tests of a general model of affective aggression. *Personality and Social Psychology Bulletin, 21,* 434–448.

Anderson, D. R., Collins, P. A., Schmitt, K. L., & Jacobvitz, R. S. (1996). Stressful life events and television viewing. *Communication Research, 23,* 243–260.

Bickham, D. S., Vandewater, E. A., Huston, A. C., Lee, J. H., Caplovitz, A. G., & Wright, J. C. (2003). Predictors of children's media use: An examination of three ethnic groups, *Media Psychology, 5,* 107–137.

Bradley, M. M., & Lang, P. J., (2000). Measuring emotion: Behavior, feeling, and physiology. In R. D. Lane, & L. Nadel (Eds.), *Cognitive neuroscience of emotion* (pp. 242–276). London: Oxford University Press.

Bryant, J., & Miron, D. (2004). Theory and research in mass communication. *Journal of Communication, 54,* 662–704.

Bryant, J., Raney, A. A., & Zillmann, D. (2002). Sports television. In B. Strauß, N. Kolb, & M. Lamus (Eds.), www.sports–goes–media.de (pp. 51–79). Schorndorf, Germany: Hofmann.

Bryant, J., & Zillmann, D. (1977). The mediating effect of the intervention potential of communications on displaced aggressiveness and retaliatory behavior. In B. D. Ruben (Ed.), *Communication Yearbook 1* (pp. 291–306). New Brunswick, NJ: Transaction.

Bryant, J., & Zillmann, D. (1984). Using television to alleviate boredom and stress: Selective exposure as a function of induced excitational states. *Journal of Broadcasting, 28,* 1–20.

Buchman, D. D., & Funk, J. B. (1996). Video and video games in the '90s: Children's time commitment and game preference. *Children Today, 24,* 12–15.

Calvert, S. L., & Tan, S. L. (1994). Impact of virtual reality on young adults' physiological arousal and aggressive thoughts: Interaction versus observation. *Journal of Applied Developmental Psychology, 15,* 125–139.

Christ, W. G., & Medoff, N. J. (1984). Affective state and the selective use of television. *Journal of Broadcasting, 28,* 51–63.

Colwell, J., Grady, C., & Rhaiti, S. (1995). Video games, self–esteem and gratifications of needs in adolescents. *Journal of Community & Applied Social Psychology, 5,* 195–206.

Colwell, J., & Payne, J. (2000). Negative correlates of computer game play in adolescents. *British Journal of Psychology, 91,* 295–310.

Crawley, A. M., Anderson, D. R., Wilder, A., Williams, M., & Santomero, A. (1999). Effects of repeated exposure to a single episode of the television program Blues's Clues on the viewing behaviors and comprehension of preschool children. *Journal of Educational Psychology, 4,* 630–637.

Davies, J. J. (2004). The effects of neuroticism, mood, and the intervention potential of media messages on selective exposure to television. Unpublished doctoral dissertation, University of Alabama, Tuscaloosa.

Detenber, B. H., & Simons, R. F. (1998). Roll 'em!: The effects of picture motion on emotional responses. *Journal of Broadcasting & Electronic Media, 42,* 113–127.

Dietz, T. L. (1998). An examination of violence and gender role portrayals in video games: Implications for gender socialization and aggressive behavior. *Sex Roles, 38,* 425–442.

Dittmar, M. L. (1994). Relations among depression, gender, and television viewing of college students. *Journal of Behavior and Personality, 9,* 317–328.

Dominick, J. R. (1984). Videogames, television violence, and aggression in teenagers. *Journal of Communication, 34,* 136–147.

Entertainment Software Association. (2004, June 30). *Top ten industry facts.* Retrieved June 30, 2004 from http://www.theesa.com/pressroom.html.

Fling, S., Smith, L., Rodriguez, T., Thornton, D., Atkins, E., & Nixon, K. (1992). Videogames, aggression, and self–esteem: A survey. *Social Behavior and Personality, 20,* 39–46.

Funk, J. B., & Buchman, D. D. (1996). Playing violent video and video games and adolescent self–concept. *Journal of Communication, 46,* 19–33.

Funk, J. B., Buchman, D. D., & Germann, J. N. (2000). Preference for violent electronic games, self–concept, and gender differences in young children. *American Journal of Orthopsychiatry, 70,* 233–241.

Funk, J. B., Hagan, J., & Scimming, J. (1999). Children and electronic games: A comparison of parents' and children's perceptions of children's habits and preferences in a United States sample. *Psychological Reports, 85,* 883–888.

Furno-Lamude, D., & Anderson, J. (1992). The uses and gratifications of rerun viewing. *Journalism Quarterly, 69,* 362–372.

Gentile, D. A., Lynch, P. J., Linder, J. R., & Walsh, D. A. (2004). The effects of violent video game habits on adolescent hostility, aggressive behaviors, and school performance. *Journal of Adolescence, 27,* 5–22.

Griffiths, M. D., & Dancaster, I. (1995). The effect of type A personality on physiological arousal while playing video games. *Addictive Behaviors, 20,* 543–548.

Griffiths, M. D., Davies, M. N. O., & Chappell, D. (2004). Online computer gaming: A comparison of adolescent and adult gamers. *Journal of Adolescence, 27,* 87–96.

Griffiths, M. D., & Hunt, N. (1995). Computer game playing in adolescence: Prevalence and demographic indicators. *Journal of Community & Applied Social Psychology, 5,* 189–193.

Griffiths, M. D., & Hunt, N. (1998). Dependence on video games by adolescents. *Psychological Reports, 82,* 475–480.

Grodal, T. (2000). Video games and the pleasures of control. In D. Zillmann & P. Vorderer (Eds.), *Media entertainment: The psychology of its appeal* (pp. 197–212). Mahwah, NJ: Lawrence Erlbaum Associates.

Helregel, B. K., & Weaver, J. B. (1989). Mood-management during pregnancy through selective exposure to television. *Journal of Broadcasting & Electronic Media, 33,* 15–33.

Kaiser Family Foundation. (2004, June 29). *New study finds children age zero to six spend as much time with TV, computers and video games as playing outside.* Retrieved June 29, 2004, from http://www.kff.org/entmedia/entmedia102803nr.cfm.

Kim, K. H., Park, J. Y., Dong, Y. K., Hak, I. M., & Ho, C. C. (2002). E–lifestyle and motives to use on–line games. *Irish Marketing Review, 15,* 71–78.

Knobloch, S., Hastall, M., Zillmann, D., & Callison, C. (2003). Imagery effects on the selective reading of Internet newsmagazines. *Communication Research, 30,* 3–29.

Knobloch, S., & Zillmann, D. (2002). Mood management via the digital jukebox. *Journal of Communication, 52,* 351–366.

Krahé, B., & Möller, I. (2004). Playing violent electronic games, hostile attributional style, and aggression related norms in German adolescents. *Journal of Adolescence, 27,* 53–69.

Larsen, R. J., & Diener, E. (1987). Affect intensity as an individual difference characteristic: A review. *Journal of Research in Personality, 21,* 1–39.

Lin, S., & Lepper, M. R. (1987). Correlates of children's usage of videogames and computers. *Journal of Applied Social Psychology, 17,* 72–93.

Mastro, D. E., Eastin, M. S., & Tamborini, R. (2002). Internet search behaviors and mood alterations: A selective exposure approach. *Media Psychology, 4,* 157–172.

McClure, R. F., & Mears, F. G. (1984). Video game players: Personality characteristics and demographic variables. *Psychological Reports, 55,* 271–276.

Meadowcroft, J. M., & Zillmann, D. (1987). Women's comedy preferences during the menstrual cycle. *Communication Research, 14,* 204–218.

Medoff, N. J. (1982). Selective exposure to televised comedy programs. *Journal of Applied Communication Research, 10,* 117–132.

Mehrabian, A., & Wixen, W. J. (1986). Preferences for individual video games as a function of their emotional effects on players. *Journal of Applied Social Psychology, 16,* 3–15.

Nelson, T. M., & Carlson, D. R. (1985). Determining factors in choice of arcade games and their consequences upon young male players. *Journal of Applied Social Psychology, 15,* 124–139.

Oliver, M. B. (2003). Mood management and selective exposure. In J. Bryant, D. Roskos–Ewoldsen, & J. Cantor (Eds.), *Communication and emotion: Essays in honor of Dolf Zillmann* (pp. 85–106). Mahwah, NJ: Lawrence Erlbaum Associates.

Perse, E. M. (1998). Implications of cognitive and affective involvement for channel changing. *Journal of Communication, 48,* 49–68.

Phillips C. A., Rolls, S. Rouse, A., & Griffiths, M. D. (1995). Home video game playing in schoolchildren: A study of incidence and patterns of play. *Journal of Adolescence, 18,* 687–691.

Potts, R., & Sanchez, D. (1994). Television viewing and depression: No news is good news. *Journal of Broadcasting & Electronic Media, 38,* 79–90.

Roe, K., & Muijs, D. (1998). Children and video games: A profile of the heavy user. *European Journal of Communication, 13,* 181–200.

Selnow, G. W. (1984). Playing videogames: The electronic friend. *Journal of Communication, 34,* 148–156.

Slater, M. D. (2003). Alienation, aggression, and sensation seeking as predictors of adolescent use of violent film, computer, and website content. *Journal of Communication, 53,* 105–121.

Sneed, C., & Runco, M.A. (1992). The beliefs adults and children hold about television and video games. *The Journal of Psychology, 126,* 273–284.

Vorderer, P. (2000). Interactive entertainment and beyond. In D. Zillmann & P. Vorderer (Eds.), *Media entertainment: The psychology of its appeal* (pp. 21–36). Mahwah, NJ: Lawrence Erlbaum Associates.

Vorderer, P. (2001). It's all entertainment—sure. But what exactly is entertainment? Communication research, media psychology, and the explanation of entertainment experiences. *Poetics, 29,* 247–261.

Vorderer, P., Klimmt, C., & Ritterfeld, U. (2004). Enjoyment: At the heart of media entertainment. *Communication Theory, 14,* 388–408.

Wakshlag, J. J., Bart, L., Dudley, J., Groth, G., McCuthcheon, J., & Rolla, C. (1983). Viewer apprehension about victimization and crime drama programs. *Communication Research, 10,* 195–217.

Wakshlag, J., Day, K., & Zillmann, D. (1981). Selective exposure to educational television programs as a function of differently paced humorous inserts. *Journal of Educational Psychology, 73,* 27–32.

Wakshlag, J., Reitz, R., & Zillmann, D. (1982). Selective exposure to and acquisition of information from educational television programs as a function of appeal and tempo of background music. *Journal of Educational Psychology, 74,* 666–677.

Wakshlag, J., Vial, V., & Tamborini, R. (1983). Selecting crime drama and apprehension about crime. *Human Communication Research, 10,* 227–242.

Weaver, J. B. III, & Baird, E. A. (1995). Mood management during the menstrual cycle through selective exposure to television. *Journalism and Mass Communication Quarterly, 72,* 139–146.

Wright, J. C., Huston, A. C., Vandewater, E. A., Bickham, D. S., Scantlin, R. M., Kotler, J. A., et al. (2002). American children's use of electronic media in 1997: A national survey. In S. L. Calvert, A. B. Jordan, & R. R. Cocking (Eds.), *Children in the digital age: Influences of electronic media on development* (pp. 35–54). London: Praeger.

Zillmann, D. (1971). Excitation transfer in communication–mediated aggressive behavior. *Journal of Experimental Social Psychology, 7,* 419–434.

Zillmann, D. (1988). Mood management: Using entertainment to full advantage. In L. Donohew, H. E. Sypher, & E. T. Higgins (Eds.), *Communication, social cognition and affect* (pp. 147–171). Hillsdale, NJ: Lawrence Erlbaum Associates.

Zillmann, D. (2000). Mood management in the context of selective exposure theory. In M. E. Roloff (Ed.), *Communication Yearbook 23* (pp. 103–123). Thousand Oaks, CA: Sage.

Zillmann, D., & Bryant, J. (1985). Affect, mood, and emotion as determinants of selective exposure. In D. Zillmann & J. Bryant (Eds.), *Selective exposure to communication* (pp. 157–190). Hillsdale, NJ: Lawrence Erlbaum Associates.

Zillmann, D., Bryant, J., Comisky, P. W., & Medoff, N. J. (1981). Excitation and hedonic valence in the effect of erotica on motivated intermale aggression. *European Journal of Social Psychology, 11,* 233–252.

Zillmann, D., Hezel, R. T., & Medoff, N. J. (1980). The effect of affective states on selective exposure to televised entertainment fare. *Journal of Applied Social Psychology, 10,* 323–339.

Zillmann, D., Knobloch, S., & Yu, H. (2001). Effects of photographs on the selective reading of news reports. *Media Psychology, 3,* 301–324.

RECEPTION AND REACTION PROCESSES

14

A Brief Social History
of Game Play

Dmitri Williams
University of Illinois at Urbana-Champaign

Who has played video games? Where have they played them? And how have games helped or hindered social networks and communities? This chapter answers these historical questions for the birthplace of video games—the United States—although many other industrialized countries have had similar patterns. In the United States, our collective stereotype conjures up an immediate image: Isolated, pale-skinned teenage boys sit hunched forward on a sofa in some dark basement space, obsessively mashing buttons. In contrast, the statistics and accounts tell a very different story—one of often vibrant social settings and diverse playing communities. Why do American conceptions of gamers diverge from reality?

The explanation is that for video game media, the sociopolitical has been inseparable from the practical. Social constructions, buttressed by the news media over the past 30 years, have created stereotypes of game play that persist within generations. This chapter will explain both the imagery and the reality. Moving from the descriptive to the analytical, it begins with the basic trends and figures: who played, when, where and why, and how changes in technology have impacted the social side of gaming. An immediate pattern appears—for both industrial and political reasons, the early 1980s were a crucial turning point in the social history of video game play. What began as an open and free space for cultural and social mixing was quickly transformed through social constructions that had little to do with content, the goals of the producers, or even demand. The legacy of that era persists today, influencing who plays, how we view games, and even how we investigate their uses and effects.

SETTING THE STAGE

Figure 14.1 gives industry revenues for home and arcade play, standardized to 1983 dollars. The data show what game historians have already presented through narratives (Herman, 1997; Herz, 1997; Kent, 2000; Sheff, 1999): A slow adoption during the 1970s led to a massive spike

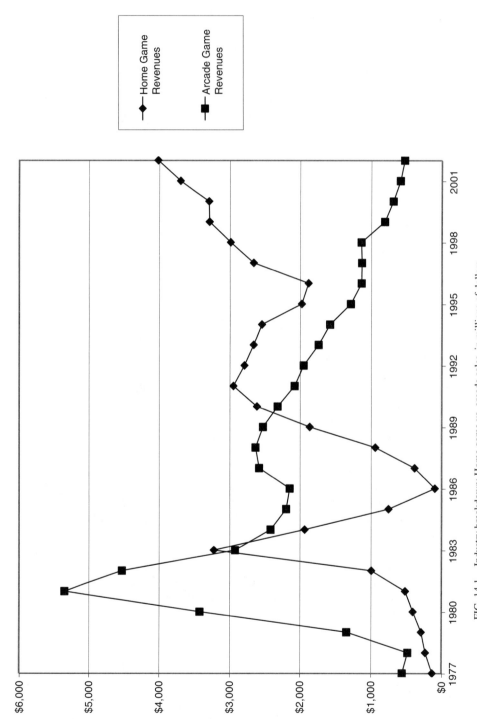

FIG. 14.1. Industry breakdown: Home game vs. arcade sales, in millions of dollars.

Note: For comparability, these data have been adjusted for inflation and standardized to their values in 1983. Unadjusted values would show higher totals in recent years. Data source: Amusement & Music Operators Association, Nintendo, PC Data, NPD, Veronis Suhler, Vending Times (1978–2001).

in popularity during the Atari heyday of the early 1980s, followed by the collapse of that company and the industry's eventual revival in the late 1980s by Nintendo. The late 1980s also saw the beginning of play moving from public to private spaces. Over the past 10 years, games have steadily become more popular to the point where they are now considered mainstream media, competing with newspapers, television, radio, and film for attention and dollars.

Aside from the early hobbyist era (Levy, 1994), commercial gaming efforts did not gain traction until the early 1970s. Long before arcades and Atari, the first mass-marketed home game machine, Magnavox's Odyssey, aimed for the mainstream family audience. Seeking to break the nascent link between gaming technology and male youth culture, Magnavox marketed their machine as the new electronic hearth, complete with advertising tableaux of families joined in domestic bliss around the new machines. Such tableaux can have powerful idealizing and norming functions (Marchand, 1985). From 1972 until the early 1980s, manufacturers tried to promote the idea of gaming as a mainstream activity. This meant convincing parents that console games could unite their families, while also convincing single adults that arcade games had sex appeal. It is a measure of the strong stereotypes about age and gaming that the following will surprise many readers: As a niche hobby, these marketing efforts had some early success. Game play in public spaces began as an adult activity, with games first appearing in bars and nightclubs before the eventual arcade boom. Then, when arcades first took root, they were populated with a wide mixing of ages, classes, and ethnicities. This mixing was quite similar to what Gabler (1999) described for turn-of-the-century nickelodeons—populist, energetic, and ultimately threatening.

There is no doubt that mainstream corporate and political forces helped to water down these public spaces (a theme to be taken up shortly). However, the single biggest cause in the decline of gaming in the 1980s was the spectacular collapse of the Atari corporation, an event so traumatic that it appeared to destroy the entire industry. But while pundits and investors alike thought Atari's collapse was proof of a faddish product and a fickle consumer, it was really no more than inept management (Cohen, 1984; Sheff, 1999). In actuality, the rise of Atari and the game industry had created a new kind of consumer, one increasingly comfortable with interactive electronic devices. The Nintendo revival of the late 1980s proved that demand had in fact not magically disappeared (it continued to flourish in Japan). Still, Nintendo's marketing and distribution solidified games as the province of children for the next 10 years.

The Crash and the New Consumer

At first, the industry's collapse was easy to explain as just another example of short attention-span American tastes: first disco, then Pet Rocks, and now *Pac-Man*. However, a closer look at the demographics and demand shows that video games helped usher in a new kind of consumer, one increasingly aware of new tools and new possibilities. Consumers were beginning to embrace home computers, compact discs, and the concept of digital systems as convenient and powerful entertainment tools.

The demand for video games should be viewed as part of a larger trend in entertainment consumption. Games' initial rise and temporary decline occurred during periods of overall increasing demand for entertainment products. Large increases in productivity and income gains have increased Americans' incentives to work even more while also giving families more discretionary income: less time, but more money. Much of this trend is due to the large-scale rise in hours worked by U.S. women (Schor, 1991). Time has become scarcer, and Americans have been steadily spending more and more of their income to enjoy it to the fullest (Vogel, 2001). In 1970, Americans were spending 4.3% of their incomes on recreation and entertainment, but by 1994, that figure had grown to 8.6% (*The national income and product accounts of the United States, 1929–1976*, 1976; *Survey of current business*, 1996). It follows that consumers have been quick to adopt digital technologies that can be enjoyed more efficiently. For example, Americans spend a great deal of time playing card games and board games (Schiesel, 2003), but it is far easier and faster (if more expensive) to play *Risk* on a computer than on a tabletop with dice.

A New Electronic Hearth: The Rise of Home Computing and Games

Throughout the 1980s, a combination of economic and technological forces moved play away from social, communal, and relatively anarchic early arcade spaces and into the controlled environments of the sanitized mall arcade (or "family fun center") or into the home. The idea of a home game machine—once confusing and new to consumers[1]—seemed less remarkable in a home with microprocessors embedding in everything from PCs to blenders. This acceptance can also be viewed as part of a general transition of technology-based conveniences away from public areas and into private ones (Cowan, 1983; Putnam, 2000).

Since their inception, video games have been harbingers of the shift from analog to digital technology for both consumers and producers. They made major portions of a generation comfortable and technoliterate enough to accept personal computers (Lin & Leper, 1987; Rogers, 1985), electronic bulletin boards, desktop publishing, compact disks and the Web, and have pushed the development of microprocessors, artificial intelligence, compression technologies, broadband networks, and display technologies (Burnham, 2001). Games functioned as stepping stones to the more complex and powerful world of home computers. Figure 14.2 shows the dual trends in adoption for home game systems and home computers. Notably, games *preceded* computers at every step of adoption and have continued to be in more homes since their arrival.

The Rise of Networks

The last major trend affecting the social site of gaming is the more recent move toward networked game play. Beginning with text-based networked games called "MUDs" and proceeding to graphical versions called "MMRPGs," online games have emerged as an important new and social game format (see Chan & Vorderer, chap. 6, this volume). But although the history of these PC-based games suggests a vibrant social universe (Dibbell, 2001, 2003; Mulligan & Petrovsky, 2003; Turkle, 1995), the casual gamer is unlikely to invest the time or money to wade into them. Such games are extremely profitable, but are still a minor part of game play (Croal, 2001; Palumbo, 1998). Instead, it is the current wave of more mainstream online game adoption that has firms investing (Kirriemur, 2002)[2] and will have social implications. This adoption is notable because it has begun to expand the game market beyond the traditionally younger, male audience that plays console games. Data from the Pew Internet and American Life Project (see Table 14.1) illustrate that higher percentages of racial minorities and women are playing than White men, and that surprisingly large numbers of older people play.[3]

The major drivers for this phenomenon are not the games themselves, but the addition of other players via the Internet (Griffiths, Davies, & Chappell, 2003; Kline & Arlidge, 2002). As one online gamer said, "[meeting new people is] the most interesting aspect of the game. This gives it a social dimension. There's another person behind every character" (Pham, 2003, p. C1). The presence of competitors and collaborators introduces a social element that has been missing from some gamers' experiences since the early-1980s heyday of the arcade (Herz, 1997).

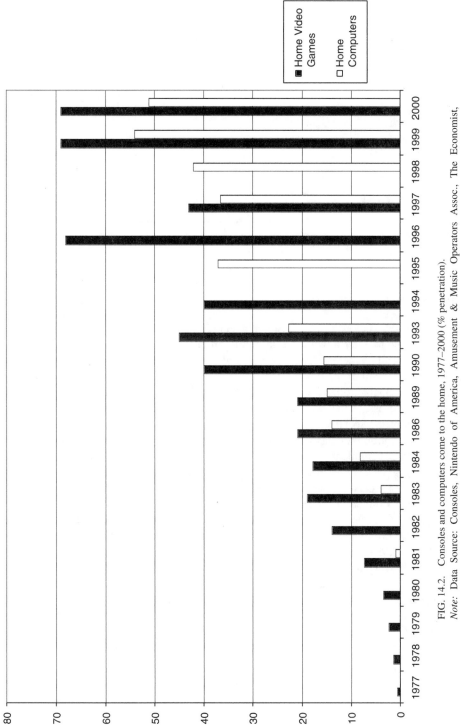

FIG. 14.2. Consoles and computers come to the home, 1977–2000 (% penetration).
Note: Data Source: Consoles, Nintendo of America, Amusement & Music Operators Assoc., The Economist, www.icwhen.com; Computers, National Science Foundation, Roper surveys, Census Bureau, Statistical Abstracts of the United States.

TABLE 14.1
Online Gaming Demographics

National adult sample, respondents who answered
yes to "ever play a game online?"

All users	37%
Men	37%
Women	38%
Whites	34%
Blacks	48%
Hispanics	54%
18–29	52%
30–49	34%
50–64	28%
65+	38%

Note: Data were collected by the Pew Internet and American Life Project in June and July, 2002.

SOCIAL PLAY

Gamers in general—but especially arcade players (Garner, 1991; Meadows, 1985; Ofstein, 1991)—were able to enter a world based purely on talent and hard work, not social status. The resulting social element of game play has always been one of the medium's strongest appeals (see Raney, Smith and Baker, chap. 12, this volume). For those who felt marginalized, unchallenged, or unable to participate in other mainstream activities, game play allowed for the contestation of issues that were less easily dealt with in everyday life. For the awkward, the underclass, or the socially restricted player, success at a game translated into a level of respect and admiration previously unavailable outside of the arcade. There was no gender or status bias in arcade competition, and the machine did not care if the player was popular, rich, or an outcast. As Herz put it, "It didn't matter what you drove to the arcade. If you sucked at *Asteroids*, you just sucked" (Herz, 1997, p. 47). Much like on playing fields, social conventions and interactions within an arcade were separate from those of "real life." Inside the arcade, gamers assumed roles separate from their outside personae and adhered to a strict set of rules governing game play, including a ban on physical aggression and a recognition of the hierarchy of skill (Ofstein, 1991).

Arcades were social magnets in the early 1980s, attracting a range of players to their populist settings. An 18-year-old-girl was quoted in *Newsweek* describing her local arcade: "Look at all these people together—blacks, whites, Puerto Ricans, Chinese. This is probably the one place in Boston where there are not hassles about race" (1981, p. 91). Class barriers were similarly low in the early years. Herz noted their similarity to pinball parlors where

> sheltered suburban teens might actually come into contact with working-class kids, high-school dropouts, down-and-out adults, cigarettes, and other corrupting influences, which made the place a breeding ground for parental paranoia, if not for crime. (Herz, 1997, p. 44)

As forbidden fruit, the appeal to video gamers was apparent. Not only could people mix with others of different ages, ethnicities and classes that they were otherwise constricted from being near, but they could also form friendships, compete, and establish an identity. Said one player, looking back on the era, "Sure, all my favorites were there, but it was the magic of the place at large, and the people there that were a major draw" (Killian, 2002).

While early arcades represented a key social site for play, consoles and PC games in homes were equally important Mitchell (1984) studied 20 families from a range of backgrounds to see what the impact of adding a console game machine was to family life. She found that family and sibling interaction increased, that no detrimental trends were found in schoolwork (there was actually a slight improvement), that none of the children or parents became more aggressive, that boys played more than girls, that girls gained a sense of empowerment, and that all of the families saw games as a bridge to personal computers. She further concluded that home video games brought families together more than any other activity in recent memory, chiefly by displacing time spent watching television (Mitchell, 1985). Murphy (1984) found that homes with video games had similar family interactions to those that did not. Instead, in nearly every case, links to deviant behavior were found to correlate with parental variables such as supervision and pressure for achievement.

THE CAUSES OF NEW MEDIA AMBIVALENCE

Games are a contentious subject in modern American society not solely because of their inherent qualities, but because they are a wholly new medium of communication, something guaranteed to provoke suspicion and ambivalence. In America and elsewhere, the advent of every major medium has been greeted with utopian dreams of democracy, but also with tales and visions of woe and social disorder or unrest (Czitrom, 1982; Neuman, 1991). This pattern has been consistent and has maintained itself dating from the telegraph (Standage, 1999), and persisting through nickelodeons (Gabler, 1999), the telephone (Fischer, 1992), newspapers, (Ray, 1999), movies (Lowery & DeFluer, 1995), radio (S. Douglas, 1999), television (Schiffer, 1991), and now with both video games and the Internet. Video games are simply the latest in a long series of media to endure criticism. Typically, the actual source of the tension lies not in the new medium, but in preexisting social issues. The tensions over new media are surprisingly predictable, in part because the issues that drive them are enduring ones such as intraclass strife and societal guilt.

Understanding how and why the medium is assigned blame can tell us a great deal about the tensions and real social problems that are actually at issue. Often, focusing attention on the medium is a convenient way of assigning blame while ignoring complex and troubling problems. Media coverage of new technology often generates a climate in which consumers of news media are terrified of phenomena that are unlikely to occur. Just as importantly, they are also guided away, purposefully or not, from complicated and troubling systemic social issues (Glassner, 1999). This is not a new trend and not particular to America. Across a wide variety of cultures, the dangers most emphasized by a society are not the ones most likely to occur, but are instead the ones most likely to offend basic moral sensibilities or that can be harnessed to enforce some social norm (M. Douglas, 1992; M. Douglas & Wildavsky, 1982). Most tragically, the guilt over mistreatment of our children can manifest itself in a painfully unjust way by casting the children themselves as the source of the problem. Resorting to the trope of the "bad seed," or blaming an external force like media, provides an excuse to ignore the primary risk factors associated with juvenile crime and violence, which are abuse from relatives, neglect, malnutrition, and, above all, poverty (Glassner, 1999).

The evidence presented so far suggests that strong social forces have shaped our reactions to games and the subsequent comfort level of certain groups to remain players. This phenomenon has a direct point of origin in the early 1980s and represents a host of social issues not directly related to games themselves. Most prominently, arcades were threatening to conservative forces. As seen in media coverage, arcades were mixing grounds for homeless children and lawyers, housewives and construction workers, and countless other socially impermissible combinations. The lashing out against arcades that followed was, according to the research, unjustified. For example, *Time* reported that children in arcades were susceptible to homosexual cruisers, prostitution, and hard liquor (Skow, 1982). It was no coincidence that the years 1981 and 1982 marked the start of the news media's dystopian frames of misspent youth or fears of injury, drug use, and the like (Williams, 2003)—and that this was precisely the period of the conservative Reagan administration's rise to power. In seeking to throw off what it perceived as the general social, cultural, and moral malaise of the 1970s, the Reagan administration campaigned on a platform that especially highlighted the culpability and irresponsibility of single "welfare queen" mothers (Gilens, 1999). This political agenda led to frames about truancy, unsupervised children, and the negative influences of electronic media—especially arcades—acting as babysitters for unconscionable working mothers.

Internet Cafés: Old Wine in New Bottles

Just as with arcades, the uncontrolled space of the Internet has predictably raised concerns about who is interacting with whom and what morally questionable activities might be taking place. The case of Internet cafés—a modern combination of anarchic arcade space and private network—shows the same patterns and concerns reoccurring.

Much like early arcades, many Internet cafés are marked by the same dark lighting and socially inclusive atmosphere (McNamara, 2003; Yee, Zavala, & Marlow, 2002). The networked game play inside the cafés is in many ways a return to the aesthetic and values of the early arcades: The spaces are morally questionable, challenging, anarchic, uncontrolled, racially diverse, and community oriented. Fears and reaction to the Internet cafés are also remarkably similar to arcades, probably owing to a parallel set of concerns and punditry centering around computer use. Table 14.2 illustrates that the same themes occur 20 years apart, suggesting that the same issues of social control and parental guilt are still operating.

TABLE 14.2
Comparing Coverage of Early Arcades With Current Coverage of Internet Cafes

	Early Arcade Coverage (1981–1982)	*Internet Café Coverage (2002–2003)*
The site operator defends the activity	"I baby-sat a bunch of kids here all summer. It may have cost them money, but they were here, they were safe, and they didn't get into trouble."*	"I think that anything that helps keep kids off the street and out of trouble is a good thing … Here there are no cigarettes, no drugs, no alcohol. Here the kids come to be with their friends."**
Conservative elders are provoked	"Taking a cue from the pool-troubled elders of the mythical River City, communities from Snellville, Ga., to Boston have recently banned arcades or restricted adolescent access."†	"It's not hard to imagine what Professor Harold Hill would have said upon entering the dim recesses of Cyber HQ in Eagle Rock. 'Trouble, with a capital T and that rhymes with C and that stands for computer game.'"**

Note. *(Skow, 1982). ** (McNamara, 2003). †(Langway, 1981).

These current concerns may or may not be valid ones, but the history of moral panics and public criticisms of public amusement gathering spaces—whether it is a nickelodeon, pinball hall, or arcade—suggests that they are likely overstated and hiding other social tensions.

AREAS OF STRUGGLE: AGE, GENDER, AND PLACE

Games, much like other new technologies, have been a means of social control. This is illustrated by presenting the everyday practices, the social construction and the framing of three issues surrounding game play: age, gender and place.

Dad, Put Down the Joystick and Back Away Slowly: Games and Age

During the mid-1980s and the 1990s, video games were constructed as the province of children. Today, as an all-ages phenomenon once again, they have begun to reenter the social mainstream. Is this adoption the result of American culture's slow and steady acceptance of gaming technology, or simply the result of an aging user base? The evidence suggests that there were both cohort and age effects (Glenn, 1977) at work over the last quarter-century. Today, youths adopt game technology at the same time as many Generation X players continue to play past adolescence. As a result, the average age of players has been rising steadily and, according to the industry, is now 30 (Top Ten Industry Facts, 2005). One cohort effect is relatively easy to isolate: the generations that ignored video games in the late 1970s and early 1980s have continued to stay away. Those who played and stopped rarely returned; by 1984, baby boomers had dramatically decreased their play, probably because of the powerful social messages they were suddenly getting about the shame and deviancy of adult gaming (Williams, 2003). Another reason may have been that the culture of games still caters primarily to adolescents, despite adults who want more mature content (Kushner, 2001; Russo, 2001).

It should be reemphasized that the popular conception of game use as a purely childcentric phenomenon did not emerge until well after games had entered the popular consciousness and home games became widespread. This is not surprising since the initial video game boom occurred in adult spaces such as bars and nightclubs. But it was not until the late 1990s that this frame finally began to dissipate, perhaps because such reporting had become so at odds with actual use. For example, Roper data showed that adult home game play was at 43% during 1993, the same year *Time* reported that grownups "don't get it" (Elmer-DeWitt, 1993). But they did "get it," and in the 1990s, adults were seemingly able to come out of the video games closet. Much of this stems from the social cache (and disposable income) that Generation X members gained upon entering independent adulthood. This trend was also likely reinforced by a transition within news magazines to younger writers for the video games beat.

Gender Gaps

The research shows a clear gender gap in video game play, but one that has only been measured for adolescents. Nearly every academic study and survey of the social impact of games, regardless of its focus, has noted that males play more often than females (Buchman & Funk, 1996; Dominick, 1984; Griffiths, 1997; Michaels, 1993; Phillips, Rolls, Rouse, & Griffiths, 1995). Some of the gender preferences may be the results of socialization and parental influence (Scantlin, 1999). Parents may have been discouraging girls at the same time they were encouraging boys to play. For example, Ellis (1984) found that parents exerted far more control over their daughters' ability to go to arcades than their sons'.

Why should this be? The explanation involves the gendering of technology. For males, technology has long been an empowering and masculine pursuit that hearkens back to the wunderkind tinkerers of the previous century. The heroic boy inventor image was first made fashionable through the carefully managed and promoted exploits of Thomas Edison and then Guglielmo Marconi (S. Douglas, 1987). Since then, technology has remained a socially acceptable pursuit for boys, one that may offer them a sense of identity and empowerment that they are not getting elsewhere (Chodorow, 1994; Rubin, 1983). One theory maintains that boys are driven to technology in large part because it helps them develop their self-identity at a time when, unlike girls, they are being forced into independence (Chodorow, 1994). Male tastes are privileged through content as well, reinforcing the choice.

This explanation fits the experience of the males, but does not fully explain the dearth of females pursuing technological interests, who continue to remain on the sidelines of science and technology. For example, women are dramatically underrepresented as engineers and scientists, despite outperforming men in science and math in high school (Seymour, 1995). The percentage of female engineering Ph.D.s who graduated in 1999 was an all-time high of only 15%.[4] If there are fewer women in technology, it must be for one of two reasons. One, women are not capable or naturally interested in technology, or two, women have been systematically socialized away from technology. Despite media framing, there is no evidence to suggest that biology plays a role. There is, however, ample evidence pointing to a social construction of science as a male pursuit (Jansen, 1989).[5] Flanagan argued that this is a direct result of the threat that female empowerment through technology poses to male power; women who use technology are not only less dependent on men but also less monitored and controllable (Flanagan, 1999). The world of video games is a direct extension of this power relationship (McQuivey, 2001). Female characters, when they do appear, tend to be objects rather than protagonists, resulting in generally negative gender stereotypes (J. Funk, 2001; Gailey, 1993; Knowlee et al., 2001). Additionally, a male, heterosexual viewpoint is assumed, with most characters playing the role of the strong, assertive man seeking glory through violence with the reward of female companionship (Consalvo, 2003). But while women experience frustration with their inability to identify with in-game characters, male designers are largely unaware of the problem (K. Wright, 2002).

The effects of such social constructions are very real: The connection between video game play and later technological interest has become a gender issue in early adolescence and persists throughout the life span. Females are socialized away from game play, creating a self-fulfilling prophecy for technology use: Girls who do not play become women who do not use computing technology (Cassell & Jenkins, 1999; Gilmore, 1999), and certainly do not aspire to make games. In my interviews with game makers over 2 years, I spoke with almost no women. It is no surprise that an industrywide masculine culture has developed in which a male point of view is nearly the only point of view. Despite the untapped sales potential of the female audience, this culture is unlikely to undergo any sea change in the near future so long as men dominate the ranks of game makers.

Place, the Final Frontier

In addition to the powerful social forces that have moderated gamers' behavior and access to the technology, changes in both technology and space have impacted play. As Spigel (1992) has shown, the introduction of a new technology or appliance into the home can have a tremendous impact on social relations within families and communities. Writing from a more community-based perspective, Putnam has argued about the negative impact that electronic media have on local conversation and sociability (Putnam, 2000). Putnam has suggested that video games are yet another media technology that is further atomizing communities by bringing individuals out of public spaces. But in this case, Putnam's line of analysis misses the actual sequence of events and presumes incorrectly that game play, regardless of location, is isolating.

The diversity of early arcade play had been drastically reduced by the mid-1980s (Herz, 1997), when games were played primarily in homes (J. B. Funk, 1993; Kubey & Larson, 1990). For play to be social, a group had to gather around a television set. Evidence suggests that in the mid-1980s, home play hit a low point for sociability (Murphy, 1984). The correlation for sociability and home console play was still positive but was not as large as for arcade play (Lin & Leper, 1987). One reason for this temporary drop was that the earliest home games usually only allowed for one or two players, as compared to the four-player consoles that became popular in the early 1990s. Once more games and console systems were made to satisfy the demand for more players, the trend reversed. By 1995, researchers were finding that play was highly social again (Phillips et al., 1995).

Some of the move toward the home was precipitated by advances in technology, and some by changes in the home itself. Over 30 years, technology has lowered the cost of processing and storage to the point where home game units are comparable to arcade units; convenience has moved games into homes. But other less obvious forces have kept game technology moving into more isolated spaces *within* homes. From 1970 to 2000, the average U.S. home size rose from 1,500 square feet to 2,200 square feet, but this space became more subdivided than ever before (O'Briant, 2001). Ten percent more homes had four or more bedrooms than in 1970, even though Americans are having fewer children ("In census data, a room-by-room picture of the American home," 2003). Consequently, there is less shared space within homes and more customized, private space for individuals. More than half of all U.S. children have a video game player in their bedroom (Roberts, 2000; Sherman, 1996). In much the same way that Putnam described televisions moving people off of communal stoops and into houses, games and computers have been moving people out of living rooms and into bedrooms and home offices.

Games, along with other mass media, may have separated families within their own houses causing less intergenerational contact, while at the same time opening up access to new social contacts of all types via networked console systems and PCs. The result is a mix of countervailing social forces—less time with known people and more exposure to new people from a broader range of backgrounds. Whether or not this virtual networking is qualitatively better or worse for social networks than in-person game play is an issue that has received little attention. However, despite the physical separation of game players, the desire to play together has remained constant.

CONCLUSION

If the social history of video games can teach us anything, it is that humans will use games to connect with each other, that technology changes the means (and thus the quality) of those connections, and that this will all generate concern. These conclusions have implications for researchers, both in how we should study gaming and in how we should consider gamers.

The Academic Agenda and a Suggestion

The political climate and news media coverage have had a direct and dramatic effect on the gaming research agenda. The most prominent figure in the U.S. health care system, Surgeon

General C. Everett Koop, was widely cited when he claimed in 1982 that video games were hazardous to the health of young people, created aberrant behavior, and increased tension and a disposition for violence (Lin & Leper, 1987). Although there was no science to back this assertion, researchers understandably went looking for it. In reviewing the resulting literature 10 years later, Funk concluded: "Despite initial concern, current research suggests that even frequent video game playing bears no significant relationship to the development of true psychopathology. For example, researchers have failed to identify expected increases in withdrawal and social isolation in frequent game players" (Funk, 1992, pp. 53–54).

The effects work on violent games and aggression has similar origins. However, simply because the motivations for the research have their roots in sociopolitical fears does not mean that the research must necessarily be invalid. Looking for effects is certainly a worthwhile activity. However, where the research can be found lacking is in its failure to incorporate social variables. The typical experiment has included bringing subjects into a laboratory and then having them play alone. Without the social context in place, it is not clear what such studies are capturing. Sherry suggested that the dominant format of laboratory studies of players playing alone against a computer may be testing for an effect that does not occur normally (Sherry, 2003; Sherry & Lucas, 2003). The long history of social play lends weight to such criticisms.

Again, this is not to say that effects will not be found. Instead, it is to suggest that social variables have been ignored in the models. In fact, some social variables may well cause stronger effects. Others may moderate or reverse them. Although many of the leading researchers on media violence have recently noted this omission (C. Anderson et al., 2003), no work has included social variables. Until experimentalists incorporate the actual circumstances of play, they will be open to criticisms of external validity. The solution does not involve creating new theories. The social learning approach used by most effects researchers is based on observational modeling and is in fact highly applicable and adaptable to social settings. The problem is that the settings and the social actors who might be modeled have been excluded.

In the fast lanes of the Information Superhighway, the speculations and initial research on gaming and online community have begun. Some researchers (Howard, Rainie, & Jones, 2001; Rheingold, 1993; Wellman & Gullia, 1999) have borrowed from Habermas (1998), Anderson (1991), and Oldenburg (1997) to explore the civic and social utopias that might be created in online spaces, including games. Others (Nie, 2001; Nie & Erbring, 2002; Nie & Hillygus, 2002) have argued that far darker outcomes are likely, especially in games (Ankney, 2002). This utopian/dystopian research agenda will be the backdrop as we enter an era of virtual gaming communities. Will these networked games help us cross social boundaries and create new relationships, or take us even farther away from our dwindling civic structures? While empirical evidence remains scant, some may wonder where the game player is in the discussion.

Players Are Not Passive

Games research is the child of mainstream U.S. social science communication research. As such, it is not particularly surprising that attention has remained focused on what games do *to* people rather than what people do *with* games. One problem with this preference is that the games-playing audience is plainly an active one. Some researchers have argued that this activity will make effects stronger (C. Anderson & Bushman, 2001). That may turn out to be true. Nevertheless, digital interactive media have made audience agency obvious to even the casual observer. These new media have destabilized the assumptions of an inactive or gullible consumer by rudely introducing media that have an inescapably active—or interactive—component.

Power, agency, and control have spread both upstream to the producer and downstream to the consumer (Neuman, 1991). It is difficult to suggest that Internet users do not have a high level of choice, agency, and activity. Likewise, video games are plainly an active medium. The starting point for video game research should be a theoretical framework that allows for active users in real social contexts. But when game players go so far as to actively participate in the creation of the content, we must consider them anew. For example, more than one quarter of *EverQuest* players say they have made some original artwork or fiction based on the game (Griffiths et al., 2003). *Counter-Strike* players create wholly new forms of self-expression within their game (T. Wright, Boria, & Breidenbach, 2002). "Modders" take the content creation tools given to them by the manufacturers and create new worlds and objectives consistent with *their*, not the producers' preferences (Katz & Rice, 2002; "PC Gamer It list," 2003). Researchers considering direct effects or limited effects models will have to come to grips with a population that takes a vigorous role in the practice and creation of their medium, not simply its consumption.

As Stephenson (1967) noted long ago:

> Social scientists have been busy, since the dawn of mass communication research, trying to prove that the mass media have been sinful where they should have been good. The media have been looked at through the eyes of morality when, instead, what was required was a fresh glance at people existing in their own right for the first time. (p. 45)

The social history of video games makes plain that we should consider not only the active ways that gamers participate in the medium, but also the long tradition of the way they play together.

ACKNOWLEDGMENTS

The author wishes to thank Kurt Squire, Constance Steinkuehler, and the editors for their comments on a draft of this chapter.

NOTES

[1] For example, the first home game machine, the Magnavox Odyssey, had trouble with consumers in part because many incorrectly assumed it would only work on a Magnavox television set (Herman, 1997).

[2] Cell phone-based games have a similar appeal, and may drive phone use (Schwartz, 1999).

[3] These data were graciously supplied to the author by Senior Research Scientist John Horrigan of the Pew Internet and American Life Project in an e-mail.

[4] Data are from the National Science Foundation's Survey of Earned Doctorates Summary report, 1999.

[5] A recent study of implicit attitudes found that both women and men see science as a male domain (O'Connell, 2003).

REFERENCES

Anderson, B. (1991). *Imagined communities: Reflections on the origin and spread of nationalism.* London: Verso.

Anderson, C., Berkowitz, L., Donnerstein, E., Huesmann, L. R., Johnson, J. D., Linz, D., et al. (2003). The influence of media violence on youth. *Psychological Science in the Public Interest, 4*(3), 81–110.

Anderson, C., & Bushman, B. J. (2001). Effects of violent video games on aggressive behavior, aggressive cognition, aggressive affect, physiological arousal, and prosocial behavior: A meta-analytic review of the scientific literature. *Psychological Science, 12*(5), 353–359.

Ankney, R. N. (2002). *The effects of Internet usage on social capital.* Paper presented at the AEJMC Annual Conference, New Orleans, Louisiana.

Buchman, D. D., & Funk, J. B. (1996). Video and computer games in the '90s: Children's time commitment and game preference. *Children Today, 24*(1), 12–15.

Burnham, V. (2001). *Supercade, a visual history of the videogame age 1971–1984.* Cambridge, MA: MIT Press.

Cassell, J., & Jenkins, H. (1999). Chess for girls? Feminism and computer games. In J. Cassell & H. Jenkins (Eds.), *From Barbie to Mortal Kombat: Gender and computer games* (pp. 2–45). Cambridge, MA: MIT Press.

Chodorow, N. J. (1994). *Feminities, masculinities, sexualities: Freud and beyond.* Lexington: University Press of Kentucky.

Cohen, S. (1984). *Zap! The rise and fall of Atari.* New York: McGraw-Hill.

Consalvo, M. (2003). Hot dates and fairy-tale romances: Studying sexuality in video games. In M. J. P. Wolf & B. Perron (Eds.), *The video game theory reader* (pp. 171–194). New York: Routledge.

Cowan, R. (1983). *More work for mother: The ironies of household technology from the open hearth to the microwave.* New York: Basic Books.

Croal, N. G. (2001, February 5). Online games get real. *Newsweek,* 62–63.

Czitrom, D. (1982). *Media and the American mind: From Morse to McLuhan.* Chapel Hill: University of North Carolina Press.

Dibbell, J. (2001). A rape in cyberspace; or how an evil clown, a Haitian trickster spirit, two wizards, and a cast of dozens turned a database into a society. In D. Trend (Ed.), *Reading digitial culture* (pp. 199–213). Malden, MA: Blackwell.

Dibbell, J. (2003, January). The 79th richest nation on Earth doesn't exist. *WIRED, 12,* 106–113.

Dominick, J. R. (1984). Videogames, television violence, and aggression in teenagers. *Journal of Communication, 34*(2), 136–147.

Douglas, M. (1992). *Risk and blame.* London: Routledge.

Douglas, M., & Wildavsky, A. (1982). *Risk and culture.* Berkeley University of California Press.

Douglas, S. (1987). *Inventing American broadcasting, 1899–1922.* Baltimore: Johns Hopkins University Press.

Douglas, S. (1999). *Listening in: Radio and the American imagination...from Amos n' Andy and Edward R. Murrow to Wolfman Jack and Howard Stern.* New York: Random House.

Ellis, D. (1984). Video arcades, youth, and trouble. *Youth & Society, 16*(1), 47–65.

Elmer-DeWitt, P. (1993, September 27). The amazing video game boom. *Time,* 67–72.

Fischer, C. S. (1992). *America calling: A social history of the telephone to 1940.* Berkeley: University of California Press.

Flanagan, M. (1999). Mobile identities, digital stars, and post-cinematic selves. *Wide Angle, 21*(1), 76–93.

Funk, J. (2001, October 27). *Girls just want to have fun.* Paper presented at the Playing by the Rules Conference, Chicago, Illinois.

Funk, J. (1993). Reevaluating the impact of video games. *Clinical Pediatrics, 32,* 86–90.

Funk, J. (1992). Commentary: Video games; Benign or malignant? *Developmental and Behavioral Psychology. 13*(1), 53–54.

Gabler, N. (1999). *Life the movie: How entertainment conquered reality.* New York: Knopf.

Gailey, C. (1993). Mediated messages: Gender, class, and cosmos in home video games. *Journal of Popular Culture, 27*(1), 81–97.

Garner, T. L. (1991). *The sociocultural context of the video game experience.* Unpublished dissertation, University of Illinois at Urbana-Champaign, Urbana-Champaign.

Gilens, M. (1999). *Why Americans hate welfare: Race, media, and the politics of antipoverty policy.* Chicago: University of Chicago Press.

Gilmore, H. (1999). Female computer game play. In M. Kinder (Ed.), *Kids' media culture* (pp. 263–292). Durham, NC: Duke University Press.

Glassner, B. (1999). *The culture of fear: Why Americans are afraid of the wrong things.* New York: Basic Books.

Glenn, N. (1977). *Cohort analysis* (Vol. 5). Newbury Park, CA: Sage.

Griffiths, M. (1997). Computer game playing in early adolescence. *Youth & Society, 29*(2), 223–237.

Griffiths, M., Davies, M. N., & Chappell, D. (2003). Breaking the stereotype: The case of online gaming. *CyberPsychology & Behavior, 6*(1), 81–91.

Habermas, J. (1998). *The structural transformation of the public sphere: An inquiry into a category of bourgeois society.* Cambridge, MA: MIT Press.

Herman, L. (1997). *Phoenix: The fall and rise of videogames.* Union, NJ: Rolenta.

Herz, J. C. (1997). *Joystick nation.* Boston: Little, Brown.

Howard, P. E., Rainie, L., & Jones, S. (2001). Days and nights on the Internet: The impact of a diffusing technology. *American Behavioral Scientist, 45*(3), 383–404.

In census data, a room-by-room picture of the American home. (2003, February 1). *New York Times,* p. A32.

Jansen, S. (1989). Gender and the information society: A socially structured silence. *Journal of Communication, 39*(3), 196–215.

Katz, J. E., & Rice, R. E. (2002). *Social consequences of Internet use: Access, involvement, and interaction.* Cambridge, MA: MIT Press.

Kent, S. (2000). *The first quarter: A 25-year history of video games.* Bothell, WA: BWD Press.

Killian, S. (2002). *The once and future arcade.* Retrieved January 11, 2002, from www.shoryuken.com/forums/ext_columns.php?f=20&t=7018.

Kirriemur, J. (2002). *The relevance of video games and gaming consoles to the higher and further education learning experience. Western Isles, Scotland*: Ceangal.

Kline, S., & Arlidge, A. (2002). *Online gaming as emergent social media: A survey*: Simon Fraser University Media Analysis Laboratory.

Knowlee, K. H., Henderson, J., Glaubke, C. R., Miller, P., Parker, M. A., & Espejo, E. (2001). *Fair play? Violence, gender and race in video games.* Oakland, CA: Children Now.

Kubey, R., & Larson, R. (1990). The use and experience of the new video media among children and adolescents. *Communication Research, 17*, 107–130.

Kushner, D. (2001, May 10). Nintendo grows up and goes for the gross-out. *New York Times*, pp. D1, 11.

Langway, L. (1981, November 16). Invasion of the video creatures. *Newsweek*, 90–94.

Levy, S. (1994). *Hackers: Heroes of the computer revolution.* New York: Penguin.

Lin, S., & Leper, M. R. (1987). Correlates of children's usage of videogames and computers. *Journal of Applied Social Psychology, 17*(1), 72–93.

Lowery, S., & DeFluer, M. (1995). *Milestones in mass communication research: Media effects.* White Plains, NY: Longman.

Marchand, R. (1985). *Advertising the American dream: Making way for modernity, 1920–1940.* Berkeley: University of California Press.

McNamara, M. (2003, January 7). Cyber cafes—new turf, same old battles. *Los Angeles Times*, p. C1.

McQuivey, J. (2001). The digital locker room: The young, white male as center of the video gaming universe. In E. T. L. Aldoory (Ed.), *The gender challenge to media: diverse voices from the field* (pp. 183–214). Cresskill, NJ: Hampton Press.

Meadows, L. K. (1985). *Ethnography of a video arcade: A study of children's play behavior and the learning process (microcomputers).* Unpublished dissertation, The Ohio State University.

Michaels, J. W. (1993). Patterns of video game play in parlors as a function of endogenous and exogenous factors. *Youth & Society, 25*(2), 272–289.

Mitchell, E. (1984). *Home video games: Children and parents learn to play and play to learn.* Paper presented at the Annual Meeting of the American Educational Research Association, New Orleans, Louisiana.

Mitchell, E. (1985). The dynamics of family interaction around home video games. *Marriage and Family Review, 8*(1), 121–135.

Mulligan, J., & Petrovsky, B. (2003). *Developing online games: An insider's guide.* Boston: New Riders.

Murphy, K. (1984). *Family patterns of use and parental attitudes towards home electronic video games and future technology.* Unpublished dissertation, Oklahoma State University.

The national income and product accounts of the United States, 1929–1976. (1976). U.S. Bureau of Economic Analysis.

Neuman, W. R. (1991). *The future of the mass audience.* Cambridge, England: Cambridge University Press.

Nie, N. H. (2001). Sociability, interpersonal relations, and the Internet: Reconciling conflicting findings. *American Behavioral Scientist, 45*(3), 420–435.

Nie, N. H., & Erbring, L. (2002). Internet and society: A preliminary report. *IT & Society, 1*(1), 275–283.

Nie, N. H., & Hillygus, D. S. (2002). The impact of Internet use on sociability: Time-diary findings. *IT & Society, 1*(1), 1–20.

O'Briant, A. (2001). *Introducing the liquid house.* Unpublished Master of Architecture Thesis, Rice University.

O'Connell, P. (2003, May 22). Online diary: Test your biases. *New York Times*, p. E3.

Ofstein, D. (1991). *Videorama: An ethnographic study of video arcades.* Unpublished dissertation, University of Akron, Akron, Ohio.

Oldenburg, R. (1997). *The great good place: Cafés, coffee shops, community centers, beauty parlors, general stores, bars, hangouts, and how they get you through the day.* New York: Marlowe.

Palumbo, P. (1998). *Online vs. retail game title economics.* Retrieved April 7, 2003, from www.gamasutra.com/features/business_and_legal/19980109/online_retail.htm.

PC Gamer "It" list. (2003, Holiday Issue). *PC Gamer, 10,* 84.

Pham, A. (2003, February 4). "Sims Online" gives creators a painful reality check. *Los Angeles Times*, p. C1.

Phillips, C. A., Rolls, S., Rouse, A., & Griffiths, M. D. (1995). Home video game playing in schoolchildren: A study of incidence and patterns of play. *Journal of Adolescence, 18*, 687–691.

Putnam, R. D. (2000). *Bowling alone: The collapse and revival of American community.* New York: Simon & Schuster.

Ray, M. (1999). Technological change and associational life. In T. Skocpol & M. Fiorina (Eds.), *Civic engagement in modern democracy* (pp. 297–330). Washington, DC: Brookings Institution Press.

Rheingold, H. (1993). *Virtual communities*. Reading, MA: Addison-Wesley.

Roberts, D. (2000). Media and youth: Access, exposure, and privatization. *Journal of Adolescent Health, 27S*(2), 8–14.

Rogers, E. M. (1985). The diffusion of home computers among households in Silicon Valley. *Marriage and Family Review, 8*(1), 89–101.

Rubin, L. (1983). *Intimate strangers*. New York: Harper & Row.

Russo, T. (2001, February). Games grow up. But is the rest of the world ready? *NextGen, 3,* 54–60.

Scantlin, R. (1999). *Interactive media: An analysis of children's computer and video game use.* Unpublished dissertation, University of Texas at Austin.

Schiesel, S. (2003, April 10). The PC generation, back to the board. *New York Times,* pp. F1, 8.

Schiffer, M. (1991). *The portable radio in American life*. Tucson: University of Arizona Press.

Schor, J. (1991). *The overworked American: The unexpected decline of leisure*. New York: Basic Books.

Schwartz, J. (1999). *Online gaming in the next century. Cambridge, MA*: Forrester.

Seymour, E. (1995). The loss of women from science, mathematics, and engineering undergraduate majors: An explanatory account. *Science Education, 79*(4), 437–473.

Sheff, D. (1999). *Game over, press start to continue: The maturing of Mario*. Wilton, CT: GamePress.

Sherman, S. (1996). A set of one's own: TV sets in children's bedrooms. *Journal of Advertising Research, 36*(6), 9–12.

Sherry, J. (2003, May 25). *Relationship between developmental stages and video game uses and gratifications, game preference and amount of time spent in play*. Paper presented at the International Communication Association Annual Conference, San Diego, California.

Sherry, J., & Lucas, K. (2003, May 27). *Video game uses and gratifications as predictors of use and game preference*. Paper presented at the International Communication Association Annual Conference, San Diego, California.

Skow, J. (1982, January 18). Games that play people: Those beeping video invaders are dazzling, fun-and even addictive. *Time,* 50–58.

Spigel, L. (1992). Installing the television set: Popular discourses on television and domestic space, 1948–1955. In L. Spigel & D. Mann (Eds.), *Private screenings: Television and the female consumer*. Minneapolis:: University of Minnesota Press.

Standage, T. (1999). *The Victorian Internet: The remarkable story of the telegraph and the nineteenth century's online pioneers*. Berkeley: University of California Press.

Stephenson, W. (1967). *The play theory of mass communication*. Chicago: University of Chicago Press.

Survey of current business. (1996). U.S. Bureau of Economic Analysis.

Turkle, S. (1995). *Life on the screen: Identity in the age of the Internet*. New York: Touchstone.

Vogel, H. L. (2001). *Entertainment industry economics: A guide for financial analysis*. Cambridge, England: Cambridge University Press.

Wellman, B., & Gullia, M. (1999). Net surfers don't ride alone: Virtual communities as communities. In B. Wellman (Ed.), *Networks in the global village*. p.331–366 Boulder, CO: Westview.

Williams, D. (2003). The video game lightning rod. *Information, Communication & Society, 6*(4), 523–550.

Wright, K. (2002). *GDC 2000: Race and gender in games*. Retrieved April 11, 2002, from http://www.women-gamers.com/articles/racegender.html.

Wright, T., Boria, E., & Breidenbach, P. (2002). Creative player actions in FPS online video games: Playing Counter-Strike. *Game Studies, 2*(2).

Yee, G., Zavala, V., & Marlow, J. (2002). On *Life & Times* [Television]. Los Angeles.

15

Video Game Uses
and Gratifications as Predictors
of Use and Game Preference

John L. Sherry
Michigan State University

Kristen Lucas
Purdue University

Bradley S. Greenberg
Michigan State University

Ken Lachlan
Boston College

Video games continue to be a highly popular form of entertainment. In 2003, over 239 million computer and video games were sold in the United States, and the video game industry reported sales of over $7 billion (Entertainment Software Association, 2004). According to an industry poll conducted by ESA (2004), 50% of U.S.-Americans play video games, the average age is 29, and 61% of players are male. An Annenberg Public Policy Center survey estimates that video game consoles are in 68% of U.S.-American homes with at least one 2- to 17-year-old and in 75% of homes with two or more children (Woodard & Gridina, 2000). These figures are expected to grow as high-speed broadband Internet access facilitates networked game play. Clearly, video games have emerged as one of the most popular forms of mass mediated entertainment in the United States among a range of people.

Despite this popularity, the study of video games is still in its infancy. To date, most video game studies have focused on traditional effects issues, particularly the effects of violent video games on aggression (see Anderson & Bushman, 2001; Funk, 1992; Mediascope, 1996; Sherry, 2001). Many of the questions found in Chaffee's (1977) 18-cell explication of media effects have gone largely unaddressed. Prominent among these questions are the reasons why people use video games and the gratifications that they receive from them. In this chapter, we explore the reasons that individuals use video games and how those reasons are translated into genre preferences and amount of time devoted to game play from a uses and gratifications perspective.

THE USES AND GRATIFICATIONS PARADIGM

Since its inception, the uses and gratifications paradigm has provided a cutting edge approach for gaining insight on the uses and impact of new communication technologies (Rubin, 1994; Ruggiero, 2000). Uses and gratifications research is rooted in the structural–functionalist systems approach (Palmgreen, Wenner, & Rosengren, 1985) to understanding the interface between biological entities and their context (e.g., Bertalanffy, 1968; Buckley, 1967; Merton, 1957; Monge, 1977). Common among these perspectives is the idea that human behavior may best be understood as a system represented as "interlinked sets of components hierarchically organized into structural wholes which interact through time and space, are self-regulating, yet capable of structural change" (Monge, 1977, p. 20). Systems theory, as applied to human behavior, places individuals, each having unique biological features expressed in both physical and mental attributes, within a multisystem context and attempts to account for the cross-system influences on behavior (Bertalanffy, 1968; Blalock & Blalock, 1959). Furthermore, humans are believed to be self-regulating; that is, individuals respond to felt needs and contextual factors (Lerner, 1987). In the case of media, an individual's media use and the effects of that media use are largely (though not completely) a function of the individual's purpose for using the media. For example, we would expect greater cognitive change resulting from reading a magazine article from an individual who perceives a deficit in knowledge on the subject (i.e., is reading for information) than from an individual who is reading to pass time (i.e., to alleviate boredom), all other traits held constant (e.g., age, intelligence, reading comprehension level, initial knowledge).

From a systems perspective, people use media to solve perceived problems in order to maintain equilibrium. Other perspectives on media use share the idea of media use to manage equilibrium such as sensation seeking, novelty, dispositional alignments, and evolutionary explanation (for summaries of these ideas, see Bryant & Miron, 2002; Sparks & Sparks, 2000). Atkin (1985) distinguished between exposure to media as being driven by the need for immediate, momentary intrinsic satisfaction (e.g., enjoyment seeking) or extrinsic utility (e.g., information). Research in mood management suggests that individuals use media to manage fluctuations in positive and negative emotional states (Zillmann & Bryant, 1985). For example, Bryant and Zillmann (1984) found that individuals who were asked to perform monotonous tasks were more likely to choose exciting media than those individuals who were under stress. Another perspective, media flow theory (Sherry, 2004), posits that exposure to media is an intrinsically rewarding experience in which media users attain the highly engaging and enjoyable flow state described by Csikszentmihalyi (1997). These equilibrium-based theories demonstrate that a complex account of media effects cannot be obtained in isolation from the reasons that individuals use media.

Human systems research requires the following set of necessary and sufficient logical conditions: (a) identification of the set of interrelated parts in the system, (b) specification of the environment in which the system operates, (c) specification of a trait or attribute of the system that is essential for the continuation of the system, (d) specification of the range within which traits must remain in order for the system to remain in operation, and (e) a detailed account of how the parts operate together to maintain the traits within acceptable ranges (Monge, 1977). The uses and gratifications paradigm advanced by Rosengren (1974) provides a theoretical explanation for the study of media effects from a systems perspective. Essentially (see Figure 15.1), basic needs (1), individual differences (2), and contextual societal factors (3) combine to result in a variety of perceived problems and motivations (4–6) to which gratifications are sought from the media (7) and elsewhere (8) leading to differential patterns of media effects (9) on both the individual (10) and societal (11) levels. Consistent with human systems research, the model is complex, multivariate, and nonrecursive (Palmgreen et al., 1985).

Historically, empirical research in the uses and gratifications tradition begins with the identification of logical conditions numbers 3 and 4; that is, the traits[1] essential for understanding the role of media in the individual's system and the range of those traits. The traits that traditionally have been studied in this paradigm are sets of motivations for media use. A number of

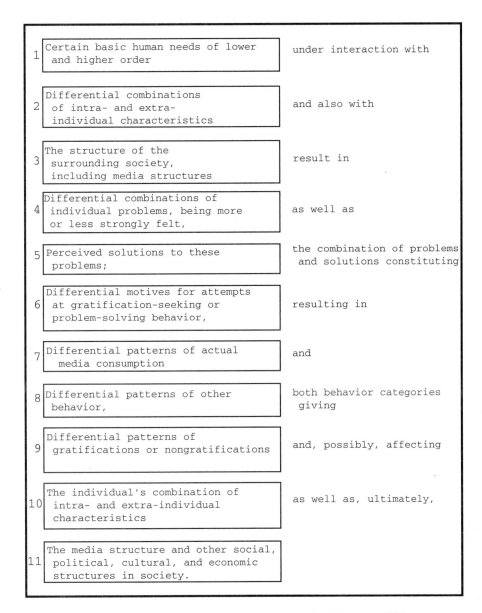

FIG. 15.1. Uses and gratifications paradigm (adapted from Rosengren, 1974).

taxonomies of these traits have emerged, sorting uses and gratifications according to entertainment genre (Abelman, 1987; Gantz, 1996), media type (Greenberg & Hnilo, 1996; Selnow, 1984), and culture of respondents (Greenberg, Li, Ku, & Tokinoya, 1991; Tokinoya, 1996; Youichi, 1996). One of the lessons of these studies is that there is no universal set of reasons for using media: Motivations vary across media, genres, and cultures. For this reason, empirical work is required at the introduction of each new medium to specify the key traits in understanding the use of that medium (Wartella & Reeves, 1985). After identification of the traits, a detailed account of how the parts of the system operate together can be specified and media effects can be explained as a function of the reasons for use.

To date, there have been few attempts at developing sets of video game uses and gratifications traits. Selnow (1984) published the first video game uses and gratifications study in which he surveyed 244 10- to 24-year-olds about the needs and gratifications met by video games. Because most video game play in the early 1980s occurred at arcades, and the home console systems that currently dominate game play did not exist, the study was limited to patterns of arcade game use. Selnow used Greenberg's (1974) television uses and gratifications scale, adding a few additional video game-specific dimensions. Selnow's exploratory factor analysis yielded five arcade video game play factors: (a) game play is preferable to human companions, (b) game play teaches about people, (c) game play provides companionship, (d) game play provides activity/action, and (e) game play provides solitude/escape. These five factors were significantly correlated with amount of game play.

A second uses and gratifications study of video games was published the following year (Wigand, Borstelmann, & Boster, 1985). Again, the focus was on arcade use, so the researchers focused largely on understanding reasons that adolescents used arcades rather than games. Based on the existing literature in the early 1980s, Wigand et al. (1985) found the main reasons for using video games at arcades were excitement, satisfaction of doing well, and tension-reduction. In 1990, Myers published a study isolating four factors of game play: fantasy, curiosity (novelty), challenge, and interactivity. All four factors were significantly related to amount of game play, with challenge consistently having the highest rankings.

More recently, researchers have included items that are similar to uses and gratifications in their projects. For example, research in the U.K. has touched on uses and gratifications of console video game play. A survey conducted by Phillips, Rolls, Rouse, and Griffiths (1995) used single-item measures of video game play motivation including: "to pass time," "to avoid doing other things," "to cheer oneself up," and "just for enjoyment." Furthermore, Griffiths' (1991a, 1991b) research on video game addicts proposed additional uses and gratifications, including arousal, social rewards, skill testing, displacement, and stress reduction. Most recently, Vorderer, Hartmann, and Klimmt (2003) have argued that competition is the chief gratification obtained by playing video games.

In order to specify a set of traits and ranges that reflect the experience of game users, the present studies were undertaken in a methodologically similar manner to Greenberg's (1974) seminal study of British children, which provided the basic set of theoretical traits for understanding television use and effects. Greenberg's study consisted of two parts: an analysis of children's essays and a survey. Children (9-, 12-, and 15-year-olds) from a London school district wrote essays on "Why I Like to Watch Television." The 180 essays collected were content-analyzed to determine reasons for watching television, resulting in nine clusters: to pass time, to forget, as a means of diversion, to learn about things, to learn about myself, for arousal, for relaxation, for companionship, and as a habit. These traits provided the basis for a 31-item scale that was included in a survey of 726 British 9-, 12-, and 15-year-olds. Subsequent television uses and gratifications scales built on Greenberg's original dimensions.

In this chapter, we followed Greenberg's multimethod approach to developing the set of theoretical traits for video game uses and gratifications. First, we conducted a series of focus groups with young adults to determine reasons for playing video games. From these responses, we developed and tested a uses and gratifications scale. Next, we conducted a survey in order to determine the range of the uses and gratifications traits and how those traits are related to amount of game use. It is important to remember that the uses and gratifications tradition assumes that individuals can accurately report their own motivations for media use. Some have argued that self-reports of media use may not be veridical to actual experience, but may represent idealized version of behavior (Bryant & Miron, 2002).

ISOLATING THE TRAITS

The purpose of Study 1 was to generate an empirically valid set of traits needed for the theoretical explication of a systems model for video game use and effects. From this set of traits, a valid and reliable instrument was created for measuring the traits.

Respondents

Eighteen- to 22-year-old U.S.-American undergraduate students were participants in a series of focus groups. This population is particularly appropriate because it is the first generational co-hort to grow up with video games both at home and in arcades. Members of this generation were born in the early 1980s, when the early home console games Atari and Mattel Intellivision were popular. As they progressed through grade school, Nintendo NES (1986), Nintendo Game Boy (1989), and Sega Genesis (1989) were introduced and rose to prominence. Their high school years saw the introduction of the Sony PlayStation system (1995) and Nintendo 64 (1996).

Focus Group Procedures and Results

Focus group interview sessions with four to eight participants each were conducted to ascertain the most common reasons from using video games ($n = 96$). Participants signed up for either "Video Game Player," "Non-Video Game Player," or "Mixed Player and Non-Player" sessions based on their own evaluation of their video game experience. Participants filled out a brief survey upon entering the interview room asking about their video game use in order to prime their memories. Once all participants were gathered and had completed the preliminary survey, a moderator conducted the interview using a standardized set of questions designed to funnel responses from general to specific and allowing for probes (Morgan, 1997). Subsequently, interviews were analyzed for repeated themes representing dimensions of video game play motivations. Analysis of focus group data resulted in six dominant dimensions of video game use that included Arousal, Challenge, Competition, Diversion, Fantasy, and Social Interaction (see Table 15.1).

Arousal. A frequently stated reason for playing video games was to stimulate emotions as a result of fast action and high-quality graphics.

Challenge. Many respondents also enjoy playing video games to push themselves to a higher level of skill or personal accomplishment. Some respondents said that the desire to solve the puzzles in order to get to the next level or beat the game can be addicting. Many of the players prefer to play a familiar set of games that they feel confident playing.

Competition. One of the most frequently cited reasons for playing video games was to prove to other people who has the best skills and can react or think the fastest. Typically, competition response came from male respondents who spoke of competing for pride or money. Hence, video game competition served the function of a dominance display among males most often seen in sports. This gratification derives its power from the reactions of others to the dominance shown by the player, establishing a relative position in the peer group's hierarchy.

Diversion. Video games are frequently used to avoid stress or responsibilities. Respondents reported playing video games to fill time, relax, escape from stress, or because there is nothing else to do.

TABLE 15.1

Examples of Dimensional Responses From Focus Groups

Arousal	"I go crazy when I'm playing video games, sometimes. I'm jumping up and down. Yelling and screaming. Things like that."
	"It was the neatest game. And I would jump up and down and swear because I would be so into the game."
	"[My friends who lived in the dorm] were all hooked up, in all the different rooms. And they were, like, *obsessed* with it. It was crazy. It was out of control, playing that game."
	"It kind of pressures you to do as much as you can, but at the same time, it's fun."
	"You can sit there and play it over and over."
Challenge	"I like it because it's a challenge and I like competition. I keep playing until I complete a level or win the game."
	"It seems to be addicting to play, you always want to do better and better."
	"The only reason why I play now is to get a better score."
	"Repeating levels bores me. I hate games that you always start at the beginning."
	"I have played with someone else, Tetris. And I didn't find too much fun in that, you know. I like to play alone. Challenge against the computer, instead of challenging the person next to you."
Competition	"We always play in all house (fraternity) tournaments. We used to put money down."
	"I love trying to beat the guys next door or brothers."
	"[Competition] is pretty much the only reason why I play. We, like, have an intercom in our house and they'll call you out and you'll have to defend."
	"It [competition among fraternity members] gets personal. It's funny."
	"When you play with someone you've never played with, and they think they're the expert, and you beat them finally, you get invited to play with them more. Because it's like, 'I'm going to beat you this time.' "
Diversion	"I like it because it's a break from studying and it's relaxing."
	"I always lose track of time while I'm playing."
	"It was so bad one year that I had to bring my Nintendo home. I was going to flunk out because all I ever did was play *Bond*."
	"When you study so much, you're just like 'uuugh, I want a break.' "
	"I don't necessarily use it for a study break. I use it more for an excuse to stop studying."
Fantasy	"It's like you're in another world. You're in a TV, you know. You're in it."
	"It takes you away from reality."
	"Because he likes the game so much, [my boyfriend] fits himself into the game. For example, in *Counter Strike* he imagines himself as the soldier. He puts himself into the picture."
	"You can do anything. It's so realistic. It's just like real skateboarding."
	"I really like shooting. [I played Duck Hunt] because you can do it without, like, playing with a real gun."
	"You get to be a Star Wars fighter pilot and you fly missions just like you are in the movies and everything."
Social Interaction	"I play with my boyfriend and friends just to have fun."
	"I like it because it's just plain fun, just being with your friends."
	" I can't play by myself. I always want someone to play with me."
	"In the dorms, it was almost, like, easier. You've got like, 50 guys there and you know that at least some of them are going to have free time to play."

Fantasy. Video games allow players to do things that they normally would not be able to do, such as drive race cars, fly, and so on. Focus group participants spoke frequently about the appeal of being able to do things they cannot do in real life.

Social Interaction. Social interaction is the main reason many individuals got involved in playing video games as a child. Respondents mentioned that the arrival of Nintendo brought sleepovers where video games were played, and the need to keep up on the games to be "cool." Many now use video games to interact with friends and learn about the personalities of others.

TABLE 15.2
Analysis of Video Game Uses and Gratifications Instrument

Item #	Item	Factor Loading	Alpha
	Competition		.86
6	I like to play to prove to my friends that I am the best.	.79	
12	When I lose to someone, I immediately want to play again in an attempt to beat him/her	.72	
16	It is important to me to be the fastest and most skilled person playing the game.	.81	
20	I get upset when I lose to my friends.	.78	
	Challenge		.80
3	I feel proud when I master an aspect of a game.	.70	
8	I find it very rewarding to get to the next level.	.84	
13	I play until I complete a level or win a game.	.64	
17	I enjoy finding new and creative ways to work through video games.	.64	
	Social Interaction		.81
7	My friends and I use video games as a reason to get together.	.83	
15	Often, a group of friends and I will spend time playing video games.	.83	
	Diversion		.89
4	I play video games when I have other things to do.	.90	
10	I play video games instead of other things I should be doing.	.90	
	Fantasy		.88
1	I play video games because they let me do things I can't do in real life.	.75	
9	Video games allow me to pretend I am someone/somewhere else.	.88	
14	I like to do something that I could not normally do in real life through a video game.	.82	
18	I enjoy the excitement of assuming an alter ego in a game.	.78	
	Arousal		.85
2	I find that playing video games raises my level of adrenaline.	.74	
5	Video games keep me on the edge of my seat.	.79	
11	I play video games because they stimulate my emotions.	.69	
19	I play video games because they excite me.	.83	

Factor analysis reduced the total instrument to 27 items (see note 1). The 27-item scale was pretested in two introductory research methodology classes ($n = 54$) where respondents were questioned about the clarity of the items. Finally, the items were subjected to an additional internal consistency and parallelism factor analysis on the survey data reported in Study 2 ($n = 550$). The final analysis resulted in a 20-item scale representing six uses and gratifications dimensions (see Table 15.2).

TRENDS IN VIDEO GAME USES AND GRATIFICATIONS

The purpose of Study 2 was to examine age and sex patterns among the set of system traits (uses and gratifications), as well as the relationship among uses and gratification traits and amount of hours playing video games. If we have identified a useful set of uses and gratifications traits, we would expect that the uses and gratifications traits will be significant predictors of amount of game use.

Method

Survey data were collected from two universities, three high schools, two middle schools, and two elementary schools in the Midwestern region of the United States. Only fully completed surveys were included, resulting in a total of 1,265 usable cases. The mean age of the university sample was 19.68 years old (range = 18 to 23). Gender distribution of respondents was 58% female ($n = 321$) and 42% male ($n = 229$). Most of the respondents were Freshmen (43%) and Sophomores (26%), while fewer were Juniors (19%) and Seniors (12%). Next, survey data were collected at two high schools, resulting in 318 usable cases. The mean age was 16.56 years old (range = 16 to 18). Gender distribution of respondents was 53% female ($n = 168$) and 47% male ($n = 150$). Third, survey data were collected from eighth graders from two Midwest middle schools. Only fully completed surveys were included, resulting in a total of 227 usable cases. The mean age was 13.84 years old (range = 13 to 16) and gender distribution of respondents was 55% female ($n = 125$) and 45% male ($n = 102$). A fourth set of survey data was collected from fourth- and fifth-grade students at two Midwest elementary schools. Only fully completed surveys were included, resulting in a total of 141 usable cases. Ages ranged from 9 to 11 years with a mean age of 10.15 years old. Gender distribution of respondents was 50% female ($n = 70$) and 50% male ($n = 71$).

The survey contained two main scales that measured: (a) amount of hours played in various dayparts during the typical week and (b) uses and gratifications of video game play. Respondents were asked to report the amount of time they spend playing video games during a typical week in the school year. In order to facilitate autobiographical memory (Menon, 1994), the respondents filled out a grid that broke the typical week first into days and then into four dayparts (before noon, between noon and 6 p.m., between 6 p.m. and midnight, and after midnight). All dayparts were summed to create a score representing the total number of hours played during the typical week. Uses and gratifications of video games were based on the 20-item scale developed in the focus group study.

Results

The average number of hours played across the total sample per week was 11.04 ($SD = 18.41$), with boys ($n = 564$) reporting playing more than twice as many hours per week ($M = 16.87, SD = 16.40$) as girls ($n = 701$) reported playing ($M = 6.34, SD = 11.31$). Among all respondents, the top reason on a scale from 1 to 7 for playing video games was Challenge ($M = 4.22, SD = 1.47$) followed by Competition ($M = 4.11, SD = 1.61$), Diversion ($M = 3.95, SD = 1.62$), Arousal ($M = 3.91, SD = 1.57$), Fantasy ($M = 3.64, SD = 1.80$), and Social Interaction ($M = 3.42, SD = 1.81$). Three additional factors were included for the fifth-, eighth-, and eleventh-grade samples, but these were not as important as the other traits. These included: playing because of the quality of the Graphics and sound ($M = 3.50, SD = 1.60$), playing because the games are Realistic ($M = 2.89, SD = 1.50$), and playing for the fantasy of being more Strong than in real life ($M = 2.57, SD = 1.58$). All means were normally distributed.

In the college, eleventh-grade, and fifth-grade samples, Challenge was the top reason for playing video games. The pattern held across sexes within each sample also. Least popular among all the gratifications in the eleventh-, eighth-, and fifth-grade samples was playing for the fantasy of being Strong (this dimension was not measured in the college sample). Males consistently ranked all gratifications higher than females did and the rank order of gratifications between males and females was fairly consistent. Among all grade samples, the rank order was consistent except that males rated Social Interaction much higher on their rank order than females did. The size of the difference between motivations was measured using Cohen's d. Cohen (1988) provides a way of interpreting his statistic: $d = .20$ is a small effect, $d = .50$ is a moderate size effect, and $d = .80$ is a large effect. Social Interaction had the largest difference between males and females as expressed as Cohen's $d > 1.0$ in all samples except eighth graders. Among the eighth graders, females ranked Challenge as the number one gratification, while males ranked Social Interaction at the top. The largest mean difference between sexes was in the fifth-grade sample (M Cohen's $d = .87$), followed by the college students (M Cohen's $d = .76$), eleventh graders (M Cohen's $d = .68$), and the eighth graders (M Cohen's $d = .59$).

Uses and Gratifications Traits as Predictor of Hours Played. The relationship between the use categories and amount of play was tested by employing stepwise linear multiple regression to determine if uses and gratifications could predict total hours played among game players. In each sample, the uses and gratifications variables were strong predictors of time spent playing video games. Among college students, 28% of the variance in game play was accounted for by uses and gratifications ($R^2 = .28$; $p < .01$, with Diversion (ß $= .29$; $p < .01$, $sr = .21$), Social Interaction (ß $= .24$; $p < .01$, $sr = .19$), and Arousal (ß $= .11$; $p < .05$, $sr = .08$) as the most important predictors of time spent playing video games per hour. The pattern was the same among 11th graders, with more variance explained ($R^2 = .36$; $p < .01$) using the same variables Diversion (ß $= .24$; $p < .01$, $sr = .19$), Social Interaction (ß $= .31$; $p < .01$, $sr = .24$), and Arousal (ß $= .16$; $p < .01$, $sr = .12$). A similar amount of variance was explained in the eighth-grade sample ($R^2 = .38$; $p < .01$) using only two of the other motivations (i.e., Diversion [ß $= .21$; $p < .01$, $sr = .18$] and Social Interaction [ß $= .47$; $p < .01$, $sr = .39$]). Uses and gratifications accounted for 28% of the variance in amount of weekly game play in the fifth grade sample ($R^2 = .28$; $p < .01$), with a different set of significant predictors: playing to be Strong (ß $= .31$; $p < .01$, $sr = .24$) and for Competition (ß $= .29$; $p < .01$, $sr = .22$).

FUTURE DIRECTIONS FOR RESEARCH

In this project, we set out to develop a nuanced understanding of the video game play experience by investigating the principal motivations for play and how these motivations relate to amount of game play. The pattern of relationships among the uses suggests that the generational cohorts studied here have developed a consistent set of reasons for playing video games. This pattern of uses is likely more purposeful and active than television use because of the costs involved in game play; in addition to owning a television set, gamers must invest approximately $200 for a gaming system and $50 for each game that they want to play. The costs are even higher for computer game players, who need to purchase a more high-powered, graphics-intensive system than the average computer user. Furthermore, unlike television viewers for whom a wide variety of content is delivered to their home either for free or for a monthly flat fee, gamers must purchase each game. As such, they are less likely to expose themselves to content that they are not sure will give them the gratification that they are seeking.

While there is evidence for use of games to equilibrate emotions, the most prominent of the motivations for game use are more social in nature. Responses in the focus groups did not center as much on gratifications derived from feelings as on relationships, resulting in only two emotional gratifications. In the survey data, diversion is the third most popular reason for using video games, while arousal ranks a very close fourth. Importantly, the two emotional gratifications are consistent predictors of game play time across age groups. While they are not the highest ranked gratifications overall, they are important predictors of amount of play.

Results show that the game experience focuses on personal and social gratifications. Players enjoy the challenge of "beating the game," but also of beating friends. For many, it is not enough to win the game; one's exploits must be known amongst one's friends. Like other contests, such as sports, game players can establish a place on a peer pecking order by being the best at a game. Focus group data suggest that competition is most acute among sports and fighter genres—games in which players compete through both agility and knowledge of the game. Unlike real-world sports and fighting, the video game world does not discriminate by physical height and strength, offering a more level playing field than is found in the real world. Like the real world, games are used as a source of social interaction, particularly for males. In fact, social interaction and diversion gratifications were the strongest predictors of time spent playing video games in the oldest three cohorts. Therefore, the diversion from life that video games provide is not necessarily diversion from other people (cf. Williams, chap. 14, this volume). Instead, gaming appears to be a type of diversion that involves other people in social interaction . This finding contradicts the idea of the solitary player isolated from social contact. In fact, frequent game play appears to be highly social with focus group participants describing the experience as being very similar to a group of friends shooting baskets at the park. The ritual is the same; only the location has changed.

Unlike findings from previous studies of television and film gratifications, video game players did not mention using video games for learning (see Lieberman, chap. 25, this volume; Ritterfeld & Weber, chap. 26, this volume). This is important for understanding the mechanisms by which video games may affect users. Bandura (1994) argued that social learning resulting from media messages occurs when users choose a role model from a media portrayal and imitate the model's behavior. Consistent with this theory, in most television uses and gratifications studies, respondents report that the major reason for using television is to learn about the world and about how they should act. Television viewers seek role models from television characters and personalities, modeling their behaviors in real life. Given the purposeful use of television for learning, social cognitive theory (Bandura, 1994) provides a logical explanation of the effects of television. In fact, the storytelling nature of television is consistent with centuries-old socialization mechanism of folklore (Levi-Strauss, 1995). However, the data in these studies suggest that video game players may not be learning by imitating video game role models as has been hypothesized by video game researchers (e.g., Calvert & Tan, 1994; Irwin & Gross, 1995). Bandura (1994) argued that during the attentional process stage, people "determine what is selectively observed in the profusion of modeling influences and what information is extracted from ongoing modeled events" (pp. 67–68). Thus, selection of role models is an active and purposeful behavior enacted via each individual's self-reflective capacity (see Bandura, 1994, pp. 64–66). If video game players were acquiring behavior via a social learning mechanism, they would be aware of and self-reflective upon their choices. However, video game users do not report using games to learn how to behave in the same manner as television and film viewers do. While there may be incidental social learning of behavior, it is more likely that video game effects will result from another mechanism such as arousal transfer, priming, or possibly desensitization.

From the early years of mass communication research, scholars have held that media use is purposeful and that the key to understanding media effects lies in understanding the reasons why people use media (Blumer, 1933; Herzog, 1944). The major theoretical contribution of these studies lies in the specification of the main reasons people use video games. As a result, we can begin to theorize more realistically about the effects of video games. However, it is important that the systematic study of the uses and gratifications of video games does not stop with the specifications of motivations for use. We must determine the antecedents of use as well as the effects of use relative to these sets of motivations. Only then will the uses and gratifications approach begun here offer an explanation of video game uses and effects as outlined in the Rosengren model (see Figure 15.1). The next step in this research project will be to determine what factors influence the decision to use one game genre over another. Do certain personality types prefer certain types of gratifications (and therefore games)? Are there age and sex differences in the use of games? These types of investigations are likely to provide a clearer picture of why people use video games and provide a basis for eventually theorizing about the effects the games may have on them.

NOTES

[1]Complete details about the analysis of psychometric properties and results tables are available from the first author.

REFERENCES

Abelman, R. (1987). Religious television uses and gratifications. *Journal of Broadcasting and Electronic Media, 31*, 293–307.

Anderson, C. A., & Bushman, B. J. (2001). Effects of violent video games on aggressive behavior, aggressive cognition, aggressive affect, physiological arousal, and prosocial behavior: A meta-analytic review of the scientific literature. *Psychological Science, 12*, 353–359.

Atkin, C. K. (1985). Informational utility and selective exposure to entertainment media. In D. Zillmann & J. Bryant (Eds.), *Selective exposure to communication* (pp. 63–91). Hillsdale, NJ: Lawrence Erlbaum Associates.

Bandura, A. (1994). The social cognitive theory of mass communication. In J. Bryant & D. Zillmann (Eds.), *Media effects: Advances in theory and research* (pp. 61–90). Hillsdale, NJ: Lawrence Erlbaum Associates.

Bertalanffy, L. (1968). *General systems theory*. New York: Braziller.

Blalock, H. M., & Blalock, A. B. (1959). Toward a clarification of systems analysis in the social sciences. *Philosophy of Science, 26*, 16–27.

Blumer, H. (1933). *The movies and conduct*. New York: Macmillan.

Bryant, J., & Miron, D. (2002). Entertainment as media effect. In J. Bryant & D. Zillmann (Eds.), *Media effects: Advances in theory and research* (pp. 549–582). Hillsdale, NJ: Lawrence Erlbaum Associates.

Bryant, J., & Zillmann, D. (1984). Using television to alleviate boredom and stress: Selective exposure as a function of induced excitational states. *Journal of Broadcasting, 28*, 1–20.

Buckley, W. (1967). *Sociology and modern systems theory*. Englewood Cliffs, NJ: Prentice-Hall.

Calvert, S., & Tan, S. L. (1994). Impact of virtual reality on young adult's physiological arousal and aggressive thoughts: Interaction versus observation. *Journal of Applied Developmental Psychology, 15*, 125–139.

Chaffee, S. H. (1977). Mass media effects: New research perspectives. In D. Lerner & L. M. Nelson (Eds.), *Communication research: A half-century appraisal* (pp. 210–241). Honolulu: University of Hawaii Press.

Cohen, J. (1977). *Statistical power analysis for the behavioral sciences*. New York: Harcourt-Brace.

Csikszentmihalyi, M. (1997). *Finding flow: The psychology of engagement with everyday life*. New York: Basic Books.

Entertainment Software Association. (2004). *Essential facts about the computer and video game industry*. Retrieved August 1, 2004, from http://www.theesa.com/EFBrochure.pdf

Funk, J. B. (1992). Video games: Benign or malignant. *Developmental and Behavioral Pediatrics, 13*, 53–54.

Gantz, W. (1996). An examination of the range and salience of gratifications research associated with entertainment programming. *Journal of Behavioral and Social Sciences, 1996*(1), 11–48.

Greenberg, B. S. (1974). Gratifications of television viewing and their correlates for British children. In J. G. Blumler & E. Katz (Eds.), *The uses of mass communications: Current perspectives on gratifications* (pp. 71–92). Beverly Hills, CA: Sage.

Greenberg, B. S., & Hnilo, L. R. (1996). Demographic differences in media gratifications. *Journal of Behavioral and Social Sciences, 1996*(1), 97–114.

Greenberg, B. S., Li, H., Ku, L., & Tokinoya, H. (1991). Affluence and mass media behaviours among youth in China, Japan, Korea and Taiwan. *Asian Journal of Communication, 2*, 87–108.

Griffiths, M. D. (1991a). The observational analysis of adolescent gambling in U.K. amusement arcades. *Journal of Community and Applied Social Psychology, 1*, 309–320.

Griffiths, M. D. (1991b). Are computer games bad for children? *The psychologist: Bulletin of the British Psychological Society, 6*, 401–407.

Herzog, H. (1944). What do we really know about daytime serial listeners? In P. F. Lazarsfeld & F. N. Stanton (Eds.), *Radio Research 1942–1943* (pp. 3–33). New York: Duel, Sloan, & Pearce.

Irwin, A. R., & Gross, A. M. (1995). Cognitive tempo, violent video games, and aggressive behavior in young boys. *Journal of Family Violence, 10*, 337–350.

Lerner, R. M. (1987). A life-span perspective for early adolescence. In R. M. Lerner & T. T. Foch (Eds.), *Biological–psychosocial interactions in early adolescence* (pp. 9–34). Hillsdale, NJ: Lawrence Erlbaum Associates.

Levi-Strauss, C. (1995). *Myth and meaning*. New York: Schocken Books.

Lewin, K. (1951). *Field theory in social science*. New York: Harper & Row.

Mediascope. (1996). The social effects of electronic interactive games: An annotated bibliography. Studio City, CA: mediascope.

Menon, G. (1994). Judgments of behavioral frequencies: Memory search and retrieval strategies. In N. Schwarz & S. Sudman (Eds.), *Autobiographical memory and the validity of retrospective reports* (pp. 161–172). New York: Springer-Verlag.

Merton, R. K. (1957). *Social theory and social structure*. Glencoe, IL: Free Press.

Monge, P. (1977). The systems perspective as a theoretical basis for the study of human communication. *Communication Quarterly, 25*, 19–29.

Morgan, D. L. (1997). *Focus groups as qualitative research*. Thousand Oaks, CA: Sage.

Myers, D. (1990). Computer game genres. *Play & Culture, 3*, 286–301.

Palmgreen, P. C., Wenner, L. A., & Rosengren, K. E. (1985). Uses and gratifications research: The past ten years. In K. E. Rosengren, L. A. Wenner, & P. C. Palmgreen (Eds.), *Uses and gratifications research: Current perspectives* (pp. 11–37). Beverly Hills, CA: Sage.

Phillips, C. A., Rolls, S., Rouse, A., & Griffiths, M. D. (1995). Home video game playing in schoolchildren: A study of incidence and patterns of play. *Journal of Adolescence, 18*, 687–691.

Rosengren, K. E. (1974). Uses and gratifications: A paradigm outlined. In J. G. Blumler & E. Katz (Eds.), *The uses of mass communications: Current perspectives of gratifications research* (pp. 269–286). Beverly Hills, CA: Sage.

Rubin, A. M. (1994). Media uses and effects: A uses-and-gratifications perspective. In J. Bryant & D. Zillmann (Eds.), *Media effects: Advances in theory and research* (pp. 417–436). Hillsdale, NJ: Lawrence Erlbaum Associates.

Ruggiero, T. E. (2000). Uses and gratifications theory in the 21st century. *Mass Communication & Society, 3*, 3–37.

Selnow, G. W. (1984). Playing videogames: The electronic friend. *Journal of Communication, 34*(2), 148–156.

Sherry, J. L. (2001). The effects of violent video games on aggression: A meta-analysis. *Human Communication Research, 27*, 409–431.

Sherry, J. L. (2004). Flow and media enjoyment. *Communication Theory, 14*, 392–410.

Sparks, G. G., & Sparks, C. W. (2000). Violence, mayhem, and horror. In D. Zillmann & P. Vorderer (Eds.), *Media entertainment: The psychology of its appeal* (pp. 73–92). Mahwah, NJ: Lawrence Erlbaum Associates.

Tokinoya, H. (1996). A typological study with media gratifications theory in Japan. *Journal of Behavioral and Social Sciences, 1996*(1), 115–137.

Vorderer, P., Hartmann, T., & Klimmt, C. (2003). Explaining the enjoyment of playing video games: The role of competition. In D. Marinelli (Ed.), *Proceedings of the 2nd International Conference on Entertainment Computing (ICEC 2003), Pittsburgh* (pp. 1–8). New York: ACM.

Wartella, E., & Reeves, B. (1985). Historical trends in research on children and the media: 1900–1960. *Journal of Communication, 35*, 118–133.

Wigand, R. T., Borstelmann, S. E., & Boster, F. J. (1985). Electronic leisure: Video game usage and the communication climate of video arcades. *Communication Yearbook, 9*, 275–293.

Woodard, E. H., IV, & Gridina, N. (2000). *Media in the home 2000: The fifth annual survey of parents and children* (Survey Series No. 7). Philadelphia: Annenberg Public Policy Center of the University of Pennsylvania.

Youichi, I. (1996). Why do people watch foreign movies and read foreign books? *Journal of Behavioral and Social Sciences, 1996*(1), 139–150.

Zillmann, D., & Bryant, J. (1985). Affect, mood, and emotion as determinants of selective exposure. In D. Zillmann & J. Bryant (Eds.), *Selective exposure to communication* (pp. 157–190). Hillsdale, NJ: Lawrence Erlbaum Associates.

16

The Role of Presence in the Experience of Electronic Games

Ron Tamborini

Michigan State University

Paul Skalski

University of Minnesota, Duluth

In the 1982 movie *Tron,* a computer-game designer named Kevin Flynn (Jeff Bridges) is literally zapped into the electronic game world he created. In this environment, he gets to play his imaginative computer games as if they were real. To many viewers at the time, the scenario envisioned by *Tron* might have been nothing more than an outlandish fantasy (which could explain the film's lack of box office success), but *Tron* has become a cult classic to many gaming enthusiasts, who applaud the film for realizing their dream of becoming totally "immersed" in the world of electronic games (Demaria & Wilson, 2002). When *Tron* was released in 1982, electronic games were in their infancy and had very limited potential to make players feel like they were engaging in real experiences. But due to the phenomenal advancements that have occurred in electronic-game technology over the past 20 years, the dream of *Tron* has almost become a reality. Many games are now being designed to create a sense of "being there" inside the game world, a feeling we call *presence.*

Presence is a relatively new concept, but its emergent academic importance is hard to overlook. It has captured the attention of philosophers, psychologists, computer scientists, and engineers along with experts from related fields. And although it has been largely overlooked by entertainment researchers, its application here is difficult to ignore. Presence can be understood as a critical concept in several areas of entertainment theory, with particular relevance to the type of interactive entertainment found in electronic games (Klimmt & Vorderer, 2003; Tamborini, et. al., 2004). Moreover, developing trends in electronic-game technology seem directly related to central dimensions of presence. Realistic graphics and sounds, haptic feedback, first-person point of view, and control devices that map natural body actions all increase the vividness and interactivity in games making them highly conducive to the sensation of presence. Presence seems central in shaping the experience of electronic games.

We begin this chapter by defining the concept of presence. Then, we provide a more detailed discussion of how developments in game technology are related to three dimensions of presence: spatial presence, social presence, and self-presence. Finally, we argue that changes in

225

presence corresponding with technological advances not only affect electronic game use, but also shape user experiences that govern the development of mental models and other outcomes of exposure.

DEFINING PRESENCE

Explaining people's experience with media has captivated the attention of scholars from fields concerned with the development of media technology to those focused on how humans interact with this technology. Until recently, media scholars did little more than acknowledge that the manner in which we experience media is important. No major undertakings described or defined the essence of media experience. This changed, however, when recent advances in "virtual reality" (VR) increased our need to distinguish experiences in virtual environments (those that can only be experienced by technology) from actual environments (those that can be experienced without technology). With technology's promise to blur the distinction between the actual and the virtual to a point where we can no longer take for granted our ability to separate the two, researchers are forced to consider subtle differences in the complex process of media experience. Although VR was once a term reserved for futuristic goggles and gloves technology, today's electronic games create virtual environments where users interact in ways hard to distinguish from interaction in actual worlds. Considerations of these subtle differences in experience are the focus of presence research.

Since Reeves (1991) directed scholarly attention to the feeling of "being there" created by media technologies, attempts to define this experience and identify its determinants have proliferated. Unfortunately, like many highly abstract concepts, efforts to develop a shared understanding of presence and its corollaries suffer from the use of different terms referring to identical concepts or use of the same term to mean slightly different things. Words like presence, telepresence, subjective presence, virtual presence, and others are used interchangeably to distinguish experience in virtual environments (Tamborini, 2000). Yet although the terminology found often differs, there is general agreement about the concept they are trying to represent. In simple and somewhat limited terms, the essence of presence is often described as the perception of nonmediation (Lombard & Ditton, 1997).[1] In this sense, presence can be understood as a psychological state in which the person's subjective experience is created by some form of media technology with little awareness of the manner in which technology shapes this perception.

Although definitions of presence are still muddled by its application in different areas of study, there seems to be general agreement that it is a multidimensional concept. In their seminal work reviewing presence in different literatures, Lombard and Ditton (1997) identified six conceptualizations including presence as social richness, presence as realism, presence as transportation, presence as immersion, presence as social actor within a medium, and presence as medium as social actor.[2] Current conceptualizations of presence generally offer typologies identifying different dimensions of presence based on domains of experience. Although different typologies have been provided to classify categories of presence (e.g., Biocca, 1997; Heeter, 1992; Lee, 2004), most schemes share certain categories in common. *Spatial presence*, *social presence*, and *self-presence* emerge prominently in this literature. We will consider the relationship of electronic games to presence in terms of these three dimensions.

Spatial presence might be understood as the sense of being physically located in a virtual environment (Ijsselsteijn, de Ridder, Freeman, & Avons, 2000) or experiencing virtual physical objects as though they are actual objects (Lee, 2004). Social presence can be thought of as the experience of virtual social actors as though they are actual social actors (Lee, 2004). Finally, self-presence has been defined as a state in which users experience their virtual self as if it were their actual self (Lee, 2004), perhaps even leading to an awareness of themselves inside a virtual environment (Biocca, 1997). Although each dimension has notable implications for understanding experience in electronic games, the concentration of research on spatial presence shows its particular importance. Focus on spatial presence seems to parallel early developments in electronic game technology designed to create a sense of "being there" and reveals the importance of spatial presence for examining outcomes from electronic-game play.

SPATIAL PRESENCE: BEING THERE IN ELECTRONIC GAMES

When Kevin Flynn enters the computer game environment in *Tron*, he moves through the virtual world as if he were interacting with the actual world. He jumps on a light-cycle, grabs its controls, and speeds his way around a virtual landscape and across the game world he perceives. Although his initial experience is shaped by an acute awareness that the objects and entities he encounters are virtual, he quickly begins to treat these virtual things as though they are actual and starts to respond as someone actually "there" in the game world. The extent to which game players feel transported to another place, as though they are physically located inside the virtual environment, can be understood as the extent to which they experience spatial presence.

Spatial presence (and related terms such as physical presence, subjective presence, and telepresence) has been identified as a critical determinant of an electronic game's affect on users. Tamborini (2000) argued that the strength of an electronic game's influence is determined by the game's ability to enhance two essential qualities of spatial presence: the feelings of *involvement* and *immersion*. He suggested that technological features associated with *interactivity* and *vividness* inherent in most electronic games heighten the user's sense of involvement with and immersion in the virtual game environment, a process Wirth et. al. (2003) explained in terms of mental models. Wirth et al. argued that situational mental models of environments are pre conditions for spatial presence. They think of spatial presence as primarily a cognitive experience. Although sensory cues can enhance the perception of spatial presence, cognitions more than cues govern this experience. People construct models of environments from the spatial cues they perceive and their memories of the spatial environment. As such, experiences of spatial presence vary from high to low as a function of individual differences in perceptions and memories (Ijsselsteijn et al., 2000) as well as characteristics of form and content found in particular media (Lombard & Ditton, 1997).

Vividness, Interactivity, and Spatial Presence in Electronic Games

Vividness refers to the technology's ability to produce a rich sensory environment (Steuer, 1992) and is defined by the manner in which information is presented to the senses. A vivid technology is high in breadth (the number of sensory channels simultaneously activated) and depth (the degree of resolution within each sensory channel). Conventional media like radio and television have somewhat limited breadth, sending signals to only the auditory and/or visual channels. Newer electronic games supplement this by adding input to haptic and orienting systems (those controlling body equilibrium), often providing simultaneous cross-model forms of redundant information by activating multiple sensory systems in ways that heighten spatial presence (Biocca, Inoue, Lee, Polinsky, & Tang, 2002). For example, a player swerving onto a rumble strip in a driving game might be met with thumping sounds plus vibrations in the game controller that match feelings in real life. The burgeoning area of haptics technology focuses on how game controllers can impart force and vibrations in response to on-screen actions — something already present in most newer game controllers and expected to be more prevalent in coming years due to advances in the understanding of neurology (Kushner, 2003). In cases of feedback to multiple sensory systems, VR technology allows user expectations based on efferent sensation to be met by afferent feedback. Thus, for example, when a player in a VR game environment turns his head, he expects to see the surrounding environment move accordingly. Spatial presence is enhanced by the ability of game technology to match expected *proprioception*, the anticipated sense of body orientation and movement.

A medium with great depth is one that delivers substantial information through each sensory channel. For example, the bandwidth transmission capabilities of the telephone provide considerably less auditory information than stereophonic systems of today's high-end games. In similar fashion, today's electronic-game graphics are dramatically more vivid than those in early games such as *Pac-Man*. Increased graphic capabilities parallel the phenomenal advancements in gaming technology. Successive home console and computer systems continue to display higher resolution and more colorful graphics. The results are electronic games with people and environments that look *real*. Although not yet equal to real-world video images, the vivid graphics of modern games have tremendous depth. Virtual reality games add to this depth by providing motion parallax through the use of head-tracking devices and stereoscopic displays that mimic binocular disparity – both considered a powerful source of spatial presence (cf. Heeter, 1992). Steuer (1992) suggested that breadth and depth interact to create spatial presence, but that cross-modal sensory activation can produce a strong feeling of presence even when signal depth is low.

Interactivity refers to the user's ability to influence the form and content of an environment (Steuer, 1992), and is considered a prerequisite of spatial presence by some (e.g., Zahorik & Jenison, 1998). Steuer (1992) identified three factors governing interactivity: *speed* (the time required for the environment to respond to input), *range* (the number of environmental attributes that can be successfully manipulated and the amount of alternatives available for each attribute changed), and *mapping* (how closely actions represented in the virtual environment match the natural actions used to change a real environment). The interactivity of electronic games has increased considerably in recent years, specifically in terms of range and mapping (Skalski, 2004). Early games were quite restricted in the number of environmental attributes that players could manipulate, probably due to limitations in the gaming hardware of the time. For example, the first entry in the popular *Mortal Kombat* series focused mainly on the moves and actions of combatants, with little attention to the surrounding environment. The newest version of the game (*Mortal Kombat: Deception*), however, features wide-ranging interaction allowing players to manipulate and destroy many aspects of the environment. In this game and other recent fighting titles like *WWE Smackdown*, players can control different game objects (e.g., steel chairs, fire extinguishers) and use them against their opponent just as they use kicks and punches. These interactions add to the sense of being in a real environment, where similar actions are possible.

Most standard game controllers do not offer considerable mapping, but mapped devices such as steering wheels and guns are available for home systems, and arcades offer additional mapping options. Arcade driving games like *Need for Speed* have steering wheels, gear shifters, and pedals that allow players' hands and feet to interact in real-time using natural actions to alter the virtual environment. When this happens instantly in substantial and meaningful ways, the interactivity should create a heightened sense of spatial presence. The extent to which interactivity is possible within an electronic game can be seen as a product of the technology;[3] however, although technology might help open the door to spatial presence, the essence of spatial presence lies in the perceiver.

Involvement, Immersion and Spatial Presence in Electronic Games

A medium's ability to focus the users' attention is a central determinant of spatial presence (Fontaine, 1992). Witmer and Singer (1998) maintained that focused attention leads to *involvement* and *immersion*, two psychological states considered by some as the essence of experiencing presence. The attention leading to involvement is generally thought to result from the meaningfulness produced by an environment's form and content; whereas for immersion, focus is largely regarded a product of the technology's ability to control the stimulus environment. When things in an environment are coherent and connected, it focuses our attention on one meaningful stimulus set (McGreevy, 1992). Witmer and Singer (1998) maintained that a medium's ability to focus user attention on a meaningful stimulus leads to *involvement*–a form of internal mental vigilance characterized as being cognitively engrossed. Meaning can occur as a result of the type of cross-model consistency discussed above, where simultaneous reception of redundant information from different sensory channels merges in coherence to produce a meaningful experience (Steuer, 1992). Perhaps as importantly, however, meaning is also drawn from the successful application of a user's mental model.

In game environments, the ability of users to apply existing mental models to the objects and events in virtual worlds allows them to make sense of the world—something widely held as critical for experiencing spatial presence (Witmer & Singer, 1998; Wirth et al., 2003). Past research indicates that spatial presence is enhanced by the game's ability to arrange objects and scenery in a manner consistent with expectations (Hoffman, Prothero, Wells, & Groen, 1998). A similar influence might be expected from the narrative structure of many games. Games that offer dramatic content and meaningful plots consistent with models from a user's actual or virtual past experience should generate a continuous flow that facilitates involvement and subsequent spatial presence (Slater & Wilbur, 1997).

Whereas involvement depends on the meaningfulness of an environment and centers on mental vigilance, immersion is determined by the environment's ability to isolate people from other surrounding stimuli. Immersion is characterized as the sense of being enveloped by and interacting with an environment — something often thought of as the degree to which a medium controls the user's access to stimuli (Witmer & Singer, 1998). Environments create immersion to the extent that they can insulate individuals from their physical environment, create the sensation that they are inside the environment instead of an outside observer, and generate a feeling that they are can interact and move within the environment in a natural manner. Witmer and Singer suggested that involvement can occur in almost any type of environment, whereas immersion is much more likely to occur in environments that isolate the user or create the perception of inclusion, natural interaction, and control. For example, an arcade racing-car game with a steering wheel that vibrates and gives the user a feeling of control might have more immersive qualities than watching the Indianapolis 500 on an IMAX screen. Similarly, the isolation produced by a head-mounted display used to create a virtual reality environment is likely to be more conducive to immersion than a playing an electronic game in the distracting environment of a busy arcade.

Wirth et al. (2003) presumed that both isolation and mental vigilance can produce the type of continuous information flow that promotes spatial presence; however, they noted that mental vigilance is more an act of volition associated with characteristics of individuals. Strong user motivations to remain involved in a virtual environment can overcome outside distractions and perpetuate strong experiences of spatial presence. As such, presence can be experienced by many individuals even when the immersive quality of technology is poor. The importance of volition to compensate for technological limitations focuses our attention on the individual's role in the experience of spatial presence.

Individual Differences in Experiencing Spatial Presence

Although most attempts to identify determinants of spatial presence deal with characteristics of specific media, some attempts have been made to uncover the characteristics of individuals that impact this experience. Several different scholars identify trait and state variables thought relevant to the experience of presence (e.g., Ijsselsteijn et al., 2000; Steuer, 1992). For example, Wirth et al. (2003) suggested that spatial presence is influenced by individual differences in user ability (information processing speed, spatial ability, absorption, need for cognition, self-efficacy, and domain-specific interest) as well as state differences like mood and user fatigue. Yet, to date there is little evidence to show how these variables impact experienced presence. We might expect spatial presence to flourish in those who are fascinated by virtual environments, and whose ability to quickly process information not only allows them to visualize imagined spatial structures but also richly elaborate on them. Skills allowing successful control of virtual environments and the confidence this brings can increase the likelihood of experiencing spatial presence, particularly in virtual environments of special interest to individual users. Tamborini et al. (2004) supported this notion by showing that previous use and preference for specific game genres predicts feelings of spatial presence experienced during game play. Meaning emerges when the user can successfully apply existing mental models to familiar scenery and thus make some sense of it. Related work on state differences suggests that situational user attributes can shape spatial presence. For example, Wirth et al. (2003) argued that fatigue can impair cognitive processing and elaboration capacities that help experiences of presence occur.

SOCIAL PRESENCE: BEING WITH OTHERS IN ELECTRONIC GAMES

Tron might have been unbelievable to most viewers in 1982, but the box-office success of more recent films delving into the world of electronic games suggests a change in the audience's willingness to accept the reality of virtual worlds. Today, the commonality of these experiences (due to the rapid diffusion of virtual technology) provides a different mental model of virtual agents for audiences. Although gaming in 1982 often involved a single human player and a number of computer-controlled characters, today's games create more social environments, the type represented by another (and fittingly more recent) movie, *Spy Kids 3D: Game Over* (2003). In this film, the spy kids, Carmen and Juni, enter an electronic game world in which they interact both with computer-controlled beings *and* other humans entering the virtual environment to play alongside them. The human-to-human interaction in the movie's virtual game world parallels the emergence of highly social electronic games in the past decade, such as Massively Multiplayer Online Games (MMOGs). Though this sensation of "being with" probably happens during many gaming experiences, Multi-User Dungeons (MUDs) and MMOGs have great potential to make players feel a sense of *social presence*. Undoubtedly, social presence is an important part of these experiences.

Though social presence might seem like a simple concept, many definitions have been advanced. In the context of mediated communication, contemporary thinking about social presence probably originated in the work of Short, Williams, and Christie (1976), who defined the concept as "the degree of salience of the other person in [an] interaction and the consequent salience of the interpersonal relationships" (p. 65). Importantly, this definition suggests that social presence is more than just a dichotomous "here or not" judgment. Instead, it exists along a continuum affected by individual perception and communication technology. The work of Short et al. on social presence has been adopted by scholars interested in determining the suitability of various media forms for different types of social interaction (e.g., Rice, 1993).

Ultimately, it involves the use of media for social purposes. Reeves and Nass (1996) have demonstrated that people have an evolutionary-based tendency to treat media like real social beings. This innate reaction should be enhanced through socially rich applications like online electronic games.

Recent work on social presence has taken a more user-centered (versus technology-centered) approach to the concept, consistent with contemporary notions of presence as a psychological state (Lee, 2004). In contrast to Short et al.'s research that focuses on user perceptions of a medium's ability to make others salient, this work examines the actual perceived salience of the other (Nowak, 2001) based on media attributes. This has led to the development of social presence "theory" focusing on two fundamental issues: (a) the *technology* question, that is, how changes in properties of media interfaces affect social presence and (b) the *psychological* question, that is, what properties of humans elicit attributions of cognitive states and other aspects of social presence to mediated representations (Biocca, Harms, & Burgoon, 2003). To help answer these questions, Biocca et al. have identified three dimensions of the concept: copresence, psychological involvement, and behavioral engagement.

Dimensions of Social Presence

Biocca et al.'s (2003) three dimensions highlight different instances in which social presence can be experienced in electronic games. Copresence is a manifestation of social presence that should happen in most electronic games, whereas psychological involvement and behavioral engagement are more likely in newer games, particularly those played by multiple users at the same time. Together, the three dimensions reveal how widespread and potentially complex the sensation of social presence can be in response to electronic games.

Copresence, in its most basic form, involves sensory awareness of an embodied other (Goffman, 1959), though it can also refer to feelings of spatial presence with another and/or a sense of mutual awareness. The seminal work by Goffman in this area emphasizes the role of the human senses in social interaction. This emphasis makes Goffman's thinking particularly applicable to media experiences that extend the senses to bodily representations (Biocca et al., 2003), including electronic game experience. Most games include visible "others" and should therefore generate some copresence, whether it is with swarms of zombies in *Diablo II* or human amusement park visitors in *Roller Coaster Tycoon*. Copresence might also share properties with spatial presence, if it is thought of as the feeling of being in the same space or location of another. This is a likely effect of more advanced VR gaming technologies that place players into virtual environments. Finally, copresence has been extended by Goffman and others to include *mutual awareness*, where the user is aware of the other and the other is aware of the user (Biocca et al., 2003). Interestingly, this type of copresence induction is missing from traditional non interactive media experiences such as television, though some viewers might perceive that television personalities are aware of them. However, most electronic games have a strong potential to create a sense of mutual awareness. For example, in games where enemy zombies attack a player character, the behavior of the computer beings (e.g., rushing over to attack) shows an awareness of the player character. Moreover, although copresence might play only a minimal role in determining the success of action/violence games, the heightened potential for copresence in new game technology seems central to the success of relationship games like *The Sims*. Although nonviolent electronic games have not sold as well customarily, we expect this to change as the potential for social presence in games increases.

Psychological involvement is a sense of access to intelligence. As Biocca et al. (2003) noted, basic sensory awareness of another (copresence) might not be enough to activate feelings of social presence when perceived intelligence is missing. Many electronic games, especially older ones, may be lacking in this regard. The second author of this chapter vividly remembers playing electronic football games in which a weakness in the defense could be exploited over and over again, with no adjustments. Though this was good for his self-esteem (he never lost), it did not do much for his sense of social presence. Because many electronic game characters are limited in intelligence (i.e., typically programmed only to "beat the player"), they might not be thought of as fully social beings. For psychological involvement to happen, the virtual body has to provide cues to its intentional states (Dennett, 1987). In electronic games, these cues have become more common in recent years due to advances in artificial intelligence (AI) programmed into computer-controlled agents. The latest versions of the *Madden* football series, for example, have been praised for their AI (e.g., Smith, 2001), and adventure games like the classic *The Legend of Zelda* include computer characters who "speak" to the player through text. These cues to intelligence should increase psychological involvement, something especially likely to happen in newer games like *The Sims* or *Singles—Flirt up your Life,* which are brimming with seemingly smart beings that foster feelings of intimacy, immediacy, and mutual understanding. These attributes are recognized components of psychological involvement (Biocca et al., 2003) and share the ability to reveal forms of intelligence that signal social presence.

The final dimension of social presence, behavioral engagement, focuses on behavioral interaction or synchronization. At this stage, entities engage in social behaviors such as talking, chatting, turn taking, eye contact, and nonverbal mirroring. Biocca et al. (2003) discussed behavioral engagement as a recent addition to research on social presence that has developed among scholars interested in the wide range of interaction channels possible in newer, high-bandwidth immersive virtual environments and advanced electronic games. For example, although the limited text-based chat of early online games surely slowed interaction and interfered with immersion, the broadband-enhanced voice chat programs available today allow levels of social presence previously unavailable. Systems like *Teamspeak* give players the ability to talk online as they are playing together. The military game *SOCOM* for Playstation 2 can be purchased with a headset through which voice commands may be issued to responsive teammates during missions. This verbal form of communication provides a more natural interface that increases perceptions of behavioral engagement. Other examples of increased synchronization can be seen in the advanced nonverbal behaviors displayed by characters in newer, graphically vivid games. Avatars and agents in games like *Halo* heighten behavioral engagement by maintaining the appearance of eye contact during interactions. Recent research on online games suggests that players use the realistic-looking nonverbal cues in today's games for a variety of purposes, including the prediction of an opponent's actions (Bracken, Denny, Utt, Quillan, & Lange, 2003). This research highlights the importance of nonverbal cues as a source of behavioral engagement.

Online Games and Social Presence

The recent surge in popularity of online games has opened up a host of new and fascinating possibilities for behavioral engagement along with other aspects of social presence. Particularly striking in this regard is the success of the MMOG genre. These games, which now include fantasy, science fiction, superhero, and even everyday life offerings, consist of vast, graphically rich electronic environments populated by a variety of social beings. Although these beings might "look" like computer-controlled entities, many are controlled by humans playing over the Internet who develop characters to take part in adventures. The motivations for playing these online games extend far beyond adventuring, however. In a survey of *Everquest* players, Griffiths, Davies, and Chappell (2003) found the favorite aspect of the game is its *social features.* In MMOGs these features can include chatting with other players, working in teams, or joining guilds and social groups (Emery, 2004), all of which are forms of behavioral engagement and psychological involvement. The social features of games can also extend to the ways in which the players are represented graphically in the environment, which can perpetuate a sense of copresence. For example, early online role-playing games were text based and perhaps less likely to remind players that other social entities were present than graphically rich newer games. As the social presence-inducing ability of these games continues to increase, these dimensions should become more important motivations for use.

SELF-PRESENCE: MANIFESTATIONS OF SELF IN ELECTRONIC GAMES

Though self-presence has received much less attention than other dimensions of presence, it can play a critical role in the understanding of electronic game experience. Some researchers consider self-presence simply as a part of spatial presence (Ijsselsteijn et al., 2000; Wirth et al., 2003), but its conceptual distinction from other dimensions seems great enough to warrant individual consideration. Moreover, the recent focus of electronic game technology on features that cultivate a player's awareness of themselves within the game environment warrants separate discussion.

Biocca's (1997) definition of self-presence identifies three "bodies" present in a virtual world: the actual body, the virtual body, and the body schema, or the user's mental model of self. He argued that when we see a graphic representation of ourselves within a virtual environment, the representation evokes mental models of our body as well as our identity. Moreover, because these mental models of self are open to change (Fisher, 1970), embodiment in a virtual world can alter both mental models, especially when the environment makes the embodied self salient. In other words, the logic argues that experiences of self-presence can alter both our self-image of our body as well as our social identity — an intriguing possibility in light of the growing popularity of games where players assume fantasy identities.

The potential influence of embodied experiences seems far-reaching. New game technology can provide redundant forms of simultaneous cross-modal sensory activation in a manner that closely maps a player's body movements. This should cultivate a mental model of being inside the game environment (Biocca, 1997). A significant advancement in this area has been the incorporation of first-person point of view (POV) into games.

First-person POV exploded onto the gaming scene with the release of *Wolfenstein 3D* in the early 1990s. In this game, the player-character walked around a labyrinth to fight Nazis and other enemies. Unlike other action offerings at the time, however, the walking was done through the eyes of the main character. Instead of the character appearing on the screen and moving, the environment moved as if the player were traveling around in it. In addition, a representation of the hands of the main character holding a weapon was included at the bottom of the screen, mainly to make the player feel more in the "space" of the game (McMahan, 2003). Today these features have been incorporated into many games as part of the highly successful 3-D first-person shooter genre. In the future we can expect even stronger perceptions of self-presence as VR technology increases its ability to completely coordinate virtual body movement with tracking devices, thereby reducing problems with proprioception that could otherwise terminate feelings of presence (Slater & Usoh, 1994).

Though discussion of self presence in this sense focuses on body schemas created in first-person POV games, we should not overlook the fact that like all forms of presence, self-presence is first and foremost a product of cognition. As such, it is not determined solely by first-person POV and game technology's ability to limit problems of proprioception. For example, some new game technologies generate representations of players inside the virtual environment without creating first-person POV. The most notable example of this type of technology is the Sony EyeToy, a small camera that captures the images of players and puts these images into games. Also notable is the ability of mapping to induce feelings of self-presence. Biocca (1997) suggested that close mapping of a virtual body to a user's actual physical body has a strong influence on both experience in VR environments and outcomes from experience. We can expect that game makers will continue to pay close attention to developing technology that creates graphic representations of the player and/or the sensation of being inside the game environment. As electronic games continue along this path, the role of self-presence as a critical factor in shaping future game experience will escalate.

UNDERSTANDING THE CONSEQUENCES OF PRESENCE

With the rapid development of presence-inducing technology and the ever-increasing growth in video game use, understanding the role of presence in shaping the outcomes from electronic-game play is a challenge of increasing consequence. Undoubtedly, many theoretical perspectives can add to our understanding of presence and electronic-game enjoyment (Klimmt & Vorderer, 2003). Still, the need for additional theories and models that can address the dual roles of witness and participant unique to interactive forms of entertainment research is apparent here (Vorderer, 2000). This section examines how presence can affect the use and enjoyment of games as well as the development of mental models influencing intended and unintended effects of game exposure.

Use and Enjoyment of Presence-Inducing Electronic Games

The ability of presence to influence the selection and use of electronic games is important for both practical and theoretical purposes. The growth in sales of presence-inducing games not only affects electronic game industry profit, but also increased game use can reduce time spent with other media and resulting advertising revenues (Powell, 2003). More central to our concern, however, is the potential for presence-inducing technology to alter the experience of game play along with its social and psychological consequence for users. Can this explain user enjoyment, the selection of particular titles, the enormous time spent with these games, or potential electronic-game dependency?

In entertainment literature, Zillmann's work on mood management and selective exposure (Zillmann, 2000) suggests that media use is often determined by its ability to serve the user's immediate affective needs, even if the needs are unknown. As such, we might expect strong and lasting states of presence in environments that serve emotional needs. This logic generally predicts that heightened presence should only occur when users in positive mood states are absorbed in virtual environments with matching valence. At the same time, however, selective exposure logic suggests that games allowing users to master feelings of fear should create a heightened sense of presence and lead to extended use. Already, research in clinical psychology suggests the value of presence-inducing technology for treating phobic disorders (Strickland, Hodges, North, & Weghorst, 1997). But whereas clinical use should be limited, entertainment theory predicts broad use for games allowing players to overcome ordinary anxieties. Evidence of this is seen in research on exposure to horror films suggesting that selection results in part from the desire of young men to master their fears (Tamborini, 2004). Signs that this occurs in electronic games might already be apparent in the popularity of survival-horror games such as *Resident Evil* and *Silent Hill* as well as first-person action games like *Doom* where players battle hordes of fierce monsters in highly immersive environments. It is not hard to imagine similar uses of electronic dating games like *Singles — Flirt up your Life*, where young men can confront their fear of rejection and other relationship-based anxieties in less interpersonally threatening game environments. Moreover, although it is hard to think of existing electronic games promoting experiences of grief, the popularity of tragic novels and film suggests the possibility that immersive games allowing users to confront these anxieties could also attract large audiences.

The connection of presence with enjoyment is hard to overlook. To some, the types of emotional experience we label as joy or delight are nothing more than pleasurable forms of what they call absorption or presence (Klimmt & Vorderer, 2003). Yet even if you do not define enjoyment as a form of presence, enjoyment is perhaps the primary outcome sought and experienced from electronic games and profitable presence-inducing technologies like IMAX films and simulator rides. As suggested by Lombard and Ditton (1997), these experiences are popular because they are fun — a concept central to literature explicating the role of presence in entertainment's appeal. Klimmt and Vorderer (2003) emphasized the need to integrate research on presence with entertainment theory, and gave examples of how the enjoyment of presence can be understood in terms of rationales related to user dispositions, mental simulation of fictional events, and the psychological mindset of play. Undoubtedly, theory on media entertainment will enhance our understanding of the role of presence in the experience of electronic games.

Mental Models as a Consequence of Spatial Presence

Logic derived from research on mental models (Roskos-Ewoldsen, Roskos-Ewoldsen, & Dillman Carpentier, 2002) and script theory (Schank & Abelson, 1977) provides the foundation for our position that the natural mapping and interactive control found in presence-inducing game technology should not only facilitate the development of more complete mental models, but should also increase their accessibility for use. Perhaps we do little more than echo the thoughts of those who note how natural mapping provides more accurate behavioral information. However, although interactivity and control are often discussed as contributors to the experience of spatial presence, we feel that important aspects of interactive control's contribution to the development of comprehensive mental models are sometimes overlooked. In this regard, the vigilance needed to control events in a virtual environment sets it apart from other media. Unlike television that continues on its own, electronic games and other interactive media require focused attention for the game to continue. The importance of such control is twofold: Not only should the vigilance required for control increase the salience of the mental models activated, but control in this circumstance also implies a decision to act.

We argue that the rehearsal of decision making is significant both because active decision making increases the sense of presence and because it adds to the structure of mental models formed through virtual experience in ways previously thought unique to actual experience. Certainly this influence is unfeasible in media that do not, by necessity, expose users to decisional cues. For example, Tamborini (2000) argued that the decision to strike or not strike an opponent made when playing violent electronic games creates a more complete mental model of aggression than that which results from watching violent television or film where decisional models are absent. Particularly with games requiring basic motor skills already part of an actor's repertoire (e.g., the act of bending your finger to pull a trigger), the virtual decision to act seems more important than the act itself in its contribution to the development of mental models. Most acts performed in an electronic game involve decision-making behaviors and triggering actions not part of other media experience. Their constant rehearsal in repeated play should make them a powerful part of the mental model.

Within this framework, presence should strengthen the influence of electronic games on a number of different behaviors, yet the greatest concerns involving electronic games emphasize the potential for unintended anti-social outcomes like those related to violence. Tamborini (2000) spoke to this issue directly by evoking an image of children playing electronic games that create worlds of virtual violence more realistic and engaging than actual experience. He argued that immersion in these games inducing players to rehearse violent plans, decisions to aggress, and the actual motor skills used to assault and kill can foster the development of mental models for aggression—models lacking many critical inhibitors associated with most actual experience. Given the preponderance of electronic games containing violence (Smith, Lachlan, & Tamborini, 2003), the special concern for unintended aggressive consequence comes as no surprise. Yet we should not overlook the potential for an equally strong influence on the development of mental models for other behaviors represented in various games. The growing popularity of electronic games like *The Sims* or, even more, those featuring highly involving intimacy behaviors like *Singles—Flirt up your Life* hold promise for great influence on related mental models. It is not hard to imagine players experiencing heightened presence in worlds of virtual desire, or to imagine these experiences shaping mental models governing decisions associated with dating behaviors.

Although we want to highlight the importance of decisional cues stemming from interactive control as a determinant of outcomes from presence-related experience, we still note the prominent role of mapping. Natural mapping should strengthen the influence of electronic games on the development of diverse mental models, both favorable and adverse. The Xbox game *Yourself!Fitness*, for example, encourages various exercises through a virtual trainer named Maya. The addition of interface technology that allows the actual physical activity to occur (e.g., a treadmill for running) would likely increase spatial presence as well as contribute to physical fitness. In this case, the physical benefits of the game result both directly from the motivated exercise and indirectly through the strengthening of more fully mapped mental models giving rise to exercise in the future. One can imagine similar applications of electronic games designed to train or simply amuse where natural mapping could engage students in dedicated behavior and make the rehearsal of relevant actions and concepts fun. Some professional race car drivers, for example, report using realistic racing video games like *NASCAR Thunder* to prepare for events. Inclusion of steering wheels, pedals, and advanced haptic feedback systems can simulate the feel of speeding around corners and allow virtual racers to see, hear, and feel what it is like to drive a particular car at different track locations. Claims that playing helps drivers learn course layouts (Emmons, 2003) suggest that the game helps build mental models of different tracks and the actions needed to navigate them for use in actual races. The strong sense of spatial presence created by this type of immersive environment provides information for use in the development of mental models unavailable to observers in traditional media environments (Skalski, Tamborini, & Westerman, 2002). Moreover, the repeated rehearsal this promotes should lead to the formation of stronger and more developed mental models.

Mental Models and Self-Presence

Closely connected with mental models originating from spatial presence is a person's mental model of self. Biocca (1997) suggested that close mapping of a user's actual physical body to a virtual body evokes a mental model of self within the virtual environment capable of changing our self-image. If representations of the self implicit in avatars really do result in distortions of body schema and social identity, predictable patterns of influence are expected from sets of avatar attributes commonly found in different game genre. For example, Anderson and Dill (2000) suggested that an electronic video game's first-person point of view and an active role in decision making help form chronically accessible mental models that change an individual's personality. Their general aggression model (GAM) submits that these attributes play a critical role in explaining how violent video games influence aggressive personality. By similar token, self-presence inducing technology in different game genres should help change other aspects of self-image, such as those related to gender and intimacy. Tamborini and Mastro (2001) argued that media play a powerful role in the development of social identities, and their ability to activate perceptual frames can govern how we see the social environment and ourselves. Such distortions should increase when the environment makes the embodied self salient. Tamborini (2000) made this point by arguing that increased identification with characters in presence-inducing electronic games strengthens effects from exposure—an experience heightened in first-person point of view games by players adopting a character's role (Laurel, 1991). Media's ability to make attributes of identity salient creates mental models that are part of the ongoing process of negotiating self-image.

Mental Models and Social Presence

Beyond the roles of spatial and self presence in the development of mental models for social behavior, other unintended outcomes are expected from the social presence created by electronic games. In addition to technology's ability to create a sense of social presence and satisfy companionship needs, the behavioral engagement and psychological involvement created by electronic gaming, particularly through online game experiences, provide unique opportunities for community building. For example, developers of the game *Yourself!Fitness* envision users creating an online community akin to a virtual gym where players use wireless headsets to talk during game play and encourage each other (Taub, 2004). One of the most interesting issues raised by this image involves the potentially broad social consequence of online gaming.

In his seminal book entitled *Bowling Alone*, Putnam (2000) claimed that as a culture we have become increasingly disconnected from each other. He argued that change over the last few decades has led to a decline in social capital, or the value we extract from the strength of our connection to family, friends, and social structures. In addition to underlying social and economic cause, developments in technology like television and computers are singled out by Putnam as contributors to this decline, alleging that these changes lead to fewer people joining organizations, connecting with their neighbors, or having a "we" mentality. Instead, more people are just "bowling alone." The image associated with this claim is a picture of isolated television viewers and alienated electronic gamers playing against agents incapable of producing the social capital Putnam values. But as the electronic game industry continues to develop along lines of social presence-inducing technology, we must ask if the social presence available here can reconnect electronic gamers through virtual communities. Just as reading groups contributed to the creation of social institutions that helped reconnect U.S.-Americans in the late 1800s, do MOGs and MMOGs help develop social networks that bring cooperation, reciprocity, trust and support even outside the game environment? Or, are most players just gaming alone? Though no research directly addresses this issue, evidence that heavy electronic-game players are more likely to spend time with their friends and feel closer to their family (Colwell, Grady, & Rhaiti, 1995) supports notions that electronic games can strengthen social capital.

CONCLUSION: PRESENCE AND GAMING SCHOLARSHIP

The phenomenal growth in electronic-game technology was overlooked for some time, but attention to it today is evident from this book. Compelled by escalating industry sales and reports of users captivated by their passion to play, scholars who once dismissed electronic games as unworthy of study are now rushing to discover the secrets of its appeal and the consequence of its use. We believe that the unique presence-inducing qualities of electronic games are central in both regards. Developments in game technology fostering presence not only increase exposure to electronic games, but also shape the form of resultant mental models that govern how we move in physical space, how we interact in social settings, and even how see ourselves. It can influence a host of outcomes ranging from the affect expressions used to pursue relational intimacy to the benefits gained from interactive learning. The vividness and interactivity of current games offer compelling mental models of spaces and people, and this will increase over time. We strive to focus attention toward the study of presence and encourage the development of related theory. Electronic games are poised to become the *ultimate* presence-inducing medium, making presence central to research exploring the experience of electronic games.

NOTES

[1] In this discussion, we purposely avoid using Lombard and Ditton's (1997) often quoted phrase that defines presence as the "perceptual illusion of nonmediation." This was done in order to preclude giving readers the impression that we define presence as a form of illusion. While debate over this issue is beyond the scope of this chapter, the authors consider presence a virtual experience that need not be identified as illusionary.

[2] Clearly, there is still disagreement concerning the dimensions that constitute the experience of presence, or even whether presence should be understood as the composite of several dimensions or a process. Some scholars question whether presence is nothing more than a special case of involvement (Klimmt & Vorderer, 2003). Other scholars think it more useful to consider presence as a process (Wirth et al., 2003). In general, current conceptualizations of presence are understood as typologies identifying different dimensions of presence based on domains of experience.

[3] Some presence scholars argue that interactivity is determined by the capability of a medium to create a mental model of possible interactions and not the ability to interact per se (Wirth et al., 2003).

REFERENCES

Anderson, C. A., & Dill, K. E. (2000). Video games and aggressive thoughts, feelings, and behavior in the laboratory and in life. *Journal of Personality and Social Psychology, 78,* 772–790.

Biocca, F. (1997). The cyborg's dilemma: Progressive embodiment in virtual environments. *Journal of Computer Mediated Communication, 3*(2). Retrieved May 29, 2004, from http://www.ascusc.org/jcmc/vol3/issue2/biocca2.html.

Biocca, F., Harms, C., & Burgoon, J. K. (2003). Toward a more robust theory and measure of social presence: Review and suggested criteria. *Presence: Teleoperators and Virtual Environments, 12*(5), 456–480.

Biocca, F., Inoue, Y., Lee, A., Polinsky, H., & Tang, A. (2002, October). *Visual cues and virtual touch: Role of visual stimuli and intersensory integration in cross-modal haptic illusions and the sense of presence.* Paper presented at the Fifth International Workshop on Presence, Porto, Portugal.

Bracken, C. C., Denny, J., Utt, K., Quillan, M., & Lange, R. (2003, October). *Sometimes I really hate coming back to this world: Presence and on-line video game playing.* Paper presented at the Sixth International Workshop on Presence, Aalborg, Denmark.

Colwell, J., Grady, C., & Rhaiti, S. (1995). Computer games, self esteem, and gratifications of needs in adolescents. *Journal of Community and Applied Social Psychology, 5,* 195–206.

Demaria, R., & Wilson, J. L. (2002). *High score! The illustrated history of electronic games.* Berkeley, CA: McGraw-Hill/Osborne.

Dennett, D. C. (1987). *The intentional stance.* Cambridge, MA: MIT Press.

Emery Jr., C. E. (2004, April). Turbine software creating a gamer's world. *The Providence Journal.* Retrieved May 1, 2004, from http://www.projo.com/technology/content/projo_20040420_turbine.127f8b.html.

Emmons, M. (2003, June). Cutting-edge video games have a place in race-day preparation. *The San Jose Mercury News.* Retrieved May 1, 2004, from http://www.mercurynews.com/mld/mercurynews/sports/6144871.htm?1c.

Fisher, S. (1970). *Body image in fantasy and behaviors.* New York: Appleton-Century-Crofts.

Fontaine, G. (1992). The experience of a sense of presence in intercultural and international encounters. *Presence: Teleoperators and Virtual Environments, 1*(4), 482–490.

Goffman, E. (1959). *The presentation of self in everyday life.* Garden City, NY: Anchor.

Griffiths, M. D., Davies, M. N. O., & Chappell, D. (2003). Online computer gaming: A comparison of adolescent and adult gamers. *Journal of Adolescence, 27,* 87–96.

Heeter, C. (1992). Being there: The subjective experience of presence. *Presence: Teleoperators and Virtual Environments, 1*(2), 262–271.

Hoffman, H. G., Prothero, J., Wells, M. J., & Groen, J. (1998). Virtual chess: Meaning enhances users' sense of presence in virtual environments. *International Journal of Human–Computer Interaction, 10*(3), 251–263.

Ijsselsteijn, W. A., de Ridder, H., Freeman, J., & Avons, S. E. (2000). Presence: Concept, determinants, and measurement. *Proceedings of the SPIE, Human Vision and Electronic Imaging V,* 3959–3976.

Klimmt, C., & Vorderer, P. (2003). Media psychology "is not yet there": Introducing theories on media entertainment to the presence debate. *Presence: Teleoperators and Virtual Environments, 12*(4), 346–359.

Kushner, D. (2003, July). With a nudge or vibration, game reality reverberates. *The New York Times.* Retrieved July 8, 2003, from http://www.nytimes.com/2003/07/03/technology/circuits/03next.html.

Laurel, B. (1991). *Computers as theatre.* Menlo Park, CA: Addison-Wesley.

Lee, K. M. (2004). Presence, explicated. *Communication Theory, 14*(1), 27–50.

Lombard, M., & Ditton, T. (1997). At the heart of it all: The concept of presence. *Journal of Computer Mediated Communication, 3*(2). Retrieved May 29, 2004, from http://www.ascusc.org/jcmc/vol3/issue2/lombard.html.

McGreevy, M. W. (1992). The presence of field geologists in Mars-like terrain. *Presence: Teleoperators and Virtual Environments, 1*(4), 375–403.

McMahan, A. (2003). Immersion, engagement, and presence: A method for analyzing 3-D video games. In M. J. P. Wolf & B. Perron (Eds.), *The video game theory reader* (pp. 67–86). New York: Routledge.

Nowak, K. (2001, May). *Defining and differentiating copresence, social presence, and presence as transportation.* Paper presented at the Fourth International Workshop on Presence, Philadelphia, PA.

Powell, C. (2003, July). Get in the game. *Marketing Magazine.* Retrieved May 29, 2004, from http://www.marketingmag.ca/magazine/current/feature/article.jsp?content=20030728_55703_55703.

Putnam, R. D. (2000). *Bowling alone: The collapse and revival of American community.* New York: Simon & Schuster.

Reeves, B. R. (1991). "Being there": Television as symbolic versus natural experience. Unpublished manuscript. Stanford University, Institute for Communication Research, Stanford, CA.

Reeves, B., & Nass, C. (1996). *The media equation.* New York: Cambridge University Press.

Rice, R. (1993). Media appropriateness: Using social presence theory to compare traditional and new organizational media. *Human Communication Research, 19,* 451–484.

Roskos-Ewoldsen, D. R., Roskos-Ewoldsen, B., & Dillman Carpentier, F. R. (2002). Media priming: A synthesis. In J. B. Bryant & D. Zillmann (Eds.), *Media effects: Advances in theory and research.* (pp. 97–120) Mahwah, NJ: Lawrence Erlbaum Associates.

Schank, R. C., & Abelson, R. P. (1977). *Scripts, plans, goals, and understanding.* Hillsdale, NJ: Lawrence Erlbaum Associates.

Short, J., Williams, E., & Christie, B. (1976). *The social psychology of telecommunications.* London: Wiley.

Skalski, P. (2004, April). *The quest for presence in video game entertainment.* Panel presentation at the Annual Conference of the Central States Communication Association, Cleveland, OH.

Skalski, P., Tamborini, R., & Westerman, D. (2002, November). *Script development through virtual worlds.* Paper presented at the Annual Conference of the National Communication Association, New Orleans, LA.

Slater, M., & Usoh, M. (1994). Body centered interaction in immersive virtual environments. In N. M. Thalmann & D. Thalmann (Eds.), *Artificial life and virtual reality* (pp. 125–148)New York: Wiley.

Slater, M., & Wilbur, S. (1997). A framework for immersive virtual environments (FIVE): Speculations on the role of presence in virtual environments. *Presence: Teleoperators and Virtual Environments, 6*(6), 603–616.

Smith, J. (2001). Madden 2002 X-Box review. *The Sports Gaming Network.* Retrieved May 5, 2004, from http://www.sports-gaming.com/football/madden_2002/review_xbox.shtml.

Smith, S. L., Lachlan, K., & Tamborini, R. (2003). Popular video games: Quantifying the presentation of violence and its context. *Journal of Broadcasting & Electronic Media, 47,* 58–76.

Steuer, J. (1992). Defining virtual reality: Dimensions determining telepresence. *Journal of Communication, 42*(4), 73–93.

Strickland, D., Hodges, L., North, M., & Weghorst, S. (1997, August). Overcoming phobias by virtual exposure. *Association for Computing Machinery, 40*(8), 34–39.

Tamborini, R. (2000, November). *The experience of telepresence in violent video games.* Paper presentation at the Annual Conference of the National Communication Association, Seattle, WA.

Tamborini, R. (2004). Enjoyment and social functions of horror. In. J. Bryant, D. Roskos-Ewoldsen, & J. Cantor (Eds.), *Communication and emotion: Essays in honor of Dolf Zillmann* (pp. 417–444). Hillsdale, NJ: Lawrence Erlbaum Associates.

Tamborini, R., Eastin, M., Skalski, P., Lachlan, K., Fediuk, T., & Brady, R. (2004). Violent virtual video games and hostile thoughts. *Journal of Broadcasting & Electronic Media, 48*(3), 335–357.

Tamborini, R., & Mastro, D. E. (2001) *Race, media, and social identity.* Paper presentation at the University of Michigan Conference on Media and Race, Ann Arbor, MI.

Taub, E. A. (2004, May). Lift and reach and hold that pose, and advance to the next level. *The New York Times.* Retrieved May 26, 2004, from http://www.nytimes.com/2004/05/24/technology/24game.html.

Vorderer, P. (2000). Interactive entertainment and beyond. In D. Zillmann & P. Vorderer (Eds.), *Media entertainment: The psychology of its appeal* (pp. 21–36). Mahwah, NJ: Lawrence Erlbaum Associates.

Wirth, W., Bocking, S., Hartmann, T., Klimmt, C., Schramm, H., & Vorderer, P. (2003). *Presence as a process: Towards a unified theoretical model of formation of spatial presence experiences.* Unpublished manuscript.

Witmer, B. G., & Singer, M. J. (1998). Measuring presence in virtual environments: A presence questionnaire. *Presence: Teleoperators and Virtual Environments, 7,* 225–240.

Zahorik, P., & Jenison, R. L. (1998). Presence as being-in-the-world. *Presence: Teleoperators and Virtual Environments, 7*(1), 78–89.

Zillmann, D. (2000). Mood management in the context of selective exposure theory. In M. E. Roloff (Ed.), *Communication Yearbook 23* (pp. 103–123). Thousand Oaks, CA: Sage.

17

The Role of Music in Video Games

Sean M. Zehnder and Scott D. Lipscomb
Northwestern University

The role of music in video games has come a long way from the "hollow ringing sound" (Winter, 2004) of *Pong* (Atari, 1975), to the symphonic orchestral scores of games like *Medal of Honor: Frontline* (*MOH: Frontline;* Electronic Arts, 2002) and the highly customizable hip-hop playlists of others like *Tony Hawk's Underground* (Activision, 2003). While we recognize that there are a wide variety of definitions of the term, we use "video games" to refer primarily to console and computer gaming. With the introduction of the latest round of game console competitors in 2000, the Sony Playstation 2 (PS2), the Microsoft Xbox, and the Nintendo GameCube, interactive gaming took large strides forward in both video and audio capabilities. MP3 compression, CD-ROM and DVD technology, as well as faster processors and larger storage devices have enabled audio in games to approach the audio-visual quality of film and television. That music plays an important role in the overall experience of video gaming is widely accepted, although there have been very few experimental or theoretical studies of the role of music in the perception of video game stimuli.

This chapter explores the contribution of music to the gaming experience from a variety of perspectives. As there are very few studies of music in video games available, we begin by reviewing literature on the role of music in the perception of film,[1] from which we draw vocabulary and a theoretical framework for the later analysis of music in video games. Next, this chapter outlines an experimental study conducted by the authors on the perception of music in the game *The Lord of the Rings: The Two Towers* (Electronic Arts, 2002). Finally, the authors propose specific areas for future research studying the role of music in this rapidly evolving medium.

LITERATURE REVIEW

The Perception of Music in Film

Almost since its inception, cinematic imagery has been accompanied by musical sound. Early presentations of films by the Lumière Brothers in the 1890s, for example, typically incorporated a piano accompaniment. Silent films of the early Hollywood film era were accompanied by a variety of musical instruments from solo piano or organ to larger ensembles, depending on venue. Originally intended simply to cover the noisy sounds of the projection equipment (Cavalcanti, 1985), the impact of including a musical score as a means of enhancing and expanding on the psychological drama of the audio-visual experience—certainly a much more significant role—was soon realized. In fact, the film score has become so integral to the cinematic experience that a movie without a musical score (e.g., *The Birds*, 1963; *Diary of a Chambermaid*, 1964), interestingly, creates tension and a sense of urgency, causing audience members to seek out other sounds (e.g., ambient noises) to fulfill the functions typically performed by the music (Lipscomb & Tolchinsky, 2005).

The number of empirical investigations into the role of the musical soundtrack has increased dramatically since the late 1980s. Early studies (Tannenbaum, 1956; Thayer & Levenson, 1983) revealed that adding a musical soundtrack enhanced the viewers' experience. Marshall and Cohen (1988) demonstrated that information provided by a musical soundtrack could affect the judgments of personality attributes assigned by research participants to each of three geometric shapes presented as "characters" in an animated film.

In 1996, an entire issue of the journal *Psychomusicology* was devoted to the "Psychology of Film Music," including both experimental and theoretical work. Thompson, Russo, and Sinclair (1996) showed that a musical score can influence the perceived closure in filmed events, concluding that the role played by music was implicit. A study by Bolivar, Cohen, and Fentress (1996) revealed that both referential and temporal aspects are important to the interpretation of an audio-visual combination. In this context, the perceived "friendliness" or "aggressiveness" of characters in the audio-visual combinations were accurately predicted based on ratings for the independent visual and audio components. Lipscomb and Kendall (1996), using excerpts from *Star Trek IV: The Voyage Home* (1986), found that the majority of research participants were able to select the composer's intended musical score for each scene from a selection of musical excerpts taken from various parts of the same movie. In addition, subject responses to a series of verbal response scales revealed highly significant differences in the ratings, depending on which musical score was paired with a given visual scene. In the context of this experimental investigation, though the authors are cautious about stating the fact too emphatically, the musical score appeared to exert *more* influence on subject ratings than the visual images. Using both verbal rating scales and open-ended questions, a study by Bullerjahn and Güldenring (1996) revealed that the musical score polarizes the emotional content of a cinematic excerpt and significantly influences understanding of the dramatic plot. Results of a study by Sirius and Clarke (1996) revealed that the relationship between simple computer-generated visual images and music was additive, that is, "music that scores highly on the evaluative dimension simply raises the evaluative scores for an audiovisual combination in which the particular music is featured, no matter what the visual material is" (p. 130). Using excerpts from 20 commercially released laserdisc recordings, Iwamiya (1996) found that, when considering matched audio-visual combinations (i.e., combined in the manner intended by the producer of the recording), the auditory component systematically influences the visual component, but the visual component did not exert such influence on the audio.

Significant research in this area has continued to the present day. Boltz and her colleagues (Boltz, 1992, 2001; Boltz, Shulkind, & Kantra, 1991) have confirmed the important role played by music and its placement within a dramatic sequence in relation to memory, especially in determining perceived duration of a scene and which specific aspects of a scene will be remembered after an initial viewing. Applying an empirical approach to a related artform, Krumhansl and Schenk (1997) investigated the relationship between Balanchine's dance choreography and the music by which it was inspired, Mozart's *Divertimento No. 15*. Vitouch (2001) asked participants, after seeing a brief film excerpt with one of two contrasting musical soundtracks, to provide a written prediction of how the plot would continue. Results revealed that anticipations of future events are "systematically influenced" by the accompanying musical sound (p. 70).

As a result of this collection of empirical studies, it is an undeniable fact that music plays a significant role in the motion picture experience. It is the belief of the present authors that the musical soundtrack plays a similarly important role in the context of video games.

Music as Communication

The essence of all music is communication, whether it is personal expression, spiritual messages, political persuasion, or commercial appeal (Sonnenschein, 2001, p. 101). According to Lipscomb and Tolchinsky (2005), "Music is a culturally-defined perceptual artifact, existing in the mind of enculturated listeners (Hood, 1982; Lomax, 1962; Merriam, 1964; Nettl, 1983) (p. 384). In order to be successful, communication must involve shared implicit as well as explicit knowledge structures (Kendall & Carterette, 1990). Many studies have investigated various aspects of musical communication as a form of expression (Bengtsson & Gabrielsson, 1983; Clarke, 1988; Clynes, 1983; Gabrielsson, 1988; Seashore, 1967/1938; Senju & Ohgushi, 1987; Sundberg, 1988; Sundberg, Frydén, & Askenfelt, 1983). A three-component communication model was proposed by Campbell and Heller (1980), including simply a composer, performer, and listener. It was upon this previous model that Kendall and Carterette (1990) expanded, elaborating upon and clearly defining the constituent parts and elucidating the specific interrelationships that exist.

In video games, as in other audio-visual media, music serves as one important component of the spectrum of sound that includes the musical score, ambient sound, dialogue, sound effects, and even silence. As Peterson (2004) argued, "Game production today is serious business, and a major part of that appeal is great audio." In games such as *MOH: Frontline*, music plays a significant part in creating a sense of immersion in the game. *MOH: Frontline* places the player in the perspective of an Allied soldier taking part in the storming of the beach at Normandy during World War II. Bullets and explosions seem to wiz past the player's head while searching for a commanding officer. *MOH: Frontline* includes an original symphonic score composed by Michael Giacchino that was performed by a live orchestra and chorus, and recorded and mixed for Dolby 5.1 surround sound. In one online review, North (2003) described a variety of contributions of the *MOH: Frontline* score, including: heightened moments of suspense, accentuation of the "horrors of war" (by the use of an adult chorus), and bittersweet passages that "brighten the beauties of war." Upon a closer look, this review suggests that a musical score can contribute to both narrative (i.e., suspense) and meta-narrative (i.e., perception of horror or beauty) aspects of the video game experience.

There is good reason to believe that composers instinctively draw on their prior experience with films when composing for video games. Don Veca, audio director for *Lord of the Rings: the Return of the King* (EA Games, 2003), argued, "We're getting to the point where we're supposed to sound like a movie (quoted in Jackson, 2004a, p. 52). Reesman (2004) argued that the two media (film and video games) require a composer to work in "grand themes and short snippets." It is striking, in fact, that many interviews with game composers include some comparison to the creative process of scoring a film (Lennertz, 2004; Sanger, 2004;

Hyde-Smith, 2004b). Music can reflect the inner feelings of a character or the outward state of the setting, and in both cases it can choose to "comment on" either state through the audiovisual contrast, one of the audio-visual relationships discussed below.

Types of Listening

A key technique in the analysis of the role of music in video games is the differentiation of various types of listening. Chion (1994) outlined three main types of listening: reduced, causal, and semantic. *Reduced* listening emphasizes the sound itself and the source or meaning of the audio. *Causal* listening refers to listening to a sound to be able to identify its cause, while *semantic* listening focuses on code systems in the audio (i.e., spoken language) that symbolize ideas, actions, or things. In the case of video game music, reduced listening would emphasize the mood of the music, causal would highlight the actions that trigger certain sounds/loops, and semantic listening would focus on the lyrical or genre-related (i.e., hip-hop vs. symphonic classical) connotations of the audio.

Some scholars have argued that the strength of a musical contribution to multimodal communication lies in its ability to convey meanings that are incommunicable via words alone. Richard Wagner claimed of 19th-century music drama, that "as a pure organ of the feeling, [music] speaks out of the very thing which word speech in itself can not speak out. . . that which, looked at from the standpoint of our human intellect, is *the unspeakable*" (Wagner, 1849/1964, p. 217). Suzanne K. Langer (1942) argued, "Music has all the earmarks of a true symbolism, except one: the existence of an *assigned connotation*" (p. 240). Royal Brown (1988) argued that it is this "unconsummatedness" of music that accounts for the predominance of the orchestral film score. As Lipscomb and Tolchinsky (2005) suggested, in order for a film to make the greatest possible impact, it is necessary for there to be an interaction between at least three modes of signification: verbal dialogue, cinematic images, and the musical score.

Gorbman (1987) proposed three methods by which music can "signify" in the context of a narrative film. *Purely musical signification* results from the highly coded syntactical relationships inherent in the association of one musical tone with another, that is, patterns of tension and release provide a sense of organization and meaning to the musical sound, apart from any extramusical association that might exist (e.g., Hanslick's [1891/1986] "absolute music"). *Cultural musical codes* are exemplified by music that has come to be associated with a certain mood or state of mind (e.g., Meyer's [1956] "referentialism"). *Cinematic codes* influence musical meaning merely due to the placement of musical sound within the cinematic context; for example, opening credits, end title music, and recurring musical themes that come to represent characters or situations within the film.

Audio-Visual Relationships

The emotional response to music and a musical soundtrack's obvious ability to establish an appropriate mood are important to the process of music communication, whether in the context of film or video games. However, there are other effective ways in which music can communicate information to the listener that are equally important. Peirce (1931–1935, vol. 2) provided a classification of three types of signs used in the communication process, differentiated by the manner in which they represent their referent. Dowling and Harwood (1986) provided the following useful explication of this delineation:

> An *index* represents its referent by having been associated with it in the past, as lightning and thunder with a storm. An icon represents through formal similarity to the referent, as

a wiring diagram represents a circuit. *Symbols* represent by being embedded in a formal system, such as language. (p. 203; italics added)

Within a musical context, both iconic and indexical signs may be considered referential means of musical communication. *Icons* provide perhaps the most concrete means of transmitting musical meaning, emulating the physiognomic structure of physical motion (Davies, 1978; Rosar, 1996). This type of signification is utilized frequently in animated cartoons. Inevitably, as a character falls toward certain destruction, the musical score will incorporate a descending melodic line or glissando.

Two instructive examples of indexical meaning can be found in Hal B. Wallis' classic film *Casablanca* (1942; musical score by Max Steiner). During the opening titles, the initial phrase of music was based on a scale consisting of semitones and augmented seconds that serves as an *index*, "pointing to" the foreign locale. Indexical meaning is utilized in the musical score throughout this film as a means of differentiating between the French citizens from the German soldiers occupying the city. A minor key rendition of "Die Wacht am Rhein," a German folk song, becomes associated with the German antagonists, while "La Marseilles" came to represent the French protagonists. Both of these musical themes are used to effectively express the tensions between these two groups during a confrontation that occurs in Rick's bar as the sound of the soldiers singing the German folk song is gradually drowned out by the refrain of the French national anthem.

Symbolic signification relies solely on the relationship between musical sounds for meaning, including such audio-visual arts as modern dance, ballet, and ice skating. In this context, the various bodily movements and musical sounds take on significance purely because of the syntactical relationship of one sound or movement to another. This relationship is directly affected by the interaction between the two modalities (i.e., sight and sound) and the alignment of their accent structures. Within the context of a motion picture or video game, varying degrees of *iconic*, *indexical*, and *symbolic* signs are used to communicate to the listener the composer's intended musical message.

REVIEW OF MUSIC IN VIDEO GAMES

With the rapid pace of technological development in the video game industry, there have been great improvements in both the audio tools for designers and composers and the possibilities for storage and playback. Twenty years ago, composers of music for games were limited to simple beeps or synthesized sound (Jackson, 2004b), 8-bit central processing units (at best!), and minimal storage space. Today composers for video games often use state-of-the-art tools and instruments to extremely high-quality soundtracks, rivaling the production quality and musical sources typical of cinema. In fact, the primary technological hurdle remaining in the arena of music for games involves the implementation of interactive music engines, within which musical tracks and loops are activated and layered "on-the-fly" based on the actions of the player in the gamespace. Designers now have the freedom to pick and choose what aspects of the audio they would like to allow the user to control—from custom playlists to the volume levels of dialogue and the customization of surround sound. Today, designers can choose whether or not they would like to license extant music performed by recording artists or to hire a composer to create a unique score for the game. In the following section, we consider the evolution of console sound technologies and the evolving role of sound within the gaming experience.

The Co-Evolution of Music-Related Technologies and Implementations

Implementations of interactive music have developed alongside music-related technologies. Belinkie (1999) described the tone of the musical sounds in the original Nintendo Entertainment System as "hardly better than that of pure sine waves," largely due to the small amount of available storage space and 7-bit playback that could only handle four simultaneous sounds. Introduced to video game consoles by Sega in 1992, most major video game consoles turned to CD-ROM media soon thereafter due to its large storage capacity, high reliability, and comparatively low cost of production. Jackson (2004a) wrote, "Everyone agrees that it was the eventual widespread adoption of the CD-ROM format for video games that allowed for the rapid improvement of game audio." Nine years later, Sony Playstation 2 introduced the first video game console to use a proprietary DVD format along with mpeg-2 standard for video compression—offering a sevenfold increase in the amount of storage capacity of a single disc compared to a CD-ROM. This dramatic increase in storage space allowed video game producers to integrate extremely high-quality audio and video recordings in video games, to the point where, today, many games include the same DVD-quality audio and video of their cinematic counterparts.

Hyde-Smith (2004b) suggested that the improvement of game audio has contributed to a shift in public perception of gaming—from the domain of geeks to mass appeal. He wrote:

> I remember when I did *The Lost World*. We used a live orchestra and as soon as you said the word "videogame" people were just uninterested because in their mind videogame music was just annoying kid stuff. But I've seen a big change since then.

Indeed, there can be striking differences between the synthetic sound of MIDI and the sound of an orchestral recording, not the least of which is that audiences have grown accustomed to the quality of sound in other entertainment media such as television and film. In fact, scholarship on the perception of music in film indicates that as audio-visual stimuli become more complex, there is a shift in the cognitive role of music from one of association to one of highlighting key aspects of the visual domain (Boltz, 2001; Lipscomb, 2005). In other words, though in the past music might have primarily served to encode the game with additional associational meaning, contemporary designers have the added possibility of using music to heighten specific aspects or emotional qualities of the visual experience.

Genres Use Music and Other Aural Components Differently

To glean a better understanding of the recent history of audio in video games, data were collected on a representative sample of video games from the years 1985–2004 ($n = 159$), across a variety of platforms and genres.[2] Each game was coded for the presence/absence of the ability of the user to control various aspects of the game's sound, including: the playlist, volume-level of the sound effects, volume-level of the music, volume-level of the in-game voice/dialogue, the presence/absence of popular or familiar music, the ability to switch between stereo and mono-aural modes, and the presence/absence of surround sound controls.

Because video games are routinely categorized into genres in a similar fashion to television and film, we considered it important to look at whether and how music is used differently between these genres. In order to do so, each video game was assigned to one of five genre categories: Action/Adventure, Racing/Driving, Sports, Role-Playing Games (RPG), and Simulation/Strategy (see Appendix). The role and functions of music within each genre were assessed. Analysis of these data, using a multivariate analysis of variance, revealed a significant main

TABLE 17.1
Post-Hoc Comparisons of Means (only statistically significant differences are shown)

Sound Option	Comparison (Means)	
Sound FX	Racing/Driving (.917)	RPG (.250)
	Simulation (.845)	RPG (.250)
	Racing Driving (.917)	Action/Adventure (.376)
	Simulation (.845)	Action/Adventure (.376)
Music Vol.	Racing/Driving (.917)	RPG (.312)
	Simulation (.845)	RPG (.312)
	Racing/Driving (.917)	Action/Adventure (.495)
	Racing/Driving (.917)	Simulation (.845)
Pop/Recognizable Music	Racing/Driving (.333)	Action/Adventure (0.0)
	Sports (.254)	Action/Adventure (0.0)
Stereo/Mono	Action/Adventure (.507)	Sports (.126)

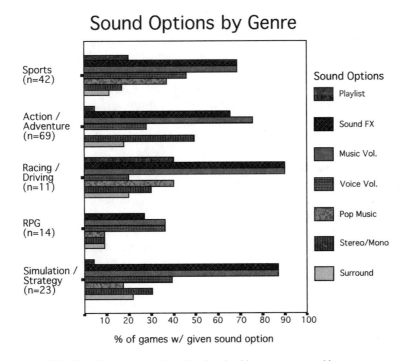

FIG. 17.1.   Comparision of sound options in video games, grouped by genre.

effect for genre ($F = 2.784; df = 4; p = .031$). Table 17.1 depicts the pairwise comparisons of means by genre ($n = 5$) and sound option ($n = 7$) that are statistically significant ($p <= .05$).

As reflected in Table 17.1, the Racing/Driving and Simulation genres have significantly higher percentage of games that allow the player to control the volume of sound effects and music than the RPG and Action/Adventure genres. Also, our findings suggest that Action/Adventure games have lower incidence of popular or familiar music than the Racing/Driving or Sports genres, as well as a higher incidence of Stereo/Mono control than sports games (see Figure 17.1).

That there is variation in the implementations of sound across video game genres is not wholly surprising. As independent game composer Darryl Duncan explained:

"Clearly if it is a medieval role playing type game, it would often call for dark dramatic orchestra based music. If the game is a Car racing game then the music would need to be more fast paced and 'driving' like Techno or Rave." (quoted in Belinkie, 1999)

What is more unexpected in our findings, however, is that genres of video games are not only differentiated by the type of music, but also through a degree of control given the player over their game experience: expressivity/customizability versus a controlled/narrated experience.

Perceptual Constraints

Along with technological constraints, composers for video games also face significant cognitive constraints on the amount of music and other sound that can be implemented effectively in a video game. For example, Stephen Rippy, composer of *Age of Mythology* (Microsoft Game Studios, 2002), stated, "You're dealing with a couple of dozen little characters on the screen at once, so it's finding a balance between hearing general mayhem and being able to identify what you're selecting and, 'Is this thing responding to what I'm telling it to do?'" (quoted in Jackson, 2004b). In order to balance what music plays when and at what levels, video game designers will have to incorporate interactive music engines that dynamically adapt the musical track to cues from the actions of the player in real-time. Additionally, the wider use of music situated in the game world will require the interactive music game engines to support the real-time spatialization of musical sound sources to complement implementations of point of view in 3-D video games.

One of the most apparent medium-specific constraints on the implementation of music in video games is that of designing music that is engaging, while not becoming intolerable after a significant amount of repetition. Many games, after all, require the user to replay levels many times while they master the skills and gather information to complete tasks and overcome obstacles. "The true test of game music is whether a player can still stand a simple theme after hearing it repeated for an hour or more," Belinkie (1999) wrote. In an environment where audio cues are given by triggers based on the player's progress through a scene or challenge, the length of time that a certain track will be looping is variable. "A kid plays his favorite game for 40 hours," Sanger (2004) wrote, "hears that hour over and over, and by the fourth hour, he's sick of the music." (p. 409) Some repetition can be helpful in establishing musical motifs, but too much can become extremely annoying.

APPLIED MEDIA AESTHETICS AND SOUND DESIGN FOR VIDEO GAMES

Adapting to technical and perceptual challenges has occupied video game composers and programmers throughout the medium's short history, and will likely continue to do so as constant innovation plays such an important role in the industry. The following section explores implementations of basic cinematic sound design techniques in video games. We contend that video game scores and incidental music serve crucial functions in the technical, aesthetic, and emotional experience of the player. In particular, music in video games can serve to enhance a sense of presence, cue narrative or plot changes, act as an emotional signifier, enhance the sense of aesthetic continuity, and cultivate the thematic unity of a video game.

Heightened Sense of Immersion

Music also serves an important role in the cultivation of the sense of perceptual or psychological presence in the video game. *Perceptual immersion* refers to a sense of having one's perceptual system submerged in a virtual environment (Biocca & Delaney, 1995), while *psychological immersion* refers to the degree to which the user feels involved in or engaged with stimuli from the virtual environment (Lee, 2004; Palmer, 1995). Tamborini and Skalski provide an overview of the components and consequences of presence in video games in their chapter "The Role of Presence" in this book.

Diegesis refers to the world of the characters and story within the film; everything that happens to the characters and in the environment portrayed on screen. In some cases, nondiegetic music (music, such as an orchestral score, that is not considered to be audible by the characters themselves) functions as an interpretive element, guiding the listener toward a certain feeling, subjectively beyond the visual elements (Sonnenschein, 2001, p. 154). As a diegetic example, music can be present as if occurring somewhere in the world of the game, such as a car radio in *Grand Theft Auto: Vice City* (Rockstar Games, 2002) or a small portable radio as in *Star Wars: Knights of the Republic* (LucasArts, 2003). In fact, in the latter example, the music is spatialized, meaning it adapts the volume-levels in the left and right stereo channels depending on the player's distance and orientation to the radio. Although they include diegetic music, *Grand Theft Auto III* (Rockstar Games, 2001) and *Grand Theft Auto: Vice City* do not have to contend with the implementation of spatial dynamics as the music plays while the player is sitting in a car. In fact, the *Grand Theft Auto: Vice City* soundtrack is not only an important part of the game play, but it has also created a profitable cross-media synergy through the commercial release of soundtrack CDs for each of the game's eight radio stations. Although it is too early to tell if this model will prove profitable for other video games or the music industry, the idea of a business model that incorporates proprietary technologies, older music properties, and cross-media synergies must be tantalizing to those in the music business who are reeling from the impact of peer-to-peer file sharing on their corporate earnings.

Narrative

Another important role of music is as a contributor to the narrative of a video game. More specifically, music can both assist in the unfolding of events in time (i.e., plot) and infuse the game experience with a heightened sense of importance, tension, or emotion.

One contribution of music to plot is as a marker of environmental changes. In games like *Final Fantasy X* (Square Electronic Arts, 2001), a change in the soundtrack is often triggered to coincide with the appearance of an enemy or triggered by a spatially located threshold to signal a change of scene. The *Lord of the Rings:The Two Towers* (EA Games, 2002) takes this one step farther by punctuating periods of silence with music cued by the sudden attack of a group of enemies (for a good example, see the marsh sequence near the end of "the Gates of Moria" level). This controlled use of silence is another important technique in sound design for video games. As Hyde-Smith (2004b) suggested, "It's actually more effective if you don't have it all the time because when it arrives it has more weight."

Music can serve as a sort of punctuation. In the "Otherworld" scene from *Final Fantasy X*, for example, the screen fades rapidly to white at the same time we hear the sound of a door slamming or the heavy metal rock music abruptly coming to an end, punctuating the end of one scene and the beginning of the next. Alternatively, sound effects and music can also be used to smooth over gaps or imperfections in the visual and sound effect layers of the experience, helping to ensure that the player is not distracted by technical irregularities/interruptions of the medium.

An extremely common archetype in both film and video games, the *hero's journey* begins with the introduction of a character who must undergo a series of trials or difficulties, often involving a descent into hell (at least metaphorically) in order to come to a revelation and return to the ordinary world (Zehnder & Calvert, 2004). This trope is easily adapted to the classic Aristotelian model of narrative typified by a rising action (desis), climax (peripeteia), and unwinding (denouement). Correspondingly, there is a long tradition in musical composition of developing themes—in the context of the common sonata form movement—that are first presented (exposition), brought into tension (development), and recapitulated (resolved) in the end (Sonnenschein, 2001).

However, the narrative structure of video games also creates a unique set of difficulties for the exposition, development, and recapitulation of leitmotifs. A *leitmotif* is a musical theme that becomes associated with a character, object, emotion, or idea within the game or motion picture. Film scores are replete with examples of such themes, for example, the "Fellowship" theme, the "Shire theme," and the "Ring" theme, to name a few of the themes composed by Howard Shore and used consistently throughout *The Lord of the Rings* trilogy. For example, in many *Action/Adventure* games, there are a few scenes that set the stage and provide the character with a narrative and navigable plane that, on completion and arrival at the goal of the journey, provide a final challenge with a "boss" or extremely powerful enemy that stands as a final obstacle before the protagonist, and the scene's denouement. Often, the music follows a narrative arc similar to the scene, where tension builds as the player makes progress navigating toward the goal, and finally reaching its climax. At this point, there is an abrupt change in the soundtrack that serves as a marker of the climax within the narrative arc. There are certainly consistencies in the musical theme that signify the new challenge as part of the preceding series of events. However, it is typical for the intensity of the piece—in rhythm and volume, for example—to signify a salient moment. This is one way in which music can simultaneously serve the purpose of marking an environmental change while also retaining important characteristics of the narrative trajectory.

Emotional Signifier

Another important role for music in video games is as an emotional signifier, or as exemplified by the quotes from of Wagner (1964) and Langer (1942) cited previously, to infuse the experience with "unconsummated symbolism." Bill Brown, composer of *Return to Castle Wolfenstein: Tides of War* (Activision Publishing, Inc., 2003), described this use of music in the following passage:

> I try to serve the subtext of what's going on in each area of the game. So I said, "Well, this is obviously a very sad thing going on, why don't we play it totally opposite and make this a really heart-wrenching piece?" It took a little bit of convincing but we finally did it and I think it worked well, while all this chaos is going on and you're trying to survive the level. We were already going to have tons of explosives and the action music would only be competing with that. I think it's a little bit more suspenseful if at times, it's the opposite. (Hyde-Smith, 2004a)

In this example, Brown described how the presence of music that contrasts the visual activity of the scene sometimes acts as a better complement to the experience as a whole. In fact, such music can heighten the sense of suspense by providing contrapuntal contrast.

Music can also infuse a scene or activity with the artist's critical perspective, although we contend that the division of labor in video game production militates against this to some degree. As Sonnenschein (2001) suggested: "Music can raise the literal to the symbolic, the ordinary to the poetic, and the particular to the universal. We feel the sense of belonging to all of mankind as we live the lives of those individuals on the screen" (p. 107). In order to enrich the experience of the World War II first-person shooter game he was producing, Steven Spielberg hired Michael Giacchino to compose an entirely orchestral score for the game *Medal of Honor* (EA Games, 1999). For this project, Giacchino composed "sad and sober pieces" that provide a powerful emotional signification throughout the continuing game series. Christopher Lennertz (2004), award-winning composer of *Medal of Honor: Rising Sun* (EA Games, 2003), asserted, "I also tend to be very visceral with my scores as well the action and tension will be on full tilt one minute then followed by a very sparse, subtle moment of plaintive drama. I think that *Medal of Honor* is the perfect canvas for that sort of thing."

EXPERIMENTAL STUDY OF THE IMPACT OF MUSIC ON IMMERSION

As stated at the outset of this chapter, there has been a surprising paucity of empirical research into the role of music in the context of gaming. Both players and, judging from all outward appearances, those involved in creating the games share the belief that the presence of music enhances the video game experience. However, our review of related literature revealed no experimental studies into this relationship. It is possible—perhaps even probable—that such research has been carried out by the companies that produce the video games, but the results— like the programming code for the games themselves—have remained proprietary. In an effort to initiate a much-needed investigation into this matter, the present authors designed and carried out an investigation into the role of music and its effect on the user's immersion in the gaming experience (Zehnder, Igoe, & Lipscomb, 2003), serving as a pilot study for the investigation reported herein. Though the purpose of this chapter is to inform the reader concerning the general role of music in video games, not to disseminate the results of experimental research, the latter is a vital—and currently absent—component in the advancement of knowledge in the field.[3]

Research Questions and Hypotheses

Several research questions were of interest: Do people perceive a game differently due to the presence of music? If so, what are the ways in which music can contribute to the aesthetic experience of the game? Does the presence of music act primarily as a complement to the visual components in that individuals who play the game with music have a similar, but more intense experience? In what ways does the audio-visual interaction contribute to the construction of narrative meaning (i.e., the player's interpretation of the scene)? We predicted that verbal responses provided by participants after exposure to one of three conditions (playing the game with music, playing the game without music, or *music-only*) would be significantly different. We also predicted that the presence of music would intensify the gaming experience, as explained below.

Method

Participants in the study included 76 university (63) and high school (13) students. This sample included 51 females and 25 males (see Table 17.2). Each subject completed a one page

TABLE 17.2
Number of Participants by Gender
and School Group

		High School	University
Gender	Male	6	19
	Female	7	44

TABLE 17.3
Adjectives for Verbal Response Scales

Not at all Pleasant	Pleasant
Not at all Active	Active
Not at all High	High
Not at all Powerful	Powerful
Not at all Loud	Loud
Not at all Good	Good
Not at all Exciting	Exciting
Not at all Annoying	Annoying
Not at all Busy	Busy
Not at all Strange	Strange
Not at all Dangerous	Dangerous
Not at all Labored	Labored
Not at all Bright	Bright
Not at all Relaxed	Relaxed
Not at all Colorful	Colorful
Not at all Cold	Cold
Not at all Gentle	Gentle
Not at all Intense	Intense
Not at all Fast	Fast
Not at all Simple	Simple
Not at all Masculine	Masculine

pre-experimental survey, in which they provided general demographic information. After being randomly assigned to one of three conditions (with music, without music, or *music-only*), each participant played three segments from *The Lord of the Rings: The Two Towers* video game for the Sony PlayStation2 (Electronic Arts, 2002). Following each segment, participants responded to a series of 21 verbal response scales, providing information about the experience (see Table 17.3). Each subject responded by moving the button on a scrollbar to an appropriate location between two terms anchoring the left and right side of the scroll (e.g., not active–active). This response was recording by the computer software as a value between 0 and 100, based on the location of the scrollbar button. After the participant completed all 21 randomly presented rating scales, they were asked to complete a free-response form containing five questions related to their perception of the musical and/or visual components of the stimulus. The participant would then move on to the next segment of game play, until all three had been completed and their responses collected.

The Stimuli

Three scenes were selected from the *Lord of the Rings: The Two Towers* video game for the purposes of this study, labeled by the investigators as: (1) "Weather Top," (2) "Moria," and (3) "Amon Hen." In all three scenes the players control the character Aragorn.

Weather Top refers to the second level of the video game (the first level after the prologue) where the player's main objective is to protect the character Frodo from the Ringwraiths—menacing-looking characters dressed in dark cloaks that emit shrill screams. The setting is atop a high mountain ruin surrounded by forest at night. In this scene the player must repeatedly relight and wield a torch in order to fend off the Ringwraiths.

Moria refers to a late portion of the second level of the video game where the player must fight through a marsh while orcs attack from the water and as arrows are shot from the surrounding cliffs. Because each participant was only asked to play the later half of this level, an experimenter cued the stimulus (with the sound off) either before the session or while the subject was busy responding to the questionnaire, depending on the randomized scene-order assignment. The scene is very dark and appears to be set at night, and goes through periods of relative quiet, punctuated by the occasional attack from an orc hidden in the marsh. The amount of activity reaches a crescendo as the player approaches the end of the level.

Amon Hen refers to the seventh level of the video game where the players must fight their way down a forest path against the fierce Uruk-Hai fighters, a stronger and larger breed of orc. The scene is set in the daytime in a forest and includes a large number of enemies in close combat. Research participants were instructed to play until they successfully reached the far side of the first bridge spanning the forest trail.

Treatment

Because the game engine for *Lord of the Rings: The Two* Towers allows players the option of turning off the music, the stimuli for Groups 1 and 2 (*game with music* and *game without music*, respectively) required no modification for the purposes of this study. However, Group 3, the *music-only* condition, required the extraction of segments of the in-game musical soundtrack, which were recorded as three separate tracks on an audio CD. Because the game does not allow for the absence of nonmusical sound (without eliminating all sound), the creation of the *music-only* stimuli required that (a) the audio of each game segment be extracted multiple times, (b) edited to eliminate the nonmusical sounds, and (c) reconstructed, keeping the music as intact as possible.[4] The lengths of the music-only stimuli were 54, 86, and 85 seconds, respectively. In comparison, scenes chosen for the game playing conditions were targeted at roughly 120 seconds each.

Results

Interestingly, the study revealed no between-subjects differences in participant responses based on the main effects of gender (female–male), age level (high school–university), or experimental condition (with music, without music, or *music-only*), nor in the interaction between these factors. However, looking closely at the subject responses as a within-subjects factor reveals some interesting differences. Not surprisingly, research participants responded differently to the various verbal scales ($F_{(20,1280)} = 22.606$, $p < .0005$). Participants from high school responded differently to the various verbal scales than the university students ($F_{(20,1280)} = 1.760$, $p = .020$). Differences emerged in the verbal responses to each game segment ($F_{(40,2560)} = 1.464$, $p = .031$), gender influence on the verbal ratings of each game segment ($F_{(40,2560)} = 1.657$, $p = .006$), and the influence of age level on the verbal ratings of each game segment ($F_{(40,2560)} = 1.622$, $p = .008$). Most important to the present study are the statistical conclusions that research participants responded differently to the verbal scales based on the experimental condition to which they were assigned ($F_{(40,1280)} = 1.522$, $p = .020$) and responded differently to the verbal scales for each scene based on the experimental condition to which they were assigned ($F_{(80,2560)} = 1.330$, $p = .028$). Therefore, though there were no significant differences between the responses of males and females, of high school and university students, or the music condition to which they were assigned, when we begin to look at more complex interactions between these factors and the verbal response scales, some interesting relationships can be identified. Most important, the ratings given to the various scales are significantly influenced by the presence or absence of music.

A detailed discussion of these empirical results is beyond the scope of the present chapter. However, it is worth noting that statistically significant differences clearly emerge as a result of the presence of music in the gaming environment. The results of this experimental investigation also suggest that gender differences and age level differences may provide fertile ground for future research into the role and functions of music in the context of video games.

AVENUES FOR FUTURE RESEARCH

It is our view that although the role of music in video games is generally respected by practitioners and fans, the foundation of scientific knowledge about its specific contribution to the perceptual salience of video games is, at present, virtually unexplored. There is a growing body of literature investigating the role of music in film, but very little experimental research into the role of music in the cultivation of a sense of immersion in the game. A significant portion of this chapter has been invested in establishing a critical vocabulary to support future research into the various roles that music plays in video games.

Of particular importance to video game designers and composers will likely be experimental research studies comparing human thresholds for discontinuity and latency in various formal features of video games, including the musical score. If, after all, music is often used to maintain a sense of continuity when visuals become jumpy or CPU load times create a momentary lag, then it would be useful to know the most effective types of music or implementation strategies that could maximize perceived continuity. Furthermore, in order to design effective interactive music engines, it would be helpful to have a better understanding of the factors influencing the cognition of musical loops and samples as accompaniment to repetitive game tasks.

There is also some important work to be done on the evolving relationship between the video game and music industries, especially with the diffusion of broadband Internet access and online gaming services. It is likely that, as more households connect to the Internet via broadband and video games are developed with more advanced online capabilities, the recording industry will increasingly seize upon the opportunity to license their content for real-time streaming and interactive sampling. Fledgling music licensing divisions like EA Trax will aggressively pursue exclusive licenses of popular music in order to maintain a creative advantage in an extremely volatile industry.

Furthermore, there is a need for historical and critical research on the co-evolution of hip-hop, rap, and video games—especially since it appears that the affinity for these styles of music cut across demographic boundaries, especially within the Sports and Action genres. As Kevin Liles of Def Jam Records explained, "Since the video game audience and hip hop demographics are clearly converging, the partnerships between our two brands is a powerful vehicle for our artists' and music" (quoted in *EA Partners With Major Music Labels and Artists on EA Trax*, 2004). Studies of this type would do well to build upon the body of research on music in contemporary culture, including research on music in music videos (Hansen & Hansen, 2000), uses and gratifications of popular music among adolescents (Gantz et al., 1978), the relationship of sex and violence in music videos (Gow, 1990), as well as early studies of the cognitive effects of gangsta rap (Hansen, 1995) and rock music (Greenfield et al., 1987).

The future of interactive gaming will undoubtedly be accompanied by an exponential improvement in sound quality and increasing role for music in the overall experience. Driven by future innovations in sound compression, spatialization, and interactive implementation as well as the diffusion of broadband Internet access, music will continue not only to enhance the narrative and emotional qualities of gaming experience, but also provide opportunities for the rich experience of live music, presented virtually.

APPENDIX—GENRE SAMPLES FOR THE "MUSICAL CAPABILITIES BY GENRE" STUDY

The following 69 games were categorized in the *Action/Adventure* genre: *Super Mario Bros, Afterburner, Karnov, Castlevania III: Dracula's Curse, Crash Bandicoot, Resident Evil, Tomb Raider, Mortal Kombat: Trilogy, 007: Goldeneye, MegaMan X4, Skull Monkeys, Star Wars: Rogue Squadron, Tomb Raider II, 007: Tomorrow Never Dies, Silent Hill, Tomb Raider: The Last Revelation, Wu Tang: Shaolin Style, Quake II, Resident Evil 2, 007: Agent Under Fire, Dead or Alive 3, Halo: Combat Evolved, Oni, Silent Hill 2: Restless Dreams, Star Wars: Obi-Wan, Blinx: The Time Sweeper, Conflict: Desert Storm, Disney's Treasure Planet, Max Payne, Medal of Honor: Allied Assault, Medal of Honor: Frontline, Monster's Inc, Robotech: Battlecry, Socom: U.S. Navy Seals, Spider-Man, Spyro: Enter the Dragonfly, State of Emergency, Super Mario Sunshine, The Getaway, Metroid Prime, Tom Clancy's Splinter Cell, Unreal Championship, Sonic Mega Collection, Freedom Fighters, Max Payne 2: The Fall of Max Payne, Prince of Persia, The Hulk, The Legend of Zelda: The Wind Waker, Unreal II: The Awakening, X2: Wolverine's Revenge, Tao Feng: Fist of the Lotus, Soul Calibur II, XIII, Half-Life, Return to Castle Wolfenstein: Tides of War, Star Wars Jedi Knight: Jedi Academy, Delta Force 2, Rayman 2: The Great Escape, Tom Clancy's Rainbow Six: Rogue Spear, Outlaws, Nam, Get Medieval, Descent, Sonic the Hedgehog, Sonic the Hedgehog 2, Sonic the Hedgehog 3, Sonic and Knuckles, Sonic 3D Blast, Sonic Spinball.*

The following 11 games were categorized in the *Racing/Driving* genre: *Cruis'n USA, Twisted Metal 2, GTA2, Midnight Club, Jet Set Radio Future, Sega GT 2002, Wreckless: The Yakuza Missions, Grand Theft Auto: Vice City, Nascar Thunder 2004, Gran Turismo 3, Grand Theft Auto: 3.*

The following 14 games were categorized in the *RPG* genre: *Dragon Warrior, Faxanadu, Parasite Eve, The Legend of Zelda: Ocarina of Time, Threads of Fate, Final Fantasy X, Harry Potter and the Chamber of Secrets, Lord of the Rings: The Fellowship of the Ring, Lord of the Rings: The Two Towers, The Elder Scrolls III: Morrowind, Lord of the Rings: The Return of the King, Star Wars: Knights of the Old Republic, Diabolo II, Dark Stone.*

The following 23 games were categorized in the *Simulation/Strategy* genre: *Metal Gear Solid 2: Sons of Liberty, Britney's Dance Beat, Mario Party 4, Mary-Kate and Ashley: Sweet 16: Licensed to Drive, Warcraft III: Reign of Chaos, Command & Conquer: Generals, The Sims Bustin' Out, Socom II: US Navy Seals, The Sims, Civilization III, Disciples, CyberStorm 2, Jagged Alliance 2, The Operational Art of War, vol 1, Armored Fist 3, Starsiege: Tribes, SimCity 2000 SE, Starfleet Command: Neutral Zone, Police Quest: Swat 2, Homeworld, Pax Imperia: Eminent Domain, Wing Commander: Prophecy, Dr. Robotnik's Mean Bean Machine.*

The following 42 games were categorized in the *Sports* genre: *Duck Hunt, NHL 99, Cool Boarders 4, Knockout Kings 2000, Grind Session, NFL Blitz: 2001, Madden NFL 2001, Tony*

Hawk: Pro Skater 2, Madden NFL 002, Amped: Freestyle Snowboarding, BMX XXX, Madden 2003, FIFA Soccer 2003, WWW Raw, Dead or Alive: Extreme Beach Volleyball, Harry Potter: Quidditch World Cup, Madden 2004, NBA Live 2004, NBA Street Vol. 2, SSX 3, Def Jam: Vendetta, Tony Hawk's Underground, World Series Baseball 2K3, NBA Inside Drive 2004, WWF Smackdown: Justin Bring It, Tony Hawk Pro Skater 3, NHL 000, Tiger Woods PGA Tour, Prime Time Starring Deion Sanders, World Series Baseball, NHLPA Hockey '93, PGA Tour Golf II, Madden 95, John Madden Football '93, NBA Live 95, FIFA Soccer 97, College Football USA 97, John Madden Football '92, Coach K College Basketball, FIFA International Soccer, PGA European Tour, NHL 95.

NOTES

[1] For the purposes of this chapter, "film" is used as an umbrella term to describe all cinematic audio-visual media without drawing distinctions between various digital and analog formats.

[2] The number and percent of games by platform were the following: NES ($n = 7$; 4.4%), Nintendo 64 ($n = 10$; 6.3%), Sega Genesis ($n = 21$; 13.2%), Sony Playstation ($n = 19$; 11.9%), Sony Playstation 2 ($n = 39$; 24.5%), Nintendo GameCube ($n = 8$; 5.0%), Miscrosoft Xbox ($n = 24$; 15.1%), and PC/CDROM ($n = 31$; 19.5%).

[3] Results of the statistical analyses are presented in an abbreviated format, providing the essential information for those to whom such figures are useful. Any reader interested in more detailed results should contact the primary author via email (s-zehnder@northwestern.edu) for details.

[4] The signal path was from the Playstation 2 audio out directly into a Motu 828. The Motu was connected via firewire to a Mac G4 Powerbook, running Logic Audio Platinum 5. The editing of the segments was done by first eliminating any unwanted nonmusical sounds from each extraction, resulting in several layers of music files with gaps of silence. These files were then combined so as to "fill in the gaps" by pasting portions of each layer together at identical zero-crossings, resulting in one seamless audio file identical to the in-game sound excluding nonmusical sounds. As a player of the game, one author (SZ) determines that these excerpts were representative of music that would accompany the gaming experience.

REFERENCES

Belinkie, M. (1999). Video game music: Not just kid stuff. Special Report for VGMusic.com. Retrieved May 15, 2004, at http://www.vgmusic.com/vgpaper.shtml.

Bengtsson, I., & Gabrielsson, A. (1983). Analysis and synthesis of musical rhythm. In J. Sundberg (Ed.), *Studies of music performance* (pp. 27–60). Stockholm: Royal Swedish Academy of Music.

Biocca, F., & Delaney, B. (1995). Immersive virtual reality technology. In F. Biocca & M. Levy (Eds.), *Communication in the age of virtual reality* (pp. 57–124). Hillsdale, NJ: Lawrence Erlbaum Associates.

Bolivar, V. J., Cohen, A. J., & Fentress, J. C. (1996). Semantic and formal congruency in music and motion pictures: Effects on the interpretation of visual action. *Psychomusicology, 13*, 28–59.

Boltz, M. (1992). Temporal accent structure and the remembering of filmed narratives. *Journal of Experimental Psychology: Human Perception and Performance, 18*, 90–105.

Boltz, M. (2001). Musical soundtracks as a schematic influence on the cognitive processing of filmed events. *Music Perception, 18*(4), 427–454.

Boltz, M., Schulkind, M., & Kantra, S. (1991). Effects of background music on the remembering of filmed events. *Memory & Cognition, 19*, 593–606.

Brown, R. (1988). Film and classical music. In G. R. Edgerton (Ed.), *Film and the arts in symbiosis: A resource guide* (pp. 165—215). New York: Greenwood Press.

Bullerjahn, C., & Güldenring, M. (1996). An empirical investigation of effects of film music using qualitative content analysis. *Psychomusicology, 13*, 99–118.

Campbell, W., & Heller, J. (1980). An orientation for considering models of musical behavior. In D. Hodges (Ed.), *Handbook of music psychology* (pp. 29–36). Lawrence, KS: National Association for Music Therapy.

Cavalcanti, A. (1985). Sound in films. In E. Weis & J. Belton (Eds.), *Film sound: Theory and practice* (pp. 98–111). New York: Columbia University Press.

Chion, M. (1994). *Audio-Vision: Sound on screen* (C. Gorbman, Trans.). New York: Columbia University Press. (Original work published in 1990).

Clarke, E. (1988). Generative principles in music performance. In J. A. Sloboda (Ed.), *Generative processes in music* (pp. 1–26). Oxford, England: Clarendon.

Clynes, M. (1983). Expressive microstructure in music, linked to living qualities. In J. Sundberg (Ed.), *Studies of music performance* (pp. 76–181). Stockholm: Royal Swedish Academy of Music.

Dowling, W. J., & Harwood, D. L. (1986). *Music cognition.* Orlando, FL: Academic Press.

Davies, J. B. (1978). *The psychology of music.* Stanford, CA: Stanford University Press.

EA Partners With Major Music Labels and Artists on EA Trax. Music4Games.com. Retrieved March 24, 2004, at http://www.music4games.net/n_eatrax.html.

Gabrielsson, A. (1988). Timing in music performance and its relations to music experience. In J. A. Sloboda (Ed.), *Generative processes in music* (pp. 27–51). Oxford, England: Clarendon.

Gantz, W., Gartenberg, H. M., Pearson, M. L., & Schiller, S. O. (1978). Gratifications and expectations associated with pop music among adolescents. *Popular Music in Society, 6,* 81–89.

Giacchino, M. (2002). Medal of Honor: Frontline (Soundtrack). Electronic Arts, Inc.

Gorbman, C. (1987). *Unheard melodies: Narrative film music.* Bloomington: Indiana University Press.

Gow, J. (1990). The relationship between violent and sexual images and the popularity of music videos. *Popular Music and Society, 14*(4), pp. 1–10.

Greenfield, P. M., Bruzzone, L., Koyamatsu, K., Satuloff, W., Nixon, K., Brodie, M., & Kingsdale, D. (1987). What is rock music doing to the minds of our youth? A first experimental look at the effects of rock music lyrics and music videos. *Journal of Early Adolescents, 7,* 315–330.

Hansen, C. H. (1995). Predicting cognitive and behovioral effects of gangsta' rap. *Basic and Applied Social Psychology, February,* 16(1–2), 43–52.

Hansen, C., and Hansen, R. (2000). Music and music videos. In D. Zillmann & P. Vorderer (Eds.), *Media entertainment: The psychology of its appeal (*pp. 175–196). Mahwah, NJ: Lawrence Erlbaum Associates.

Hanslick, E. (1986). *On the musically beautiful: A contribution towards the revision of the aesthetics of music* (8th ed., G. Payzant, Trans.). Indianapolis, IN: Hackett Publishing. (Original work published in 1891).

Hood, M. (1982). *The ethnomusicologist.* Kent, OH: Kent State University Press.

Hyde-Smith, A. (2004a). Interview with Composer Bill Brown. *Music4Games.com.* Retrieved March 23, 2004, at http://www.music4games.net/f_billbrown.html.

Hyde-Smith, A. (2004b). Interview with Medal of Honor Composer Michael Giacchino. *Music4Games.com.* Retrieved March 23, 2004, at http://www.music4games.net/f_moh_mgiacchino.html.

Iwamiya, S. (1996). Interactions between auditory and visual processing when listening to music in an audio visual context: 1. Matching 2. audio quality. *Psychomusicology, 13,* 133–153.

Jackson, B. (2004a). Game sound: Audio at electronic Arts. *Mix,* March. Retrieved July 7, 2005, from http://mixonline.com/recording/interviews/audio_audio-electronic_arts/.

Jackson, B. (2004b). Game sound: Ensemble Studios on gods, trolls and minotaurs. *Mix,* at http://mixonline.com/recording/applications/audio-ensemble-studios-goods/.

Kendall, R. A., & Carterette, E. C. (1990). The communication of musical expression. *Music Perception 8 (2),* 129–163.

Krumhansl, C. L., & Schenck, D. L. (1997). Can dance reflect the structural and expressive qualities of music? A perceptual experiment on Balanchine's choreography of Mozart's *Divertimento No. 15. Musicae Scientiae, 1*(1), 63–85.

Langer, S. K. (1942). *Philosophy in a new key: A study of symbolism of reason, rite, and art* (3rd ed.). Cambridge, MA: Harvard University Press.

Lee, K. M. (2004). Presence, explicated. *Communication Theory, 14*(1), February, 27–50.

Lennerz, C. (2004). Interview with New Medal of Honor Composer Christopher Lennertz. *Music4Games.com.* Retrieved March 23, 2004, at http://www.music4games.net/f_moh_christopherlennertz.html.

Lipscomb, S. D. (2005). The perception of audio-visual composites: Accent structure alignment of simple stimuli. *Selected Reports in Ethnomusicology, 11,* Winter, 37–67.

Lipscomb, S. D., & Kendall, R. A. (1996). Perceptual judgment of the relationship between musical and visual components in film. *Psychomusicology, 13*(1), 60–98.

Lipscomb, S. D., & Tolchinsky, D. E. (2005). The role of music communication in cinema. In D. Miell, R. MacDonald, & D. Hargreaves (Eds.), *Music communication.* Oxford, UK: Oxford University Press, 283–404.

Lomax, A. (1962). Song structure and social structure. *Ethnology, 1,* 425–451.

Marshall, S. K., & Cohen, A. J. (1988). Effects of musical soundtracks on attitudes toward animated geometric figures. *Music Perception, 6,* 95–112.

Merriam, A. P. (1964). *The anthropology of music.* Evanston, IL: Northwestern University Press.

Meyer, L. B. (1956). *Emotion and meaning in music.* Chicago: University of Chicago Press.

Nettl, B. (1983). *The study of ethnomusicology.* Urbana: University of Illinois Press.

North, A. (2003). Medal of Honor: Frontline (Soundtrack Review). Retrieved May 10, 2004, at http://www.film-tracks.com/titles/medal_honor3.html. Last modified: July 27, 2003.

Palmer, M. T. (1995). Interpersonal communication and virtual reality: Mediating interpersonal relationships. In F. Biocca & M. R. Levy (Eds.), *Communication in the age of virtual reality* (pp. 277–302). Hillsdale, NJ: Lawrence Erlbaum Association.

Peirce, C. S. (1931–1935). *Collected papers* (Vols. 1–6), C. Hartshorne & P. Weiss (Eds.). Cambridge, MA: Harvard University Press.

Peterson, G. (2004). From the editor: Not just a game! *Mix*, March. Retrieved 7/27/05, from http://mixonline.com/mag/audio-not-game/

Reesman, B. (2004). Music for games: Bob Daspit. *Mix*, March. Retrived 7/27/05, from http://mixonline.com/recording/interviews/audio-music-game/

Rosar, W. H. (1996). Film music and Heinz Werner's theory of physiognomic perception. *Psychomusicology, 13*, 154–165.

Sanger, G. A. (2004). *The fat man on game audio: Tasty morsels of sonic goodness.* Indianapolis, IN: New Riders.

Seashore, C. E. (1938). *Psychology of music.* New York: McGraw-Hill. (Reprinted, New York: Dover, 1967).

Senju, M., & Ohgushi, K. (1987). How are the player's ideas conveyed to the audience? *Music Perception, 4*, 311–323.

Sirius, G., & Clarke, E. F. (1996). The perception of audiovisual relationships: A preliminary study. *Psychomusicology, 13*, 119–132.

Sonnenschein, D. (2001). Sound design: The expressive power of music, voice, and sound effects in cinema. Studio City, CA: Michael Wiese Productions.

Stam, Robert (2000). *Film theory.* Oxford, England: Blackwell.

Sundberg, J. (1988). Computer synthesis of music performance. In J. A. Sloboda (Ed.), *Generative processes in music* (pp. 52–69). Oxford, England: Clarendon.

Sundberg, J., Frydén, L., & Askenfelt, A. (1983). What tells you the player is musical? An analysis-by-synthesis study of music performance. In J. Sundberb (Ed.), *Studies of music performance* (pp. 61–67). Stockholm: Royal Swedish Academy of Music.

Tannenbaum, P. H. (1956). Music background in the judgment of stage and television drama. *Audio-Visual Communications Review, 4*, 92–101.

Thayer, J. F., & Levenson, R. W. (1983). Effects of music on psychophysiological responses to a stressful film. *Psychomusicology, 3*, 44–54.

Thompson, W. F., Russo, F. A., & Sinclair, D. (1996). Effects of underscoring on the perception of closure in filmed events. *Psychomusicology, 13*, 9–27.

Vitouch, O. (2001). When your ear sets the stage: Musical context effects in film perception. *Psychology of Music, 29*, 70–83.

Wagner, R. (1964). *Wagner on music and drama.* New York: Da Capo Press, Inc.

Winter, David. Atari Pong—the first steps. Retrieved April 20, 2004, at http://www.pong-story.com/atpong1.htm.

Zehnder, S., & Calvert, S. (2004). Developmental differences in older and younger adolescents' understanding of heroic depictions. *Journal of Communication Inquiry*, April, 122–137.

Zehnder, S., Igoe, L., & Lipscomb, S. D. (2003). Immersion factor—sound: A study of the influence of sound on the perceptual salience of interactive games. Paper presented at the conference of the Society for Music Perception & Cognition. Las Vegas, NV, June 2003.

18

Narrative and Interactivity in Computer Games

Kwan Min Lee, Namkee Park, and Seung-A Jin
University of Southern California

According to a recent industry report, the U.S. computer-game industry (including console-based, P.C. or Mac based, and arcade games) grew 8% to a record-breaking $6.9 billion in 2002, surpassing the box-office movie industry (ISDA, 2003a). In addition, about 60% of U.S.-Americans aged 6 or older, or approximately 145 million people, regularly play computer games. (By computer games, we mean all types of electronic games including console-based games [PlayStation, Nitendo, and Gameboy], CD-ROM games, online games, and arcade games.) Quite surprisingly, as of 2000, nearly half (43%) of these game players were female (Burke, 2000). The average age of game players is 29 as of 2003 (ISDA, 2003b). In South Korea, a country well-known for MMORPG, or massively multi-player online role-playing games, NCsoft, the biggest South Korean online games firm, earned an operating profit of 77 billion won ($62 million) on revenues of 155 billion won—a margin of nearly 50% in 2002 (Economist, 2003). Clearly, computer games are becoming a mainstream entertainment medium.

One major characteristic of entertainment media is its focus on delivering a compelling narrative to the audience. Old entertainment media such as books, radio, TV, and film have each developed unique conventions to maximize this narrative power. Novels, plays, and movies are constructed according to certain styles and formats in order to create compelling narratives (Bal, 1985; Onega & García Landa, 1996; Vorderer & Knobloch, 2000). In fact, they are judged by the quality of the narratives they convey to the audience. However, despite the differences in style and format of narrative creation and delivery, all of these entertainment media share one major characteristic—*a linear relationship between the creator and the audience*. Narratives always flow from the creator (e.g., book author, TV producer, film director) to the audience. Although we acknowledge the existence of a claim that the reader (i.e., the audience) psychologically participates in the narrative construction in his or her mind while reading a book (Oatley, 2002), we think that narratives delivered by old entertainment media such as books or movies can never be *physically* changed or reconstructed by the audience. That is, the physical linear relationship between the creator and the audience cannot be changed.

Computer games as entertainment media are fundamentally different from old entertainment media in various ways. Many scholars argue that interactivity is the major factor that sets computer games apart from old entertainment media (Grodal, 2000; Vorderer, Knobloch, & Schramm, 2001). In fact, they contend that it is one of the major reasons why computer games have become so popular in such a short time (Murray, 1997; Plowman, 1996). Thanks to the interactive characteristic of this new entertainment medium (Brody, 1990, p. 103; Vorderer, 2000, p. 23), game players can actively interact with simulated environments, A.I. (Artificial Intelligence)-based software agents, or networked game players. As a result, the linear relationship between the creator and the audience of a narrative has been blurred.

With regard to narratives, computer games are different from old entertainment media in two ways. First, for some (usually old) types of computer games, the creation and delivery of narratives have never been a major concern. For example, simple games such as *Tetris, Pinball,* and *Solitaire* did not focus on narrative creation and delivery. These games mainly focused on engaging players in fun and usually repetitive mental and/or physical exercises. Most prototypical arcade games such as *Space Invaders* (1978) and *Pac-Man* (1980) also did not position themselves as narrative media. Even though they provided some rudimentary tacit narratives (e.g., protecting the Earth, rescuing a princess, and so forth), the major gratification derived from playing these games was not based on emotional engagement in the narratives but on the simple perceptual (e.g., music, graphics) or motivational (e.g., earning the highest point) engagements in game modalities and rules.

Second, even when computer games focused on creating and delivering compelling narratives, their narratives were somewhat different from those of the traditional media. As indicated by many scholars, they provided nonlinear narratives that blur the traditional one-way relationship between creators and the audience of entertainment media. Computer-game players could change the course of the narrative or even construct a new narrative by changing their behaviors and performance during a game. Therefore, unlike book readers or movie watchers, players of narrative-intensive computer games such as *Myst* (1993), *Tomb Raider* (1996), and *Max Payne* (2002) can encounter *physically* different narratives even when they play the same game. This difference between computer games and old entertainment media is more important than the first one discussed above, because narratives are increasingly becoming an integral part of contemporary computer games (Wolf, 2001).

Although we do not predict the total demise of nonnarrative or simple arcade games, we believe that these games are not—and will not be—the mainstream computer games. As with other types of entertainment media, narratives will become a killer application of computer games and computer games will gradually become another medium for storytelling. However, because of the interactive nature of computer games, narratives in computer games will be somewhat different from those in other media.

Despite the increasing use of narratives in computer games and the enormous efforts made into developing compelling narratives for games, there have been few academic discussions on the *impacts of narratives* on computer games. This is due to two reasons. First, there has been a lack of academic interest in the study of computer games, even though computer games have existed for 40 years (for a similar claim made to the study of entertainment, see Vorderer, 2003; for a detailed discussion on the history of computer games and game playing, see chapters by Williams and Lowood in this book). Consequently, game studies have been very slow in achieving the status of an independent academic area. For example, almost all existing academic associations (e.g., Digital Game Research Association), conferences (e.g., Electronic Entertainment Expo, Digital Game Research Conference), and journals (e.g., *Game Studies*) on computer games were created only within the last decade. Second, existing literature on computer games has mainly focused on two factors—interactivity as a form factor and violence as a content factor—and neglected other issues such as narratives (see Peng & Lee, 2004). That is, the majority of existing literature has basically tried to examine the effects of interactivity on aggression-related variables (aggressive behaviors, attitudes, thoughts, and affects) when people play violent computer games. The effects of narratives in computer games, thus, have received little academic interest.

In this chapter, we provide a theoretical explanation of what *narrative* is and why it is still important in *interactive* media such as computer games. We start our discussion by explicating (for an explication procedure, see Chaffee, 1991) the two major concepts—*interactivity* and *narrative*. Based on the explications, we examine two opposing theoretical arguments on the relationship between the two concepts. After the examination, we will look at how *interactivity* and *narrative* have been actually integrated together in computer games by examining the film-adaptation trend in the computer-game industry and by categorizing various types of narratives currently employed in various computer games. Finally, we explain the reasons why a narrative is important in interactive media such as computer games.

INTERACTIVITY, EXPLICATED

Early Definitions of Interactivity

Interactivity has been defined in various ways (Biocca, 1998; Chaffee, Rafaeli, & Lieberman, 1985; Ha & James, 1998; Haeckel, 1998; Heeter, 1989; Rafaeli, 1988; Rafaeli & Sudweeks, 1997; Rogers, 1995; Steuer, 1992; Vorderer, 2000; Walther & Burgoon, 1992). We categorize three main views of interactivity—technology-oriented, communication setting-oriented, and individual-oriented views—from the previous literature.

First, technology-oriented approaches have regarded interactivity as a characteristic of new technologies (usually containing a computer as a component) that makes an individual's participation in a communication setting possible and efficient (Biocca, 1998; Heeter, 1989; Steuer, 1992; Walther & Burgoon, 1992).

In contrast, Chaffee et al. (1985), Rafaeli (1988), and Rafaeli and Sudweeks (1997) considered interactivity as a process-related characteristic of a communication setting. Rafaeli (1988) defined interactivity as "an expression of the extent that in a given series of communication exchanges, any third (or later) transmission (or message) is related to the degree to which previous exchanges referred to even earlier transmissions" (p. 111).

Finally, individual-oriented views have defined interactivity from a technology user's point of view. For example, Rogers (1995) defined interactivity as "the degree to which participants in a communication process can exchange roles and have control over their mutual discourse" (p. 314). Ha and James (1998) also conceptualized interactivity from a participant's perspective. However, since they were more interested in the outcome of interactive communications, they defined interactivity in relation to persuasiveness and goal achievement, such as fulfilling consumer needs. Table 18.1 illustrates the early definitions of interactivity as categorized by the three different approaches.

Problems of Early Definitions

Although technology-oriented definitions are the most popular, they have inherent problems in that even the same medium can be considered as having a different degree of interactivity according to the way it is actually used by an individual. For example, if a user has competence in using every command of a computer game, the user may feel much more interactivity in playing the game than a novice user. Therefore, the notion of interactivity as a characteristic of a particular medium is null in that a medium itself cannot have a fixed degree of interactivity.

TABLE 18.1
Early Definitions of Interactivity

Main Approach	Researchers	Definitions
Technology-oriented approach	Biocca (1998) Steuer (1992) Walther & Burgoon (1992)	"Interactivity is the name given to properties of a medium that simulate the properties of human interaction with the physical world and/or other humans (intelligent beings)" (Biocca, 1998, p. 5). "Interactivity is the extent to which users can participate in modifying the form and content of a mediated environment in real time" (Steuer, 1992, p. 84).
Communication setting-oriented approach	Chaffee et al. (1985) Rafaeli (1988) Rafaeli & Sudweeks (1997)	"A relationship between interactants A and B over time is interactive if an act by A at one time is constrained by an act of B at an earlier time that was constrained by an act of A yet an earlier time" (Chaffee et al., 1985, p. 9). "... interactivity is an expression of the extent that in a given series of communication exchanges, any third (or later) transmission (or message) is related to the degree to which previous exchanges referred to even earlier transmissions" (Rafaeli, 1988, p. 111).
Individual-oriented approach	Ha & James (1998) Rogers (1995)	"... interactivity should be defined in terms of the extent to which the communicator [Web page providers] and the audience respond to, or are willing to facilitate, each other's communication needs [such as playfulness, choice, connectedness, information collection, and reciprocal communication]" (Ha & James, 1998, p. 461). "... the degree to which participants in a communication process can exchange roles and have control over their mutual discourse" (Rogers, 1995, p. 314).

Communication setting-oriented definitions, which emphasize mutual exchanges of information, also contain problems in that they have unrealistic assumptions that every participant of a communication setting wants the same level of information exchange and that the actual exchange of information is the first condition for interactivity. However, a symmetrical flow of information is rare and interactivity can occur without the actual exchange of information. For example, in a massive multi-player game such as *The Sims Online*, a participant may feel a high level of interactivity by merely observing and reading other players' behaviors and text messages, even when he or she is not actively participating in virtual social interaction. Most fundamentally, these definitions cannot explain the fact that interacting participants of the same communication setting could have different levels of interactivity. That is, they cannot explain why players of the same computer game can feel different levels of interactivity even when they are part of the same communication setting.

For existing individual oriented definitions, the limitation is that they cannot be successfully applied to human–computer interaction situations, because they were made in the context of human-to-human interactions (i.e., mediated communications) in which at least two interacting humans are required (e.g., Ha & James, 1998; Rogers, 1995). Therefore, a possible interaction between a game player and an autonomous software agent inside the game world cannot be easily explained by existing individual-oriented definitions (for the claim that people interact with software agents in a similar way as they would interact with real people, see Reeves & Nass, 1996; Nass & Moon, 2000).

An Alternative Definition of Interactivity

In order to solve the various problems of the early definitions, we should consider interactivity as a *perceived* characteristic of a communication *act*, which varies according to a communicating actor's perception. That is, interactivity is not just a given characteristic of a particular medium or a communication setting but a constructed characteristic of a communication act according to an individual's perception.

Interactivity has two Latin roots—"*inter-*" that broadly means (a) between or among and (b) mutual or reciprocal, and "*-act*" that means "do" (Webster's Third New International Dictionary, 1976). Together, "interact" means to act upon each other and to have reciprocal effects or influences. Therefore, interactivity can mean "the state of acting upon each other and having reciprocal effect or influence." From this nominal definition, we can see that interactivity requires two fundamental conditions: (a) at least two interacting participants (usually two persons, or in the case of human–computer interaction, at least, a person and a computer [or agent]); and (b) reciprocal effects of an interacting participant's act on the communication process. Based on the two fundamental conditions raised above, we propose the following conceptual definition:

Interactivity is a perceived degree that a person in a communication process with at least one more intelligent being can bring a reciprocal effect to other participants of the communication process by turn-taking, feedback, and choice behaviors.

The term "perceived" means that the interactivity defined here is more concerned about the individual's subjective perception than objective technology characteristics. The term "degree" means that interactivity will be quantitatively defined according to the level of individual perception, which is more or less subjected to objective factors such as turn-taking possibility, feedback mechanisms, and the quantity and quality of user choices available in the system. "Communication process" means a type of communication acts such as interpersonal conversation, TV watching, playing computer games, using a computer, and so forth. "Intelligent being" can be a person, an avatar, a software agent, a computer, or a robot. "Reciprocal effect" means that an act by an interacting participant in a communication process should bring a result of visible changes to the acts of other interacting participants in that communication process. For example, game players' killing of enemies (which are either other players or software agents) in computer games should bring a visible change to the elimination of the dead players or agents. "Turn-taking" means the change of the role between the sender (or the creator) and the receiver (or the audience). It is different from feedback in that it is usually longer than feedback and requires an interacting participant to formally take the other role. "Feedback" means a receiver's reaction to a sender about a received message (or an observed behavior). "Choice" means a user's choice/selection of something from among various sets of options or possibilities (e.g., choosing a movie in VOD system, choosing a particular behavior for his or her game character).

This definition of interactivity can be applied to broad areas of communication settings, including interpersonal, mediated, and even human–computer interactions and communications. This is a psychologically oriented concept. The level of analysis is individual. The unit of analysis is an individual who participates in a communication process. It varies quantitatively across individuals and each given communication process. Therefore, using the same technology or being in the same communication setting does not guarantee the same level of interactivity. For instance, players of *The Sims Online* can feel different levels of interactivity depending on their experience and behaviors during the game. The level of interactivity becomes higher as an interacting participant acquires more confidence about the process through his or her own efforts and experiences. For example, a game player will feel a higher level of interactivity as he or she gains higher computer-game self efficacy after repeated plays.

TABLE 18.2
A Representative List of Early Definitions of *Narrative* (in Chronological Order)

Author	Definition
Aristotle (as cited by Riessman, 1993)	Narrative is a story that has a beginning, a middle, and end (in *Poetics*).
Labov & Waletzky (1967)	Narrative is "a story which follows a chronological sequence. The order of events moves in a linear way through time and the order cannot be changed without changing the inferred sequence of events in the original semantic interpretation" (p. 21).
Labov (1972)	Narratives are stories about specific past events.
Fisher (1984)	Narratives are "stories we tell ourselves and each other to establish a meaningful life-world" (p. 6).
Brown (1987)	Narrative is "an account of an agent whose character or destiny unfolds through actions and events in time" (p. 143).
Young (1987)	Narrative is a consequential sequencing. One event causes another in the narrative, although the links may not always be chronological.
Lowe (2000)	Narrative is a recounted story (p. 18).
Mallon & Webb (2000)	"Narrative is one of the oldest constructs humans use for understanding and giving meaning to the world" (p. 270). Causality, temporality, and linearity are the key features of narrative.
Ryan (1997, 2001)	Narrative can mean a discourse reporting a story, as well as the story itself.
Abbott (2002)	"Narrative is the representation of a series of events" (p. 12).

NARRATIVE, EXPLICATED

Previous Definitions of Narrative

Narrative has as long a history as that of humankind. For example, Barthes (1966, p. 1) stated that:

> *Narrative is present in every age, in every place, in every society; it begins with the very history of mankind and there nowhere is nor has been a people without narrative. All classes, all human groups, have their narratives, enjoyment of which is very often shared by men with different, even opposing, cultural backgrounds. Caring nothing for the division between good and bad literature, narrative is international, trans-historical, trans-cultural; it is simply there, like life itself.*

Mallon and Webb (2000) also argued that "narrative is one of the oldest constructs humans use for understanding and giving meanings to the world" (p. 270).

As we can expect from the long history of narratives, there have been many theoretical explorations into the conceptual understanding of narrative. They inevitably produce considerable disagreements on the definition of narrative (Riessman, 1993; see Table 18.2 for the list of early definitions). For example, Aristotle (as cited by Riessman, 1993) saw narrative as a story that has a beginning, a middle, and an end. Labov and Waletzky (1967) defined narrative as "a story that follows a chronological sequence" (p. 21). According to Fisher (1984), narratives are "stories we tell ourselves and each other to establish a meaningful life-world" (p. 6). For Ryan (1997, 2001), narrative is a sign with a signifier (discourse) and a signified (story, mental image, semantic representation). The signifier can have many different semiotic manifestations, such as verbal acts of storytelling or gestures and dialogues performed by actors. He argued that a narrative is medium-free because "narrativity" is located in the level of the signified (story), not the signifier. Lowe (2000) simply put narrative as "a recounted story" (p. 18). Most recently, Abbot (2002) defined narrative as "the representation of an event or a series of events" (p. 12).

Whatever the definition, there seldom has been an objection to the statement that a narrative has two major components—"story" and "plot." Different terms are sometimes used to represent the concept of "story" and "plot." For example, Russian structuralists use the term "fabula" for "story" and "sjuzet" for "plot." Abbott (2002) used the term "narrative discourse" instead of "plot." Even though we acknowledge the subtle differences raised by Abbott (2002), we use the terms "story" and "plot" here in order to follow the major tradition in narratology. "Story" is an event or a sequence of events involving entities, whereas "plot" is the way the story is conveyed. Stories can be delivered by a variety of media (e.g., books, film, radio, TV, computer games) with sometimes dramatically diverse plots. A plot plays a significant role in story delivery by "organizing events and actions into a logically unfolding development" (Brown, 1987, p. 143). Therefore, a story is always constructed by a plot (Abbott, 2002, p. 19).

Problems of Applying Early Definitions to Interactive Media

The early definitions have several limitations when we try to employ them to explain narratives in interactive media. First, most of the definitions derived from the Russian structuralism tradition viewed narrative as a given or predetermined structure. They tended to neglect the possible cognitive role an audience plays in understanding a given narrative or in forming a narrative. Thus, these traditional definitions focused heavily on structural analyses of a narrative, rather than psychological processing of a narrative. This might work in traditional media, but the negligence of the active user participation in narrative interpretation cannot work in interactive media such as computer games (Vorderer, Knobloch, & Schramm, 2001).

Second, the early definitions of narrative were conceptualized with a specific medium in mind (usually either film or books). Therefore, they cannot be appropriately applied to the study of new media whose technological properties are quite different from traditional media (e.g., Nass & Mason, 1990).

Finally, the early definitions of narrative focused on a stable narrative given by a noninteractive medium. Nonfixed narratives made possible by interactive media such as computer games cannot be effectively studied under these definitions. Sticking to the traditional definitions of narrative, therefore, has prevented researchers from reconciling interactivity and narrative in emerging interactive media environments.

An Alternative Definition of Narrative

To overcome the limitations of previous definitions of narrative, we propose the following definition, which can be easily applied to interactive media:

Narrative is a representation of events that provides a cognitive structure whereby media users can tie causes to effects, convert the complexity of events to a story that makes sense, and thus satisfy their primitive urges to understand the physical and social worlds.

We believe that, for a narrative to be realized, it requires any type of tangible technology or medium—from primitive media such as nonverbal (e.g., body language) and verbal languages (e.g., text) to advanced media such as radio, TV, film, and computer games. In other words, a narrative needs a sensible medium or technology through which either fictional or non fictional events are represented. We also believe that human beings have an innate mechanism for narrating events and understanding the narrated events (Abbott, 2002). We assume that the innate schemata have evolved to satisfy human beings' desire to make sense of their physical and social worlds (e.g., Tooby & Cosmides, 2001). In short, people try to understand the causal relationship of the world through the lenses of narratives (e.g., Mithen, 1996). Therefore, unlike previous definitions, we do not regard narrative as a simple recounted story (Lowe, 2000, p. 18). Rather, we consider it a cognitive structure whereby causal understanding of an event can take place.

Instead of structure analyses of narrative forms and units (which is usually the case in early definitions made for structural analyses of narrative units), we propose a psychological approach to narratives (e.g., Oatley, 2002). Defining narratives from a psychological perspective has two major benefits. First, with regard to the level of analysis, defining narratives at a psychological/individual level gives us a fundamental understanding of human cognition and motivation, contributing to explanation of "why" narrative is important in human understanding of an event. Second, in terms of applicability to other media, defining narrative at the psychological level makes it easier to apply the definition to new interactive media such as computer games. That is, unlike previous medium-specific definitions which focused on either books or film, our current definition can cut across various media, as long as a medium conveys a story that satisfy our demand for the causal understanding of an event.

THE RELATIONSHIP BETWEEN NARRATIVE AND INTERACTIVITY IN COMPUTER GAMES

There are two opposing views on the relationship between interactivity and narrative in computer games. One view claims that interactivity and narrative cannot go together easily. This view argues that the two concepts have an inverse relationship in which an increase in one concept will result in a decrease in the other concept (Aarseth, 1997; Adams, 1999; Eskelinen, 2001; Frasca, 2003; Juul, 2001). According to this perspective, the interactive nature of computer games disturbs the flow of narratives. That is, unlike film viewers or book readers who passively follow a narrative predetermined by a director or an author, game players constantly interrupt and change the shape of a narrative (if there is any), which has been preprogrammed by game designers. For instance, when a game character narrates, game players may skip the narration (even though it is vital for the construction of a compelling narrative), as the narration might prevent them from fully enjoying the game's interactive features (e.g., Smith, 2002). This view claims that designers of interactive media products should focus on maximizing interactive potentials of the media. According to this view, trying to build a compelling narrative in interactive media is futile, because a narrative is not what users of interactive media are looking for (Forrester, 1996). To sum up, this view argues that games are designed "not to tell a story but to stimulate and create an environment for interactive experimentation" (Frasca, 2003, p. 225).

The opposing view, however, proposes that interactivity and narrative can coexist and should be integrated in interactive media environments. This view has its root in the studies on interactive educational media, which tried to show how well-organized narratives in interactive media can enhance learning. Scholars like Laurel (1993), Murray (1997), Plowman (1996), and Wolf (2001) have developed this view. For example, in the early 1990s when computer and games were making inroads into the mass entertainment market, Laurel (1993) viewed interactive media as an extension of drama. In a similar fashion, Murray (1997) emphasized the effects of narratives in game situations. Plowman (1996) also suggested that interactivity does not have to lead to the absence of narratives because using familiar narrative conventions enables the players to feel "imaginary coherence" in playing interactive games. She claimed that interactive media with good narratives increase comprehension and enjoyment. Finally, Wolf (2001) argued that interactivity may work together with multiple lines of narratives. For him, interactivity provides a variety of narrative possibilities, because there is no single fixed or predetermined narrative in interactive media contexts. Players can shape the way in which a story line is constructed by freely interacting with environments and agents in a game. Through this procedure, "interactive narrative" is made possible. Wolf (2001) further argued that interactive narratives are becoming more widespread in interactive media and are beginning to provide new opportunities for the study of narratives.

Between the two views, we agree with the second view, which accepts the possibility of coexistence of interactivity and narrative and predicts "interactive narrative" as the future of narratives. Why? We explain the reason by examining the trend of narrative integration in the game industry. The examination will show how interactivity and narrative have already been integrated together in contemporary computer games. We will also review various ways of narrative integration by listing four types of computer-game narratives. The categorization will give a sense of the breadth of narrative integration in contemporary computer games.

NARRATIVES IN COMPUTER GAMES: INDUSTRY TRENDS

Games developed nowadays are significantly different from earlier games. In addition to much more advanced graphics, sounds, and interactive functions, *narrative* is becoming an integral part of contemporary games. The game industry has been putting considerable amount of efforts into the development of compelling narratives. Huge successes of character-driven games (e.g., *Tomb Raider, Hit Man*), background story-based games (e.g., *Myst, Maxpayne, Age of Empire*), and movie or drama sequel games (e.g., *Star Wars, Matrix, Spider-Man, Harry Potter*) clearly indicate that narrative is obviously a killer application in the computer game industry. In the case of recent blockbusters such as *Matrix* and *Spider-Man2*, movies and computer games are developed and marketed almost at the same time. This practice of Hollywood tie-ins with the computer-game industry has already proved to be very successful and will increasingly become a standard strategy for game development.

The importance of narratives in computer games can also be confirmed by the fact that there are abundant cases of successful computer games that have been adopted from successful movies such as *Star Wars*. Examples of movies that were later transformed into computer games are, among others, *48 Hours, 9 to 5, E.T. the Extra-Terrestrial, Excalibur, Friday the 13th, Ghostbusters, Ghostbusters II, Gremlins, Jaws, Star Trek II: The Wrath of Khan, Star Trek III: The Search for Spock, Lion King,* and *Mission Impossible*. In addition, there are also many computer games that were based on popular television shows—*Jeopardy, Knight Rider, M*A*S*H, M*A*S*H II, Magnum PI,* and *Wheel of Fortune*. In fact, among the top 20-selling computer and video games software in 2002, a considerable number of games are based on winning movies such as *Spider-Man, The Lord of the Rings, Harry Potter & The Sorcerer's Stone,* and *Star Wars: Jedi Knight II* (IDSA, 2003a). The transformation of movies into computer games seems to have continued. In 2003, the leading companies in the industry planned to develop or market computer games based on hit movies such as *Matrix Reloaded, Finding Nemo,* and *Terminator 3* (Pham, 2003). Many game developers have assumed that game players can easily understand the rules and logics of a movie-adapted computer game, thanks to their prior exposure to the movie, which provides the platform for understanding the basic story line or background information of the game.

In contrast to the popular and successful conversion of movies into computer games, the transformation of games into movies is less common and usually yields disappointing box office revenues. Examples are *Super Mario Bros.* (game, 1985; movie, 1993), *Street Fighter* (game, 1985; movie, 1993), *Mortal Kombat* (game, 1992; movie, 1995), and *Tomb Raider* (game, 1996; movie, 2001). It may be because game players might feel disappointed to see the discrepancy between their own constructed narratives and the narratives build by film directors. This is a very similar reason why most book-turned-into-movies receive unfavorable comments from the audience who have already read the book (except for *Lord of the Rings*). In other words, the dynamic and interactive characteristic of games, which can be spurred by gamers' creativity, may be easily diminished in the case of game-adapted movies. In addition, virtually unlimited action events in a computer game would have been shrunken into only dozens of pre-selected events in a movie. As a result, a movie that was adapted from a game might have been felt like limited game sessions played by limited protagonists with limited predetermined outcomes (Juul, 2001). This makes it difficult for games to be successfully adapted into movies and renders their success in the box office questionable, despite the earlier huge successes in their game title sales.

To sum up, the industry practice and data suggest that the integration of a compelling narrative into a computer game positively affects the evaluation of the game and its sales. So, then, what are the various ways of integrating narratives in computer games? The next section categorizes four types of narratives in current computer games. This will help us examine the various ways of narrative integration in modern computer games.

TYPES OF NARRATIVES IN COMPUTER GAMES

Computer games can provide much more diverse types of narratives than traditional entertainment media. Following Jenkins (2004), we categorize four types of narratives in computer games—*evoked narratives, enacted narratives, embedded narratives,* and *emergent narratives*.

Evoked narratives are narratives that provide only simple broad outlines coming from previously existing stories. Therefore, players of games with evoked narratives have a task of painting the details through their own playing. For instance, users who play the game, *Star Wars*, may well know the story of the film, *Star Wars* (Jenkins, 2004). However, the game, *Star Wars*, does not provide an identical story with the film (Jenkins, 2004). The game just offers a similar background environment of the film, and hence, players can creatively shape their own stories based on their performance in the game. As discussed in the previous section on industry trends, evoked narratives have been a popular device to ensure the success of a computer game. When the alteration of existing details of a narrative is poorly done, however, evoked narratives can backfire, as suggested by the occasional failures of film-adapted video games.

Enacted narratives are narratives that present broadly defined goals and conflicts in games and provide game players with limited choices of paths a main character in a game can take. Unlike evoked narratives, enacted narratives are not based on pre existing narratives before a game. A story in an enacted narrative is constructed around a character's (and thus, a player's) actions in the game environment, and the features of the environment may hinder or facilitate the story's trajectory. Thus, the organization of the story depends on both how the geography of the imaginary game world is designed and the choices players make during the game. For example, players of *Tomb Raider* may choose a route in a cave among several paths and the choice will lead to localized incidents that have been pre-determined by programmers. Jenkins (2004) called the localized incidents "micronarratives," which are the foundation for the creation of a large-scale master narrative. Different master narratives are made during a game, depending on upcoming events in the specific route players choose.

Embedded narratives are narratives that are discovered only when players deeply process information inside the game world. Embedded narratives demand more than the simple choice behaviors required by enacted narratives. In games with embedded narratives, players are asked (in a sense, *required*) to actively seek and process textual information or environmental cues inside games in order to fully develop the stories. Thus, the comprehension of embedded narratives is a process by which players actively set up and test hypotheses about plausible story developments after their possible choices (Jenkins, 2004). Without active and deep information processing, players of games with embedded narratives can neither enjoy the game nor discover the hidden narratives. *Myst* is a good example of an embedded narrative. In this game, players cannot move further unless they actively process environmental or textual cues in order to solve the problem that they are facing. Narratives in this category include detective or conspiracy stories in which game players' activities of deciphering codes and tracing clues or hidden connections between past and current events are important. Hence, it is necessary for the players to identify, speculate, investigate, and interpret the pieces of a linear story in order to construct a narrative (Jenkins, 2004).

Finally, *emergent narratives* are the ones that provide a rich array of possible narratives, making it possible for players to actively build stories (Jenkins, 2004). Although emergent narratives are not totally unorganized or unstructured, they provide the players with much more latitude so that the players can define their own goals and write their own stories. Unlike the other types of narratives, emergent narratives are truly open-ended structures that provide almost unlimited narrative possibilities to players. Players of a game with an emergent narrative, thus, can get deep emotional gratification from their own story building while they are playing the game. The creators of emergent narratives do not and cannot know all the possible narrative paths and outcomes, as opposed to the creators of the other types of narratives, who clearly pre-envision all possible narrative paths and outcomes when they design a game. *The Sims Online* is probably the best current example of an emergent narrative. It should also be noticed that story building in emergent narratives cannot be established from a vacuum (Jenkins, 2004). That is, creators of emergent narratives should design the characters' (or agents') behaviors in such a way that the characters have their own wills and motivations. For example, Will Wright, the creator of *The Sims Online*, programmed the characters' behaviors of the game based on Maslow's hierarchy of needs.

So far we have examined four different types of narratives in computer games, which vary with regard to the extent to which the room for the players' own story building is provided and the intensity of narrative integration. In the next section, we examine the major reasons why narrative is important in computer games.

WHY IS NARRATIVE IMPORTANT IN COMPUTER GAMES?

The most fundamental reason for the importance of narrative is the human being's intrinsic preference for narratives. The ability and tendency to enjoy narratives in media seems to be a cross-culturally universal phenomenon. This universality implies that human attraction to vicarious experiences through narratives is one way of adaptation (Tooby & Cosmides, 2001). Otherwise, the tendency to enjoy narratives would have been selected out through evolution. Although the engagement in narrative worlds may seem to be nonfunctional and even excessively nonutilitarian at first glance, it has huge survival and reproductive advantages (Cosmides & Tooby, 2000a; 2000b; Pinker, 1997; Tooby & Cosmides, 2001; see also

Steen et al., this chap. 21, volume). Why? The most fundamental reason is that narratives provide surrogate experiences. The worlds that human beings can experience directly are very limited. While direct experiences provide survival-critical lessons such as "Run away when something bigger than you approach you" or "Do not work with cheaters," they are very limited, and the cost for a failure can be as enormous as to lead to one's extinction. Therefore, human beings have to build a mental mechanism to experience unexplored worlds in a safe way. The mental engagement in narrative worlds (along with the ability to pretend and imagine) provides the solution. Thanks to the engagement, human beings can have the secondary experience of the narrative worlds both by transporting themselves into the narrative worlds (Hilgard, 1979) and by activating psychological processes of participatory responses to characters in narratives (e.g., self-identification, side participation, simulation of autobiographic memories; see Oatley, 2002; Polichak & Gerrig, 2002). As a consequence, people benefited from narrative engagement in that the depicted experiences become their own, not those of the character(s). Through this process, people can be enriched by and adapt with what they have *not* actually experienced (Tooby & Cosmides, 2001). The enjoyment people get from narrative engagement, thus, is natural and critical for their continuing survival and reproduction. As a result of long-term adaptation, people prefer rich narratives to pure information, even though the latter takes less time to consume. Thanks to this intrinsic preference for narratives, human beings have developed dramas, plays, novels, television programs, films, and other types of narrative-delivery arts throughout history (Tooby & Cosmides, 2001). The push for rich narratives in computer games should be understood from this evolutionary perspective.

In addition to the intrinsic human preference for narratives in entertainment media, narratives are becoming important in computer games because narratives significantly reduce cognitive loads when people try to understand virtual worlds inside games. It has been argued that narratives are critical for understanding virtual worlds depicted in old media such as novels and films (Forrester, 1996; McLellan, 1993). Human beings have evolved to possess specialized cognitive mechanisms that allow them to get into, and participate in, virtual worlds with some guidance by narratives (Tooby & Cosmides, 2001). Narratives facilitate quick understanding of imagined or unexplored worlds, because they make evident causal relationships of physical (e.g., why objects are moving in certain ways) and social worlds (e.g., who is the enemy or the partner). Knowledge stored in narrative forms can be easily retrieved without much cognitive effort, as human beings are very good at remembering episodic memory. In fact, most of the knowledge we use in our day-to-day lives is stored in our memory structures as narratives (Schank & Abelson, 1995). In interactive media such as computer games, the reduction of cognitive loads is even more important, because users of interactive media need to mentally focus on the tasks at hand (e.g., killing enemies) in order to fully enjoy the interactivity. If they are constantly confused about the way physical and social environments inside the game world work, they cannot fully enjoy themselves during game playing. With narratives, game players can easily focus on their task at hand with the intuitive understanding of the physical and social environments inside the game world. Without narratives, game players need to go through complicated cognitive processes in order to figure out what they are supposed to do inside the game world (Laurillard, 1998; Mandler & deForest, 1979; Plowman, 1991, 1992; Stratfold, 1994). The extent and range of guidance provided by narratives may vary according to the types of narratives, as explained in the previous section.

Third, narratives in computer games provide strong motivation to play games. Narratives create specific situations where appropriate actions by game players are required to finish a story. Unlike users of old media, game players observe the direct consequences of their actions because they are playing the main character's role (Bielenberg & Carpenter-Smith, 1997). As Biocca (2002) said, "to become an interactor in a narrative experience is to have a role"

(p. 119). Actually, interactive narrative has its origins in role-playing games such as Dungeons and Dragons (Cook, Tweet, & Williams, 2000). With compelling narratives, game players feel a strong need and pressure to enhance their performances. Players feel motivated to do well when they are psychologically involved with the game character they are role-playing (Klimmt & Vorderer, 2003). Moreover, because it is hard for players to get to the final stage of a game in the first place except for a few experts, narratives provide strong incentives to players to continue to play the game. The desire to see the destiny of their roles or simple curiosity about the ending of a story gives game players a strong motivation to finish the game.

Finally, narratives make it possible for game players to feel a strong presence while playing games (for more detailed discussion on the concept of presence, see Lee, 2004a, 2004b). The concept of presence—*"a psychological state in which virtual objects are experienced as actual objects in either sensory or non-sensory ways"* (Lee, 2004a, p. 27); or simply *"the perceptual illusion of nonmediation"* (Lombard & Ditton, 1997; Lombard, Reich, Grabe, Bracken, & Ditton, 2000, p. 77)—has great relevance to the design and evaluation of media products and computer interfaces, especially in entertainment (e.g., movies, reality television programs, computer and video games, conversational interfaces). As technologies for simulating interaction with people and places become more sophisticated, computer scientists, psychologists, and communication scholars pay even greater attention to this concept (e.g., Held & Durlach, 1992; Kim & Biocca, 1997; Loomis, 1992; Sheridan, 1995; Whitmer & Singer, 1998). Narratives can be regarded as a classic device to bring feelings of presence, especially the sense of being transported into a virtual world depicted by narratives. For example, readers often feel strong feelings of presence when they read a good novel. Skillful storytellers can induce a very compelling sense of presence to their listeners. Sometimes good narratives can induce stronger feelings of presence than a highly sophisticated 3-D virtual world. While it is clear that narratives are the key in *noninteractive* media such as books, TV and movies, little is known about the role of narratives in *interactive* media such as computer games. Recent studies about the effects of narratives (Lee, Jin, Park, & Kang, 2004; Schneider, Lang, Shin, & Bradley, 2004) fill the gap. These studies show that narratives in computer games increase feelings of presence. The increased feelings of presence yield a positive evaluation of the game. The studies provide compelling empirical evidence that narratives are important for user gratification in interactive as well as noninteractive media.

FINAL REMARKS

In this chapter, we explicated two important concepts—interactivity and narrative—in interactive entertainment media such as computer games. After the careful explication and new conceptual definition of each concept, we examined two competing views on the relationship between interactivity and narrative, and concluded that interactivity and narrative can coexist and should be integrated in interactive media environments such as computer games. In order to support our conclusion, we examined both the industry trend of successful integration of movies and computer games. Finally, we explained the reasons why narratives are important not only in noninteractive media but also in interactive media such as computer games. We provided four major reasons for the importance of narratives in computer games—human being's intrinsic preference for narratives, reduction of cognitive loads through the use of narratives, increased motivation by narratives, and narrative impacts on feelings of presence. Even though the current chapter provides strong theoretical reasons why narratives are important, future studies should empirically test the effects of narratives on various psychological variables such as enjoyment, memory, cognition, motivation, and feelings of presence.

REFERENCES

Aarseth, E. (1997). *Cybertext: Perspectives on ergodic literature*. Baltimore: Johns Hopkins University Press.

Abbot, H. P. (2002). *The Cambridge introduction to narrative*. New York: Cambridge University Press.

Adams, E. (1999). *Three problems for interactive storytellers*. Retrieved June 12, 2004, from http://www.gama-sutra.com/features/designers_notebook/19991229.htm.

Bal, M. (1985). *Narratology: Introduction to the theory of narrative*. Toronto: University of Toronto Press.

Barthes, R. (1966). Communication 8, "Introduction á l'analyse structurale des récits" essay in Recherches Semi-ologiques. L'Analyse Structurale du Récit. (S. Heath, Trans.). (1977). *Introduction to the structural analysis of narratives*. Essay within Image-Music-Text. London: Fontana.

Bielenberg, D. R., & Carpenter-Smith, T. (1997). Efficacy of story in multimedia training. *Journal of Network and Computer Applications, 20*, 151–159.

Biocca, F. (1998). *Interactivity explained*. Unpublished manuscript.

Biocca, F. (2002). The evolution of interactive media. In M. C. Green, J. J. Strange, & T. C. Brock (Eds.), *Narrative impact: Social and cognitive foundations* (pp. 97–130). Mahwah, NJ: Lawrence Erlbaum Associates.

Brody, E. W. (1990). *Communication tomorrow: New audiences, new technologies, new media*. New York: Praeger.

Brown, R. H. (1987). *Society as text: Essays on rhetoric, reason, and reality*. Chicago: University of Chicago Press.

Burke, K. (2000). *Sixty percent of all Americans play video games, contributing to the fourth straight year of double-digit growth for the interactive entertainment industry*. Retrieved May 14, 2004, from http://www.isda.com/releases/4-21-2000.html.

Chaffee, S. (1991). *Explication*. Newbury Park, CA: Sage.

Chaffee, S., Rafaeli, S., & Lieberman, S. (1985, May). *Human computer interactivity: A concept for communication research*. Paper presented at the annual meeting of International Communication Association, Chicago, IL.

Cook, M., Tweet, J., & Williams, S. (2000). *Dungeons and dragons players handbook*. San Francisco: Wizards of the Coast.

Cosmides, L., & Tooby, J. (2000a). Consider the source: The evolution of adaptations for decoupling and metarep-resentation. In D. Sperber (Ed.), *Metarepresentations: A multidisciplinary perspective* (pp. 53–115). New York: Oxford University Press.

Cosmides, L., & Tooby, J. (2000b). Evolutionary psychology and the emotions. In M. Lewis & J. M. Haviland-Jones (Eds.), *Handbook of emotions* (pp. 91–115). New York: Guilford.

Economist (2003, December 11). Invaders from the land of broadband. *Economist*. Retrieved August 10, 2004, from http://www.economist.com/business/displayStory.cfm?story_id=2287063.

Eskelinen, M. (2001). The game situation. *Game Studies, 1*(1). Retrieved May 22, 2004, from http://www.game-studies.org/0101/eskelinen/.

Fisher, W. R. (1984). Narration as a human communication paradigm: The case of public moral argument. *Communication Monographs, 51*, 1–22.

Forrester, M. (1996). Can narratology facilitate successful communication in hypermedia environments? *Intelligent Tutoring Media, 7*, 11–20.

Frasca, G. (2003). Simulation versus narrative: Introduction to ludology. In M. J. Wolf & B. Perron (Eds.), *The video game: Theory reader* (pp. 223–235). New York: Routledge.

Grodal, T. (2000). Video games and the pleasure of control. In D. Zillmann & P. Vorderer (Eds.), *Media entertainment: The psychology of its appeal* (pp. 197–213). Mahwah, NJ: Lawrence Erlbaum Associates.

Ha, L., & James, E. L. (1998). Interactivity reexaminied: A baseline analysis of early business Web sites. *Journal of Broadcasting & Electronic Media, 42* (4), 457–474.

Haeckel, S. H. (1998). About the nature and future of interactive marketing. *Journal of Interactive Marketing, 12*(1), 63–71.

Heeter, C. (1989). Implications of new interactive technologies for conceptualizing communication. In J. L. Salvaggio & J. Bryant (Eds.), *Media use in the information age: Emerging patterns of adoption and consumer use* (pp. 217–236). Hillsdale, NJ: Lawrence Erlbaum Associates.

Held, R. M., & Durlach, N. I. (1992). Telepresence. *Presence: Teleoperators and Virtual Environments, 1*(1), 109–112.

Hilgard, E. R. (1979). *Personality and hypnosis: A study of imaginative involvement*. Chicago: University of Chicago Press.

IDSA (2003a). *Essential facts about the computer and video game industry: 2003 sales, demographics and usage data*. Retrieved May 14, 2004, from http://www.theesa.com/EFBrochure.pdf.

IDSA (2003b). *Top ten industry facts*. Interactive Digital Software Association. Retrieved May 14, 2004, from Inter-active Digital Software Association Web Site: http://www.idsa.com/pressroom_main.html.

Jenkins, H. (2004). Game design as narrative architecture. In N. Wardrip-Fruin & P. Harrington (Eds.), *First person* (pp. 118–130). Cambridge, MA: MIT Press.

Juul, J. (2001). Games telling stories? A brief note on games and narratives. *Game Studies, 1* (1). Retrieved June 12, 2004, from http://www.gamestudies.org/0101/juul-gts/.

Kim, T., & Biocca, F. (1997). Telepresence via television: Two dimensions of telepresence may have different connections to memory and persuasion. *Journal of Computer-Mediated-Communication, 3*(2). Retrieved January 5, 2004, from http://www.ascusc.org/jcmc/vol3/issue2/.

Klimmt, C., & Vorderer, P. (2003). Media psychology "is not there": Introducing theories on media entertainment to the Presence debate. *Presence: Teleoperators and virtual environments, 12*, 346–359.

Labov, W. (1972). The transformation of experience in narrative syntax. In W. Labov (Ed.), *Language in the inner city: Studies in the Black English vernacular* (pp. 354–396). Philadelphia: University of Pennsylvania Press.

Labov, W., & Waletzky, J. (1967). Narrative analysis: Oral versions of personal experience. In J. Helm (Ed.), *Essays on the verbal and visual arts* (pp. 12–44). Seattle: University of Washington Press.

Laurel, B. (1993). *Computers as theater*. Reading, MA: Addison-Wesley.

Laurillard, D. (1998). Multimedia and the learner's experience of narrative. *Computers and Education, 31*, 229–242.

Lee, K. M. (2004a). Presence, explicated. *Communication Theory, 14*(1), 27–50.

Lee, K. M. (2004b). Why presence occurs: Evolutionary psychology, media equation, and presence. *Presence: Teleoperators and Virtual Environments, 13,* 494–505.

Lee, K. M., Jin, S. A., Park, N., & Kang, S. (2004, November). *Effects of narrative on feelings of presence in computer/video games*. Paper presented at the annual meeting of National Communication Association, Chicago: IL.

Lombard, M., & Ditton, T. (1997). At the heart of it all: The concept of presence. *Journal of Computer-Mediated Communication, 3*. Retrieved June 14, 2004, from http://www.ascusc.org/jcmc/vol3/issue2/lombard.html.

Lombard, M., Reich, R., Grabe, M., Bracken, C., & Ditton, T. (2000). Presence and television: The role of screen size. *Human Communication Research, 26*, 75–98.

Loomis, J. M. (1992). Distal attribution and presence. *Presence: Teleoperators and Virtual Environment, 1*(1), 113–119.

Lowe, N. J. (2000). *The classical plot and the invention of Western narrative*. New York: Cambridge University Press.

Mallon, B., & Webb, B. (2000). Structure, causality, visibility, and interaction: Propositions for evaluating engagement in narrative multimedia. *International Journal of Human-Computer Studies, 53*, 269–287.

Mandler, J. M., & deForest, M. (1979). Is there more than one way to recall a story? *Child Development, 50*, 886–889.

McLellan, H. (1993). Hypertextual tales: Story models for hypertext design. *Journal of Educational Multimedia and Hypermedia, 2*, 239–260.

Mithen, S. (1996). *The prehistory of the mind: The cognitive origins of art and science*. New York: Thames & Hudson.

Murray, J. H. (1997). *Hamlet on the Holodeck: The future of narrative in cyberspace*. New York: Free Press.

Nass, C., & Mason, L. (1990). On the study of technology and task: A variable-based approach. In J. Fulk & C. Steinfeld (Eds.). *Organization and communication technology* (pp. 46–67). Newbury Park, CA: Sage.

Nass, C., & Moon, Y. (2000). Machines and mindlessness: Social responses to computers. *Journal of Social Issues, 56*(1), 81–103.

Oatley, K. (2002). Emotions and the story worlds of fiction. In M. C. Green, J. J. Strange, & T. C. Brock (Eds.), *Narrative impact: Social and cognitive foundations* (pp. 39–69). Mahwah, NJ: Lawrence Erlbaum Associates.

Onega, S., & García Landa, J. A. (1996). Introduction. In S. Onega & J. A. García Landa (Eds.), *Narratology: An introduction* (pp. 1–41). London: Longman.

Peng, W., & Lee, K. M. (2004, May). *What do we know about computer and video games?: A comprehensive review of the current literature*. Paper presented at the annual meeting of the International Communication Association, New Orleans, LA.

Pham, A. (2003, May 12). Game makers are playing it safe. *Los Angeles Times*, pp. C1, C4.

Pinker, S. (1997). *How the mind works*. New York: Norton.

Plowman, L. (1991). *An investigation of design issues for group use of interactive video*. Unpublished doctoral dissertation, Brighton Polytechnic, Brighton, UK.

Plowman, L. (1992). An ethnographic approach to analysing navigation and task structure in interactive multimedia: Some design issues for group use. In A. Monk, D. Diaper, & M. D. Harrison (Eds.), *People and computers VII* (pp. 271–287). Cambridge, UK: Cambridge University Press.

Plowman, L. (1996). Narrative, linearity and interactivity: Making sense of interactive multimedia. *British Journal of Educational Technology, 27*, 92–105.

Polichak, J. W., & Gerrig, R. J. (2002). Get up and win! Participatory responses to narrative. In M. C. Green, J. J. Strange, & T. C. Brock (Eds.), *Narrative impact: Social and cognitive foundations* (pp. 71–95). Mahwah, NJ: Lawrence Erlbaum Associates.

Rafaeli, S. (1988). Interactivity: From new media to communication. In R. P. Hawkins, J. M. Wieman, & S. Pingree (Eds.), *Advancing communication science: Merging mass and interpersonal processes* (pp. 110–134). Newbury Park, CA: Sage.

Rafaeli, S., & Sudweeks, F. (1997). Networked interactivity. *Journal of Computer-Mediated Communication, 2*(4), Retrieved November 20, 2003, from http://www.cwis.usc.edu/dept/annenberg/vol2/issue4/rafaeli.sudweeks.html.

Reeves, B., & Nass, C. (1996). *The media equation: How people treat computers, televisions, and new media like real people and places*. New York: Cambridge University Press.

Riessman, C. K. (1993). *Narrative analysis*. Newbury Park, CA: Sage.

Rogers, E. M. (1995). *Discussion of innovations* (4th ed.). New York: Free Press.

Ryan, M-L. (1997). Interactive drama: Narrativity in highly interactive environment. *Modern Fiction Studies, 43*(3), 677–707.

Ryan, M-L. (2001). *Narrative as virtual reality: Immersion and interactivity in literature and electronic media.* Baltimore: Johns Hopkins University Press.

Schank, R. C., & Abelson, R. P. (1995). Knowledge and memory: The real story. In R. S. Wyer (Ed.), *Advances in social cognition* (Vol. VIII, pp. 1–85). Hillsdale, NJ: Lawrence Erlbaum Associates.

Schneider, E. F., Lang, A., Shin, M., & Bradley, S. D. (2004). Death with a story: How story impacts emotional, motivational, and physiological responses to first-person shooter video game. *Human Communication Research, 30* (3), 361–375.

Sheridan, T. B. (1995). Teleoperation, telerobotics and telepresence: A progress report. *Control Engineering Practice, 3*(2), 205–214.

Smith, G. M. (2002). Computer games have words, too: Dialogue conventions in Final Fantasy VII. *Game Studies, 2* (2). Retrieved April 5, 2004, from http://www.gamestudies.org/0202/smith/.

Steuer, J. (1992). Defining virtual reality: Dimensions determining telepresence. *Journal of Communication, 42*(4), 73–93.

Stratfold, M. (1994). *Investigation into the design of educational multimedia: Video, interactivity and narrative.* Unpublished doctoral dissertation, Open University, UK.

Tooby, J., & Cosmides, L. (2001). Does beauty build adapted minds? Toward an evolutionary theory of aesthetics, fiction, and the arts. *SubStance, Issue 94/95, 30*(1), 6–27.

Vorderer, P. (2000). Interactive entertainment and beyond. In D. Zillmann & P. Vorderer (Eds.), *Media entertainment: The psychology of its appeal* (pp. 21–36). Mahwah, NJ: Lawrence Erlbaum Associates.

Vorderer, P. (2003). Entertainment theory. In J. Bryant, D. Roskos-Ewoldsen, & J. Cantor (Eds.), *Communication and emotion: Essays in honor of Dolf Zillmann* (pp. 131–153). Mahwah, NJ: Lawrence Erlbaum Associates.

Vorderer, P., & Knobloch, S. (2000). Conflict and suspense in drama. In D. Zillmann & P. Vorderer (Eds.), *Media entertainment: The psychology of its appeal* (pp. 59–72). Mahwah, NJ: Lawrence Erlbaum Associates.

Vorderer, P., Knobloch, S., & Schramm, H. (2001). Does entertainment suffer from interactivity?: The impact of watching an interactive TV movie on viewer's experience of entertainment. *Media Psychology, 3*, 343–363.

Walther, J. B., & Burgoon, J. K. (1992). Relational communication in computer-mediated interaction. *Human Communication Research, 19*(1), 50–88.

Webster's Third New International Dictionary (1976). Springfield, MA: Merriam.

Whitmer, B. G., & Singer, M. J. (1998). Measuring presence in virtual environments: A presence questionnaire. *Presence: Teleoperators and Virtual Environment, 7*(3), 225–240.

Wolf, M. J. (2001). *The medium of the video game*. Austin: University of Texas Press.

Young, K. G. (1987). *Taleworlds and storyrealms: The phenomenology of narrative*. Boston: Martinus Nijhoff.

19

Realism, Imagination, and Narrative Video Games

Michael A. Shapiro, Jorge Peña-Herborn, and Jeffrey T. Hancock
Cornell University

Many video games are narratives. They have a plot that takes place over time, characters that interact, a setting, and some form of conflict. Even first-person shooters are structured as stories in which the player is the hero involved in a conflict with others trying to kill the player—usually in an exotic setting. Other narrative games include adventure games, simulation games, and online multiplayer games. These stories range greatly in sophistication. Some are little more than high-tech games of tag with audio, visual, and sometimes haptic effects and a thin story line that stitches together the multiple stages that constitute the game. However, some adventure games are starting to develop more elegant stories. Some of the more recent multiplayer games (e.g., *EverQuest*) create immersive worlds in which real people interact in a complex world (see Chan & Vorderer, chap. 6, this volume; Peña & Hancock, in press; Yee, 2001). Narratives in video games may have important effects. In one recent study, adding a story to a video game increased identification with characters and made the game more enjoyable and arousing (Schneider, Lang, Shin, & Bradley, 2004).

One important element of video game stories is their *realism*. Faster processors and increased bandwidth allow video games to provide a more and more realistic sensory experience (see Frauenfelder, 2001). Currently, most of the effort to increase the realism of computer-mediated narrative entertainment focuses on enhancing the sensory experience—particularly better graphics (Newman, 2002) and sounds (Kramer, 1995). Efforts to create a sensory experience that is more and more "lifelike" are likely to continue.

While the sensory experience of playing a video game contributes to a feeling of realism, considerable evidence indicates that by middle childhood perceived realism becomes increasingly influenced by the conceptual and the abstract as well as by sensory representations (Ang, 1985; Austin, Roberts, & Nass, 1990; Davies, 1997; Dorr, 1983; Downs, 1990; Flavell, 1986; Hall, 2001; Morison, Kelly, & Gardner, 1981; Wright, Huston, Reitz, & Piemyat, 1994; Wright, Huston, Truglio, Fitch, Smith, & Piemyat, 1995). Knowledge about the world and social knowledge contribute more and more to judgments of realism as children mature. In addition, many video game narratives are about worlds that never existed (e.g., "Nosgoth" in *Legacy of Kain 2: SoulReaver*) or events that the player will probably never experience (e.g., racing cars on the streets of downtown Hong Kong). On average video game players are adults—mean age 29 (Entertainment Software Association, 2004). Thus, the ability to imagine things that have not or cannot be experienced is also an important element in perceiving the realism of video games.

The current state of video game storytelling is equivalent to a novel in which almost all the effort went into describing the setting—ignoring what the reader knows, feels, and can infer. In part, this is a technical issue. Game designers know a great deal more about how to make something *look* real than they do about how to make us *think* it is real. But before designers can start meeting the challenge, they need to know something about what other factors influence perceptions of realism.

Some video game designers are starting to consider these nonsensory cues. For example, the designers of *The Sims Online* reportedly used Maslow's hierarchy of needs and architecture theory to enhance the realism of the game (Levine, 2002; Thompson, 2003). Also, computer scientists have developed agents endowed with "humanness" to pit against human players (Laird & Duchi, 2000). Developers are also working on believable agents to develop video game characters exhibiting rich personalities, emotions, and humanlike strategies (Mateas & Stern, 2000), and the ability to form synthetic social relations (Tomlinson, 2002). Others have tried to implement known psychological processes to better model social interactions between players and game agents (Mao & Gratch, 2003).

This chapter reviews the research in *perceived realism* and how it might apply to narrative video games. The focus will be on both the sensory and the thinking aspects of realism judgments. Overall, our claim is that the ability to imagine what a setting, character, or conflict would be like if it happened is a key ability in realism judgments. We start by examining why realism is important and what people mean when they say something is realistic, showing that imagination plays an important role. We then examine the impact of some story elements on perceived realism. We also look at how our social judgments and our knowledge of the world are applied to judgments about realism. Finally, we discuss how this applies to current and future video games.[1]

PERCEIVED REALISM

Perceived realism is an important psychological characteristic of stories. Perceived realism may influence mental processing, beliefs, attitudes, and behavior (Potter, 1988), forming parasocial relationships with media characters (Rubin, 1979; Rubin, Perse, & Powell, 1985; Perse, 1990), aggressiveness (Atkin, 1983; Geen, 1975; Geen & Rakosky, 1973), and the effectiveness of health messages (Andsager, Austin, & Pinkleton, 2001; Bahk, 2001). For dramas, there is at least some evidence that increased realism leads to increased enjoyment (Shapiro & Chock, 2003). Most information about video game realism is anecdotal; however, one recent study found that game players strongly prefer realistic sound effects and realistic graphics—particularly males (Wood, Griffiths, Chappell, & Davies, 2004). Males also strongly preferred realistic settings. Both males and females were interested in character development over time.

What Does Realistic Mean?

One problem we need to address is what a person means when he or she says one story is more realistic than another. One possibility is that a person compares a story to what he or she knows of daily life. That is certainly part of any realism judgment. It works better for some stories than others. For example, family dramas often deal with relatively familiar people, events, and issues. Even for those stories, we run into a problem. Almost by definition, an engaging story deals with topics that are more amusing or dramatic than usual. While the characters and the relationships on *Little House on the Prairie* seemed relatively familiar, the setting and many of the tasks of farm life in that era were unfamiliar. Using experience (both direct and indirect) to gauge realism becomes even more problematic as we look at other kinds of stories. For instance, few of us spend much time dealing with the issues, settings, and conflicts of a typical legal or police drama. While many viewers have direct or at least indirect experience of some dramatic aspects of these stories, such as dating or divorce, fewer have experience of criminal trials, murders, drug raids, or police interrogations. The only experience a viewer or player has of the fantasy worlds in science fiction and horror movies or the typical video game is the program or game itself. Of course, elements of fantasy stories parallel the real world (interpersonal relationships, for example), but such stories also contain people, events, settings, and perhaps conflicts that have never happened and perhaps cannot happen. Yet it is not difficult for people to rank order the realism of science fiction stories. A theory of perceived realism must be able to explain how we do that.

Imagine a convict who has spent most of his life in prison. He (or she) knows (in brutal detail) what prison life is like. When he watches a prison drama on television, he can draw on his experience to make detailed comparisons to his reality. On the other hand, when he watches a drama about life outside prison his experience inside prison is of limited usefulness. If he is smart, he draws on what he does know to imagine what life would be like outside prison and compare what he is seeing to what he imagines the real world is like.[2] His imaginings may be inaccurate, but they probably have some relationship with reality.

Most of us are in the reverse position when we judge the realism of a prison drama. We can call on some indirect knowledge about prison life (mostly from the media), but in the end we have to compare what we are seeing to what we *imagine* prison to be like. Rather than being a limitation, this ability to use our imagination in judging realism enables us to make reasonable, although perhaps not totally accurate, realism judgments far beyond our experience.

In fact, it requires a certain degree of maturity to apply imagination to realism judgments. This is evident in the changes in what frightens children. In early childhood, young children tend to be frightened by what looks scary rather than what acts scary but is actually harmless or friendly (Hoffner & Cantor, 1985). As children mature, they start to recognize that something that looks scary (e.g., a dragon) may be a fantasy and is not really something to fear (Cantor, 2001). At the same time, children start to be more frightened by real dangers they could not conceptualize at an earlier age and by depictions of abstract concepts or hypotheticals. For example, older children were more frightened by a television movie about the aftermath of a nuclear war than were younger children (Cantor, Wilson, & Hoffner, 1986). This particular movie had only a few graphic scenes. The movie's impact came from an abstract understanding of what had happened.

By middle and late childhood, children still use physical characteristics, but perceptions of realism are increasingly associated with more abstract, conceptual, and even hypothetical aspects of reality (Shapiro & McDonald, 1992). By adulthood, understanding of realism is complex. It is not surprising that people usually rate news as more real than drama. However, people know that news is not entirely real. Adolescents know that television news presents events as less complex, more intense, and more solvable than people know them to be in real life (Adoni, Cohen, & Mane, 1984; Cohen, Adoni, & Drori, 1983). Judgments of characters are very complex and include knowledge of the world, knowledge of social relationships, physical characteristics, behaviors, emotional states, and knowledge of the specific character (Hoffner & Cantor, 1991; Livingstone, 1989).

Several recent studies make a distinction between *absolute perceived realism* and a more *relative (imaginative) perceived realism* (Shapiro & Chock, 2003; Shapiro & Weisbein, 2001).

Absolute perceived realism is defined as the judged likelihood that a depicted event could happen in the real world. This is indexed by the event's likelihood in real life or likelihood to happen to the viewer. By that standard, most media events should seem relatively unrealistic because most of what we see on television, in movies, in video games, and other entertainment is rare. Such absolute realism judgments seem to be part, but not all, of what people mean by realistic. The realism decisions made by adults and older children seem at least in part to include the more abstract, conceptual, and hypothetical. Viewers/users commonly make a relative judgment about how realistic an event is if that sort of event were to happen. Thus, earthquakes are absolutely rare in most parts of the United States. But if an earthquake were to happen, we can judge depicted behaviors and events as more or less realistic in that circumstance. This is true even if the viewer/user has no direct experience with the events. This hypothetical sense of realism seems particularly critical to making judgments about the realism of video games, fantasy, and science fiction. A viewer/user can think—IF an extraterrestrial species existed, would it look and act as depicted? Even for games based on the real tactics and equipment of military commandos such as the war game *SOCOM: U.S. Navy SEALS* (see Werde, 2004), players must imagine what it would be like to be in that setting, with those characters, living through those events.

In summary, people make judgments about the realism of media stories as they view them. While the likelihood of an event in the real world enters into judgments of realism, people can also judge the realism of unlikely events and even events that never happened by imagining what they would be like if they did happen. For older children and adults, we would expect the realism of video games to depend on two broad types of information: (a) sensory information such as the look, sound, and feel of a game, and (b) the more inferential and imaginative elements that factor into the judgment of "IF this were to happen, would it be like this?" We begin our examination of perceived realism in video game contexts by examining the first category, the role of sensory development in the perceived realism of video games.

STORY ELEMENTS

Look, Sound, and Feel

Enhancing the look and sound of video games is important to the perception of realism. Indeed, more than two decades ago Loftus and Loftus (1983) suggested several reasons why improved graphic capabilities in video games are desirable. First, higher image resolution looks better and makes characters appear more lifelike. Second, higher image resolution should result in smoother and more realistic appearance of motion.

Buchman and Funk (1996) noted that early video games, though compelling at the time, were graphically modest in comparison to newer titles. The sensory aspects of video games have substantially improved since the days of *Pong* and *PacMan* (see Grodal, 2000). Obviously, the evolution of computer technologies, such as increased memory capacity, processing power, and graphics abilities, has played a major role in the development of video games. Developers advanced the sensory realism of video games primarily by enhancing their graphical and aural features and, to a much lesser extent, haptic features such as joysticks that vibrate in response to features in the game. Improved technology and software developments have led to graphics engines that can now render 3-D images in real-time with high resolutions. For example, a recent racing video game used millions of pictures to simulate, in accurate detail, the photorealistic look of Moscow and Florence (*Project Gotham Racing 2*). The effort involved in (re)creating realistic settings is certainly not wasted, as the majority of game players rate a realistic setting as important to their enjoyment of the game (Wood et al., 2004).

The *graphics* of recent video games also simulate real-time physics phenomena, such as explosions, gravity, running water, and the actions and reactions of inanimate objects (Feldman, O'Brien, & Arikan, 2003; Frauenfelder, 2001). According to Will Wright, lead designer of video games such as *The Sims* and *SimCity*, "you'll be able to interact with more and more of the surroundings, to the point where you can pick up a crowbar and pry a nail out of the wall" (Frauenfelder, 2001, p. 119). Indeed, the motion of objects and characters has been carefully modeled with the use of body motion capture technology, which scans and records an actor's or an object's movements to accomplish more true-to-life motion (Ryan, 2004). However, this technique usually requires additional work by skilled animators. Advances in animation have made human and nonhuman characters look increasingly photorealistic so that they look more like a video of a character than an animation. Some recent games, such as *James Bond 007: Everything or nothing,* include the facial and body likeness of the actual movie cast (Thompson, 2004).

There is some evidence that *audio* fidelity influences attention, memory, and liking more than video fidelity (for a discussion, see Reeves & Nass, 1996). The audio aspects of video games have also evolved enormously. Sound effects are now projected into the 3-D space surrounding the player, and the sound effects are drawn from increasingly sophisticated sound libraries, which can range from the clip-clop of galloping horses to the electric clash of two light sabers to the rustling of the leaves in the wind. Playing a spooky game in the dark, alone at home, with the sound turned up, can be a truly haunting experience. This and other experiences while playing are in part due to the contributions of professional composers and musicians who develop scores and soundtracks for specific video game titles (Tanner, 1997), and the voices of professional actors who bring video game characters to life (Thompson, 2004).

Music is often used to create a mood or emotion. Soundtracks that include songs associated with particular settings can also play an important role in enhancing the realism of a video game. For example, a video game that attempts to portray a certain era or aesthetic can use representative soundtracks that include songs from that era or setting to enhance the sense of that period of time or environment (e.g., *Grand Theft Auto 3: Vice City; Tony Hawk's ProSkater 4*). Overall, one of the primary goals of sounds is to enhance the realism of video games and other virtual realities (Kramer, 1995; Zehnder & Lipscomb, chap. 17, this volume).

Why does better sensory information make something seem more real? One answer is that it does not always. No matter how realistic they may make it, children playing let's pretend games (or grown-up versions like *Majestic* that are embedded in everyday life) know that the game is not "real." But another answer is that there does seem to be a tendency for better sensory information to enhance the realism of a game. One reason may be evolution. A dark shadow in the woods could just be a dark shadow. But one that sounds, smells, and moves like a bear is almost certainly a bear. As Reeves and Nass (1996) pointed out, there has not been enough time for people to evolve specific psychological strategies for dealing with media. We use the same mental strategies for media (including video games) that we use for the real world. Many of these mental strategies make reflexive judgments based on sensory information. So, reflexively at least, something that sounds and moves like a bear may seem more realistic than something that just looks like a bear.

Although tremendous advances have been made in the sensory development of video games, not all technological advances have enhanced realism, and some may even diminish it. For example, in one study, rumbling joysticks reminded players they were in front of a screen, which appeared to reduce their immersion in the game (Wood et al., 2004). As noted above, factors other than sensory components also play an important role in the perception of realism. Indeed, some authors have criticized the current focus on the sensory aspects of video games and have lamented the fact that the sensory aspects of video games have far outstripped conceptual factors, such as story elements. Newman (2002), for example, argue that video games must overcome the current emphasis on "visualism," contending that the pleasure of playing video games is more related to the way the game "feels" rather than how it "looks."

Setting

The effect of setting on realism is complex. Familiar settings are likely to benefit from familiar detail. So a game racing through the streets of Chicago (e.g., *The Getaway*) is likely to seem more realistic if all of the local landscapes are accurately represented. However, as pointed out earlier, the setting does not have to be familiar to seem realistic. Players are likely to be much more forgiving of atypical objects in a setting if the setting is exotic or fantastic (Shapiro & Chock, 2004) or if the story is comic (Shapiro & Chock, 2003).

However, developers are working to enhance realism based on previous (indirect) knowledge of settings. For instance, one recent strategy of game manufacturers is to mix high-end graphics with current news. In *Kuma: War* players enact missions based on real-life events and scenarios, such as trying to capture Saddam Hussein's sons, Uday and Kusay, and Saddam Hussein himself (Werde, 2004). The capture of the Hussein brothers is set in a virtual re-creation of the Mosul villa where they were killed (CNN.com, 2004a), and programmers re-created Saddam's capture using satellite photos and declassified military intelligence (Werde, 2004). *Call of Duty* puts gamers into virtual versions of battles during World War II as a member of a unit. The effect on realism of this strategy remains to be seen, but the games have received good reviews.[3] Attention to contextual detail is likely to enhance realism judgments in virtual environments (Shapiro & McDonald, 1992).

Pacing

At any given moment, there is a limit to how much a person can mentally process. In a video game, some of that limited capacity is automatically allocated to processing the game's sensory features. That can have an impact on judgments of realism. Lang (2000) pointed out that for television both the meaning and the features have to be processed simultaneously. These features can include a variety of sensory features including audio and visual elements (Bolls, 2002; Bolls & Lang, 2003; Lang, 2000). Some features, particularly scene changes, voice changes and arousing material, are processed automatically and initially increase attention to and processing of audio-visual material. However, if such material comes at the viewer too fast, it exceeds limited capacity and other processing aspects of the stimulus can suffer.

The mental capacity available for processing information may also influence perceived realism—although research in this area is in its early stages. First, it appears that judgments about physical aspects of the story are more automatic and less influenced by other processing than conceptual aspects. For example, if other tasks limit the amount of capacity available for processing a story, conceptual judgments are slowed but physical realism judgments are unaffected (Shapiro & Shen, 2003).

There is probably a complex relationship between pacing and perceived realism in general. Contrary to the traditional notion that audiences actively suspend disbelief when viewing fiction, some social scientists believe that viewers first accept a story as real. Only if a person has adequate mental capacity at a given moment can he or she devote that mental capacity to thinking about whether a story seems unreal at that moment (Gilbert, 1991; Reeves & Nass, 1996). However, several studies indicate that formal features of entertainment media that limit the mental capacity available at a given moment (such as scene changes) have a complex effect on perceived realism. There appear to be effects based on the kind of language used (Bradley & Shapiro, 2004), the topic of the story, and perhaps initial assumptions about the perceived realism of a particular type of story (Shapiro & Shen, 2003; Shapiro, Shen, & Weisbein, 2002).

Also, the effect of available capacity may be curvilinear (Bradley & Shapiro, 2004; Shapiro & Shen, 2003). In some cases, the moderately paced stories were perceived most realistic.

A reasonable guess is that fast pacing tends to enhance the realism of aspects of the story that seem less real with thought and diminish the realism of aspects of the story that require thought to seem real. Hence, a video game about criminal activity in an exotic environment is probably enhanced by increasing the pace. In contrast, designers should probably give users a chance to think about realistically depicted interpersonal relationships.

Conceptual Elements

Typicality. As people move through the world, they judge what they see in a number of ways. If we witness a minor traffic accident, we expect the drivers to be tense not calm, but not out of control. If a driver is hysterical, we would consider that atypical. We might also conclude that driver does not handle stress well. Of course, how we judge a driver's behavior might also depend on whether we had seen or experienced an accident before. The same factors enter into our judgments about media characters and events in video games.

Our judgments of what is typical in a particular situation seem to have a powerful effect on perceived realism. Several studies have shown that manipulating the typicality of story elements strongly influences what we think is real (Shapiro & Chock, 2003). However, in some cases this is modified by other factors, such as the familiarity of the context of the information (whether the story takes place in a familiar environment such as the United.States. or an unfamiliar one such as Brazil; Shapiro & Chock, 2004). In particular, people seem more willing to accept atypical events in unfamiliar settings.

Character Types. In general, in addition to the player's own character, there are two other types of characters currently possible in narrative video games: (a) characters controlled by other humans, often called avatars, and (b) nonplayer characters (NPCs or bots) that are computer-controlled agents. An important recent trend in video games is the increasing availability of multiplayer options for games in which the player interacts in a video game setting with other avatars, such as *Star Wars Galaxies: An empire divided* and *The Sims Online*. Multiplayer games can range from a few friends playing *Quake III* at "LAN parties," when players agree to meet at a location and play online from different terminals while being co-present, to thousands of distributed players interacting simultaneously in a virtual space composed of a number of Internet servers that maintain a persistent, nonstop virtual reality (e.g., *EverQuest*). Multiplayer games are at the height of their popularity (see CNN.com, 2004b), and the majority of players polled in a recent survey consider multiplayer capability to be an important feature of a video game (Wood et al., 2004).

What makes playing with other avatars so important, and does having other humans controlling characters in a video game enhance the perceived realism of the video game? A thought experiment may be useful here. Assume for the moment that NPCs can pass the Turing test, and that a player could not distinguish between an NPC and an avatar. Would a player's judgments of realism change with the knowledge that a character was controlled by the computer or by a human? The Social Response to Communication Technology (SRCT) perspective (Nass, Fogg, & Moon, 1996; Nass & Moon, 2000; Nass, Moon, & Carney, 1999; Reeves & Nass, 1996) argues that people respond to social cues coming from humans or media acting like humans in much the same way. The SRCT perspective does not suggest, however, that people do not know they are interacting with a nonhuman—only that we tend to respond socially to social cues. One can speculate, however, that the more social cues that are provided and the more natural those cues, the more likely players will respond with social responses. Below we discuss in more detail the deceptively complicated notion of what it means for a character to act as socially expected.

Such an ideal video game environment is not likely to exist in the near future. Bringsjord (2001) asserted that current agents do not originate behaviors or artifacts: They just do what they have been designed to do. Today's NPCs still lack "personhood," as they have no creativity or free will. However, NPCs are increasingly able to perceive information in the game and execute accurate, relevant, and believable actions based on that information. van Lent and colleagues (1999) argued that these types of intelligent characters, who can "perceive," "think," and "act," can play an important role in the perceived realism of video games. Researchers have begun to analyze the believability, in terms of humanness, of NPCs based on different strategies and abilities allocated to NPCs. For example, MacInnes (2004) observed that NPCs that used a variety of strategies modeled after human playing were perceived as more human than NPCs that used more simple rule-based systems (e.g., "If player moves around corner, then follow").

Similarly, Laird and colleagues (Laird & Duchi, 2000; van Lent et al., 1999) have created video game bots that can sense information from the game, select relevant knowledge, and carry out internal and external actions. For example, Laird and Duchi (2000) created several bots for *Quake II* with different skill levels and assessed the perceived "humanness" of NPCs. Their characters varied across skill levels according to decision time, number of tactics, aiming skill, and aggressiveness. The characters' skills were tested against highly experienced human players, and "humanness" was measured using a variation of the Turing test. Their results indicate that both decision-time and aiming skill played a role in judgments about the humanness and performance of the character. In particular, characters with the best aiming skill were judged as unlikely to be human. Instead, characters with a decision time similar to humans and medium aiming skill were judged to be more humanlike (Laird & Duchi, 2000).

Currently, players know who the avatars are and who the NPCs are. Even in the ideal computer environment posited above, the mere knowledge that a character represents another human or is an NPC may affect a player's behavior, intentions, and attributions toward the character. Indeed, recent research by Shechtman and Horowitz (2003) suggested that while people do display some social responses during conversations with a computer program, as suggested by the SRCT perspective, they did not display as many of the relational and attributional behaviors that establish the interpersonal nature of a relationship as when they believed they were conversing with a human. These data suggest that there may be something unique about how we interact and form relationships with characters we believe to be human versus those that we do not believe to be human. In particular, people seem more willing to engage in relational behaviors and to make attributions about characters they believe to be human. If this is the case, then what features must a character exhibit in order for players to engage in interpersonal attributions or to evoke interpersonal, relational behavior? That is, what elements are required for a video game user to perceive a character to be real enough to engage with relationally, such as the case of online players expressing socioemotional and task communication (see Peña & Hancock, in press)? Social psychology and communication research has only begun to examine the underlying dimensions of relational dynamics in video game contexts.

Character Judgments.   One element is suggested by research on interpersonal attributions, which suggests that perceptions of realism can be influenced by the judgments of the cause of a person or character's actions, but it depends on whether the events are good or bad. People frequently make judgments about whether other people's behavior results from characteristics of that person (dispositional attribution) or from the situation that person finds him or herself in (situational attribution; for a review see Eagly & Chaiken, 1993). Under most circumstances, people spontaneously generate dispositional explanations for bad things happening to other people and spontaneously generate situational explanations for good things happening to other people (Gilbert & Malone, 1995; Gilbert, Pelham, & Krull, 1988; Krull & Erickson, 1995; Lee & Hallahan, 2001; Lupfer, Clark, & Hutcherson, 1990; Uleman, 1987; Winter & Uleman, 1984). In other words, if it is bad it is because the person has a bad character. If it is good it is because of circumstances. How do such judgments influence the realism of story characters?

One possibility is that if the story is congruent with what a viewer generates spontaneously, it will seem more real. As reasonable as that sounds, the opposite is true: The story seems more real when it contains information incongruent with what a person generates spontaneously (Shapiro, Barriga, & Beren, 2004). If a character is dismissed from her job, viewers are likely to spontaneously generate dispositional information explaining why she was fired. Adding situational information (e.g., the company was doing badly) seems to enrich the story and increase realism. The explanation seems to be that when the information in the story adds to what the viewer spontaneously generates, it makes the story seem richer and more real.

Whether avatar or NPC, players are likely to make causal attributions about game characters. Designers should be careful to provide information about the character that complements the spontaneous attributions the player is likely to make. For example, in an adventure game such as *The Longest Journey*, if a character responds rudely the player is likely to spontaneously generate a dispositional attribution. If situational information is provided at that point (e.g., the character has been asked that question too many times), it might enhance perceived realism.

Emotion.   Another question is whether something similar happens with emotion—that unexpected emotions seem more real. Recent research suggests that this does not seem to be the case. Shapiro and Chock (2003) noted that items identified as atypical often contained incongruent emotional reactions. More atypical items seemed to be perceived as less real. Recent data gathered in the first author's laboratory seem to confirm this. Stories with unexpected emotions seemed less real than those with expected emotions. So, if a character betrays a friend, expressing pride seems less realistic than expressing the expected emotion shame.

In summary, the judgments we make about story characters do seem to influence perceived realism. When bad things happen to a character, people tend to spontaneously supply dispositional information. Situational information in those stories makes them richer and more realistic. Conversely, when good things happen to a character, people tend to spontaneously supply situational information. In that case, supplying dispositional information enhances realism. Preliminary data indicate that an unexpected and unexplained emotion makes a story seem less real. However, future research may indicate situations in which this effect does not hold.

Designers may want to take this into consideration by supplying situational information when bad things happen to a character and dispositional information when good things happen. It also seems wise to make sure that emotional expressions are congruent with what the user would expect.

CONCLUSION

Recently, one of the authors was watching two children play a highly visually realistic football simulation game. Considerable effort had gone into making the players look and move realistically and to capture the sound of a real game. The screen resolution was actually higher than nondigital television, and he joked that the game looked more real than a televised football game. But that was not really true. The game did not look real no matter how good the sensory information was because one of the children knew little about football, and ran plays that were inconsistent with the situation. Ultimately, the perceived realism of the game depended on knowledge as well as fancy graphics and sound.

Until recently game designers, and to a lesser extent researchers, have focused on sensory immersion as a path to video game realism (e.g., Tamborini, 2000; see also Tamborini & Skalski, chap. 16, this volume). Of course, there is nothing wrong with using sensory information in judgments of realism. Players tend to rate realistic graphics and sound to be important in their judgments of video games (Wood et al., 2004). Children appear to develop skill using sensory information early and continue to use sensory information as they add the ability to judge realism using conceptual and imaginative processes as they mature. Video game designers need to consider all aspects of realism and researchers need to investigate these more conceptual aspects of realism judgments for stories in general and for video games in particular.

Designing Conceptual Realism

Based on current research, how do you construct a narrative video game that thinks real as well as looks real? If the story is a drama that reflects familiar real-world or historical settings, events, and people, then the designer needs to be concerned with several things. First, the player's direct and indirect experience with those settings, events, and people may color realism perceptions. Events that seem improbable or atypical may make the game seem less real. Typical contextual detail should enhance realism. In the familiar setting, those contextual details should be what the player would expect in that setting. Also important, however, is a player's imagination about what the people, events, and settings are like.

This imaginative aspect of realism judgment is critical if the events, people, and settings are unfamiliar, fantastic, or comic. In this case, players are much more likely to accept atypical events, locations, and people as reasonable. Fantasy settings can include fantasy contextual detail, as long as it makes sense within the set of assumptions associated with that fantasy setting. In either case, the story must make sense within that set of assumptions. The player's imagination of what the game situation would be like should match the game experience.

The design of both real-world and fantasy games needs to consider the developmental stage and the sophistication of realism judgments of the game's intended users. For example, a simplistic plot with stereotyped characters may have less appeal to older players, while a complex story with richly detailed characters may not appeal to younger players.

While processing sensory cues seems to be relatively automatic, thinking about people, settings, and events takes a bit more mental capacity. Designers should consider when players need a little time to be thoughtful about the realism of a story and when it is okay to speed up the pace. Indeed, designers might consider using fast-paced action to conceal aspects of a game that may have good sensory values but that do not necessarily make sense if you think about them. For example, action adventures often are so fast-paced a viewer has little time to think about the absurdity of the plot. But fast pacing may be a mistake if the player will appreciate the realism of a social situation only after a few seconds' thought. If a the actions of a character reveal something important about that character, a player may have difficulty reaching that conclusion if he or she is in the middle of a fast-paced chase.

Players are likely to make social judgments about game characters dependent on whether a human (avatar) or the computer (NPC) controls the character. Two of these judgments, causal attributions about behavior and emotional congruity have been studied with respect to realism judgments. Designers should treat them quite differently. Characters who express emotions congruent with the situation are judged more real than those who express emotion incongruent with the situation. Causal attributions are more complex. People tend to think bad things happen to people because of their dispositions (what they are like) and good things happen to other people because of the situation, if something bad happens to a character, the player is likely to spontaneously generate a dispositional explanation. That is that the bad thing happened because of what the character is like. If the game provides the missing situational information, it is likely to seem more real. The reverse is true if something good happens to a character. Designers should keep in mind that the act of providing this information may change subsequent attributions. For example, if a character's bad act is justified by information about the situation, then that may change the attributions that follow. Also liking for or identification with a character may make situational attributions more likely after bad acts and dispositional attributions more likely after good acts. Many other characteristics of game characters are likely to influence realism judgments—for example, personality type or the congruence between personality type and behavior. More research is needed to explore the impact of these factors on social judgments and realism judgments.

Another topic for further investigation is any differences in how realism judgments are made about characters controlled by the game and those controlled by other humans. At the current level of technology, experienced players are likely to know which other characters are avatars and which are NPCs (MacInnes, 2004). While considerable research indicates that people make social judgments about computers that act social, players are more likely to engage in relational behavior with avatars than with NPCs, especially as player experience increases. However, as designers pay more attention to creating NPCs that respond in realistic ways, that may change. Designers should also be careful that NPCs have skills that are comparable to humans at different stages of the game.

Although these suggestions only begin to scratch the surface of the complex task of making video games both look and think more real, they address an important goal shared by many current game designers, and it may match consumers' expectations of increasingly better video games. Clearly, many more factors than those listed above affect the perceived realism of video games, which highlights the fact that much more research is required to examine the factors that underlie the psychological perception of realism in the context of video games. In particular, while the sensory enhancement of realism has made tremendous strides over the last several decades, designers and researchers need to consider the conceptual and more abstract elements that go into the judgment: "IF this were to happen, would it happen like this?"

Limitations and Future Opportunities

One limitation of the research discussed here is that most of it is research on other media, particularly television and movies. That raises some questions and some opportunities for research that enhances our understanding of realism judgments. We anticipate that the principles outlined here will apply to video games, but that relationship should be tested. Perhaps more important is to identify the unique aspects of video games that change the way players think about realism (see Vorderer, 2000). Video games are a participatory medium. A player does not just observe the action; he or she is to a certain extent in control of the action (Grodal, 2000). In the case of some first-person shooters, that participation is often simple and brutal. But it still raises the question of how this kind of direct participation influences realism. Adventure games and online multiplayer games provide the opportunity for the player to play a role and to develop unique relationships with other characters—sometimes real, sometimes computer-generated— in a fantasy setting that does not exist outside cyberspace. This complex immersion in an environment raises multiple questions about how realism judgments are made about the game and about other characters.

Immersion, the feeling of being part of the story, may also be a factor. This feeling is likely to be complex, multi dimensional, and originate in a number of psychological phenomena (Klimmt & Vorderer, 2003; Tamborini & Skalski, chap. 16, this volume). There is recent evidence that story-based video games are more immersive than nonstory games (Schneider et al., 2004). Some research shows a relationship between immersion and perceived realism (Shapiro & Weisbein, 2001). It is unclear if realism creates more of a feeling of immersion or if immersion creates more of a feeling of realism. One possibility is that the feeling of "flow" in an immersive environment contributes to realism (Csikszentmihalyi, 1991).

Video games are an interesting platform to test various aspects of realism judgments. A variety of variables can be manipulated with considerable control. It would help researchers if computer designers could provide a set of easy-to-use tools that would enable them to design a narrative video game while manipulating variables that influence conceptual judgments about realism. For example, a tool that allowed researchers to create multiple versions of simple stories, settings, and characters would help investigators look at how emotional congruity, attribution, and other aspects of a game might influence realism.

Although regarded as toys by many, video games have the potential to revolutionize how we tell stories. Participating as a character in a story is now an everyday possibility for players. Authors have a new tool for telling stories that is still not well understood. Video games also offer an opportunity to understand how people mentally process stories, including how people make judgments about the realism of a story.

NOTES

We would like to thank the editors of this volume as well as William Tomlinson for their many thoughtful comments and suggestions on this chapter.

[1] We recognize that philosophy, literary theory, and film theory have a great deal to say about realism; however, the focus of this chapter is on the relevant social psychological research.

[2] Jerzy N. Kosinski makes a similar point (along with many others) about the consequences of being totally dependent on television for knowledge about the world in his novel *Being There*.

[3] "Good" reviews refers to the entertainment value of the game. It seems fair to also note that these games have been criticized by some for glorifying war, racism, and for representing only a narrow political viewpoint.

REFERENCES

Adoni, H., Cohen, A. A., & Mane, S. (1984). Social reality and television news: Perceptual dimensions of social conflicts in selected life areas. *Journal of Broadcasting, 28*(1), 33–49.

Andsager, J. L., Austin, E. W., & Pinkleton, B. E. (2001). Questioning the value of realism: Young adults' processing of messages in alcohol-related public service announcements and advertising. *Journal of Communication, 51*(1), 121–142.

Ang, I. (1985). *Watching Dallas: Soap opera and the melodramatic imagination*. New York: Methuen.

Atkin, C. (1983). Effects of realistic TV violence vs. fictional violence on aggression. *Journalism Quarterly, 60*(4), 615–621.

Austin, E. W., Roberts, D. F., & Nass, C. I. (1990). Influences of family communication on children's television-interpretation. *Communication Research, 17*(4), 545–564.

Bahk, C. M. (2001). Perceived realism and role attractiveness in movie portrayals of alcohol drinking. *American Journal of Health Behavior, 25*(5), 433–446.

Bolls, P. (2002). I can hear you, but can I see you? The use of visual cognition during exposure to high-imagery radio advertisements. *Communication Research, 29*(5), 537–563.

Bolls, P., & Lang, A. (2003). I saw it on the radio: The allocation of attention to high-imagery radio advertisements. *Media Psychology, 5*(1), 33–56.

Bradley, S. D., & Shapiro, M. A. (2004). Parsing reality: The interactive effects of complex syntax and time pressure on cognitive processing of television scenarios. *Media Psychology, 6*, 307–333.

Bringsjord, S. (2001). Is it possible to build dramatically compelling interactive digital entertainment (in the form, e.g., of computer games)? *Game Studies: The International Journal of Computer Game Research, 1*(1). Retrieved March 15, 2004, from http://www.gamestudies.org/0101/bringsjord/.

Buchman, D. D., & Funk, J. B. (1996). Video and computer games in the '90s: Children's time commitment and game preference. *Children Today, 24*(1), 12–15.

Cantor, J. (2001). The media and children's fears, anxieties, and perceptions of danger. In D. G. Singer & J. L. Singer (Eds.), *Handbook of children and the media* (pp. 207–221). Thousand Oaks, CA: Sage.

Cantor, J., Wilson, B. J., & Hoffner, C. (1986). Emotional responses to a televised nuclear holocaust film. *Communication Research, 13*(2), 257–277.

CNN.com. (2004a). Game: Try your hand at nabbing Hussein's sons. Retrieved February 7, 2004, from http://www.cnn.com/2004/TECH/fun.games/02/04/cnna.saddam.games/.

CNN.com. (2004b). Tale of two gaming worlds: Online consoles soar while PCs stumble. Retrieved April 4, 2004, from http://www.cnn.com/2004/TECH/fun.games/04/02/two.virtual.worlds.ap/.

Cohen, A. A., Adoni, H., & Drori, G. (1983). Adolescents' perceptions of social conflicts in television news and social reality. *Human Communication Research, 10*(2), 203–225.

Csikszentmihalyi, M. (1991). *Flow: The psychology of optimal experience.* New York: Harper Collins.

Davies, M. M. (1997). *Fake, fact, and fantasy: Children's interpretations of television reality.* Hillsdale, NJ: Lawrence Erlbaum Associates.

Dorr, A. (1983). No shortcuts to judging reality. In J. Bryant & D. R. Anderson (Eds.), *Children's understanding of television: Research on attention and comprehension* (pp. 199–220). New York: Academic Press.

Downs, A. C. (1990). Children's judgments of televised events: The real versus pretend distinction. *Perceptual and Motor Skills, 70*(3, Pt. 1), 779–782.

Eagly, A. H., & Chaiken, S. (1993). *The psychology of attitudes.* New York: Harcourt Brace.

Entertainment Software Association. (2004). *Essential facts about the computer and video game industry.* Retrieved July 1, 2004, from http://www.theesa.com/EFBrochure.pdf.

Feldman, B. E., O'Brien, J. F., & Arikan, O. (2003, July). *Animating suspended particle explosions.* Paper presented at the ACM SIGGRAPH 2003, San Diego, CA.

Flavell, J. H. (1986). Development of children's knowledge about the appearance–reality distinction. *American Psychologist, 41*(4), 418–425.

Frauenfelder, M. (2001, August). Smash hits. *WIRED, 9.08,* 116–121.

Geen, R. G. (1975). The meaning of observed violence: Real vs. fictional violence and consequent effects on aggression and emotional arousal. *Journal of Research in Personality, 9*(4), 270–281.

Geen, R. G., & Rakosky, J. J. (1973). Interpretations of observed aggression and their effect on GSR. *Journal of Experimental Research in Personality, 6*(4), 289–292.

Gilbert, D. T. (1991). How mental systems believe. *American Psychologist, 46*(2), 107–119.

Gilbert, D. T., & Malone, P. S. (1995). The correspondence bias. *Psychological Bulletin, 117*(1), 21–38.

Gilbert, D. T., Pelham, B. W., & Krull, D. S. (1988). On cognitive busyness: When person perceivers meet persons perceived. *Journal of Personality and Social Psychology, 54*(5), 733–740.

Grodal, T. (2000). Video games and the pleasures of control. In D. Zillmann & P. Vorderer (Eds.), *Media entertainment: The psychology of its appeal* (pp. 197–213). Mahwah, NJ: Lawrence Erlbaum Associates.

Hall, A. (2001, May). *Evaluating media realism: Components of audiences' judgments of the relationship between representation and reality.* Paper presented at the International Communication Association, Washington, DC.

Hoffner, C., & Cantor, J. (1985). Developmental differences in responses to a television character's appearance and behavior. *Developmental Psychology, 21*(6), 1065–1074.

Hoffner, C., & Cantor, J. (1991). *Perceiving and responding to mass media characters.* Hillsdale, NJ: Lawrence Erlbaum Associates.

Klimmt, C. & Vorderer, P. (2003). Media psychology "is not yet there": Introducing theories on media entertainment to the Presence debate. *Presence: Teleoperators and Virtual Environments, 12*(4),346–359.

Kramer, G. (1995). Sound and communication in virtual reality. In F. Biocca & M. R. Levy (Eds.), *Communication in the age of virtual reality* (pp. 259–276). Hillsdale, NJ: Lawrence Erlbaum Associates.

Krull, D. S., & Erickson, D. J. (1995). Judging situations: On the effortful process of taking dispositional information into account. *Social Cognition, 13*(4), 417–438.

Laird, J. E., & Duchi, J. C. (2000, November). *Creating human-like synthetic characters with multiple skill levels: A case study using the Soar Quakebot.* Paper presented at the AAAI 2000 Fall Symposium Series: Simulating Human Agents, North Falmouth, MA.

Lang, A. (2000). The limited capacity model of mediated message processing. *Journal of Communication, 50*(1), 46–70.

Lee, F., & Hallahan, M. (2001). Do situational expectations produce situational inferences? The role of future expectations in directing inferential goals. *Journal of Personality and Social Psychology, 80*(4), 545–556.

Levine, R. (2002, November). The Sims online. *WIRED, 10.11,* 176–179.

Livingstone, S. M. (1989). Interpretive viewers and structured programs: The implicit representation of soap opera characters. *Communication Research, 16*(1), 25–57.

Loftus, G. R., & Loftus, E. F. (1983). *Mind at play: The psychology of videogames.* New York: Basic Books.

Lupfer, M. B., Clark, L. F., & Hutcherson, H. W. (1990). Impact of context on spontaneous trait and situational attributions. *Journal of Personality and Social Psychology, 58*(2), 239–249.

MacInnes, W. J. (2004). Believability in multi-agent computer games: Revisiting the Turing test. Paper presented at the *Proceedings of the Conference on Computer-Human Interaction,* Vienna, Austria.

Mao, W., & Gratch, J. (2003, September). The social credit assignment problem.Paper presented at the 4th International Working Conference on Intelligent Virtual Agents, Kloster Irsee, Germany.

Mateas, M., & Stern, A. (2000, November). Towards integrating plot and character for interactive drama. Paper presented at the Working Notes of the Social Intelligent Agents: The Human in the Loop Symposium. AAAI Fall Symposium Series, Menlo Park, CA.

Morison, P., Kelly, H., & Gardner, H. (1981). Reasoning about the realities on television: A developmental study. *Journal of Broadcasting, 25*(3), 229–242.

Nass, C., Fogg, B. J., & Moon, Y. (1996). Can computers be teammates? *International Journal of Human Computer Studies, 45*(6), 669 –678.

Nass, C., & Moon, Y. (2000). Machines and mindlessness: Social responses to computers. *Journal of Social Issues, 56*(1), 81–103.

Nass, C., Moon, Y., & Carney, P. (1999). Are people polite to computers? Responses to computer-based interviewing systems. *Journal of Applied Social Psychology, 29*(5), 1093–1110.

Newman, J. (2002). The myth of the ergodic videogame: Some thoughts on player-character relationships in videogames. *Game Studies: The International Journal of Computer Game Research, 2*(1). Retrieved March 15, 2004, from http://www.gamestudies.org/0102/newman/.

Peña, J., & Hancock, J. T. (in press). An analysis of socioemotional and task communication in a online multiplayer videogame. *Communication Research.*

Perse, E. M. (1990). Media involvement and local news effects. *Journal of Broadcasting and Electronic Media, 34*(1), 17–36.

Potter, W. J. (1988). Perceived reality in television effects research. *Journal of Broadcasting & Electronic Media, 32*(1), 23–41.

Reeves, B., & Nass, C. (1996). *The media equation: How people treat computers, television, and new media like real people and places.* Stanford, CA: CSLI Publications.

Rubin, A. M. (1979). Television use by children and adolescents. *Human Communication Research, 5,* 109–120.

Rubin, A. M., Perse, E. M., & Powell, R. A. (1985). Loneliness, parasocial interaction, and local television news viewing. *Human Communication Research, 12*(2), 155–180.

Ryan, S. (2004). Jet Li comes to life in new video game. Retrieved March 10, 2004, from http://edition.cnn.com/2004/TECH/03/10/rise.to.honor/.

Schneider, E. F., Lang, A., Shin, M., & Bradley, S. D. (2004). Death with a story: How story impacts emotional, motivational, and physiological responses to first-person shooter video games. *Human Communication Research, 30*(3), 361–375.

Shapiro, M. A., Barriga, C., & Beren, J. (2004, May). Tell me something I didn't know about why you did that: Attribution and perceived reality. Paper presented at the International Communication Association Conference, New Orleans.

Shapiro, M. A., & Chock, T. M. (2003). Psychological processes in perceiving reality. *Media Psychology, 5*(2), 163–198.

Shapiro, M. A., & Chock, T. M. (2004). Media dependency and perceived reality of fiction and news. *Journal of Broadcasting & Electronic Media, 48*(4), 675–695.

Shapiro, M. A., & McDonald, D. G. (1992). I'm not a real doctor, but I play one in virtual reality: Implications of virtual reality for judgments about reality. *Journal of Communication, 42*(4), 94–114.

Shapiro, M. A., & Shen, F. (2003, May). *The effect of limited capacity, social judgment, and advertising topic on perceived reality.* Paper presented at the International Communication Association, San Diego, CA.

Shapiro, M. A., Shen, F., & Weisbein, L. (2002). The effect of cognitive load on perceived reality. Paper presented at the Association for Education in Journalism and Mass Communication, Miami, FL.

Shapiro, M. A., & Weisbein, L. (2001, May). Only thinking can make it false: Limited capacity, presence and perceived reality of television. Paper presented at the International Communication Association, Washington, DC.

Shechtman, N., & Horowitz, L. (2003, April). Media inequality in conversation: How people behave differently when interacting with computers and people. Paper presented at the CHI 2003, Ft. Lauderdale, FL.

Tamborini, R. (2000, November). The experience of telepresence in violent video games. Paper presented at the 86th annual convention of the National Communication Association, Seattle, WA.

Tanner, M. (1997, August). Musicians lured to videogames. Retrieved April 26, 2004, from http://www.wired.com/news/culture/0,1284,6041,00.html.

Thompson, C. (2003, October). Suburban rhapsody. *Psychology Today, 36,* 32–40.

Thompson, C. (2004). The game's the thing: Why are Hollywood actors starring on your PlayStation? Retrieved March 18, 2004, from http://slate.msn.com/id/2097296/.

Tomlinson, W. M. (2002). *Synthetic social relationships for computational entities.* Unpublished doctoral dissertation, Massachusetts Institute of Technology, Cambridge.

Uleman, J. S. (1987). Consciousness and control: The case of spontaneous trait inferences. *Personality and Social Psychology Bulletin, 13*(3), 337–354.

van Lent, M., Laird, J. E., Buckman, J., Hartford, J., Houchard, S., Steinkraus, K., et al. (1999, July). Intelligent agents in computer games. Paper presented at the *Proceedings of the National Conference on Artificial Intelligence,* Orlando, FL.

Vorderer, P. (2000). Interactive entertainment and beyond. In D. Zillmann & P. Vorderer (Eds.), *Media entertainment: The psychology of its appeal* (pp. 21–36). Mahwah, NJ: Lawrence Erlbaum Associates.

Werde, B. (2004, March). The war at home. *WIRED, 12.03,* 104–105.

Winter, L., & Uleman, J. S. (1984). When are social judgments made? Evidence for the spontaneousness of trait inferences. *Journal of Personality and Social Psychology, 47*(2), 237–252.

Wood, R. T., Griffiths, M. D., Chappell, D., & Davies, M. N. (2004). The structural characteristics of video games: A psycho-structural analysis. *CyberPsychology & Behavior, 7*(1), 1–10.

Wright, J. C., Huston, A. C., Reitz, A. L., & Piemyat, S. (1994). Young children's perceptions of television reality: Determinants and developmental differences. *Developmental Psychology, 30*(2), 229–239.

Wright, J. C., Huston, A. C., Truglio, R., Fitch, M., Smith, E., & Piemyat, S. (1995). Occupational portrayals on television: Children's role schemata, career aspirations, and perceptions of reality. *Child Development, 66*(6), 1706–1718.

Yee, N. (2001). The Norrathian Scrolls: A study of EverQuest (version 2.5). Retrieved October 12, 2002, from http://www.nickyee.com/eqt/report.html.

20

Playing Online

Ann-Sofie Axelsson
Chalmers University of Technology Göteborg, Sweden

Tim Regan
Microsoft Research Cambridge, UK

Maður er manns gaman.
Man is the joy of man.
—Hávamál (from the Elder or Poetic Icelandic Edda)

Over the past 10 years, there has been a major rise in online gaming. While a modern stand-alone computer provides an advanced and multiskilled playmate to compete against, or a fantastic virtual environment in which the player can test his or her skills and powers, a networked computer provides further opportunities, connecting the single player to a vast number of Web sites where he or she can meet hundreds of thousands of coplayers to play with and against in everything from checkers to racing competitions to massively multiplayer online role-playing games (MMORPGs). But how do online games provide and encourage social participation between players, and is this participation at the cost of other communication modes?

The focus of this chapter is on people's social activities in online games, describing and analyzing *what* people do, *with whom* they do it, and, perhaps most interesting, *why* they do what they do in online games. Because chapter 6 in this volume provides a comprehensive overview of online games and their variety as well as their common characteristics, we narrow the focus in this chapter and deepen the analysis to one specific game and its social aspects. It is informative to compare the success of social gaming in some genres with its failure for games like *The Sims Online,* which is examined in chapter 21.

We provide a broad picture of the social aspects of online games, but in order to make it not only descriptive but also analytical, we give continuous references to our empirical study of a MMORPG called *Asheron's Call* (Turbine Entertainment Software Corp.). The material discussed and analyzed in the chapter was collected in a large study concerned with the social aspects of *Asheron's Call.* The study was conducted during the summer of

2001, and the aim of the study was to obtain an understanding of the social activities within and outside the game and how these various activities interrelate and constitute the community of *Asheron's Call* (*AC*). The empirical study is described in greater detail in the next section.

The section that follows after that describes the demographics of online computer games with reference to the empirical study of *AC*. We describe *who* plays online computer games, *with whom* they play, and why. In the following section we discuss the relationship between what we call *technical features* and the social effects of these. Examples of technical features are group communication systems, contact lists, game-related Web sites, social groups, and out-of-game communication technologies. In this section we also bring in the view of the developers of *AC* on in-game social issues.

The fourth section briefly concludes the previous sections.

THE STUDY: THE SOCIAL ASPECTS OF *ASHERON'S CALL*

The study on which this chapter is based was conducted during the summer of 2001 within the *Social Computing Group* at Microsoft Research. The aim of the study was to obtain an understanding of the social activities within and outside *AC* and how these various activities interrelate and constitute the community of *AC*. The social interaction in the game was hypothesized to be related to specific features of the game that are implemented in order to increase this social interaction between players and to foster good behavior. Two of these features, the allegiance and the fellowship systems, and their influence on social interaction have been described elsewhere (Axelsson & Regan, 2002).

Within the larger study we carried out interviews with the game developers, observation and log studies within the game, as well as an extensive online survey directed toward players of *AC*. In this chapter we report mainly on parts of this study, namely the interviews and the online survey.

In order to get a sense of what the social action in *AC* is like and, of even more interest, what directs and shapes the social interaction from the systems side, we conducted two interviews with program managers of *AC*.

In the online survey, players of *AC* were invited to answer questions concerning their engagement in the game, including activities carried out in the game as well as outside the game (either online but in another context than the game context, or offline). During the 3 weeks the survey was online, we received 7,364 entries. We base the chapter on a subsample consisting of the first 2 weeks' participants. The total number of entries during the first 2 weeks was 5,587, and after excluding multiple and incomplete entries the total valid number is 5,064 participants.

PLAYERS, COPLAYERS, AND MOTIVES IN ONLINE GAMES

Who Plays Online?

There is a social aspect to almost all game play. People want to play games that are (or will be) popular with their friends. Bulletin boards are set up to swap game experiences, compare levels reached, and so on. But online role-playing games have social aspects as the key driver to the game play itself. In these games much of the effort of the creators has gone into crafting the social interplay between players. Hence, this attracts a different kind of gamer (or at least gamers in a distinctly social mood). There are distinctions to be made amongst the approaches to social gaming. Some players revel in the role-play, others enjoy pitting their wits against an opponent far more skillful than an in-game AI, while others want to wreak destruction on their peers. Some want a social way to spend a few hours a week, while others use it as a primary mode of communication with friends.

Early pundits of online social gaming predicted that the new genre would bring in droves of new players who had been uninterested in offline gaming. This has not happened, but we do see a diversity of players. For example, the nationalities represented in our survey range from Afghanistan to Switzerland and include Costa Rica, Greece, Mexico, and Russia, though the numbers of returns are severely skewed toward the United States (3,928) rather than elsewhere (960).

The age range of respondents to our survey was also varied, and the frequencies followed those we would expect from offline adventure gaming, that is, centered on the 19–38 group. We found that respondents reported a mean age of 30 with a standard deviation of 0.13 (Fig. 20.1). Gender gave a similar result with the proportion coming out at 72% male.

Our survey respondents were very keen on *Asheron's Call*, playing an average of 21 reported hours a week. This does not mean that people are retiring from social contact, as we shall now show.

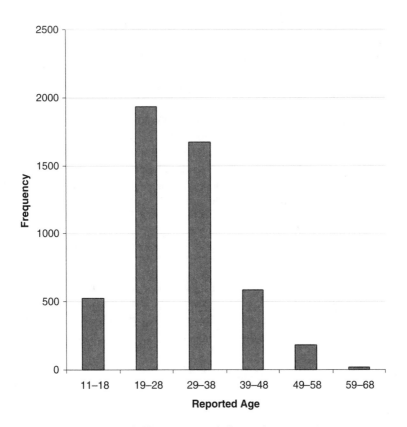

FIG. 20.1.   Age spread of respondents.

With Whom Do People Play?

Within *Asheron's Call*, the people with whom one plays can become acquaintances, friends, or even closer. Ninety percent of our respondents have become acquainted with other players through the game, and an average of 13 friends and 3 close friends have been made through the game. Moreover, a number of our 5,064 respondents had developed friendships beyond that: 144 were dating, 25 were engaged, and 12 were married to partners they met through *Asheron's Call*.

Sixty percent of our respondents have contacted other *Asheron's Call* players outside the game. This contact is often by e-mail, but non-PC-based communication is also popular: 46% of these respondents had used the phone for such contacts, and 8% had used paper letters. (NB: These are overlapping figures; e.g., 82% had used e-mail.) Twenty five percent of respondents met face-to-face with friends they had met through *Asheron's Call*. The reasons people gave for meeting in real life were varied. Many met because they developed online friendships within *Asheron's Call* with people whom they found to live nearby. For example, one respondent commented: "Found out a neighbor was a player on same server—we had a good laugh and then role-played in game." Another said: "Found out folks I played with lived close by. They needed computer support so I met up with them to help them out."

There were also deep relationships that started online in *Asheron's Call*, as this comment illustrates: "I went because I had fallen in love with my patron and wanted to meet him. He was as wonderful as I had hoped he'd be."

Inevitably, meeting in real life is not part of the role-play of *Asheron's Call*, though vestiges may persist. For example, the previously quoted responder used the word *patron* to describe someone who has become a close friend. The following reply also shows echoes of the game-world coming into the real: "[we met] for fun, to compare notes, to go see a movie...I guess the same reasons comrades would gather in taverns in fantasy novels."

We also found people playing with existing friends, colleagues, and family members. Just as e-mail, chat, phone calls, text messaging, and so on offer technological ways to maintain relationships over distance, so does playing *Asheron's Call*.

Though we have talked mostly about networked meetings, people often play games in the same physical space as other people, either formally in, for example, a LAN game, or just by random juxtaposition, for example, by playing on the family room PC while other family members watch TV. We asked respondents how they most often play *Asheron's Call*. Most respondents sat alone to play (75%), but there were still significant numbers of respondents (12% and 13%, respectively) who mainly sat with other players or others not playing as they played.

Why Do People Play Online?

We found a variety of reasons why different people choose to play an online game. Let us deal with the most surprising of these first. There are some players for whom the online aspect is secondary. This minority uses the game mainly for solo hunting and for solo questing. But even these users often augment their game play with social interaction. For example, one responder noted of his typical game session: "I don't interact unless there's someone who needs help or a dispute over creature 'ownership'; normally, I just hunt." Another pointed out: "I am a solo hunter almost exclusively. [...] If there are people I know online when I jack in, I will usually but not always chat with." So for this player *Asheron's Call* provides the chat channel, and perhaps common context, for discussion with friends, but the game itself is played solo.

Of those whose play is inherently social, there are three main types of play: helping, cooperating, and hostility, although these classes overlap substantially. A social motivation for some players is to help others. One responder described his typical session as "I hunt along the beach in Eastham. If I see any newbies or players who are having trouble, I help them out with a heal or a buff."

Cooperative play involves joining a group (either formally through a fellowship and/or allegiance) and hunting or questing together. This forms the most common reason given by our respondents for enjoying *Asheron's Call*.

The final reason, hostility, can take two forms. There is hostility between characters that is encouraged by the game design, and hostility that is not. Consider the following quote: "I harass other PKers, recomp, kill people, loot them, taunt them, recomp, go power level." *Asheron's Call* provides a world, Darktide, where players can kill one another (so-called player killers, or PKers). Strangely, this is perhaps the most social world because the value of one's allegiance is at its most important. So the killing described in the quote is typical of the structured play in Darktide, and many respondents enjoyed questing with fellow allegiance members within that world. Taunting and other activities whose sole enjoyment is the annoyance caused to others are not encouraged, but they are still social. Taunting is only enjoyable if one imbues the receiver with human characteristics, and in online games the receiver is human.

We have seen that *Asheron's Call* is used by a variety of people as a technological adjunct to their social lives. In the next section we investigate how players use technical features in and outside the game, for what reasons they do so, and what social implications this might have.

TECHNICAL FEATURES—SOCIAL EFFECTS

This section looks at some of the game functions and features of *AC* and discusses how they influence the social interaction of the players.

First, we provide a more general view of the relationship between technological functions and features and social interaction.

A number of recent studies on computer-based environments (that is, online computer games and shared virtual environments, SVEs) have highlighted the relationship between the technology and the social interaction, that is, the computer-generated surroundings in which users, players and others move around and with which they interact, and what people do in and with the environment.

Some recent studies have explored the influence of virtual appearance (avatar appearance) on social interaction in MMORPGs and SVEs, discussing issues like social inhibition and interpersonal distance (Bailenson, Blascovich, Beall, & Loomis, 2003; Slater & Steed, 2002; Becker & Mark, 2002) and social stratification and group affiliation (Axelsson, 2002; Taylor, 2002a). In these studies both the appearance of the virtual bodies (e.g., virtual clothing, expression of gender, etc.) and the function of these bodies (e.g., the ability to walk through other people's virtual bodies, to use eye gaze, etc.) are discussed.

Other studies have investigated the influence of the existence of virtual objects (e.g., buildings and avatars) on social interaction in the same kind of environments, discussing ownership issues and vandalism (Hudson-Smith, 2002; Taylor, 2002b). Still other studies have looked at the technology *outside* the SVEs, the hardware and the input and output devices, discussing how different technologies provide different interaction possibilities and examining issues such as leadership and social inhibition (Slater, Sadagic, Usoh, & Schroeder, 2000; Axelsson, 2002).

In the interviews with the program managers of *AC*, it was found that they stressed the importance of promoting interaction between the players with help from technology, by introducing single objects to "give people interesting things to talk about; swords, quests etc." as well as by introducing group structures into the game and encouraging people to join these structures. In *AC I*, there were two types of group structures in the game, fellowships and allegiances; and to *AC II* (which was released in November 2002) a third group structure, *the Kingdom*, was added. Apart from enhancing interaction between players in general, these social structures (engaging their members both inside and outside the game) are tools that the developers intentionally use to create socially active players. Other features and functions in *AC* that have been used by the developers for creating and maintaining a socially vibrant community are, for example, contact lists (contact lists for friends), communication systems (fellowship and allegiance chat systems), and fan Web sites (player sites dedicated to *AC* issues).

In the following three sections, the technical features and functions are discussed in relation to the social interaction they bring about. In the next section we highlight in-game features that are used for finding and keeping contact with people in MMORPGs over time: contact lists, communication systems, and social structures like *allegiances* and *fellowships*. Then, we emphasize features that some gamers use and/or create in order to enhance the gaming experience: out-of-the-game communication media and fan Web sites. In the final section we will go into depth discussing the role of role-playing in MMORPGs, how players look upon their own and other players' role-play, and also to what extent players gender-play and why they do so.

FINDING PEOPLE TO PLAY WITH: CONTACT LISTS, COMMUNICATION TECHNOLOGIES, AND SOCIAL STRUCTURES

In the 1990s, there were no MMORPGs around, other than MUDs (Multi-User Dungeons or Domains; see Curtis, 1992). Researchers were particularly interested in the anonymity offered by the Internet medium and in the interesting ways such anonymity could be taken advantage of: "On the Internet, no one knows you're a dog" was an often-cited cartoon quote. People online, researchers argued, take every opportunity to swap identity and appear as somebody else, and it is not uncommon that they have several online characters going on at the same time, each identity reflecting a personality trait which otherwise would have been neglected (Turkle, 1997). However, more recent studies show that people's identity online seems to be much more stable than has previously been held, having one or two characters and not, as was believed earlier, multiple (Schiano & White, 1998; Schroeder & Axelsson, 2000). Also, the reason for many online players or users of SVEs having more than one character seems to be mainly pragmatic, allowing people to obtain some privacy, away from friends who would otherwise see that they were online and contact them (Schroeder & Axelsson, 2000). This applies especially to long-term users who are well-known in the community by a large number of people and who will therefore often be contacted.

In *AC* there is an upper limit of 40 characters one player is allowed to have on his or her game account. However, as *AC* consists of altogether 8 worlds, and there is a world limit of 5 characters, players do not have 40 characters in the same social space (as the worlds are nonpermeable, not allowing players to take their characters from one world to another), but only up to 5. The data analysis of the questionnaire data ($n = 4,685$) shows that players have on average 6.41 characters in the game as a whole, with no significant difference in relation to gender. However, in relation to time spent playing the game, one can find a significant difference such that the longer people have played the game, the more characters they have. However, even if the average number of characters seems to be rather high ($M = 6.41$), the analysis shows that 92.6% use one character more than the others. This finding also indicates that *AC* players have a persistent identity but have additional characters for occasional and possibly instrumental use. One hypothesis, deduced from game observations, is that players use their additional characters mainly in two senses: either for storing things like weapons, food, or medicine to pick up when needed (a player does not want his or her character in play to be overburdened as the load might hinder the player in fights, etc.) or as complementary characters to be called upon when the situation is critical. If a player's main character, for example, is specialized in warfare, he or she might log out and log in with his or her healer character in case a fellow player is wounded. This description might also explain the fact that more experienced players have more characters than inexperienced players, as both explanations require a certain amount of time in the game, in order to collect or buy items, or to become skilled enough to be able to log in and out during critical situations. What seems to be the case, nevertheless, is that the vast majority of *AC* players use mainly one character in the game, maintaining a stable identity in the game and thereby facilitating contact initiatives from other players.

Most other online games and SVEs (and all online chat communities) allow players and users to set up contact lists in order to stay in touch easily with other players. The list usually shows the names of the contacts and whether these are online at the moment or not. It varies, but in some systems one can only communicate synchronously, that is to say, one cannot send messages to contacts that are *not* logged on, while in other systems this is possible.

What is a condition, however, in many systems for being able to create a contact list and for being put on someone else's list, is that the person is a paying player or user of the system. The unequal possibilities of establishing a social network in these systems, depending on whether you are a regular and paying user or a visitor only, and the social consequences of such technological solutions, have been discussed elsewhere (Axelsson, 2002). What can be said about the contact list system in *AC* is, of course, that players cannot play the game at all without purchasing the game and paying the monthly fee. However, being paying players of the game, people are still not allowed to set up a contact list just for communication with people they get to know. In order to be able to do so, the player must create a new or join an already existing social group: either a fellowship—a short-term group (only for one playing session) of two to nine members—or an allegiance—a long-term group with a minimum of two members and no upper limit.

The players of the same allegiance or fellowship are automatically placed on each others' contact lists. It is not uncommon, however, that a player is a long-term member of an allegiance and also joins, occasionally, fellowships for a single play session. An analysis of the survey sample showed that the vast majority in the survey sample have all or some of their characters (including their main character) in an allegiance and also usually join a fellowship when playing (85.1%). Also, players who have *all* or *some* of their characters (including the main character) in an allegiance have significantly *more* characters ($M = 6.51$) on average than players with *no* or *some* characters (*not* including the main character; $M = 4.77$) in an allegiance. The explanation for this could be that players who are more engaged in allegiances make more use of their additional characters in the instrumental ways mentioned above, for storing and/or for assisting fellow players, than players less engaged in allegiances. An additional explanation is that even though players are members of an allegiance, they do not rely heavily on the contact list function of the allegiance to stay in touch with people (and therefore they do not need to keep one stable identity), but that they have other means of keeping in touch, for example, deciding out-of-game or offline (using other communication media or communicating face-to-face) that they should log on and play the game together at a specific time. In the next section, use of other communication media in *AC* as complementary communication tools will be discussed.

Players also join both long- and short-term groups in order to have access to a communication system that makes it possible to easily keep track of and communicate with other players. An analysis showed that, when asking players about the reasons for joining a group structure (fellowship or allegiance), a fairly large number of people playing with others in fellowships (22.2%) reported that access to the group communication system was one reason, while a much lesser number of people belonging to an allegiance (1.0%) reported that as a reason. To be noted, however, is that this option, "joining a group in order to access the communication feature," was not a predefined alternative that could be chosen in the questionnaire, but something the respondents reported as a free response in a text box (in connection to the alternative "Other"). Had this been a predefined option in the questionnaire, the number of players reporting this as a reason for joining and playing in groups might have been even higher.

For example, one respondent, who occasionally plays in fellowships, said: "The color in radar is a big deal to me, but I like the fellowship chat. If you're in a huge group of people all the chat looks the same except your fellowship" (male, 32), while another player reported that one reason was: "[S]o we can chat on the fellowship channel and not broadcast our conversations. More for being considerate to the other players around us not being in our fellowship" (female, 30).

As shown in the comments above, players appreciate the chat function available to the fellowship members for a variety of reasons: orientation and recognition, privacy, and concern about other players not in the fellowship.

ENHANCING THE EXPERIENCE: COMMUNICATION MEDIA AND FAN WEB SITES

The previous section analyzed and discussed *in-game* communication systems and contact lists in relation to the use of multiple characters in MMORPGs and SVEs. However, online activities do not only take place *online*. Recent studies of SVEs and MMORPGs show that an important part of the interaction between players or users takes place *offline* or *out-of-game*, either face-to-face or via other, nongame communication media (see, for example, Taylor, 2002b). This section will therefore look at the use of various *out-of-game* communication media used by *AC* players, particularly in relation to membership in online social groups (allegiances and fellowships), gender, and time they have played the game. We also describe and discuss fan Web sites with *AC* content. Who creates the fan Web sites and why, who visits them, and for what reason?

The main motivation for investigating these issues is the importance of widening the perspective on gaming activities also to include certain offline activities or out-of-game activities. Two such activities that most likely have implications for gaming activities, and therefore should be studied, are the use of complementary media (out-of-game) and the creation and use of fan Web sites. Both activities can be seen as social activities, strengthening the relationships between the players in the game and making the sense of belonging to a gaming community even stronger.

Communication Media

The online survey contained, as we have already seen, a number of questions about the players' relationships with each other: with whom they tend to play, how many and how close friends they have in the game, and so on. A question in relation to these, regarding social relationships with other players, is of course whether players use additional communication tools that are not in the game, such as telephone, e-mail, chat, ICQ, or other media, in order to enhance the play sessions.

Analyses showed some interesting results. A large number of players reported that they use one to three other communication media to enhance the play session (81.6%), while only a few (10.6%) reported that they use four to six media. Most common is to use email (38.9%) and telephone (35.9%). The reasons for using complementary media vary. For example, one respondent commented that he used them in order "to coordinate meeting times. Also to send messages, since *AC* doesn't support personal mail. Sometimes, there are six of us hunting and using Gamevoice is easier than typing out everything"(male, 28) while another player said that she uses "[t]elephone—to talk to another person while we're both playing the game, or to get directions on how to do something from my husband (who also plays) while I am in game and he's at work. E-mail—to correspond with guild mates and other friends that play" (female, 22).

The two quotes exemplify some common uses of additional communication media in *AC* that respondents report: out-of-game planning and organizing of play, facilitating play sessions online, and obtaining out-of-game advice while playing online. The examples indicate what has been suggested earlier: that players of online (!) games do not perceive what is going on in the game and out of the game or even offline as separate worlds, but only as different sides of the game.

The analysis also showed that females report a significantly higher number of additional communication media used than males, reporting $M = 1.96$ and $M = 1.78$ media, respectively. There is also a significant correlation between engagement in the game, and use of additional communication media, such that the longer players have played the game and the more social groups they belong to, the more additional communication media they use. This might be explained by the fact that the more socially absorbed a player is online, the more he or she wants to contact other players. Conversely, the more socially engaged one is, the more people there are to contact him or her. This explanation would be consistent with the studies of online personal networks (e.g., distance education groups, work groups) showing the importance of the social context for the use of technologies, such as that people who belong to dense social networks seem to use a larger number of communication technologies and to a higher degree than do people belonging to social networks with fewer and weaker ties (Garton & Wellman, 1995; Haythornthwaite, 2000). The more social people are, the more media they use, and the more social they become. Also, when looking at how people become acquainted with new communication technologies and start to use them, it seems that the social network (e.g., family, friends, coworkers, and coplayers) is highly important (Haddon, 2004). In that sense *AC* is a perfect place for *warm experts*, that is, people in one's social network who are knowledgeable about new technologies and who would be happy to introduce oneself to these technologies (Bakardjieva, 2001).

There were also a fairly large number of respondents in the survey who reported that they had decided *not* to use any of the additional media mentioned in parallel with *AC* (21.7%). The reasons for choosing not to use complementary media varied. The most common reasons reported as free responses were either of technical or social kinds. The technical reasons were specified as either the lack of computer or network capacity to have other programs than *AC* running on the computer (e.g., use of modem would hinder simultaneous use of telephone), lack of computer skills to use other programs simultaneously (including even communication technologies like an ordinary telephone), or that it was a disturbance to "alt-tab" out of the game to, for example, chat via another program. The social reasons were described as either not wanting to break the role-play and the immersive illusion by using out-of-game technologies, not needing to have an online communication channel at all with other players as the coplayer(s) was/were usually in the same physical room, that the respondent regarded him or herself as a solo player, or as not wanting one's offline life to become involved in the online activities.

Fan Web Sites

We now turn to the next area of interest in this section, fan Web sites. What will be presented and discussed in this section about fan Web sites in relation to *AC* are mainly two aspects: first, the meaning of fan Web sites (i.e, what they are and what they are for) and, second, the use of them (i.e, who creates them and who uses them).

Fan Web sites can be defined as Web pages with contents mainly related to one specific phenomenon, for example, a pop band, a soccer club, or an online computer game made and maintained by individuals dedicated to this phenomenon. That is on the more practical level. When it comes to the meaning of fan Web sites, one explanation could be Jenkins' (1992) description of fandom in relation to television (which could easily be replaced by other phenomena) as "an institution of theory and criticism, a semi-structured space where competing interpretations and evaluations of common texts are proposed, debated, and negotiated and where readers speculate about the nature of mass media and their own relationship to [them]" (p. 86). Even though this and most other analyses of fandom look at communities of a different kind, communities mainly occupied with discussing the doings of characters in a soap opera or the line-up of a soccer team, the term is still adequate as the main activity also of the *AC* fan Web sites is to exchange information, criticism, and interpretations regarding a publicly available fantasy environment where people's doings are in focus. Here, engaged players exchange information about how to do certain things in the game (e.g., how to dye virtual clothes, the best way to deal with a specific monster, why bother cooking in the game), publish texts about and screenshots from the game, and express their ideas about what changes should be made to the game by the developers. The *AC* fan Web sites can thus be seen as an asynchronous communication and information technology, connecting players and distributing information within the community, but also an informal communication channel to the developers, expressing the players' thoughts about the game.

An analysis of the survey data showed that a large percentage (52.5%) of the *AC* players frequently visit "nonofficial *AC* Web sites" (Web sites hosted by others than MS), while a smaller percentage (45.9%) frequently visit "official *AC* Web sites." There was a significant difference between male and female respondents, such that there were more male respondents (59.6%) reporting that they frequently visit nonofficial Web sites than female respondents (47.4%), while there was no significant gender difference regarding official *AC* Web sites. However, in relation to time spent playing the game as well as engagement in an allegiance, one can find a significant difference such that the longer people have played the game and the more they are engaged in allegiances, the more they visit *AC*-related Web sites.

We also asked the respondents about reasons for visiting *AC*-related Web sites (official and nonofficial), and the reasons were found to be appreciated in the following order: find information (77.4%), view own and others' texts and screenshots (25.2%), participate in online chat (22.4%), and other (5.4%). A difference was found such that players who were more engaged in allegiances reported significantly higher on all reasons apart from "finding information" (on official and nonofficial *AC* Web sites) and "other" (on official *AC* Web site). We also found a gender difference, such that males reported to a significantly higher extent than females that they visit *AC* Web sites to "view own and others' texts and screenshots" (official and nonofficial Web sites) and "find information" (only nonofficial Web sites). Female respondents, on the other hand, scored significantly higher on "find information" on official *AC* Web sites.

A significant correlation between Web site hosting and time in the game was found, such that people reporting long time in the game also reported, to a higher degree, that they hosted an *AC* Web site. The same pattern was shown regarding allegiance membership, such that players who have more characters in an allegiance are also more likely to host an *AC* Web site. This is not a surprising finding, as *AC* fan Web sites are usually hosted by one or several members of an allegiance and used for organization of allegiance activities, distribution of information, and the like. Of the respondents reporting that they host an *AC* fan Web site ($n = 777$), 18.0% reported that they hosted it for the benefit of the allegiance.

What can be concluded from the survey results reported above is that a large number of *AC* players seem to use official and/or nonofficial Web pages mainly for finding information and that quite a few players host *AC*-related Web sites themselves. Web sites thus seem to be important to *AC* players, something that is also reported in the survey. Only 2.5% of the respondents found *AC* Web sites unimportant or very unimportant to their engagement in *AC*, while as many as 73.6% found Web sites important or very important. There was no significant gender difference or difference between players more and those less involved in allegiances. However, a significant correlation between Web site hosting and time in the game was found, such that people reporting long time in the game also reported that they find *AC* Web sites more important for their engagement in the game.

Two aspects emerging from free-response answers were considered important: (a) the community aspect, bringing people together for sharing experiences and knowledge; and (b) the information aspect, being the main information channel, the newspaper of *AC*, collecting and distributing all sorts of information about the game to its players.

The conclusion that can be drawn from the results regarding additional media use and fan Web sites is that many players seem to feel a need for complementary media (including Web sites related to the game) in order to be able to play the game in a way that they feel is enjoyable. The game itself does not, according to the survey respondents, provide a sufficient communication system, nor does it include all necessary information about the game. Therefore, a need for additional, out-of-game media has arisen, generating perhaps not only, to use social-network theoretical terms (Turner, 1991, pp. 550–553), *more* ties to *more* people, as discussed above, but also *stronger* ties between people, consisting of an intense exchange of information via multiple communication channels. More social interaction and creating a Web community was also the intention of the developers: "*Not* providing help or providing a difficult quest, or so, can actually increase interaction." And, "We try to foster a web community by supporting fan sites."

SHARED FANTASY: ROLE-PLAY AND GENDER-PLAY

Role-play and gender-play are activities in online computer games that some but not all players are engaged in. How many (and which) *AC* players role-play their characters? How common is it that players have characters with a different gender than their real-life gender? And, what does all this mean? Game technology and design make these activities possible, but why do people do it? And, what might these activities bring about in the game?

Gender-play or gender swapping is a much-talked-about and researched Internet phenomenon, apparently taking place in such different online environments as chat groups, MUDs, and online computer games (see, for example, Bruckman, 1993; Turkle, 1997; Yee, 2004). Anonymity was, as already mentioned, much discussed in the early days of the Internet and part of the fascination was unquestionably the possibility to interact with other people acting as someone else, as being of another gender and sometimes even of no gender at all. Lorber and Farrell (1991) described gender as a social institution, and gender characteristics as means according to which we define ourselves and others. Therefore, to break loose from gender (if ever possible) in an online environment would be to become emancipated from the norms controlling the real-life gender and to enjoy, for a moment, the freedom of the opposite sex. Or not?

In the following we will present the results regarding the extent to which *AC* players perform gender-play, that is, whether they have characters with another gender than their real-life gender, and why they do so. We also try to draw some conclusions about how gender-play as well as role-play might influence the social interaction in the game on a more general level. However, first let us look at role-playing in *AC*.

Role-Play

The question is: Does the game qualify as a MMORPG or is it just a MMOG? To find out to what extent *AC* players role-play and their attitude toward other players' role-play, we presented four statements to the respondents in the survey to either agree or disagree with, rating their answers on a 5-point scale: (a) "I role-play the majority or all of my characters," (b) "People who role-play extensively bother me," (c) "I get annoyed when people drop out of the role-play and bring up out-of-the-game issues (e.g., other games, movies, work/school)," and (d) "I sometimes drop out of role-play myself."

An analysis showed that 36.6% of the respondents reported that they agree or strongly agree with the first statement, meaning that a good third of the sample seems to be engaged in role-play. One third, 33.4%, reported that they disagree or strongly disagree with the first statement, and the remaining 30% that they neither agree nor disagree.

Regarding the second statement, a majority of the respondents (87%) reported that they do not mind people role-playing. Only 13% agree or strongly agree with the statement.

The third statement, asking how annoying the respondents think it is that other players break the role-playing illusion by talking about out-of-the-game issues, received similar answers. A majority of the respondents (88.4%) reported that they do not mind people dropping out of the role-play.

The fourth and last role-play statement, asking whether the respondents sometimes drop out of role-play themselves, gave similar results. A good majority of the respondents (91%) report that they agree, strongly disagree, or neither with the suggested statement, meaning that people seem to frequently drop out of role-play, and without any major concerns.

The only significant gender difference in relation to role-play was found in relation to the third statement, which addressed annoyance about other players dropping out of the role-play. Female respondents reported a lower sense of annoyance than male respondents. Concerning the relationship between group membership, time in the game, and role-play we found significant differences, showing that players who are more engaged in groups (allegiances) seem to be more tolerant toward other players dropping out of the role-play, and more often drop out themselves, than players less engaged in group play. Almost the same pattern appeared when looking at whether time in the game has an influence on how much players role-play and their attitude toward role-playing in general. The longer time a player has played the game, the more characters are role-played, but less seriously it seems, as the attitude toward role-play dropouts seems to become more relaxed as time goes on.

The findings regarding one's own and other players' role-play thus suggest that a relatively large number of *AC* players seem to be role-playing most of their characters and do not mind other players' role-play. However, what was also found, and may seem a bit contradictory, is that *AC* players do not seem to take role-playing terribly seriously. Dropping out of the game is not considered a serious problem, either when they do it themselves or when other players do it. The results show that time in the game has an influence on role-play, such that more experienced players role-play more than inexperienced players. Experienced players are also more tolerant toward role-play dropout, an attitude they share with highly socially engaged players. These results have certain similarities with findings from a study of language meetings in an Internet-based SVE, showing that experienced users are more tolerant toward language-play (dropout from the normal language use) than less experienced users (Axelsson, Abelin, & Schroeder, 2003).

One can thus conclude about role-play that a large number of *AC* players seem to be engaged in the shared fantasy of role-play, but that they also seem to be tolerant toward players that, for one reason or another, disrupt this fantasy. One interpretation of these results could be that the game is highly engaging, making people lose themselves in the virtual world, at the same time as the offline world is never far away.

Gender-Play

To find out to what extent *AC* players play the game using characters with another gender than their real-life gender, we asked the following question: "Do you have any characters that have a gender different from your real-life gender?" followed by a multiple-choice question ("If yes, why?") asking for additional information about the gender-play phenomenon. The predefined options were the following: (a) "I wanted to try out what it would feel like being seen and treated as a woman/man by other players," (b) "I thought I would obtain certain advantages in the game by choosing a female/male character," (c) "For role-play purposes," (d) "I had no real reason for my choice," and (e) "Other." The respondents were also given the possibility to state, as free responses, what kind of advantages they wanted to obtain as well as what other reasons they had for their gender-play.

An analysis showed that a large part of the respondents (47.9%) report that they have characters with the other gender (male characters if female player and vice versa). There was a significant gender difference, such that male players reported to a higher extent (48.7%) that they gender-play than female players (42.3%). The reasons for gender-play reported by the respondents varied. A larger proportion of female respondents (14.7%) than male respondents (11.9%) reported that they wanted to try out how it would feel to be seen and treated as the opposite sex. On all other reasons, male respondents reported slightly higher, apart from the last one ("Other"), which 28.8% of the females chose and only 20.9% of the males.

The two most common additional free response reasons were that the respondent had been fond of a specific name and had to create a suitable character to fit that name, and that a family member or friend occasionally used their account and wanted a character that suited their real-life gender.

However, when analyzing the free responses describing the advantages that the respondents stated that they wanted to obtain by using the opposite gender, we found several interesting differences between male and female respondents. First, while most male respondents said that they played females in order to attract other male players in order to receive their help (e.g., gifts, money, advice), most females said that they played males in order to *avoid* attracting male players. One male respondent explained his reason as follows: "By making a pretty looking female character i thought i could appeal to male personality and swear to male characters and get free stuff from them then break and tell them how much they got screwed and stuff" (male, 20). One female respondent described her choice, playing a male character, like this: "People pester me a lot and/or come onto me as a female character. Playing a male character gives me a bit of a break from that" (female, 28).

However, it is not only the *material* advantages that male respondents report that they are after in playing female characters: They also want to be approached differently and treated differently by the other male players, something they believe will happen if they interact with others by using a female character. One male respondent stated it very strongly, as follows: "females get more respect!!" (male, 15).

Female respondents, on the other hand, see certain interactional advantages with playing a male: "Individuals in game treat you differently as a male. I receive more recognition as a male character than a female and people allow me to take a more active role as a male" (female, 23).

As discussed elsewhere (Schroeder, 1997; Taylor, 2002a; Axelsson et al., 2003), the use of jargon (a typical in-group language understood and spoken by only a few) by a small group of users/players in a SVE or an online computer game strengthens the ties between the group members. Because jargon typically is a common element in role-play, one could assume that the role-play in *AC* also has some community-strengthening implications, at least for the players participating in the role-play. As should perhaps be emphasized more often, there is a flip-side to the coin of inclusive group elements like jargon, namely that if some people are being *included*, others are being *excluded*.

Gender-play, on the other hand, can be interpreted more as a managing activity in the game that players (many but not all) carry out in order either to gain obvious material advantages or to redirect the attention (and actions) of other players in a direction that is beneficial to the player him or herself. However, these activities, role-play and gender-play, would never be carried out unless the game itself made them possible by providing the players with flexible character creation and with anonymity of real-life identity and gender.

DISCUSSION OF THE RESULTS AND CONCLUSIONS

Through the analysis of *Asheron's Call* players, we have seen that there are a variety of reasons that people choose to play online role-playing games as opposed to their offline counterparts. We have seen how people's reasons could be grouped into helping, cooperation, and hostility.

Asheron's Call provides many different social technologies to support these aspects of people's play. We found that fellowships provide mechanisms like a private chat channel and a distinct color for fellows on the radar. We found that allegiances provide longer-term benefits with a small proportion of experience points gained by vassals accruing further up the allegiance hierarchy. The provision of Darktide, a world where players can kill each others' characters, provides a social outlet (albeit in some sense negative) for those players who enjoy hostility. The technologies encourage beneficial social behavior and try to contain destructive or antisocial behavior.

Chan and Vorder's chapter on (MMOGs) (see chap. 6) provides a sense of the breadth of these games, and we have shown how one subgenre can successfully provide players with meaningful social exchanges. Unlike *The Sims Online* (see chap. 21), the technology provided in *AC* fitted in with the participants' varying styles of play and online relationships.

Gamers in our survey used technologies outside the game to support their online game relationships; instant messaging, email, and the phone were some of the additional technologies mentioned. The social features of the game can be seen as a catalyst for these relationships between players, but the diversity of people within a friendship group and the affordances of communications technologies lead to a diversity of these technologies. Although there is a temptation to try to provide sufficient communications features within the game and keep all communications in-game (as advised, for example, in *Habitat*), we found that the range of requirements and preferences, up to and including face-to-face meetings, makes that impossible. In fact, we have seen that different communications technologies demark different cliques within a friendship group so that people choose a specific out-of-game technology to communicate with a friend purely because they know that the friend prefers that technology.

Using out-of-game communication technologies could be considered a threat to the role-play aspects of the game. But our study shows that many groups who consider the role-play important are also tolerant of those who drop in and out of role.

We have presented these results through the quantitative and qualitative analysis of a large survey of *Asheron's Call* players, and we would like to end with a few of their summaries of whether *Asheron's Call* had a positive or a negative effect on their lives. Many of the respondents echo the themes we have laid out in this chapter, highlighting their enjoyment of *Asheron's Call* as one of the communication media they use with their old and new friends. We will let them have the last words: "It's fun to meet new people from different places." "*Asheron's Call* offers entertainment that my wife and I both love and can share." "It is something to share with my friends."

REFERENCES

Asheron's Call. Turbine Entertainment Software Corp., at http://ac.turbinegames.com.

Axelsson, A. S. (2002). The digital divide: Status differences in virtual environments. In R. Schroeder (Ed.), *The social life of avatars: Presence and interaction in shared virtual environments* (pp. 188–204). London: Springer.

Axelsson, A. S., & Regan, T. (2002). *How belonging to an online group affects social behavior—A case study of Asheron's Call* (MSR-TR-2002-07). Available online at www.research.microsoft.com/scripts/pubs/view.asp?TR_ID=MSR-TR-2002-0.

Axelsson, A. S., Abelin, Å., & Schroeder, R. (2003). Anyone speak Spanish? Language encounters in multi-user virtual environments and the influence of technology. *New Media and Society, 5,* 475–498.

Bailenson, J. N., Blascovich, J., Beall, A. C., & Loomis, J. M. (2003). Interpersonal distance in immersive virtual environments. *Personality and Social Psychology Bulletin, 29,* 1–15.

Bakardjieva, M. (2001). Becoming a domestic Internet user. *Proceedings of the 3rd International Conference on Uses and Services in Telecommunications,* 2001, Paris, June 12–14.

Becker, B., & Mark, G. (2002). Social conventions in computer-mediated communication: A comparison of three online shared virtual environments. In R. Schroeder (Ed.), *The social life of avatars: Presence and interaction in shared virtual environments* (pp. 19–39). London: Springer.

Bruckman, A. S. (1993). Gender swapping on the Internet. *Proceedings of INET '93.* Reston, VA: The Internet Society. Available at http://www.cc.gatech.edu/~asb/papers/gender-swapping.txt.

Curtis, P. (1992). Mudding: Social phenomena in text-based virtual realities. *Proceedings of Directions and Implications of Advanced Computing, DIAC'92,* Berkeley, California, no pp.

Garton, L., & Wellman, B. (1995). Social impacts of electronic mail in organizations: A review of the research literature. *Communication Yearbook, 18,* 434–453.

Haddon, L. (2004). *Information and communication technologies in everyday life: A concise introduction and research guide.* Oxford, UK: Berg.

Haythornthwaite, C. (2000). Online personal networks: Size, composition and media use among distance learners. *New Media and Society, 2,* 195–226.

Hudson-Smith, A. (2002). 30 days in Active Worlds—Community, design and terrorism in a virtual world. In R. Schroeder (Ed.), *The social life of avatars: Presence and interaction in shared virtual environments.* (pp. 77–89). London: Springer.

Jenkins, H. (1992). *Textual poachers: Television fans and participatory culture.* New York: Routledge.

Lorber, J., & Farrell, S. A. (1991). *The social construction of gender.* Newbury Park, CA: Sage.

Schiano, D., & White, S. (1998). The first noble truth of cyberspace: People are people (even when they MOO). *Proceedings of CHI 98, Los Angeles, USA,* April 18–23, 352–359.

Schroeder, R. (1997). Networked worlds: Social aspects of networked multi-user virtual reality technology. *Sociological Research Online, 2.* Available online at http://www.socresonline.org.uk/2/4/5.html.

Schroeder, R., & Axelsson, A. S. (2000). Trust in the core: A study of long-term users of Active Worlds. Paper presented at *Digital Borderlands: A Cybercultural Symposium,* Norrköping, Sweden, May 12–13.

Slater, M., Sadagic, A., Usoh, M., & Schroeder, R. (2000). Small group behavior in a virtual and real environment. *Presence: Teleoperators & Virtual Environments, 9,* 37–51.

Slater, M., & Steed, A. (2002). Meeting people virtually: Experiments in shared virtual environments. In R. Schroeder (Ed.), *The social life of avatars: Presence and interaction in shared virtual environments* (pp. 146–171). London: Springer.

Social Computing Group, Microsoft Research Corp. http://research.microsoft.com/scg.

Taylor, T. L. (2002a). Living digitally: Embodiment in virtual worlds. In R. Schroeder (Ed.), *The social life of avatars: Presence and interaction in shared virtual environments* (pp. 40–62). London: Springer.

Taylor, T. L. (2002b). Whose game is this anyway?: Negotiating corporate ownership in a virtual world. In F. Mäyrä (Ed.), *Computer Games and Digital Cultures Conference Proceedings*. Tampere: Tampere University Press, no pp.

Turkle, S. (1997). *Life on the screen: Identity in the age of the Internet*. London: Phoenix.

Turner, J. (1991). *The structure of sociological theory* (5th ed.). Belmont, CA: Wodsworth.

Yee, N. (2004). *The Norrathian Scrolls*. Available online at http://www.nickyee.com/eqt/home.html.

21

What Went Wrong With *The Sims Online*: Cultural Learning and Barriers to Identification in a Massively Multiplayer Online Role-Playing Game

Francis F. Steen, Patricia M. Greenfield, Mari Siân Davies, and Brendesha Tynes

University of California, Los Angeles

When *The Sims Online* was launched in mid-December 2002, expectations were skyhigh. "The Sims games are always the best," a beta tester wrote in October (pizan36, 2002). "I think this is the best sim game ever made," another chimed in, screaming; "Dude I MEAN WHATS BETTER THAN PLAYING A GAME WERE YOU LIVE AND STUFF ONLINE" (snake72, 2002). Journalists also struggled to find the right superlatives. *Time* called it a "daring collective social experiment that could tell us some interesting things about who we are as a country" and proclaimed, "We're about to witness the birth of Simulation Nation" (Grossman, 2002). *Newsweek* issued a special report on "The next frontiers" with *The Sims Online* on the cover, writing, "America's hottest PC game is moving to the Net, where thousands of players will interact and live virtual lives. Is this the future of home entertainment?" (Croal, 2002). Subscriber numbers, however, never reached into the hundreds of thousands seen in other successful multiplayer online games (Woodcock, 2005). In the following, we examine the technical, psychological, and social dynamics of *The Sims Online* to understand why the game has not lived up to its expectations.

Creating Culture Through Social Interaction and Communication

One of us (Greenfield) observed the spontaneous rise of a series of collaborative cultural developments in the aftermath of the Northridge earthquake that hit Southern California on January 17, 1994 (Greenfield, 1997). It looked like human beings were adapting to new physical and social conditions produced by the violent earthquake by co-creating new cultural practices and meanings. She wondered whether *The Sims Online*, a massively multiplayer online role-playing game in which each screen agent (that is, a graphically represented character) is controlled by a real person, would also manifest rapid cultural evolution as thousands of players prepared to begin "living" in the simulated Sims world, a world of houses, clothing, jobs, and entertainment. As *Newsweek* noted, "*The Sims Online* stands out because social interaction is the game's *raison d'etre*, not an afterthought" (Croal, 2002, p. 52). With the collaboration of Brendesha Tynes and the research team at Children's Digital Media Center, UCLA, she initiated data collection on *The Sims Online* from the first month that the game was open to the general public. The intention was to have a laboratory in which processes of cultural evolution could be observed. By starting participants at the game's beginning, Greenfield hoped to observe not just adaptation to an existing culture, but the actual creation of a culture from scratch.

Because the essence of culture is a shared world of meanings, activities, norms, and a constructed physical environment, human beings create culture by a process of co-construction, that is, through processes of social interaction and communication (Greenfield, 1997). Hence, the initial research plan was to look for evidence of culture formation by studying the conversational discourse that would take place in the game. Greenfield and Tynes planned to make the qualitative study of conversation among players a key method of study. Indeed, at the outset of the Sims study, they were in the process of utilizing this methodology to analyze the culture of teen chat rooms (Greenfield & Subrahmanyam, 2003; Subrahmanyam, Greenfield, & Tynes, 2004; Tynes, Reynolds, & Greenfield, 2004). When *The Sims Online* first became open to the public, they heard from a video-computer technician who was helping us set up the study that *The Sims Online* was simply a 3-dimensional chat room. Although he put forth this characterization as a complaint, his description made the game sound perfect for their purposes. However, as the reader will see, the chat analogy did not turn out to be an accurate one. When Steen and Davies joined the project in early 2004 and we began to examine the assembled materials, the team quickly determined that the original analytic strategy had to be very much altered.

Background: *The Sims* and *The Sims Online*

In their press release at the launch, Electronic Arts, the publisher of *The Sims Online*, noted that "The Sims is the top-selling PC game of all time," selling more than 20 million units (Electronic Arts, 2003); according to insiders, they initially projected 400,000 subscribers to the online version the first year (Woodcock, 2005). Over the Christmas season, subscriptions immediately rocketed to 80,000; this was the point at which our study began.

The number of subscriptions is, of course, economically important to the game manufacturer. However, it is also critical to the game play itself. As a Massively Multiplayer Online Role-Playing Game (MMPORG), *The Sims Online* permits thousands of subscribers to play at once and interact with each other. This feature distinguishes it from other types of online gaming (see Chan & Vorderer, chap. 6, this volume, for a typology of online gaming).

Avatars and Player Control

As in *The Sims Online* (*TSO*), the offline Sims game offers an electronically generated, 3-dimensional world in which human figures, called sims, eat, sleep, build houses, work, and have roommates. While this may sound exotic, it is merely a virtual parallel to playing with "action figures" or "dolls." As in many facets of real life, in both versions of *The Sims*, the precise sequence of events is determined by the player or players, not by the program.

In the highly popular offline version of *The Sims*, the player controls all the virtual characters. Using the words of the electronic gaming community, the player "plays God" in *The Sims* world. The control, however, is not direct. In contrast to most action games, there is no one-to-one correspondence between controller movement and screen character movement (Kirk, 2004). Instead, the player instructs the sim to perform a particular act by selecting from a contextual menu of options, as if commanding a robot.

The Sims Online also utilizes this robotic form of character control. However, in the online version, the perspective is not the omniscient perspective of God. Instead, each screen character is an *avatar*; that is, it is controlled by and "channels" its own unique human being. Interaction is not between robots controlled by a single omnipotent, godlike player. Instead, character interaction takes place between human beings channeled by individual onscreen avatars. How would the interaction of these avatars create cultural activities, norms, and meanings? That was the question with which the research began.

Environment and Building: A Different Type of Player Control

The basic environment of the offline Sims is an invariant scene, a suburban subdivision on which the game invites the player to build one or several houses, and to move ready-made sim families into them. In *TSO*, the player chooses from nearly a dozen housing developments, virtual towns located in attractive artificial geographies and inhabited by thousands of avatars. Alphaville is the first and largest of these, situated on a coastal plain and archipelago and home to some 7000 sims. A player can maintain up to 3 simultaneous avatars, each in a different city.

The core of the offline Sims was the construction and decoration of house and grounds, and this code was inherited by *TSO*. While the avatar inhabits the virtual world, controlled indirectly by robotic instructions, the player has direct control over the act of building the house and landscaping the property. Construction materials are purchased from virtual shopping windows and placed on the building site at the click of a mouse. The task of construction itself is thus direct rather than robotic, relying on the familiar click and drag system. There is a one-to-one relation between hand movement and screen action (Kirk, 2004). In sum, game control is robotic (i.e., indirect) for the "living" avatars and direct for inanimate building materials.

This simple contrast is emblematic of the game psychology as a whole. Indeed, our analytic journey took us beyond culture and into more psychological issues. Initial issues included identity, identification, and social exploration. Later analysis led us to focus on the psychological processes elicited by the game that appeared to stand in the way of cultural evolution and cultural creativity. After we realized how many cultural norms were already programmed into the game, we turned our attention from cultural evolution to cultural learning.

As will become clear below, our first finding was that *The Sims Online* was failing to fulfill the initial high expectations of popularity. We ultimately traced this failure to the differing requirements of player–character identification and strategic control in the offline and online versions. Player control was our analytic key to understanding what went wrong with *The Sims Online*. We will make the argument that, whereas robotic control seems psychologically ideal in a stand-alone virtual society, direct control would work much better, both psychologically and socially, in a multiplayer online virtual society.

METHOD

Participants

The participants were a 23-year-old woman, a 2nd-year graduate student (KM), and a 28-year-old man (SH), a freelance producer and part-time actor. They knew each other before they started playing and occasionally arranged to meet each other in Alphaville. Both participants had grown up playing electronic games and were experienced players of MMORPGs. In terms of past Sims experience, SH noted that he had played Sim City and the offline, stand-alone

Sims, "but not avidly." KM described herself as a recovered addict of the offline version of *The Sims*. Both were curious to try the online version. We take our participants to be representative of the majority of *TSO* subscribers in age, SES, and general game experience. As we had intentionally built into our design, they also represented a range of experience with the offline version of *The Sims*.

Procedure and Data

The participants were asked by Greenfield and Tynes to record their play over a year's period with a camcorder-PC set-up provided for them; they were also asked to keep a handwritten journal of their play detailing their goals, gaming activities, strategies, and general reactions to aspects of game, other players, or gaming session overall. They received instructions, shown in the Appendix. Children's Digital Media Center, UCLA, under whose aegis this research was conducted, bought their subscription. Participants received $100 per month for playing and $1 per page for their journal entries. Tynes collected their journals every 3 months.

From KM we received 17 tapes of play capture over the period of a year from December 30, 2002 through December 29, 2003. KM's total recorded game play comprised 15 hours and 53 minutes. From SH we received 15 tapes of play capture over the period of a year from December 19, 2002 through December 24, 2003. SH's tapes comprised 8 hours 9 minutes of play time. A third player, LC, played once and then had to stop because of her college commitments (she was a freshman). We are not using her session in the present analysis. One of us (Davies) supplemented the first two players' data by playing herself and keeping notes from January through March 2004.

In addition to these primary sources of data on game play, we have benefited from the extensive coverage of *The Sims Online* and its design provided by the online gaming publications GameSpot.com, GamePro.com, and Game-Revolution.com, as well as from the player reports and conversations on threads devoted to *TSO* beta testing at Gamers.com.

ANALYSIS

Our data collection methods were planned with qualitative data analysis in mind. We checked the journals against the game capture videos at several points in time for each primary participant and found that the journals were an accurate reflection of what had transpired in the game. In addition, the journals provided the players' interpretations of their game play and a record of their changing motivations.

We began our analysis by going through the journals and Davies' notes to see what issues were coming up for the participants. In line with the original motivation of the study, we were on the lookout for evidence of cultural evolution. However, the journals provided more material relevant to the psychosocial issues of social motivation, social interaction, identity, and identification.

We also used the journal notes to understand some trends within the game as a massive multiplayer phenomenon. As it became clear that our participants' gaming experience (and that of thousands of other players) had been very different from what we had anticipated, we focused in on the surprising dynamics of avatar interaction. Finally, we were interested in the process by which a new player learns how to play the game, understood primarily as a process of cultural learning.

We analyzed the discourse of the journals and the transcriptions of game play from the video recordings, supplemented by the online discussions of other *TSO* players, in order to gain evidence relevant to all of these issues.

INITIAL FINDINGS AND NEW QUESTIONS

Our Participants and Their Multiplayer World: The Primacy of Social Interaction

In the months after Christmas 2002, subscriptions to *The Sims Online* peaked just above 100,000, not an unimpressive number, yet far short of predictions (Woodcock, 2005) and not sufficient to justify the financial investment. Electronic Arts noted in their fiscal 2003 report to shareholders "a pre-tax charge of $67 million as a result of impaired assets and restructuring costs in the online division" (Electronic Arts, 2003, p. 12). By April of 2004, the company estimated 57,500 subscribers (Glassman, 2004); at the end of the year, in spite of an international marketing campaign, the number had likely dropped to around 35,500 (Woodcock, 2005). Within the gaming community, there is a widespread expectation that the game will be terminated within another year.

This generally negative response to the game was reflected in our participants' declining interest in playing. Recorded sessions gradually became shorter and less frequent. From December 2002 through July 2003, all of KM's taped sessions comprised at least 1 hour of game play. From September through December, recorded tapes were all under 42 minutes. For SH, from December through May, 4 of his first 6 recorded video tapes were 1 hour or longer. After that, all his recordings comprised less than 1 hour of tape. While the instructions implied that we wanted the players to record their play at least once a month, both players skipped months. SH experienced some difficulties recording, which may explain part of his missed 3 months. KM skipped 1 month. Even paying them to play was not enough to get them to play without fail. The apparent fall-off in interest suggested new questions: What had gone wrong? Was something very different from the offline version of the game? Why was the online version apparently less engaging than the stand-alone version of the game?

In an interview just before the release, Will Wright, lead designer and creator of all the Sims games, told Gamespot.com, "I wanted to make parts of the game boring so you'd be encouraged to talk with others" (Keighley, 2002). An initial look at our journals indicated that Wright had succeeded in making much of the game boring. However, for some reason the game's interactive features did not have the intended effect of spurring a rich conversation. We therefore utilized our participant journals, video captures of their game play, and researcher notes to document the nature, quantity, and motivations for social interaction and communication.

Cultural Evolution Versus Cultural Learning

In this material, we did not find evidence of the original topic of interest, cultural evolution. It turned out that too much "culture" had already been programmed into the game by the designers. In an interview during the alpha stage of the design process, Will Wright spoke of "an economy of motives" (Green, 2002) intended to shape player behavior. A similar economy of motives in the offline version of *The Sims* reminds the player when her sims need rest, company, food, and so on, mimicking the nurturing needs of virtual "human pets." Applied to a channeled avatar in a massively multiplayer game, however, these imposed "motives," along with other incentive systems within the game, constrained the individual player's agency. They tended to crowd out the possibility of the formation of new cultural norms and values (but see Steen, Davies, Tynes, and Greenfield, 2005). Thus, in *TSO*, the rules of the game to a significant extent stood in for and functioned as a shared culture.

Further reducing the in-game possibility space of our participants was the fact that thousands of alpha and beta testers had already been playing the game for months, building up a large number of properties and creating a modified cultural environment into which newcomers, or "newbies," had to fit. An early review spelled out the emerging culture of the game:

> Since a player can earn money simply by enticing other players to congregate on their property, and because all the other players truly want to do is earn money, the object of the game is reduced to building—not a "house" in which your Sim will live, but a labor camp in which other Sims will come to earn money. Providing beds, showers, food and a pool table persuades your guests to stay longer and spend more of the money they are earning, owing their souls to the company store, so to speak, and never truly needing a place of their own. The result is a "city" in which nearly every house is a sweatshop. (White, 2003)

In case this cultural code was not immediately apparent to a new player, Electronic Arts provided personal guidance. As evidence of the way the "culture" was programmed into the game, consider this phone message from the game manufacturer that was spontaneously left for Greenfield when the research project initially enrolled in *The Sims Online*:

> "Hi, This is a message from *The Sims Online*, with some tips for playing the game.... Here are some helpful tips. Be sure to choose a skill to specialize in early, and find a job object that maximizes payout based on that skill."

For our players, then, the game was less an opportunity for creating new culture than a fully determined, preexisting culture to which they had to adapt. We therefore shifted our analytic focus from culture creation to cultural learning.

Identity, Identification, Avatar Control, and Perspective

Our player data suggested that the psychological processes of player–avatar identification interacted with the technicalities of game control, described earlier, to raise some interesting issues. Identity, identification, and game control therefore became key topics in our analysis. We adopt the general perspective that in *The Sims Online*, as the name suggests, the player is invited to treat her own avatar, the other avatars she encounters, and the entire computer-generated world within the game as a simulation of social reality. In contrast to a novel or a film, where the course of action is determined by the author or filmmaker, in *The Sims Online*, each player controls one of the characters within the simulation. In describing the emotional and imaginative engagement of the player in the fortunes of her and others' avatars, it is therefore not appropriate to apply the respondent-as-witness model, which explains such involvement in fictive scenarios in terms of empathy (Zillmann, 1994).

Instead, we utilize a concept of identification-as-simulation along the lines proposed by Oatley (1992, 1994) and applied to fictive narratives, although explicitly modeled on computer games. In identifying with a character, the player treats the avatar, in certain limited ways, as if it were herself. In the present context, we focus on the player's willingness to treat the avatar's movements on the screen as if they were her own bodily movements in space. We also examine the player's willingness to adopt the avatar's epistemic, emotional, and intentional states by integrating the avatar's goals, obstacles, and resources into her own planning system (Oatley, 1994; Steen & Owens, 2001; Steen 2005).

In the following, we first examine the issue of identity by assessing the characteristics and our participants' interpretations of the avatars they created to play the game. Second, we look at the degree to which they identified with these avatars during game play. The theory and evidence that we will develop posits that greater identification occurs with direct rather than robotic control, and with a first-person rather than an omniscient or "godlike" perspective.

Interaction and Communication

The tapes and journals indicated that interaction and communication, the *sine qua non* for cultural co-construction, were much more rare and limited than anticipated. The reasons for this appear to be linked to the ironic nonessentiality of social communication and social learning built into the incentive structure of *The Sims Online*. We will try to make the case that the player control mechanism also contributed to this problem: Robotic control created a time and operational lag that was unfavorable to social interaction. In addition, we analyze the game's incentive structure and other rules aimed at generating sociality, and show that they had the unintended effect of discouraging a generative culture based on interpersonal relationships and player communication.

RESULTS: IDENTITY, SOCIAL INTERACTION, AND CULTURAL LEARNING

The First Newbie's Perspective

The Sims Online begins with the creation of an avatar. From a large array of alternatives, a player custom-builds his or her sim by selecting a gendered body with a particular skin tone, a head, and a set of clothes. Our first participant, KM, began play on December 30, 2002 by creating an *alter ego*:

> I created this character based on myself. It took me a while to go through all of the hair and outfits to pick one that I thought resembled me. (KM, p. 16)

This initial choice indicates that the player expects the avatar to function as an effective channel for her own identity. As the reader will discover, this common initial expectation was eventually violated for both our players.

KM entered the game with an extensive offline Sims game background, and had realistic expectations about the incentive structure and rules for getting ahead in the game. On her first day of play, she wrote:

> Today the goal for my character was to explore the game and figure out what was going on. Since I had played the sims before I knew that the two most important things to advance are money and skill points. I knew I had to get more skill points so I could earn some more money. (KM, pp. 16–17)

KM correctly assumed that the rules of the offline game had been ported to and inherited by the online version, effectively forming its cultural norms.

In spite of this knowledge, KM was surprised and explicit in reacting to the lack of social interaction and conversation within a supposedly "interactive" online game:

> I said "hi" to the other person playing chess, but little conversation happened. I noticed that no one in this house was talking everyone was just earning skill points. (KM, p. 17)

Six months later, KM realized that even conversation could appropriately be treated as a commodity in *The Sims Online*. On July 10, 2003, she wrote:

> Last I went to [illegible] P Clymat to talk to other people in order to increase my social meter. (KM, p. 75)

This is how the incentive system is designed to work: Players are encouraged to socialize. Yet the encouragement operates by converting the players' real-life desire to communicate with others into an avatar's mechanical "motive" to exchange speech bubbles, potentially devoid of any meaning or engagement, for points—the points acquired are the same with or without meaningful conversation. It is well known that extrinsic motives, such as rewards, reduce intrinsic motivation (Greene & Lepper, 1974); and this process could well have reduced motivation to converse. The reduction of a genuine human desire to converse and the increase of the motive to mechanically create speech bubbles makes the avatar less like a human being and less like oneself, therefore exposing the game to a disidentification of avatar and player.

KM early on encountered a related dehumanization of game avatars. Observing the paucity of conversation at a skilling scene, she wrote:

> It made me wonder if anyone was actually working the characters or if someone just plopped their Sim at a computer to get more skills and then left. (KM, p. 19)

Her suspicion that many of the avatars had been left unattended was strengthened when she realized it was a logical outcome of the structure of the game. Skill points and money were made by time-consuming and boring robotic sequences that required no attention:

> After about 10 minutes of playing chess, and seeing how long it took for the skill meter to go up I could see why no one was talking—probably no one was there! (KM, p. 19)

Here, then, was a weakness in the design of the online "chat room" of *TSO*: Many avatars are only intermittently attended, operating on autopilot while skilling and working. This creates the need, on the part of attending players, to weaken the link of identification between other players and their avatars, as the presence of one frequently does not entail the presence of the other.

Robotic control, plus omniscient player perspective, also makes conversation and social interaction unnatural, by opening up dissociations between player and avatar perceptual and epistemic states. The player has an omniscient visual perspective—an overview of the whole Sims world. However, the Sims avatars appear as though they have first-person perspectives (cf. O'Keefe & Zehnder, 2004); for example, the avatars can face in different directions. This situation leads to some social anomalies. We analyzed a recorded conversation at Lucky Luc's Slots, where the AJ, the proprietor's roommate, gives KM instructions on how to play a particular gambling game. During most of this extended exchange, the two avatars have their backs to each other. While a player can in fact see everything, he or she is also likely to utilize standard social meanings and interpret an avatar that talks to her with her back to her as disrespectful. Yet players do not actually control his or her avatar's orientation, nor does avatar perspective influence player perspective. By failing to face the person you are speaking with, you violate social convention and imply a lack of interest in the other—although it was clear from the content of the conversation that this would have been an incorrect inference.

Several times during the conversation, AJ made reference to gaming results that neither her nor her interlocutor's avatar could observe. This behavior breaks a cardinal rule of mental-state attribution that "seeing is knowing" and opens up a gap between the embedded, in-game perspective of the avatar and the panoptic perspective of the player.

A similar dissociation can be witnessed between intentional state and mechanical behavior, even resulting in a player leaving her avatar "body"! In one session of recorded game play, we observed KM giving instructions to her avatar to play the guitar, and then adopt a passive stance while the avatar played. Further into the same scene, KM says to the other players, "I am going to go too, cant stand listening to my own music." Here KM announces her intention to leave the room, and in fact zooms out of the building to end up in a bird's eye view of Alphaville, used to decide where to go next. Meanwhile, until it is commanded otherwise, her avatar stays behind and continues to play the guitar. Clearly, this kind of dissociation between intended action and mechanical behavior, not to mention the ability to leave one's on-screen body behind, seemed likely to result in a disidentification of player with avatar.

Indeed, these dissociations appeared to have the unintended effect of weakening the identification of the player with her own avatar. On June 19, 2003, after nearly 6 months of play, KM wrote:

> I started a new character today because I was bored of the old one and I wanted to try something different. I made a character that is nothing like myself—I made my character very weird looking, and it is a male. (KM, p. 63)

The creation of a new character may simply indicate a desire to try something new—as explained in Turkle's (1995) pathbreaking work, KM was using the game to experiment with different identities. At the same time, the fact that this character was "weird," of the opposite sex, and "nothing like myself" implies that KM's expectation of a high degree of avatar identification was no longer present; we might call the effect electronic alienation. This process of disidentification, which we will revisit in our second participant, may be caused in part by the common practice of going *afk* ("away from keyboard"), resulting after 15 minutes of no activity in *ato* ("avatar timed out"). At that point, the avatar is electronically removed from the scene.

In KM's diaries, we begin to understand why so little conversation takes place in *The Sims Online*: Typical game play is characterized by long absences from the keyboard, as the robotic work of skilling and making money is itself experienced as boring. Our analyses of her game play captured on video indicate that this boredom forms part of a series of dissociations between players and avatars. These dissociations act cumulatively not only to weaken the bond of identification between self and avatar, but also to reduce the perception of other avatars as representing real people, capable of meaningful interaction.

The Second Newbie's Progress

The diary of our second participant, SH, reveals a series of psychological stages of identity, motivation, and cultural learning. He began to play on December 19, 2002, within days of the release of the game. While he recorded his first game, he did not start his diary until a month later. In the first recording, we observe the creation of an avatar with an insect head. He is given the name Fred Mandible and wanders around in Alphaville, steadfastly ignored by other players in spite of his efforts to attract attention. SH clearly was not aware of the widespread practice of leaving your avatar at work while afk ("away from keyboard").

> On January 19, 2003, SH writes that Fred Mandible was an experimental character I developed in order to understand the game. This character has been retired today. I didn't really develop the character fully because I knew he wasn't going to last. (SH, p. 5)

It is unclear what he means by "developing the character fully"—is this a statement about his own emotional involvement, that is, a lack of identification with his insect-headed character? Or is it, more prosaically, about gaining skill points? At this point in the game, SH may be uncertain and open-ended about what it means to develop a character, but judging from his next remark, his conceptions are extremely rich. He continues the previous entry by introducing his new character Sammar:

> I chose to develop this character because he is the closest thing to my alter ego. I needed an outlet for that ego in order to help myself in real day to day life. I'm hoping that I'll be able to learn from my other self and take those characteristics that I feel I lack and forge them into my real life. (SH, p. 9)

This statement suggests that SH expects the game to provide the opportunity for a personal and social learning experience, one in which his virtual life will allow him to explore and to cultivate modes of being and responding to the world that he can subsequently incorporate into his own life in a selective manner. It is this goal that motivates his progression from a sim very dissimilar from himself, a character with an insect head, to a character that he will identify with, one that very explicitly represents himself. He also provides an update on the first experiences of his new avatar:

> So far "Sammar" has done fairly well in getting to know his way around. He has visited a few different places. He's gotten a roommate, "KM," who also is a beginner. And he's made a few extra bucks making pizza. Not too bad for a first day. His frustrations are easing too with all knew [sic] knowledge and info he gains. His main goal at present is to make enough money to build a party pad by the beach. (SH, pp. 9, 11)

The first sentence indicates an awareness of cultural learning—he must learn his way around this constructed, virtual world. Acquiring a roommate—none other than our other participant KM—provides welcome company. SH makes it clear that, at this point, social interaction is a key motive for him and lies at the heart of what he expects from the game. He envisions his short-term goal as building a house in an aesthetically attractive location in order to provide a venue for himself to host parties and thus engage socially with other players.

SH is highly cognizant of and positive about the quality of his social encounters, even though the video recordings show a very modest level of interaction with others:

> A lot of the people I've visited at their properties have been exceptionally nice. I imagine it has to do with their visitor bonus. The people I met in the pizza place are not nearly as friendly. It's amazing what greed will do. (SH, p. 13)

There is a mild irony in the conjunction of his warm appreciation of the owner's friendliness and his dismay at the unfriendly and impatient behavior of the pizza makers, given that he realizes both are caused by the same greed. Yet he is clearly learning the system of incentives that constitute the built-in rules of *TSO* culture.

On January 22, SH expresses satisfaction at his avatar's development:

> This character has made huge strides in his skill levels. Interaction between "Sammar" and the other characters has increased dramatically. His logic skill is up 2 notches and his cooking skill is up a notch and change. His interaction and friendship with people is increasing mostly due to his helpfulness in cleaning and other household duties. (SH, p. 17)

SH approaches the game with an anticipation that he will encounter a friendly and collaborative environment. He expects to be liked and appreciated by contributing to a common good. Although he speaks of his avatar in the third person, his level of identification is high:

> Sammar is feeling accepted in this community. He is still figuring out the finer details but it's coming along well. He aspires to make his skills at their peak and make as much money as possible. (SH, p. 19)

The attribution is striking: "Sammar is feeling accepted in this community." By reporting an emotional engagement in his avatar's social position, SH indicates a subjective, in-game experience of moving toward membership within a larger community. The implication is that Alphaville is a community, that this community either accepts or rejects newcomers, and that such acceptance or rejection influences the success of a sim. At the same time, he is acquiring the tacit culture that has been programmed into the game, realizing that success depends straightforwardly on gaining skills and making money.

The next day, it is this pragmatic level of cultural acquisition that predominates:

> He's figured his way around, and his skill levels are constantly increasing as is his money levels are increasing [sic]. He's building a friendship base that's making him money and skill. (SH, p. 23)

He has now started to think of friendships in passive, instrumental terms: They are "making him money and skill," and are not necessarily pursued for their own sake. This subtle transition is critical to understanding why we observed very little interaction in our participants' Sims Online videos. *TSO* has been set up to function as an economy in which earning money is the only means to succeed at the game, and casual chitchat is a waste of time.

An underlying reason that the conversational opportunities so carefully engineered into The Sims Online remain underutilized is that the game lacks a sufficient strategic complexity to make information a scarce commodity. *TSO* characters have little information to share, because their world is structured in such a manner that information is not a critical element of gaming success or, perhaps more importantly, gaming fun. As in real-life communities, the need for vital information transfer, provoking collaborative energy, may be a critical ingredient for the rise of generative culture, and may be a reason such culture is absent in *TSO*.

Although SH's diary demonstrates he is learning this foreign culture by immersion, its dynamics continually disappoint his formulated motivational goals of exploratory social identity development and interaction. Indeed, while his diary has been upbeat and positive, he does not return to the game for more than 3 weeks.

> When he returns to the game on March 18, 2003, SH reports: "Sammar" has built his skill levels, mostly mechanical & logical, and is making a decent amount of money making gnomes. He has a home now and is in the process of building it up to be a place where other sims can come to relax and make money. (SH, p. 25)

He now shows little emotional involvement with his avatar, whose activities in this session are directed not toward forming relationships, but on building skills and making money. However, he sees these activities as a temporary means to a more attractive goal of building a house. The purpose of this house is still to provide a place for others, but he no longer imagines they will come to party. Instead, they will come to his house to hang out and to make money. His cultural learning has deepened; he recognizes that social interaction for its own sake is a dispreferred mode of exchange in the cultural environment of *The Sims Online*. Nevertheless, his own motivations remain strongly altruistic:

> My characters main goal at present is to be a viable and successful character who can help other Sims in their money and skill earning endeavors. (SH, pp. 25, 27)

SH wants to contribute to the success of other sims, but first his sim must become a "viable and successful character." We now see the first acknowledgments that the game is not intrinsically motivating; rather, it is a means to an end projected further into the game. At this point, playing has become more like work:

> The game part of the Sims is somewhat boring because character development is almost in real time unlike other Sims games where I can fast forward through time. (SH, p. 29)

He is beginning to critique the game: It is not fun, and reaching the part of the game that would be fun takes too long. His previous assumptions about the community structure and social nature of the game have fragmented into an acceptance that moving forward in *The Sims Online* means chasing skill points and money rather than developing relationships in social interactions. Starting from ignorance, false assumptions, and misguided aspirations, SH is slowly making the culture programmed into *The Sims Online* his own.

Continuing his March 18 entry, SH reflects on the lack of player interaction. He feels dead-ended by limitations in the design of the game and begins making recommendations for improvements:

> I think an offshoot room where people can seperate from the game and chat or have some other activity like a Sims poker room would make the game far more interesting. (SH, p. 29)

At this point, rather than playing *The Sims Online*, SH is inclined to "separate from the game" and just play some other game with people online. *TSO* itself is "somewhat boring" for not encouraging social interaction:

> The game would be more condusive to chatting if email were accessible while playing to swap pics and personal info. A real possibility of meeting these people off line would get the place buzzing. (SH, p. 39)

The subtext here is that avatar encounters in the game do not really feel like real encounters. They are not emotionally satisfying or engaging and thus do not draw people in. His suggestions of introducing e-mail, swapping pictures, and meeting people offline indicates that he experiences the on-screen characters as poor representatives of the players' social selves and social agency: Identification has become unattractive and the channeling role of the avatars is failing.

After recording a little more than 4 hours of play over a period of about 2 months, SH lost interest in his "alter ego." His hopes of using *TSO* to explore and practice new character traits and behaviors in a social context were effectively quashed. He stopped playing for a few weeks, until on April 9, 2003 he began a new sim:

> I started Freakstick today to retry The Sims.com. Freakstick is an odd looking character. He has a skinless body. He sorta looks like the anatomy figure from highschool biology class. And his head is a mask that's part tribal part the big blue character in Monsters, Inc. played by John Goodman. I chose to develop this character to express my off the wall personality, as opposed to my other character "Sammar" who is whom I'd like to be in real life, an alter ego, if you will. (SH, pp. 33–37)

The player's goal has now been reformulated: To "express my off the wall personality" (p. 37) rather than to "learn from my other self and take those characteristics that I feel I lack and forge them into my real life" (p. 9). The goal retains an element of sociality, but now as a playful exhibitionism. Like KM, SH is using the game to experiment with different identities (cf. Turkle, 1995). Nonetheless, like KM's second avatar, Freakstick is explicitly a distanced identity, a sharp contrast to the *alter ego*.

While *Newsweek* thought sociability was the point of the game (Croal, 2002, p. 52), the cultural code programmed into the game contains a very different reward structure. It is clear from the way in which SH's entry continues that the goal of sociability has receded from view:

> Now that I've learned the main tricks and tips in succeeding in The Sims, I have a new way of going about things. I plan on amassing large amounts of mechanical and logical skill. Those skills have the greatest amount of financial profitability with the least amount of constant attention. (SH, pp. 37–39)

SH has learned that profit, not sociability, is the highest goal in *The Sims Online*. In spite of having understood the culture, however, he persists on May 7, 2003 in his own desire to just socialize:

> In a perfect world a seperate area just for chatting would be great. A setup with regional room choice would be optimal. (SH, p. 43)

Far from experiencing a desire to live in a virtual world, SH is looking for people he can meet in real life to motivate him to play on. SH still wants to exit the game to chat, get away from the skilling and working—in spite of the fact that the game has excellent built-in chat features. Two and a half months go by with no further game play.

On July 27, 2003 he goes on for a brief session to build skills. There were few people online, so the effort did not pay off as much as he had hoped. "Maybe there will be more people the next time I log on," he comments (SH, p. 45). He is now speaking of people as a simple means to speed up the gaining of skill points; he is no longer interested in socializing or meeting friends. The incentives of the game appear to have ground sociality out of him.

On August 4, 2003, SH feels the need to explain why he bothers to keep playing with Freakstick:

> I'm continuing this character because I have invested time to build his skills up and his money a little bit. (SH, p. 47)

His relation to his avatar has become increasingly detached—he is no longer interested in trying out new traits or even showing off. The avatar is spoken of as disposable and identification is marginal. His goal is subtly reconceptualized, even as he represents it to himself as unchanged:

> My ultimate goal, still, is to gain enough skill and money to build the ultimate house where I won't have to work at making money. Rather I earn money by collecting the revenues given to me by the Sims for visitors coming to my house. Also I will get residuals for every dollar that my guests make. (SH, pp. 47, 49)

SH is now fully converted to the cultural code of the game: It is not about human relationships at all, or about good intentions moving you forward within a community of shared norms and meanings, as in his earlier desire to help others. The cultural imperative of *The Sims Online* is to live off the labor of others, to become a "sim lord" (SH, p. 41), reducing the primary significance of human relationship to economic dominance through virtual sweatshops.

Having finally made the tacit culture programmed into the game explicit, however, SH proves unable to maintain interest in its values. Over the next couple of months, complaining that the "time it takes to build skill is a little overwhelming, not mention boring" (26 August 2003, p. 51), his game play trails off and he abandons the game.

CONCLUSIONS

What went wrong with *The Sims Online*? "We expected *The Sims Online* to be our flagship online subscription," the company confessed (Electronic Arts, 2003, p. 28). Gamers and the press alike anticipated success. "Finally it's here!" aquafan1 wrote on the beta tester board September 12, 2002. "An online game where you don't just kill things. The Sims series has to be the best set of games known to man. I mean what can be better than playing God? And now you can play with people all over the world! What more can you ask for?" (aquafan1, 2002).

The experience of our players provides some answers. Drawing on KM's diaries and recorded game play, we have documented the gradual discovery and internalization of five player–avatar dissociations. First, the dissociation of controller (mouse) movement from sim behavior was familiar to her from the offline version of the game. Second, in the online version, because money-making and skill-acquiring activities are so boring, this basic dissociation develops into a cultural practice of going afk, or "away from keyboard." Third, we witnessed KM's dissociation of her avatar's mechanical "need" to interact with other avatars from her own desire to socialize. Fourth, KM experienced the dissociation of her own intentions and her avatar's behavior, as the avatar persisted in playing the guitar after she had tired of it. This incident also involved a dissociation of player from avatar "body." Lastly, in her conversation with AJ at Lucky Luc's Slots, KM experienced the dissociation between the player's panoptic perspective and the avatar's embodied social presence. The cumulative effect of these dissociations, we propose, is to weaken the identification of the player with his or her avatar.

Using SH's diary, we chronicle how the player's own goals gradually are supplanted by the tacit cultural imperatives of the game. His initial expectations are of an open and free virtual community, in which he can experiment with new modes of being and behaving, where he expects to be accepted and appreciated for his altruistic behaviors, where avatars effectively and faithfully channel real people, and social interaction is a goal in itself. Starting by being ignored by other players, he is soon faced with the mind-numbingly boring necessity of skilling. Session by session, the realization mounts that social interaction in *TSO* is instrumental and devoid of intrinsic value. The last remnants of sociality pounded out of him, he finally adopts the game's cultural value of becoming a sim lord, sustaining himself off the labor of others.

Both our players progress from an *alter ego* avatar early in the game to a preference for a "weird" or alienated avatar toward the end. This disidentified avatar, we suggest, may function as the solution to the cognitive and emotional dissonance generated by the combined force of the five dissociations and the persistent, steamrolling pressure of the game's cultural imperatives. Gradually acquired, these slowly expel any desire for genuine social interaction from the player's own in-game aspirations. For our participants, what remained was not sufficient to motivate continued play.

The problems we discovered can be traced back to an imperfect transition from *The Sims* to *The Sims Online*. Robotic control, enjoyable when the player controls all the characters, as in the stand-alone Sims, generated a series of dissociations that hindered effective player–avatar identification in the massively multiplayer situation. The godlike power of *The Sims* could not be ported to *TSO* with its many interacting players. This analysis suggests that many problems could have been solved and the affordances for natural conversation and social interaction better provided if player control in *The Sims Online* had been direct rather than robotic. What we do not know is whether this key adaptation to the multiplayer environment would have been sufficient to overcome the chilling effect on social interaction—the *raison d'etre* of the game—of the materialistic cultural code, a code that was well accepted by players of stand-alone *Sims*.

Our study complements Axelsson and Regan's interesting chapter (chap. 20, this volume). Using survey methodology to study players of one particular online game, *Asheron's Call*, they confirm the strong social motivations of MMORPG players. Their findings emphasize the importance of appropriate technologies to support the players' desires, such as multiple channels of communication and the ability to collaborate in short-term fellowships as well as enduring allegiances.

Massively multiplayer online role-playing games are still young. The ongoing interplay of game designers and gamers create unique experiments at the intersection of individual psychology, social dynamics, hardware infrastructure, and computer code. A careful analytic study of these games may help locate some of the most fruitful and interesting neighborhoods in the emerging landscape of persistent online worlds.

ACKNOWLEDGMENTS

We thank the participants for making a year-long commitment to participating in this study. We also thank the National Science Foundation for funding the Children's Digital Media Center, UCLA, under whose auspices the research was conducted. Last but not least, Patricia Greenfield would like to thank Frank Evers for planting the idea that massive multiplayer online games are a major phenomenon that would be a highly significant topic of investigation; she is also grateful for his helpful guidance at the early stages of the research.

APPENDIX. INSTRUCTIONS GIVEN TO PARTICIPANTS

Sims Online Study

Introduction and Journal/Recording Instructions

We want to thank you in advance for your participation in the Sims Online study.

For those of you who aren't familiar with the Sims, this is your opportunity to participate in an online multiplayer game that allows you to create/recreate yourself in a cybersociety.

You will be paid to play! All we ask of you is that you keep a journal and record your play for one week of every month. We'd like for you to begin to play as soon as your game is installed in your computer. The amount of time you play is up to you.

Journals

Your participation in the Sims Study will require that you journal your activities after your play. Please turn in your journals at the end of each month along with the tape of your play (see below). We will make a copy of your entries and give it back to you until it is full. Once you've completed each journal we will give you a new one.

The first page of your journal should be a brief biography including your age, occupation, family background and prior experience with video games. Please note whether you have played any of the Sims games before and whether it was "stand-alone" or on-line. Please note whether you have played any other multi-player on-line game.

When journaling please be sure to do the following:

1. Start each new entry on a new page
2. Write only on the front of each journal page
3. Skip every other line
4. Write clearly and use a pen
5. Don't tear pages out or put loose pages into the journal.

Each entry should include the following:

1. The date, start time, and finish time
2. The name of your character
3. Why you chose to develop or continue to play this character
4. Perspective of your character (e.g., goals and feelings)—Please be as detailed as possible
5. Any thoughts or feelings you may have about what your character is doing.
6. Which places you go to online and why
7. Any feelings you might have about other characters or the game in general
8. If you use the website, describe what you did there.
9. Have you contacted anyone you met in the game outside the game? If so, why and how?

Please journal whenever you play, even if you are not recording.

Recording your play

We ask that you record your play for one entire week each month. *Before you start playing, please do a short test to make sure you are recording!* Whether you play for two hours or ten hours that week, we ask that you record your play from start to finish. Each mini DV records 90 minutes. Be sure to put a new DV tape in when the one you are using runs out before you play more. Label each DV with your name, the date, and the time of play for that DV. Indicate on the label how many DVs there are for that session (e.g., 1/2, 2/2). Put your recorded DVs in a safe place until they are picked up.

We are going to loan you a mini-DV camera and some mini-DVs. Once you have recorded a week of play, we will pick up the DVs from you. Please call Brendesha Tynes at xxx-xxx-xxxx or email her at btynesb@ucla.edu when you have completed the recording and she can come to pick up the data.

In some cases we are loaning you a computer that belongs to the Children's Digital Media Center. You are responsible for all the equipment that we have loaned you.

Scheduling your play

You are free to play whenever you want. However, the recording will be limited to one week per month. Note above that you will journal even when you are not recording.

Payments

You will be paid $300 for every three months of play (to be paid at the end of the 3rd month) as well as $1 per page of the journal. We have paid for the software and your subscription for three months. Initially, you will be asked to leave a credit card. They will charge your credit card $9.95 per month and this amount will be added to your $300 stipend every third month.

Troubleshooting

In case you have problems when you're trying to play, there is a 24-hour help-line. The number is 1 866-543-5435. There is also a website: EA.com.

If you are having technical difficulties with the equipment, please call Dom Alvear, xxx-xxx-xxxx. This is his cell phone.

REFERENCES

aquafan1. (2002). Online discussion group posting. Available at http://www.gamers.com/game/1016135/reviews/userpreview.

Croal, N. (2002). Sims family values. *Newsweek,* November 25, 46–53.

Electronic Arts. (2003). *Annual Report,* March 31. Available at http://www.ccbn26.mobular.net/ccbn/7/266/277/.

Glassman, M. (2004). Braving bullying hecklers, simulants run for president. *New York Times,* April 1. Available at http://www.nytimes.com/2004/04/01/technology/circuits/01sims.html.

Green, J. (2002). *The Sims Online*: Indulging your weirdo. Preview at Gamers.com, May 1. Available at http://www.gamers.com/game/1016135/previews?page=1.

Greene, D., & Lepper, M. R. (1974). Intrinsic motivation: How to turn play into work. *Psychology Today, 8*(4), 49–54.

Greenfield, P. M. (1997). Culture as process: Empirical methods for cultural psychology. In J. W. Berry, Y. Poortinga, & J. Pandey (Eds.), *Handbook of cross-cultural psychology: Vol. 1. Theory and method* (pp. 301–346). Boston: Allyn & Bacon.

Greenfield, P. M., & Subrahmanyam, K. (2003). Online discourse in a teen chat room: New codes and new modes of coherence in a visual medium. *Journal of Applied Developmental Psychology, 24*(6), 713–738.

Grossman, L. (2002). Sim nation. *Time,* November 25. Available at http://www.time.com/time/magazine/article/subscriber/0,10987,1101021125-391544,00.html.

Keighley, G. (2002). The endless hours of *The Sims Online*. Gamespot.com, November 28. Available at http://gamespot.com/gamespot/features/pc/simsonline.

Kirk, C. (2004). Culling external sensory response: How we feel videogames. Unpublished manuscript, Department of Communication Studies, UCLA.

Oatley, K. (1992). *Best laid schemes. The psychology of emotions.* New York: Cambridge University Press.

Oatley, K. (1994). A taxonomy of the emotions of literary response and a theory of identification in fictional narrative. *Poetics, 23:* 53–74.

O'Keefe, B. J., & Zehnder, S. (2004). Understanding media development: A framework and case study. *Journal of Applied Developmental Psychology. Special Developing Children, Developing Media: Research from Television to the Internet from the Children's Digital Media Center. A Special Issue Dedicated to the Memory of Rodney R. Cocking, 25*(6), 729–740.

pizan36. (2002). Online discussion group posting. Available at http://www.gamers.com/game/1016135/reviews/userpreview.

snake72. (2002). Online discussion group posting. Available at http://www.gamers.com/game/1016135/reviews/userpreview.

Steen, F. F., Davies, M. S., Tynes, B., & Greenfield, P. M. (2005, in press). Digital dystopia. Player control and strategic innovation in *The Sims Online*. In R. Schroeder & A. Axelsson (Eds.), *Avatars at work and play*. London: Springer.

Steen, F. F. (2005). The paradox of narrative thinking. *Journal of Cultural and Evolutionary Psychology, 3*(1), 87–105.

Steen, F. F., & Owens, S. A. (2001). Evolution's pedagogy: An adaptationist model of pretense and entertainment. *Journal of Cognition and Culture, 1*(4), 289–321.

Subrahmanyam, K., Greenfield, P. M., & Tynes, B. (2004). Constructing sexuality and identity in an Internet teen chat room. *Journal of Applied Developmental Psychology. Special Developing Children, Developing Media: Research from Television to the Internet from the Children's Digital Media Center. A Special Issue Dedicated to the Memory of Rodney R. Cocking, 25*(6), 651–666.

Turkle, S. (1995). *Life on the screen: Identity in the age of the Internet.* New York: Simon & Shuster.

Tynes, B., Reynolds, L., & Greenfield, P. M. (2004). Adolescence, race, and ethnicity on the internet: A comparison of discourse in monitored vs. unmonitored chat rooms.*Journal of Applied Developmental Psychology. Special Developing Children, Developing Media: Research from Television to the Internet from the Children's Digital Media Center. A Special Issue Dedicated to the Memory of Rodney R. Cocking, 25*(6), 667–684.

White, A. A. (2003). Chatting for dummies. Game-Revolution.com, January. Available at http://www.game-revolution.com/games/pc/sim/sims_online.htm.

Woodcock, B. S. (2005). An analysis of MMOG subscription growth. Version 18.0. July. Available at http://www.mmogchart.com.

Zillmann, D. (1994). Mechanisms of emotional involvement in drama. *Poetics, 23,* 33–51.

EFFECTS
AND CONSEQUENCES

What Do We Know About Social and Psychological Effects of Computer Games? A Comprehensive Review of the Current Literature

Kwan Min Lee and Wei Peng

University of Southern California

Despite the exponential growth of computer games (including console-based video games, arcade games, online games, and stand-alone computer games) as entertainment media, empirical studies on games are somewhat limited. As early as 1982, U.S. Surgeon General C. Everett Koop lamented the lack of scientific evidence on the effects of video games on children (Selnow, 1984). Even now, more than 20 years since the surgeon general's lament, there is still a cry about the lack of scientific and theoretical studies on computer games (Dill & Dill, 1998; Villani, 2001; Vorderer, 2000). One of the main factors contributing to this continuing dissatisfaction is the lack of a comprehensive review on existing game literature. Even though some studies provide meta-analyses of the effects of violent games on aggression (Anderson & Bushman, 2001; Griffiths, 2000; Gunter, 1998; Sherry, 2001), computer game literature in general has never been comprehensively reviewed. In order to advance our understanding of this relatively new form of entertainment medium, a comprehensive review on existing game literature is needed. Two lines of research traditions—effect studies and uses and gratifications approaches—are the major paradigms adopted in the game research. Because chapter 16 covers the uses and gratifications approach, in the current chapter, we provide only an extensive review of almost 30 years of computer game studies on effects. Positive or negative effects of diverse game contents will be reviewed individually.

Research on the social and psychological effects of game playing focuses on three aspects: (a) to test negative consequences (effects) of violent entertainment games, (b) to demonstrate the utility of educational (training) games, and (c) to examine general effects of entertainment games and game playing without specifying particular content types. In the 1980s and early 1990s, most of the research mainly focused on the negative effects of entertainment games. Almost all the studies on negative effects of games have focused on *violence* (including both explicit and implicit manifestations) embedded in games. Positive effects were mainly associated with educational games. Only recently have scholars begun to realize the potential positive effects of entertainment games. Consequently, studies on game effects can

TABLE 22.1
Effects of Computer Game Playing

	Consequences (Effects)	
	Negative Effects	*Positive Effects*
Violence entertainment games	Affect (hostility, anxiety) Aggressive behaviors Arousal Empathy toward others Physiological responses (heart rate, blood pressure, skin conductance) Priming of aggressive thoughts Prosocial behavior	Catharsis
Nonviolent entertainment games	Addiction or game dependency Gender stereotyping Physical health problems	Training Sociability Academic performance Therapy Spatial visualization Cognitive abilities
Educational games		Learning Motivation Retention memory Utility for special groups (attention-deficit children, patients)

be effectively mapped into a 2 (Consequences [Effects]: negative vs. positive) by 3 (Content Types: violent entertainment vs. nonviolent entertainment vs. education [training]) table (see Table 22.1). In this chapter, we will elaborate on four of the six cells in the table—(a) negative effects of violent entertainment games, (b) negative effects of nonviolent entertainment games, (c) positive effects of nonviolent entertainment games, and (d) positive effects of educational games.

Positive effects of violent entertainment games and negative effects of educational games will not be reviewed in this chapter for the following reasons: First, even though the catharsis theory proposes that violent computer games can generate positive outcomes to their users by providing a safe outlet to exercise violence (see Sherry, 2001), little evidence has been found to support this argument (see Bushman, Baumeister, & Stack, 1999; Gunter, 1994). Second, there has been little research on negative effects of educational games. As a result, we were unable to find sufficient research on these two issues.

NEGATIVE EFFECTS OF VIOLENT ENTERTAINMENT GAMES

In a similar way that research on violence in television has been the main concern for media scholars for the last 5 decades, negative effects of violent games have been the prime focus of most empirical studies on computer games. Three theoretical perspectives—social cognitive theory (Bandura, 1997, 2001), excitation transfer theory (Tannenbaum & Zillmann, 1975;

Zillmann, 1988), and priming effects (Berkowitz, 1984; Berkowitz & Rogers, 1986)—have been applied to explain possible detrimental effects of violent games. Based on social cognitive theory, scholars hypothesize that symbolic violence explicitly *justified* (Funk & Buchman, 1996) during game playing is easily internalized by players and can be substantially transferred to the real world, because players tend to identify themselves with the game characters (Chambers & Ascione, 1987; Graybill, Strawniak, Hunter, & O'Leary, 1987; Schutte, Malouff, Post-Gordon, & Rodasta, 1988; Winkel, Novak, & Hopson, 1987). Another popular explanation for the effects of violent games on aggression is the excitation transfer model by Zillmann (1988). According to this model, residual excitement from previous game playing may serve to intensify a later emotional state of a game player (Anderson & Ford, 1986; Ballard & West, 1996; Calvert & Tan, 1994; Sherry, Curtis, & Sparks, 2001; Silvern & Williamson, 1987; Winkel et al., 1987). This model, therefore, does not necessarily predict the valence of game players' emotional state. Rather, it is about the intensity of game players' emotional state. The third theoretical explanation of negative effects of violent games is based on cognitive priming. According to this explanation, playing violent games increases accessibility to a subset of cognitions specifically related to violence and aggression, which later can be transferred to real-world aggressive behaviors (Anderson & Dill, 2000; Calvert & Tan, 1994; Chory-Assad & Mastro, 2000; Tamborini et al., 2001).

By incorporating these three theoretical perspectives into a single theoretical framework, the general aggression model (GAM; Anderson & Dill, 2000; Anderson & Bushman, 2001) tries to explain both short-term effects of violent media on aggressive cognitions, affects, behaviors, and physiological arousal and long-term effects of violent games on aggressive attitudes, schemata, personality, and aggression desensitization. Violent media increase short-term aggression by teaching users how to aggress, by increasing arousal and aggressive affective states, and by priming aggressive cognitions. Repeated playing of violent games reinforces aggression-related cognitive structures, aggressive perceptual schemata, and aggressive behavioral scripts. Most importantly, repeated playing of violent games increases the aggressive personality of a game player, which then leads to changes in the player's environment (e.g., new peer groups that are more aggressive). The combination of the newly intensified aggressive personality and the new aggressive environment leads to the use of more violent media, which then increases another short-term aggression. Long-term and chronic aggressive behaviors are thus formulated and reinforced during the above process.

Empirical studies regarding negative effects of violent games on aggressive affects, behaviors, thoughts, physiological arousal, and other social and psychological variables (e.g., empathy, prosocial behaviors, and school performance) reveal mixed results. We summarize them one by one.

Aggressive Affects

Violent video games such as *Mortal Kombat* cause more intense feeling of aggression than nonviolent video games such as *Corner Pocket,* a billiard game (Ballard & West, 1996). Violent games also induce a higher level of anxiety than nonviolent ones, at least temporarily (Anderson & Ford, 1986). In a survey of college ($N = 307$) and high school ($N = 82$) students, daily video game use was found to be highly correlated with general hostility and anger among those students (Chory-Assad & Mastro, 2000). In a survey of 355 students in sixth through eighth grades, Abel-Cooper (2001) found that playing games of any categories (i.e., from most violent to less violent) was a weak predictor of the anger state. Anderson and Ford (1986), however, did not find any significant effects of playing a violent game such as *Zaxxon* on hostility. In Anderson's later study (Anderson & Dill, 2000), he found similar null effects. In an experiment with 210 college students, no significant effect of playing a violent game on hostility was observed, even though a much more graphically violent game (*Wolfenstein 3D*) was used. Scott (1995) also failed to find a significant relationship between playing violent games and aggressiveness as measured by the Buss-Durkee Hostility Inventory and the Eysenck Personality Questionnaire. As explained before, no significant effect has been found with regard to the impact of violent games on positive affects (Fleming & Rickwood, 2001). Based on the results of mediation analyses, Anderson and his colleague (Anderson & Dill, 2000) suggested that playing violent video games affects violent behaviors through a *cognitive* path, not through an affective path. That is, it is aggressive thoughts that cause aggressive behaviors, not aggressive affects.

Aggressive Behaviors

Again, the results are mixed. In some experiments, researchers were able to find significant causal relationships between playing violent games and various measures of postgame aggressive behaviors, such as (a) the duration (Anderson & Dill, 2000) and intensity (Cohn, 1996) of a noxious noise blast to opponents, (b) attacking Bobo doll (Schutte et al., 1988), (c) toddlers' (aged 4–6 years) aggressive behaviors in a free-play setting (Silvern & Williamson, 1987), and (d) time spent playing with an aggressive toy (a spring-release fist that fires darts; Cooper & Mackie, 1986). However, some researchers found no effect of violent games on other measures of violent behaviors, such as withholding money from another (Winkel et al., 1987), pushing buttons that could punish or reward others (Graybill et al., 1987), and suggesting punishment or reward to friends (Cooper & Mackie, 1986; Kirsh, 1998). Survey results also provide mixed findings, suggesting both significant and nonsignificant relationships (for a short review, see Goldstein, 2001).

Aggressive Thoughts

Unlike results on aggressive affects and behaviors, consistent significant effects of violent games on aggressive thoughts have been reported. As predicted by the cognitive priming hypothesis, playing violent games such as *Wolfenstein 3D* (as opposed to a nonviolent game such as *Myst*) increases accessibility to aggressive thoughts, as measured by reaction time speed to aggressive words (Anderson & Dill, 2000). Calvert and Tan (1994) also found that players of a violent game listed more aggressive thoughts than simple observers of the same violent game. Playing a violent game increased the aggression-attribution bias of third- and fourth-grade children, which was measured by the negative interpretation of an ambiguous situation (Kirsh, 1998). With regard to the long-term effects of violent games on aggressive cognition, it has been proposed that continuous exposure to violent games makes aggressive thoughts more chronically accessible to players (Bushman, 1998). After continuing exposure to violent games, aggressive thoughts can be fully internalized into players' minds. These internalized aggressive thoughts are usually measured by Gerbner's mean world syndrome index (Chory-Assad & Mastro, 2000).

Physiological Arousal

With regard to effects of violent games on physiological arousal, consistently significant effects have been reported with one exception (Winkel et al., 1987). Playing violent games increases heart rate (Ballard & West, 1996; Griffiths & Dancaster, 1995; Fleming & Rickwood, 2001). The increased heart rate, however, was found to be only temporary: It returned to its baseline 15 minutes after play (Griffiths & Dancaster, 1995). One study found a gender effect, with boys reporting less arousal change than girls (Fleming & Rickwood, 2001). In addition to heart rate, violent games increase systolic blood pressure more than nonviolent games (Ballard & West, 1996). In contrast to the above results, Winkel et al. (1987) found no significant effect of violent game play on heart rate in an experiment with 56 eighth graders.

Other Social and Psychological Variables

In addition to the negative consequences listed above, other negative effects of violent games on prosocial behaviors, delinquency, self-perception, and school performance have been found. In an experiment with 160 (80 third and fourth graders and 80 seventh and eighth graders) children, Chambers and Ascione (1987) found that playing violent games significantly reduced the amount of monetary donation to a charity. In a survey of 278 children (aged 10–14) in the Netherlands, Wiegman and van Schie (1998) found that children with a high preference for violent games, and especially boys, showed significantly less prosocial behaviors than those with a low preference for violent games. In a survey of 227 college students (Anderson & Dill, 2000), playing violent games predicted delinquent behaviors such as drinking alcoholic beverages and destroying school property. And, in a survey of 364 children in fourth- and fifth-grade levels, a significant association between a high preference for violent games and low self-perceptions of behavioral conduct was found for both boys and girls (Funk, Buchman, & Germann, 2000).

Research Syntheses

There have been a series of attempts to synthesize literature on negative effects of violent games. Surprisingly, the results are mixed. Though two quantitative research syntheses based on meta-analysis (Anderson & Bushman, 2001; Sherry, 2001) provide a conclusion that violent games have small but significant negative effects on various social and psychological outcomes, some scholars (Griffiths, 2000; Gunter, 1998) hesitate to draw such a conclusion based on critical reviews of existing literature. After conducting a meta-analysis of 35 research reports, Anderson and Bushman (2001) concluded that playing violent games significantly increases aggressive affects ($r = 0.18$), behaviors ($r = 0.19$), cognition ($r = 0.27$), and physiological arousal ($r = 0.22$) and significantly decreases prosocial behaviors ($r = -0.17$). In their analysis, gender, types of research (survey vs. experiment), and age (children vs. adults) do not moderate the negative effects of violent games. Based on a meta-analysis of 25 empirical studies, Sherry (2001) also acknowledged the existence of a small yet significant effect of violent game play on aggression-related measures ($r = 0.15$). Some scholars, however, argue that the state of current game literature does not warrant any conclusion due to two main methodological limitations: (a) no measurement of long-term effects (Griffiths, 1999) and (b) few observations of actual aggression rather than simulated or pretended aggression (Gunter, 1998).

NEGATIVE EFFECTS OF ENTERTAINMENT GAMES IN GENERAL

Aggressive behaviors triggered by violent content are the main concerns of the studies on the negative effects of entertainment games. Besides violence, there are other possible detrimental effects from general game playing, such as poor academic performance, social isolation, addiction or computer game dependency, gender stereotyping, and vision and other physical health problems. Though school performance and social isolation were thought to be serious negative effects of games in the late 1980s and early 1990s, recent studies indicate that playing games has no significant correlation with suboptimal school performance and social isolation. Actually, some surveys even show that computer game players score *better* than nonplayers in several measurements, including self-concepts of mechanical and computer skills, family closeness, and attachment to school (Durkin & Barber, 2002). Thus, in this section, only addiction or game dependency, gender stereotyping, and vision and other physical health problems will be discussed.

Addiction or Game Dependency

It is widely believed that playing computer games will result in addictive behaviors and thus bring a number of detrimental outcomes, such as irrational spending of money and time on playing and decrease in healthy leisure activities. While there is some anecdotal evidence of game addiction, such as the recent tragedy at August 8th, 2005, in S. Korea where an adult gamer died after 50 hours of continuous playing, the current literature does not provide a consistent support for the game-addiction hypothesis. A general review of studies on computer game addiction by Tejeiro (2001) reveals that most of them are based on either a general survey with unrepresentative samples or an experiment usually focusing on a particular type of nonpopular games. Therefore, generalizability of these studies is somewhat limited. Some researchers correctly suggest not using the word "addiction" but rather "dependency," which applies to a particular category of persons for whom playing games is not simply a preoccupation, but also serves special social and psychological functions in their lives (Shotton, 1989). Adopting the concept of "dependency" rather than "addiction," Griffiths and Hunt (1998) conducted a survey with 387 adolescents (aged 12–16 years). The analysis indicated that one in five adolescents was currently "dependent" on computer games. Boys played significantly more regularly than girls and were more likely to be classified as "dependent." It appeared that the earlier children began playing computer games, the more likely they were to be playing at "dependent" levels.

Gender Stereotyping

When the organization "Children Now" surveyed 1,716 characters of video games on the market, male human characters totaled 1,106 (64%) whereas female human characters numbered only 283 (17%). The remaining 19% were characters with no explicit gender. On average, 17 male characters appeared in each game compared to only four female characters. Moreover, among the game characters, female characters were even less likely to be player-controlled characters with which players usually identify themselves. In this study, of the 874 player-controlled characters, 635 (73%) were males and only 107 (12%) were females (Children Now, 2001). The frequency of the gender of characters is just one indicator of gender stereotyping of computer games. When it comes to the traits of the game characters, it is even more obvious. Female and male characters are portrayed in stereotypical ways. Female characters are often very sexy, with either very thin or very voluptuous bodies. Male characters are usually hypermasculinized. Female characters are usually victims and male characters are the heroes who usually rescue female characters. The concern is that the portrayals of females in games will not only affect the self-image of young girls but also boys' expectations of and attitude toward females (Cesarone, 1994). For more discussion about the content of computer games, refer to chapters 4 and 5.

Physical Health Problems

There are some anecdotal reports of children having seizures while playing video games. The issue is whether this is a mere coincidence provoked by fatigue or stress in a particular individual or a systematic phenomenon related to specific colors, movements, and electronic signals of a particular game. In a study of 387 patients who were extremely sensitive to electronic visual simulation, Kasteleijn-Nolst and colleagues (Kasteleijn-Nolst et al., 1999) found that patients became more sensitive when playing games than when simply viewing them. Interestingly, the patients were more sensitive to *Super Mario* than any other standard games. Other potential negative effects are that game players tend to sit in front of the television or computer for long periods of time. It is not only harmful to vision; the sedentary habits also substitute for outdoor activities, which will eventually affect game players' physical health and development.

POSITIVE EFFECTS OF ENTERTAINMENT GAMES IN GENERAL

When positive effects of games are discussed, most often it is in the context of educational games. When it comes to entertainment computer games, usually it is the negative effects such as alienation, addiction, and violent behaviors that are discussed. However, a growing number of empirical studies have indicated that nonviolent entertainment games can also produce positive outcomes. In this section, positive effects of nonviolent entertainment games on improving training, spatial skills, cognitive abilities, academic performance, adolescents' sociability, and therapy will be examined.

Training

Because the purpose of educational games is to facilitate educational outcomes, they should produce learning of intended skills. For instance, educational computer games and simulations have long been used for training in military sectors. However, entertainment games, which are not specifically designed for instruction or training, can also result in some positive training effects. For example, the Marine Corps Modeling and Simulation Management Office modified and used the game *Doom* to teach combat tactics. Recently, the Marine Corps awarded a contract to MaK Technologies for a high-level architecture (HLA)-compliant PC game for operational training of Marine commanders and staffs (Coleman, 2001). The recently released game *America's Army* provides civilians with insights into soldiering, from the barracks to the battlefields, and has been used for Army recruitment. Conservatively speaking, violence is involved in the game, *America's Army*. However, considering the context of the training project, which is for the Army, this kind of violence is probably necessary to simulate the real battlefield.

Spatial Skills

Research consistently shows that computer games can facilitate spatial skills. Playing computer games was found to facilitate the development of spatial skills for 3-dimensional mental rotation in fifth-, seventh-, and ninth-grade students (McClurg & Chaille, 1987). For 2-dimensional mental rotation, a positive effect was also found among seventh and eighth-graders (Miller & Kapel, 1985). Though it has long been acknowledged that girls are weaker than boys in spatial skills, computer games can improve spatial skills for girls and boys equally. Subrahmanyam and Greenfield (1996) and De Lisi and Wolford (2002) found that video game practice or computer-based instructional activity could significantly improve spatial skills for both girls and boys. Another experiment with kindergarten children in Israel observed a similar result (Perzov & Kozminsky, 1989). Two studies conducted by Pepin and Dorval (1986) assessed the effects of playing video games on spatial visualization of college students ($N = 70$) and seventh-grade students ($N = 101$). The first study with college students produced significant results indicating that both men and women gained equally from playing the video game. The second study with seventh-grade students, however, did not reveal any significant improvement. One possible explanation is that the college student sample had little prior experience with video games whereas the adolescent sample had some previous experience with games, which probably introduced more noncontrollable variables to the study. Another plausible explanation is that age is an important factor determining the effect of computer games for improving spatial skills. Empirical results on mediating or moderating effects of age on gaining spatial skills through computer games are mixed. For example, studies found mild, mixed, or no effects of game playing on spatial skills among elderly participants (Gagnon, 1985; Pepin & Dorval, 1986). Although it is still not very clear why playing computer games can improve spatial skills, it is plausible that the 3-D visualization effect of the video game facilitates spatial perception, mental rotation, and spatial visualization—the three most important spatial skills (Linn and Petersen, 1985).

Cognitive Abilities

Playing computer games demands that the users acquire certain cognitive skills, such as proactive and recursive thinking, systematic organization of information, interpretation of visual information, general search heuristics, means–ends analysis, and so forth (Pillay, 2003). Thus, it is hypothesized that playing computer games can help children develop cognitive skills. Considerable empirical evidence supports the claim that cognitive skills obtained in playing computer games can be transferable to other tasks.

Greenfield and her associates found that exposure to computer games—either in the long term through the natural experience of playing games or in the short term via the use of games as part of experimental manipulation—was positively correlated with better cognitive skills in understanding and interpreting scientific and technical information presented graphically on the computer screen (Greenfield, Brannon, & Lohr, 1994; Greenfield et al., 1994). It was also found that playing computer games facilitated flexibility in dealing with knowledge structures to overcome functional fixedness (Doolittle, 1995). In this experiment, students who played computer games and solved computer riddles were more likely to generate a wide variety of alternative hypotheses for a problem situation.

Some evidence also suggests that computer games enhance inductive reasoning (Camaioni, Ercolani, Perrucchini, & Greenfield, 1990; Honebein, Carr, & Duffy, 1993) and facilitate the development of complex thinking skills related to problem solving (Keller, 1992), strategic planning (Jenkins, 2002; Keller, 1992), and self-regulated learning (Rieber, 1996; Zimmerman, 1990). Computer games also enable the development of different learning styles, as the speed and the level of difficulty can be adjusted according to the player's skill level (Jenkins, 2002).

Academic Performance

One of the most disturbing concerns about entertainment games is that they might interfere with players' academic performances by offering a more attractive option than doing homework. Research findings with regard to the effects of playing computer games at home on academic performance are mixed. In his testimony before the U.S. Congress, David Walsh (2000), president of the National Institute on Media and the Family, suggested that a strong preference for violent games is associated with poor school performance among teens. Yet, for nonviolent entertainment games in general, a positive relationship was found between time spent on entertainment computer games and a child's intelligence in a survey of 346 seventh and eighth graders from seven elementary schools (van Schie & Wiegman, 1997). Durkin and Barber

(2002) also found that children who played games in moderation had higher GPAs than children who did not play games at all. However, excessive playing of games does deteriorate academic performance. For instance, in a large-scale study of 10- and 11-year-olds, Roe and Muijs (1998) found that heavy use of computer games was related to negative outcomes such as low self-esteem, poor academic achievement, and less sociability.

Sociability

The popular hypothesis is that game play has a negative effect on children in terms of social adjustment because children who play games just stay home alone with their console or computer and thus have little social interaction with their peers. However, some recent surveys reveal that the reality is much different. Three surveys among elementary school children showed that the frequency of video game use had no correlation with children's popularity among classmates (Sakamoto, 1994). In contrast to the above results, a Japanese study found that children who played console games developed *higher* sociability than children who were nonplayers (Shimai, Masuda, & Kishimoto, 1990). Similarly, heavy video game players were more likely to see their friends outside school and had a need to see their friends on a regular basis than nonplayers (Colwell, Grady, & Rhaiti, 1995). Frequent players were also found to enjoy just as many friendship and contacts with friends as less frequent players (Philips, Rolls, Rouse, & Griffiths, 1995). The most recent survey conducted by Durkin and Barber (2002) in Australia demonstrates similar results. The survey examined the relationship between game play and several measures of adjustment or risk taking in a sample of 1,304 sixteen-year-old high school students. They classified players into three categories, based on the frequency of playing video games: high, low, and nonplayers. No evidence was obtained of negative outcomes among game players. Actually, high players scored higher on several measurements, including self-concepts of mechanical and computer skills, family closeness, and attachment to school. Low players scored higher on most measurements, such as lower depressed mood, lower aggression, lower disobedience, higher self-esteem, and higher GPAs. Surprisingly, nonplayers did not score high on any measure. Thus, the authors concluded that computer game play could be a positive feature of a healthy adolescent. One of the reasons that playing video games does not result in isolation but rather improves sociability might be that when adolescents play video games, they do not just play by themselves but with friends and family. They also exchange their gaming experiences with peers, not only face-to-face but also through the Internet.

Therapy

Computer games have the potential for therapeutic treatment for psychological as well as physical problems. Lynch (1981) used video games as a training aid for certain cognitive and perceptual-motor disorders and various types of mental disorders (e.g. stroke patience). Gardner (1991) concluded that the application of video games in his psychotherapy sessions was more successful than the traditional technique, because video games provided common grounds between himself and his clients and facilitated excellent behavioral observation opportunities. Computer games can also alleviate feelings of anxiety (Naveteur & Ray, 1990), trigger motivation to exercise and increase metabolic activity during wheelchair use (O'Connor, Fitzgerald, Cooper, Thorman, & Boninger, 2001), divert attention from side effects of cancer chemotherapy and reduce the feeling of pain (Redd et al., 1987), treat disabled children with speech difficulties (Horn, Jones, & Hamlett, 1991), rehabilitate a child with palsy (Krichevets, Sirotkina, Yevsevicheva, & Zeldin, 1995), rehabilitate cognitive problems (Larose, Gagnon, Ferland, & Pepin, 1989), and retard memory decline among the elderly (Drew & Waters, 1986;

Dustman, Emmerson, Steinhaus, & Dustman, 1992; Goldstein et al., 1997; see Griffiths, 1997, for a general review of using computer games for clinical treatment).

POSITIVE EFFECTS OF EDUCATIONAL GAMES

As early as 1980s, scholars have begun to notice the potential of computer games for education and the learning process. It is believed that children engaged in computer game playing may acquire more general strategies for "learning to learn" in novel environments (Stowbridge, 1983). By playing games, kids growing up in the digital age learn the rules of processing multimedia information, which is fundamentally different from how information in the printing age was presented and processed. They learn how to learn in a nonlinear way using the aid of abundant hypertextual and visual cues. These skills learned during game playing may be applied in instructional settings (Malone, 1981) and help develop other important skills, such as inductive discovery and problem solving through trial-and-error learning (Greenfield, 1983) and eye–hand coordination and spatial visualization (Pepin & Dorval, 1986). For a long time, the military has been aware of the potential of computer games for simulative flight training (Kennedy, Bitter, & Jones, 1981; Lintern & Kennedy, 1984) as well as cognitive skills, such as rapid information processing and the ability to think about a number of things at the same time (Trachtman, 1981). Though academic research on the positive effects of educational computer games and simulations has been done since the 1980s, studies on this issue are scarce compared with the studies done on the negative effects of violent computer games. Existing studies on this issue have mainly focused on the positive effects of computer games on learning, motivation, retention memory, and utility for special groups such as attention-deficit children, the elderly, or patients. The following section will elaborate on each area. We begin this section with several theoretical approaches that explain why computer games can generate such positive effects.

Theoretical Models

The theoretical explanation of why computer games can facilitate positive learning outcomes is still not very clear. Currently, three theoretical concepts offer possible explanations—immersion (or presence), flow, and intrinsic motivation.

According to Hubbard (1991), the learning process that results from playing video games is due to the immersion effect. This immersion effect creates an environment in which the players submerge themselves and progressively increase their attention and concentration on a goal. This theoretical approach could be used to explain the positive effects of games on memory retention and their utility for special groups. Computer games can engage players deeply in the learning environment, which makes the players very attentive to the educational materials embedded in the environment. The concept of immersion is similar to the concept of "presence," which has been discussed extensively in other domains including virtual reality, computer-mediated communication, and human–computer interaction. Following Lombard and Ditton (1997) and Lee (2004a), we believe that *presence*—a "psychological state in which virtual objects are experienced as actual objects" or "perceptual illusion of nonmediation"— lies at the heart of virtual experiences mediated or created by communication and/or computer technologies (for a general review, see Lee, 2004a, 2004b). Computer games can create more intense feeling of presence than other media due to the interactive nature of game playing (see Steuer, 1992). As demonstrated by Lee and Nass (2004), the feeling of presence will mediate computer games' various psychological effects on users.

Similar to Hubbard's idea of immersion is Csikszentmihalyi (1990)'s concept of flow (see Klimmt & Vorderer, 2003 for a detailed differentiation between the two concepts). Flow is the state of optimal experience whereby a person is so engaged in an activity that self-consciousness disappears, time becomes distorted, and the person engages in complex, goal-oriented activities not for external reward, but simply for the exhilaration of doing. Using Csikszentmihalyi and Larson's (1980) discussion of "flow," Bowman (1982) analyzed *Pac-Man* players and illustrated the appeal of video games as their ability to place users in "flow states." When using computer games in the learning environment, students learn in a flow state where they are not just passive recipients of knowledge but active learners who are in control of the learning activity and are challenged to reach a certain goal. However, the flow state is not necessarily easy to attain. It rests on certain skills of the learner and the condition of the challenge. Because every game has certain rules and requires certain skills of the user in order to perform well, when the learning process is embedded in the game environment, the skills of the learner should match with the challenge of the game so as to gain the optimal enjoyment. If the game is too challenging, the learner will not have the sense of control and cannot gain pleasure while playing the game.

The third concept is intrinsic motivation adopted by Garris, Ahlers, and Driskell (2002) in their input–process–outcome game model, which tries to elucidate the learning process during the game. The input elements involve the instructional content and the game characteristics. The effective process is what the authors called the "game cycle," in which certain characteristics of games trigger intrinsic motivation of the users and then generate the repeated cycles of user judgment (e.g., enjoyment), behavior (game play), and feedback. The game cycle engages the user in repetitive play and the user continually returns to the game activity over time. When the user is engaged in playing game continuously, experiential learning based on intrinsic motivation can occur. The characteristics of computer games that trigger intrinsic motivation are fantasy, rules/goals, sensory stimuli, challenge, mystery, and control. When instructional contents are successfully paired with appropriate game features, the game cycle results in recurring and self-motivated game play. Malone (1981) also identified three aspects of computer games that trigger intrinsic motivation: challenge, fantasy, and curiosity. The assumption of this model is based on the experiential learning approach of Dewey (1938) and Kolb, Boyatzis, and Mainemelis (2000). According to the experiential learning paradigm, people do learn from active engagement with the environment. Coupled with some instructional support, the active engagement during the game can produce an effective learning environment.

Learning

Research findings with regard to the effectiveness of educational computer games on learning are mixed. Randel, Morris, and Wetzel (1992) conducted a literature review to compare the instructional effectiveness of electronic simulations or games to conventional classroom instruction. This review produced the following results: 56% of the studies found no difference, 32% found differences favoring simulations/games, 7% favored simulations/games but raised questions about their experimental design, and the remaining 5% found differences favoring conventional instruction. Randel et al. (1992) also investigated the effectiveness of using games to deliver knowledge of different subject matters, including social sciences, math, language, arts, logic, physics, and biology. Not all subject matters demonstrated beneficial effects of using games. Math was the subject with the greatest percentage of results favoring games. Thirtythree out of 46 social science games/simulations exhibited no difference from classroom instruction. A more recent empirical study found that computer games and multimedia instruction had reliable and positive effects on achievement in mathematics problem solving, reading comprehension, and word study, yet the same study found no reliable effects on mathematics procedure and reading vocabulary (Blanchard, Stock, & Marshall, 1999). In addition, visual and interactive components in educational business games did not improve the knowledge in the specific domain, though the games were enjoyed more (Sedbrook, 1998). Computer games and simulation also produced no significant improvement in basic electricity and electronics training in the military (Parchman, Ellis, Christinaz, & Vogel, 2000). Overall, computer games and simulations are effective compared with the traditional instructional mode, but their effectiveness depends on specific subject/content domains.

Investigations have also found that educational computer games are beneficial in the areas of teaching strategic management (Hsu, 1989; Wolfe, 1997), statistical concepts (Lane & Tang, 2000), scientific discovery learning (de Jong & van Joolingen, 1998), language learning (Jordan, 1992; Hubbard, 1991; Kovalik & Kovalik, 2002), skill-based learning (Gopher, Weil, & Bareket, 1994), health education (Dorman, 1997) such as safe sex education (Cahill, 1994; Kashibuchi & Sakamoto, 2000) and juvenile diabetes self-care (Brown et al., 1997), and medical education (Boreham, Foster, & Mawer, 1989).

Besides the subject/content factor, characteristics of a particular computer game also matter with regard to instructional effectiveness. Malone (1981) identified three aspects of computer games that triggered intrinsic motivation of users to be the main features of successful instructional games: fantasy, challenge, and curiosity. Fantasies can make the instructional environment more interesting and facilitate focalization of attention. An emotionally appealing fantasy or metaphor that is related to game skills will more easily engage the learner in the learning process. Rieber (1996) amended Malone's approach by differentiating endogenous and exogenous fantasies. According to Rieber, only endogenous fantasy, which could weave the content into the game, will produce positive learning effects. Just adding fantasy context that has nothing to do with the learning material (exogenous fantasies) does not help. In order for an instructional environment to be challenging, it must provide goals whose attainment is uncertain. The goals should be obvious and personally meaningful to players. Players also need to receive feedback on whether they are achieving their goals or not.

Another issue concerning the use of computer games for education is whether noncomputer-based games provide the same effectiveness as computer-based games. In other words, the question is whether the effectiveness results from the particular delivery medium—a computer—or from the content embedded in the computer game environment. Wiebe and Martin (1994) investigated the impact of a computer-based adventure game involving geography content on students' recall of geography facts and their attitudes toward studying geography. They found no differences between "non-computer classroom games and activities" and "computer-based adventure games for reinforcing geography facts and student attitudes." Similarly, Antonietti and Mellone (2003) conducted an experiment to explore a computer-based version and a traditional version of *Pegopolis*, a solitaire game. These two versions of games were the same, except that they were played by moving pieces either on a real board or on a virtual computer-presented board. No significant difference was found between conditions with respect to the performance and strategies followed during the game. Though two studies are not enough to ensure a conclusion, it is plausible to make a tentative proposal that it might be the pure characteristics of game, not the medium of computer, that makes the learning process of computer game-based education more effective.

Motivation

Different from the mixed findings on effectiveness of educational computer games on learning, it is commonly agreed that educational computer games have positive effects on motivation. Randel et al. (1992) summarized in their literature review that students reported more interest in simulation and game activities than in conventional classroom instruction. A study conducted with learning-disabled students of sixth through eighth grades ($n = 25$) found that game features produced higher levels of continuing motivation (Malouf, 1987). A similar experiment with intermediate-level students ($n = 41$) with learning disabilities also showed that game format had a facilitative effect on continuing motivation of students with low initial attitudes toward mathematics (Okolo, 1992). Many theoretical explanations of why computer games can enhance learning are usually based on enhanced motivation. Nevertheless, enhanced motivation does not fully guarantee more effective learning. For example, though Parker and Lepper (1992) found that a fantasy game increased both children's motivation and their actual learning, Druckman (1995) found no convincing evidence that enhanced motivation by games increases actual learning. Some other mediating variables are probably needed for a fuller explanation.

Retention Memory

Randel et al. (1992) found that educational computer games and simulations produced greater retention over time than conventional classroom instruction in 12 out of 14 studies. In a military training context, participants assigned to the game condition scored significantly higher on a retention test compared to pretest performance. Furthermore, participants assigned to the game condition scored significantly higher on a retention test than participants assigned to the text condition (Ricci, Salas, & CannonBowers, 1996). In a comparison of learning outcomes between a computer game environment and a multimedia environment, Moreno and Mayer (2000) found that personalized rather than neutral messages produced better retention performance in the computer game condition and better problem solving in both the computer game and the multimedia environments. A possible explanation for why games can improve retention might be that users are more attentive to the media stimuli when playing games than when using other types of media. As psychologists have long established, this intense attention during games will increase memory retention (Nelson, 1995; Norman, 1976). It is also proposed that greater engagement during learning, which is the result of intrinsic motivation triggered by the game features, leads to longer retention of information (Hannafin & Hooper, 1993). However, no scholar has offered a specific mechanism and thus we cannot draw any conclusion now.

Utility for Special Groups

Educational computer games also serve as useful tools for special groups, such as children with learning disabilities and the cognitively impaired elderly. Educational computer games improve motivation for children with learning disabilities (Malouf, 1987; Okolo, 1992). They have also been used to help children with attention-deficit/hyperactivity disorders. For example, Pope and Bogart (1996) developed a video game that becomes more difficult to play when the brainwaves of a player indicate waning attention. The player can succeed at the game only by maintaining an adequate level of attention. Educational games are also used to help elderly people who are often cognitively impaired. For example, a game called Memory of Goblins was used to study its effects on elderly people's memory ability and life satisfaction. Though no statistically significant result was obtained due to the small sample size, the study suggests some evidence for the positive impact of computer games on the cognitive ability of the elderly population (see Farris, Bates, Resnick, & Stabler, 1994). Another experiment with 22 noninstitutionalized elderly people (aged 69 to 90) investigated the effects of video game playing (*Super Tetris*) on reaction time, cognitive/perceptual adaptability, and emotional well-being. The video game playing group had faster reaction times and felt a more positive sense of well-being compared to their nonplaying counterparts (Goldstein et al., 1997).

CONCLUSION AND SUGGESTIONS
FOR FUTURE RESEARCH

Based on our extensive review of the current literature, we find that the media effects paradigm still dominates the academic research on computer games. As a result, there has been a paucity of research on the nature of game playing as an entertainment *experience*. This is a perplexing situation, because studies on consequences of something can be significantly enhanced by the understanding of its intrinsic nature. In fact, studies on social consequences of game play would have been more systematically done if they were based on theoretical understanding of the nature of game experience. We believe that if we want to have a fuller understanding of this new form of media entertainment, we need to know what users actually experience while they are playing games.

In addition, the existing game literature usually focuses on the effects of media *contents* (predominantly violent or educational contents) and neglects the impacts of media *forms*. According to Reeves and Nass (1996), media forms such as size, fidelity, cuts, synchrony, and movements are equally important factors for determining psychological impacts of media. In fact, early studies on television effects during the 1970s have confirmed that formal features (e.g., loud noises, unusual camera effects, fast action) of television are at least partly responsible for television's effects on children's aggressive behaviors (Lowery & DeFleur, 1995, p. 360). More importantly, media forms and contents interact with each other. For example, violence depicted on a small screen with poor audio fidelity might have less impact on aggressive thought (Anderson & Dill, 2000) and cardiovascular activities (Ballard & West, 1996) than the same violence depicted on a large screen with high-fidelity audio. In fact, most researchers are increasingly concerned about the potential harmful effects of the newer generation of violent games due to the increasing realism in games made possible by new form factors such as high-fidelity video and audio, lifelike display size, and seamless interactivity (see Anderson & Dill, 2000; Ballard & West, 1996; Calvert & Tan, 1994; Dill & Dill, 1998; Provenzo, 1991). Therefore, studies on the main effects of computer games' form factors (e.g., cut, motion, 3-D, fidelity, audio, and so on) and possible interaction effects between the form factors and the content types (e.g., violence, sex, humor, sports, and so on) are needed in order to get a fuller understanding of game effects.

REFERENCES

Abel-Cooper, T. B. (2001). The association between video game playing, religiosity, parental guidance and aggression, in sixth through eighth grade students attending Seventh-Day Adventist schools. *Dissertation Abstracts International, 61*, 3910.

Anderson, C. A., & Bushman, B. J. (2001). Effects of violent video games on aggressive behavior, aggressive cognition, aggressive affect, physiological arousal, and prosocial behavior: A meta-analytical review of the scientific literature. *Psychological Science, 12*, 353–359.

Anderson, C. A., & Dill, K. E. (2000). Video games and aggressive thoughts, feelings, and behavior in the laboratory and in life. *Journal of Personality and Social Psychology, 78*, 772–790.

Anderson, C.A., & Ford, C. M. (1986). Affect of the game player: Short-term consequences of playing aggressive video games. *Personality and Social Psychology Bulletin, 12*, 390–402.

Antonietti, A., & Mellone, R. (2003). The difference between playing games with and without the computer: A preliminary view. *Journal of Psychology, 137*, 133–144.

Ballard, M. E., & West, J. R. (1996). Mortal Kombat™: The effects of violent videogame play on males' hostility and cardiovascular responding. *Journal of Applied Social Psychology, 26*, 717–730.

Bandura, A. (1997). *Self-efficacy: The exercise of control*. New York: Freeman.

Bandura, A. (2001). Social cognitive theory: An agentic perspective. *Annual Review of Psychology, 52*, 1–26.

Berkowitz, L. (1984). Some effects of thoughts on anti- and pro-social influence of media events: A cognitive neoas-sociationist analysis. *Psychological Bulletin, 95*, 410–427.

Berkowitz, L., & Rogers, K. H. (1986). A priming effect analysis of media influences. In J. Bryant & D. Zillmann (Eds.), *Perspectives on media effects* (pp. 57–81). Hillsdale, NJ: Lawrence Erlbaum Associates.

Blanchard, J., Stock, W., & Marshall, J. (1999). Meta-analysis of research on a multimedia elementary school curriculum using personal and video-game computers. *Perceptual and Motor Skills, 88*, 329–336.

Boreham, N. C., Foster, R. W., & Mawer, G. E. (1989). The Phenytoin Game: Its effect on decision skills. *Simulation & Games, 20*, 292–299.

Bowman, R. F. (1982). A "Pac-Man" theory of motivation: Tactical implications for classroom instruction. *Educational Technology, 22*, 14–16.

Brown, S. J., Lieberman, D. A., Gemeny, B. A., Fan, Y. C, Wilson, D. M., & Pasta, D. J. (1997). Educational video game for juvenile diabetes: Results of a controlled trial. *Medical Informatics, 22*, 77–89.

Bushman, B. J. (1998). Priming effects of media violence on the accessibility of aggressive constructs in memory. *Personality and Social Psychology Bulletin, 24*, 537–545.

Bushman, B. J., Baumeister, R. F., & Stack, A. D. (1999). Catharsis, aggression, and persuasive influence: self-fulfilling or self-defeating prophecies? *Journal of Personality and Social Psychology, 76*, 367–376.

Cahill, J. M. (1994). Health works: Interactive AIDS education videogames. *Computers in Human Services, 11*, 159–176.

Calvert, S., & Tan, S. L. (1994). Impact of virtual reality on young adults' physiological arousal and aggressive thoughts: Interaction versus observation. *Journal of Applied Developmental Psychology, 15*, 125–139.

Camaioni, L., Ercolani, A. P., Perrucchini, P., & Greenfield, P. M. (1990). Video games and cognitive ability: The transfer hypothesis. *Italian Journal of Psychology, 17*, 331–348.

Cesarone, E. (1994). *Video games and children.* (Report No. EDO-PS-94-3). Washington, DC: Office of Educational Research and Improvement.

Chambers, J. H., & Ascione, F. R. (1987). The effects of prosocial and aggressive video games on children's donating and helping. *Journal of Genetic Psychology, 148*, 499–505.

Children Now. (2001, December). *Fair play? Violence, gender, and race in video games. Children and the media.* Retrieved August 31, 2003, from http://www.childrennow.org/media/video-games/2001/#gender.

Chory-Assad, R. M., & Mastro, D. E. (2000, November). *Violent videogame use and hostility among high school students and college students.* Paper presented at the Annual Conference of National Communication Association (NCA), Seattle, WA.

Cohn, L. B. (1996). Violent video games: Aggression, arousal, and desensitization in young adolescent boys. *Dissertation Abstracts International, 57*(2-B), 1463.

Coleman, D. S. (2001). PC gaming and simulation supports training. *Proceedings of United States Naval Institute, 127*, 73–75.

Colwell, J., Grady, C., & Rhaiti, S. (1995). Computer games, self-esteem, and gratification of needs in adolescents. *Journal of Community and Applied Social Psychology, 5*, 195–206.

Cooper, J., & Mackie, D. (1986). Video games and aggression in children. *Journal of Applied Social Psychology, 16*, 726–744.

Csikszentmihalyi, M. (1990). *Flow: The psychology of optical experience.* New York: Harper Perennial.

Csikszentmihalyi, M., & Larson, R. (1980). Intrinsic rewards in school crime. In K. Baker & R. J. Rubel (Eds.), *Violence and crime in the schools.* Lexington, MA: DC Health.

de Jong, T., & van Joolingen, W. R. (1998). Scientific discovery learning with computer simulations of conceptual domains. *Review of Educational Research, 68*, 179–201.

De Lisi, R., & Wolford, J. L. (2002). Improving children's mental rotation accuracy with computer game playing. *The Journal of Genetic Psychology, 163*, 272–282.

Dewey, J. (1938). *Experience and education.* New York: Macmillan.

Dill, K. E., & Dill, J. C. (1998). Video game violence: A review of the empirical literature. *Aggression and Violent Behavior, 3*, 407–428.

Doolittle, J. H. (1995). Using riddles and interactive computer games to teach problem-solving skills. *Teaching of Psychology, 22*, 33–36.

Dorman, S. M. (1997). Video and computer games: Effect on children and implications for health education. *The Journal of School Health, 67*, 133–138.

Drew, B., & Waters, J. (1986). Video games: Utilization of a novel strategy to improve perceptual motor skills and cognitive functioning in the non-institutionalized elderly. *Cognitive Rehabilitation, 4*, 26–31.

Druckman, D. (1995). The educational effectiveness of interactive games. In D. Crookall & K. Arai (Eds.), *Simulation and gaming across disciplines and cultures: ISAGA at watershed* (pp. 178–187). Thousand Oaks, CA: Sage.

Durkin, K., & Barber, B. (2002). Not so doomed: Computer game play and positive adolescent development. *Journal of Applied Developmental Psychology, 23*, 373–392.

Dustman, R. E., Emmerson, L., Steinhaus, D., & Dustman, T. (1992). The effects of videogame playing on neuropsy-chological performance of elderly individuals. *Journal of Gerontology, Psychological Sciences, 47*, 168–171.

Entertainment Software Association. (2002, May). *Top ten industry facts.* Retrieved August 12, 2003, from http://www.theesa.com/pressroom.html.

Farris, M., Bates, R., Resnick, H., & Stabler, N. (1994). Evaluation of computer games' impact upon cognitively impaired frail elderly. *Computers in Human Services, 11*, 219–228.

Fleming, M., & Rickwood, D. (2001). Effects of violent versus nonviolent video games on children's arousal, aggressive mood, and positive mood. *Journal of Applied Social Psychology, 31*, 2047–2071.

Funk, J. B., & Buchman, D. D. (1995). Video game controversies. *Pediatric Annals, 24*, 91–94.

Funk, J. B., & Buchman, D. D. (1996). Playing violent video games and adolescent self-concept. *Journal of Communication, 46*, 19–32.

Funk, J. B., Buchman, D. D., & Germann, J. (2000). Preference for violent electronic games, self-concept, and gender differences in young children. *American Journal of Orthopsychiatry, 70*, 233–241.

Gagnon, D. (1985). Video games and spatial skills: An exploratory study. *Educational Communication and Technology Journal, 33*, 263–275.

Gardner, J. E. (1991). Can the Mario Bros. help? Nintendo games as an adjunct in psychotherapy with children. *Psychotherapy, 28*, 667–670.

Garris, R., Ahlers, R., & Driskell, J. E. (2002). Games, motivation, and learning: A research and practice model. *Simulation & Gaming, 33*, 441–472.

Goldstein, J. H. (2001, October). *Does playing violent games cause aggressive behavior?* Paper presented at Playing by the Rules: The Cultural Policy Challenges of Video Games Conference. Cultural Policy Center, University of Chicago, Chicago, IL.

Goldstein, J. H., Cajko, L., Oosterbroek, M., Michielsen, M., van Houten, O., & Salverda, F. (1997). Video games and the elderly. *Social Behavior and Personality, 25*, 345–352.

Gopher, D., Weil, M., & Bareket, T. (1994). Transfer of skills from a computer game trainer to flight. *Human Factors, 36*, 387–405.

Graybill, D., Strawniak, M., Hunter, T., & O'Leary, M. (1987). Effects of playing versus observing violent versus nonviolent video games on children's aggression. *Psychology: A Quarterly Journal of Human Behavior, 24*, 1–8.

Greenfield, P. M. (1983). Video game and cognitive skills. In S. S. Baughman & P. D. Claggett (Eds.), *Video games and human development: A research agenda for the 80s* (pp. 19–24). Cambridge, MA: Harvard Graduate School of Education.

Greenfield, P. M., Brannon, G., & Lohr, D. (1994). Two-dimensional representation of movement through three-dimensional space: The role of video game expertise. *Journal of Applied Developmental Psychology, 1*, 87–103.

Greenfield, P. M., Camaioni, L., Ercoloni, P., Weiss, L., Lauber, B. A., & Perrucchini, P. (1994). Cognitive socialization by computer games in two cultures: Inductive discovery or mastery of an iconic code. *Journal of Applied Developmental Psychology, 15*, 59–85.

Griffiths, M. D. (1997). Video games and clinical practice: Issues, uses and treatments. *British Journal of Clinical Psychology, 36*, 639–641.

Griffiths, M. D. (1999). Violent video games and aggression: A review of the literature. *Aggression & Violent Behavior, 4*, 203–212.

Griffiths, M. D. (2000). Video game violence and aggression: Comments on "Video game playing and its relations with aggressive and prosocial behavior" by O. Wiegman and E. G. M. van Schie. *British Journal of Social Psychology, 39*, 147–149 (Part 1).

Griffiths, M. D., & Dancaster, I. (1995). The effect of Type A personality on physiological arousal while playing computer games. *Addictive Behaviors, 20*, 543–548.

Griffiths, M. D., & Hunt, N. (1998). Dependency on computer games by adolescents. *Psychological Reports, 82*, 475–480.

Gunter, B. (1994). The question of media violence. In J. Bryant & D. Zillmann (Eds.), *Media effects: Advances in theory and research* (pp. 163–212). Hillsdale, NJ: Lawrence Erlbaum Associates.

Gunter, B. (1998). *The effects of video games on children: The myth unmasked.* Sheffield, UK: Sheffield Academic Press.

Hannafin, M. J., & Hooper, S. R. (1993). Learning principles. In M. Fleming & W. H. Levie (Eds.), *Instructional message design: Principles from the behavioral and cognitive sciences* (pp. 191–231). Englewood Cliffs, NJ: Educational Technology Publications.

Honebein, P. C., Carr, A., & Duffy, T. (1993). The effects of modeling to aid problem solving in computer-based learning environments. In M. R. Simonson & K. Abu-Omar (Eds.), *Annual proceedings of selected research and development presentations at the national convention of the Association for Educational Communications and Technology* (pp. 373–406). Bloomington, IN: Association for Educational Communications and Technology.

Horn, E., Jones, H. A., & Hamlett, C. (1991). An investigation of the feasibility of a video game system for developing scanning and selection skills. *Journal of the Association for Persons with Severe Handicaps, 16*, 108–115.

Hsu, E. (1989). Role-event gaming simulation in management education: A conceptual framework and review. *Simulation & Games, 20*, 409–438.

Hubbard, P. (1991). Evaluating computer games for language learning. *Simulation & Gaming, 22*, 220–223.

Jenkins, H. (2002). Game theory. *Technology Review, 29*, 1–3.

Jordan, G. (1992). Exploiting computer-based simulations for language-learning purposes. *Simulation & Gaming, 23*, 88–98.

Kashibuchi, M., & Sakamoto, A. (2000). "POMP & CIRCUMSTANCE": The effectiveness of a simulation game in sex education. *International Journal of Psychology, 35*, 156.

Kasteleijn-Nolst, D. G., da Silva, A. M., Ricci, S., Binnie, C. D., Rubboli, G., Tassinari, C. A., & Segers, J. P. (1999). Video-game epilepsy: A European study. *Epilepsia, 40*, 70–74.

Keller, S. M. (1992). *Children and the Nintendo*. (ERIC Access No: ED405069).

Kennedy, R. S., Bitter, A. C., & Jones, M. B. (1981). Video game and conventional tracking. *Perceptual & Motor Skills, 53*, 510.

Kirsh, S. J. (1998). Seeing the world through Mortal Kombat-colored glasses: Violent video games and the development of a short-term hostile attribution bias. *Childhood: A Global Journal of Child Research, 5*, 177–184.

Klimmt, C., & Vorderer, P. (2003). Media psychology "is not yet there": Introducing theories on media entertainment to the Presence debate. *Presence: Teleoperators and Virtual Environments, 12*(4), 346–359.

Kolb, D. A., Boyatzis, R. E., & Mainemelis, C. (2000). Experiential learning theory: Previous research and new directions. In R. J. Sternberg & L. F. Zhang (Eds.), *Perspectives on cognitive, learning, and thinking styles* (pp. 227–247). Mahwah, NJ: Lawrence Erlbaum Associates.

Kovalik, D. L., & Kovalik, L. M. (2002). Language learning simulation: A Piagetian perspective. *Simulation & Gaming, 33*, 345.

Krichevets, A. N., Sirotkina, E. B., Yevsevicheva, I. V., & Zeldin, L. M. (1995). Computer games as a means of movement rehabilitation. *Disability & Rehabilitation, 17*, 100–105.

Lane, D. M., & Tang, Z. H. (2000). Effectiveness of simulation training on transfer of statistical concepts. *Journal of Educational Computing Research, 22*, 383–396.

Larose, S., Gagnon, S., Ferland, C., & Pepin, M. (1989). Psychology of computers: XIV. Cognitive rehabilitation through computer games. *Perceptual & Motor Skills, 69*, 851–858.

Lee, K. M. (2004a). Presence, explicated. *Communication Theory, 14*, 27–50.

Lee, K. M. (2004b). Why presence occurs: Evolutionary psychology, media equation, and presence. *Presence: Teleoperators and Virtual Environments, 13*, 494–505.

Lee, K. M., & Nass, C. (2004). The multiple source effect and synthesized speech: Doubly disembodied language as a conceptual framework. *Human Communication Research, 30*, 182–207.

Linn, M. C., & Petersen, A. C. (1985). Emergence and characterization of sex difference in spatial ability: A meta analysis. *Children development*, 1479–1498.

Lintern, G., & Kennedy, R. S. (1984). Video game as a covariate for carrier landing research. *Perceptual and Motor Skills, 58*, 167–172.

Lombard, M., & Ditton, T. (1997). At the heart of it all: The concept of presence. *Journal of Computer-Mediated-Communication, 3*. Retrieved October 21, 2002, from http://www.ascusc.org/jcmc/vol3/issue2/lombard.html.

Lowery, S., & DeFleur, M. (1995). *Milestones in mass communication research: Media effects* (3rd ed.). New York: Longman.

Lynch, W. J. (1981, August). *TV games as therapeutic interventions*. Paper presented at the American Psychological Association, Los Angeles, CA.

Malone, T. W. (1981). Toward a theory of intrinsically motivating instruction. *Cognitive Science, 4*, 258–277.

Malouf, D. B. (1987). The effect of instructional computer games on continuing student motivation. *Journal of Special Education, 21*, 27–38.

McClurg, P. A., & Chaille, C. (1987). Computer games: Environments for developing spatial cognition? *Journal of Educational Computing Research, 3*, 95–111.

Miller, G. G., & Kapel, D. E. (1985). Can non-verbal puzzle-type microcomputer software affect spatial discrimination and sequential thinking skills of 7th and 8th graders? *Education, 106*, 160–167.

Moreno, R., & Mayer, R. E. (2000). Engaging students in active learning: The case for personalized multimedia messages. *Journal of Educational Psychology, 92*, 724–733.

Naveteur, J., & Ray, J. C. (1990). Electrodermal activity of low and high trait anxiety subjects during a frustration video game. *Journal of Psychophysiology, 4*, 221–227.

Nelson, C. (1995). *Attention and memory: An integrated framework*. New York: Oxford University Press.

Norman, D. A. (1976). *Memory and attention: An introduction to human information processing* (2nd ed.). New York: Wiley.

O'Connor, T. J., Fitzgerald, S. G., Cooper, R. A., Thorman, T. A., & Boninger, M. L. (2001). Does computer game play aid in motivation of exercise and increase metabolic activity during wheelchair ergometry? *Medical Engineering & Physics, 23,* 267–273.

Okolo, C. M. (1992). The effect of computer-assisted instruction format and initial attitude on the arithmetic facts proficiency and continuing motivation of students with learning disabilities. *Exceptionality: A Research Journal, 3,* 195–211.

Parchman, S. W., Ellis, J. A., Christinaz, D., & Vogel, M. (2000). An evaluation of three computer-based instructional strategies in basic electricity and electronics training. *Military Psychology, 12,* 73–87.

Parker, L. E., & Lepper, M. R. (1992). Effects of fantasy context on children's learning and motivation: Making learning more fun. *Journal of Personality and Social Psychology, 62,* 625–633.

Pepin, M., & Dorval, M. (1986, April). *Effect of playing a video game on adults' and adolescents' spatial visualization.* Paper presented at the annual meeting of the American Educational Research Association, San Francisco, CA.

Perzov, A., & Kozminsky, E. (1989). The effect of computer games practice on the development of visual perception skills in kindergarten children. *Computers in the School, 6,* 113–122.

Philips, C. A., Rolls, S., Rouse, A., & Griffiths, M. D. (1995). Home video game playing in schoolchildren: A study of incidence and patterns of play. *Journal of Adolescence, 18,* 687–691.

Pillay, H. (2003). An investigation of cognitive processes engaged in by recreational computer game players: Implications for skills of the future. *Journal of Research on Technology in Education, 34,* 336–349.

Pope, A. T., & Bogart, E. H. (1996). Extended attention span training system: Video game neurotherapy for attention deficit disorder. *Child Study Journal, 26,* 39–50.

Provenzo, E. F. (1991). *Video kids: Making sense of Nintendo.* Cambridge, MA: Harvard University Press.

Randel, J. M., Morris, B. A., & Wetzel, C. D. (1992). The effectiveness of games for educational purposes—a review of recent research. *Simulation & Gaming, 23,* 261–276.

Redd, W. H., Jacobsen, P. B., Dietrill, M., Dermatis, H., McEvoy, M., & Holland, J. C. (1987). Cognitive-attentional distraction in the control of conditioned nausea in pediatric cancer patients receiving chemotherapy. *Journal of Consulting and Clinical Psychology, 55,* 391–395.

Reeves, B., & Nass, C. (1996). *The media equation.* New York: Cambridge University Press.

Ricci, K. E., Salas, E., & CannonBowers, J. A. (1996). Do computer-based games facilitate knowledge acquisition and retention? *Military Psychology, 8,* 295–307.

Rieber, L. P. (1996). Seriously considering play: Designing interactive learning environments based on the blending of microworlds, simulations and games. *Educational Technology Research and Development, 44,* 43–58.

Roe, K., & Muijs, D. (1998). Children and computer games: A profile of the heavy user. *European Journal of Communication, 13,* 181–200.

Sakamoto, A. (1994). Video game use and the development of sociocognitive abilities in children: 3 surveys of elementary-school students. *Journal of Applied Social Psychology, 24,* 21–42.

Schutte, N., Malouff, J., Post-Gordon, J., & Rodasta, A. (1988). Effects of playing video games on children's aggressive and other behaviors. *Journal of Applied Social Psychology, 18,* 451–456.

Scott, D. (1995). The effect of video games on feelings of aggression. *Journal of Psychology, 129,* 121–132.

Sedbrook, T. A. (1998). Visual-interactive business games: Design and pedagogical effects. *Journal of Computer Information Systems, 38,* 33–40.

Selnow, G. W. (1984). Playing videogames: The electronic friend. *Journal of Communication, 34,* 148–156.

Sherry, J. L. (2001). The effects of violent video games on aggression: A meta-analysis. *Human Communication Research, 27,* 409–431.

Sherry, J. L., Curtis J., & Sparks, G. (2001, May). *Arousal transfer or desensitization? A comparison of mechanisms underlying violent video game effects.* Paper presented at the 51st annual convention of the International Communication Association, Washington, DC.

Shimai, S., Masuda, K., & Kishimoto, Y. (1990). Influences of TV games on physical and psychological development of Japanese kindergarten children. *Perceptual & Motor Skills, 70,* 771–776.

Shotton, M. (1989). *Computer addiction? A study of computer dependency.* London: Taylor & Francis.

Silvern, S. B., & Williamson, P. A. (1987). The effects of video game play on young children's aggression, fantasy, and prosocial behavior. *Journal of Applied Developmental Psychology, 8,* 453–462.

Steuer, J. (1992). Defining virtual reality: Dimensions determining telepresence. *Journal of Communication, 42,* 73–93.

Stowbridge, M. D. (1983). Becoming a better student with computer games. *Journal of Learning Skills, 2,* 35–43.

Subrahmanyam, K., & Greenfield, P. M. (1996). Effect of video game practice on spatial skills in girls and boys. In P. M. Greenfield, R. R. Cocking, & R. Rodney (Eds.), *Interacting with video* (pp. 95–114). Westport, CT: Ablex.

Tamborini, R., Eastin, M., Lachlan, K., Skalski, P., Fediuk, T., & Brady, R. (2001, May). *Hostile thoughts, presence and violent virtual video games.* Paper presented at the 51st annual convention of the International Communication Association, Washington, DC.

Tannenbaum, P. H., & Zillmann, D. (1975). Emotional arousal in the facilitation of aggression through communication. In L. Berkowitz (Ed.), *Advances in experimental social psychology: Vol. 8* (pp. 149–192). New York: Academic Press.

Tejeiro, R. (2001). La adiccion a los videojuegos [Video games addiction: A review]. *Adicciones, 13*, 407–413.

Trachtman, P. A. (1981). Generation meets computers—and they are friendly. *Smithsonian, 12*, 50–61.

van Schie, E. G., & Wiegman, O. (1997). Children and videogames: Leisure activities, aggression, social integration, and school performance. *Journal of Applied Social Psychology, 27*, 1175–1194.

Villani, S. (2001). Impact of media on children and adolescents: A 10-year review of the research. *Journal of the American Academy of Child and Adolescent Psychiatry, 40*, 392–401.

Vorderer, P. (2000). Interactive entertainment and beyond. In D. Zillmann & P. Vorderer (Eds.), *Media entertainment: The psychology of its appeal* (pp. 21–36). Mahwah, NJ: Lawrence Erlbaum Associates.

Walsh, D. (2000, March). *Interactive violence and children: Testimony submitted to the Committee on Commerce, Science, and Transportation, United States*, March 21, 2000. Retrieved November 22, 2002, from http://www.mediaandthefamily.org/press/senateviolence-full.shtml.

Wiebe, J. H., & Martin, N. J. (1994). The impact of a computer-based adventure game on achievement and attitudes in geography. *Journal of Computing in Childhood Education, 5*, 61–71.

Wiegman, O., & van Schie, E. G. M. (1998). Video game playing and its relations with aggressive and prosocial behavior. *British Journal of Social Psychology, 37*, 367–378.

Winkel, M., Novak, D. M., & Hopson, H. (1987). Personality factors, subject gender, and the effects of aggressive video games on aggression in adolescents. *Journal of Research in Personality, 21*, 211–223.

Wolfe, J. (1997). The effectiveness of business games in strategic management course work. *Simulation & Gaming, 28*, 360–374.

Zillmann, D. (1988). Cognition–excitation interdependences in aggressive behavior. *Aggressive Behavior, 14*, 51–64.

Zimmerman, B. (1990). Self-regulated learning and academic achievement: An overview. *Educational Psychologist, 25*, 3–17.

23

Aggression and Violence as Effects of Playing Violent Video Games?

René Weber
Michigan State University

Ute Ritterfeld and Anna Kostygina
University of Southern California

On June 3, 2003, the United States 8th Circuit Court of Appeals decided that video games, including violent video games, must be considered free speech (Interactive Digital Software Association v. St. Louis County, 2003). Hence, like literature, violent video games are protected by the First Amendment of the U.S. Constitution. Emphasizing the "small number of ambiguous, inconclusive, or irrelevant (conducted on adults, not minors) studies" (p. 6), the court pointed out that "when the government defends restrictions on speech it must do more than simply posit the existence of the disease sought to be cured. [The court] believes that the County must demonstrate that the recited harms are real, not merely conjectural, and that the regulation will in fact alleviate these harms in a direct and material way" (p. 6). *In dubio pro reo* seems to be the conclusion of the court. Nevertheless, the violent video game *Grand Theft Auto* is currently the subject of a $246 million lawsuit after a fatal shooting by two teenagers who later told the investigators that "they got the rifles from a locked room in their home and decided to randomly shoot at tractor-trailer rigs, just like in the video game *Grand Theft Auto III*" (Mansfield, 2003, p. 1). Could this just be an excuse to avoid taking responsibility for their gruesome actions?

In fact, experts describe accused adolescents increasingly shifting their responsibility for violent behavior to the mass media (Kunczik, 2002). Correspondingly, the scientific community predominantly agrees that mass media and violent video games are only one of many factors (e.g., poverty, poor parental care) that may explain aggression and violence in a society and are surely not the most important ones (Potter, 2003; Anderson, 2003). However, the National Center for Injury Prevention and Control (2001) stated that, although youth violence has dropped in recent years, it remains unacceptably high in the United State Besides poor parental supervision, the report names exposure to violence and beliefs supportive of violence as key risk factors. This raises the question of whether exposure to violent mass media in fact contributes to aggressive thoughts and behavior.

To shed light on this question, we will first introduce more recent violent games and, secondly, apply common theories to explain if and why playing those games may affect aggression and violence in real life. After discussing empirical evidence, the chapter closes with a description of methodological pitfalls, challenges, and innovations in research on violent video games.

VIOLENCE IN RECENT GAMES

Whereas violence in first-generation video games consisted of fighting animated, cartoonlike characters, modern games offer a much more realistic representation of violence. In *Grand Theft Auto: Vice City*, for example, the player may be rewarded for having sex with a prostitute and kicking her to death to avoid payment. In *Mortal Kombat*, a player earns extra points by ripping out the opponent's heart or decapitating him. In some versions of the game *Manhunt*, the player must simply approach an opponent from behind and press a button. The longer the player holds down the button, the more gruesome the killing appears. The homocides are shown in detailed and realistic graphics. Such M-rated games (*M*ature, which means for individuals age 17 and older only) may contain "sexual violence," "blood and gore," or "intense violence" as listed on the Internet homepage of the Entertainment Software Rating Board (http://www.esrb.org/index.asp). But even numerous T-rated games (for *T*eens, which means for individuals age 13 and older) are listed under the same content keywords.

Based on recent games research (Potter & Tomasello, 2003) as well as a number of studies on TV violence (see Gunter, Harrison, & Wykes, 2003; National Television Violence Study, 1999), the context of violence and the viewers' interpretations are particularly important in predicting negative effects. Hearold's (1986) meta-analysis reveals that the portrayal of rewarded, justified, and realistic violence (e.g., Atkin, 1983) committed by sympathetic characters shows the largest effects on aggression; unrewarded/punished, unjustified violence by unappealing characters reduces the influence of the portrayal. A comprehensive study on the amount *and* the context of violence in state-of-the-art video games was provided by Smith, Lachlan, and Tamborini (2003) and also by Thompson and Haninger (2001). Smith et al. analyzed violence in 60 of the most popular video games. The basic template for aggression in such games is "a human perpetrator engaging in repeated acts of justified violence involving weapons that results in some blood shed to the victim" (2003, p. 60). They found that the amount and context of violence presented in state-of-the-art video games rated T or M (and even E games, rated for *E*veryone) clearly pose risks for negative effects such as the development of aggressive scripts for social problem solving (see also Haninger, Ryan, & Thompson, 2004; B. Smith, chap. 4, this volume; S. Smith, chap. 5, this volume). It may be the content of those games that brought Grossman (2000) to the pointed title of his essay: "Violent video games are mass-murder simulators."

Furthermore, it is questionable whether these negative effects may be even more intense in video games than they ever were in TV or film, simply because of the interactive nature of games. Unlike TV or movies, video games involve players as virtual participants in a violent social setting (Grodal, 2000; Vorderer, 2000; see also Lee, Park, & Jin, chap. 18, this volume; Ritterfeld & Weber, chap. 26, this volume). The player may also identify with an aggressive game character more easily than TV viewers usually do (Zillmann, 1994). The expectations, preferences, fears, or desires of a player allow him or her to take on a role provided by the game. This role, however, is free of real-life sanctions for conduct and can therefore be played with apparent disregard for consequences. While there is some evidence for the assumption that interactive environments have a stronger effect on aggressive thoughts (e.g., Calvert & Tan, 1994), more traditional media seem to still pose the greater risk of negative effects (Sherry, 2001).

The amount and context of violence in state-of-the-art video games could be of minor importance if only a few mature people would play them. However, best practice surveys reveal: In 2003 more than 239 million computer and video games were sold, that is, almost two games for every household in America (Entertainment Software Association, 2004); most U.S. children and adolescents (more than 90%) play video games for an average of 30 minutes daily (Kaiser Family Foundation, 2002); "the motion picture and electronic game industries have acted far more responsibly in improving their self-regulatory programs, yet continue to allow advertising of R-rated movies and M-rated games in venues that attract large numbers of teens" (Federal Trade Commission, 2002, p. 1); consequently, games rated M are extremely popular with preteen and teenage boys who report no trouble buying the games (National Institute on Media and the Family, 2002; Federal Trade Commission, 2000, 2002). In a nutshell, violent video games are among the most popular entertainment products for teens and adolescents, especially for boys (cf. Klimmt, 2004).

THEORETICAL FOUNDATIONS

A substantial body of theory in the fields of psychology, communication, and sociology has been developed to explain the processes by which exposure to violence could cause both short- and long-term increases in human aggressive and hostile behaviors (for an overview: cf. Anderson & Bushman, 2002a; Dill & Dill, 1998; Griffiths, 1999). Early psychological theories used the notions of aggressive instinct, catharsis, and frustration to explicate potential origins of human aggression (Bushman, 2002; Anderson & Huesmann, 2003). Recent theorizing, however, explains the long-term effects of media violence on aggression as originating from long-term observational learning of cognitions related to aggressive behavior (Bandura, 1973, 2001, 2002; Berkowitz, 1993; Huesmann, Moise-Titus, Podolski, & Eron, 2003; Krahé & Moeller, 2004), desensitization or emotional habituation (Funk, Baldacci, Pasold, & Baumgardner, 2004; Rule & Ferguson, 1986), and cultivation processes (Gerbner, Gross, Morgan, Signorielli, & Shanahan, 2002; van Mierlo & van den Bulck, 2004), whereas short-term effects may be based on processes like priming and imitation (Huesmann, Moise-Titus, Podolski, & Eron, 2003; Jo & Berkowitz, 1994) as well as arousal processes and excitation transfer (Zillmann, 2000, 2003). The general aggression model from Anderson and Bushman (2002a) can be considered as an integration of different theories (except catharsis theory) explaining effects of media violence (see also Buckley & Anderson, chap. 24, this volume). The following sections provide an outline of the aforementioned theories.

Symbolic Catharsis Hypothesis

In response to the criticisms of violent video games, some researchers have claimed creative and prosocial applications of such games. It has been argued that the aggressive content of video games could allow players to release their stress and aggression in a nondestructive way and, in fact, would have the effect of relaxing them (Bowman & Rotter, 1983; Kestenbaum & Weinstein, 1985). These arguments are based on the idea of symbolic catharsis—discharging aggression by purging aggressive emotions or emotional purification (Golden, 1992; Scheff, 1979; Scheele, 2001).

In psychoanalytic theory, sexuality and aggression serve as the two major drives (Ellis, 2002). Both drives arrogate for release that may happen in reality or in fantasy, the latter called symbolic catharsis. Therapeutic ideas on emotional catharsis "form the basis of the hydraulic model of anger. The hydraulic model suggests that frustrations lead to anger and that anger, in turn, builds up inside an individual, similar to hydraulic pressure inside a closed environment until it is released in some way. If people do not let their anger out, but try to keep it bottled up inside, it will eventually cause them to explode in an aggressive rage" (Bushman, 2002, p. 725).

The notion of catharsis was brought to the field of media violence research by Feshbach (1955), who investigated the impact of exposure to television violence on aggression. According to this line of thinking, the exposure to violence in video games would permit viewers to engage in fantasy aggression, thus discharging their hostility in a satisfactory way and reducing the need to carry out aggression in a behavioral realm (Sparks & Sparks, 2002). However, so far there is hardly any empirical support for a catharsis effect or the existence of the hydraulic anger system (Bushman, 2002). The fact that the evidence for the symbolic catharsis hypothesis is very sparse (see also Griffiths, 1997) may be due to the flaws in the conceptualization of cathartic effects (Scheele, 2001). Particularly, the restriction of the meaning of catharsis to the process of purging emotional excitement through fantasy is arbitrary as originally the term was introduced by Aristotle to refer to the processes of both emotional purgation and purification or "intellectual clarification" (Scheele, 2001; Scheff, 1979). The restricted interpretation might not fully account for the effects of violence in video games and other media. Furthermore, only inferior research designs have been applied in order to test the catharsis hypotheses with respect to violent video games. Subjects must feel aggressive and angry before they become confronted with different game genres or other media (mainly TV and movies) that they may select and use any way they like. Under these conditions, a controlled experiment (with a control group that is also in an aggressive state but has no media choices) could reveal if aggression really decreases as a result of playing a violent video game.

Contemporary Theories Explicating Long-Term Effects

Cognitive Neoassociation Theory.   One helpful theory in exploring long-term effects is the cognitive neoassociation theory (Berkowitz, 1993). Contrary to the theory of catharsis, this theory suggests that discharge of aggressive emotions, for instance, through behaving aggressively against inanimate objects, should increase rather than decrease angry feelings and aggressive behaviors. Virtual aggression or violence is supposed to prime individuals' aggressive thoughts and feelings and enable real-life violence (cf. Bushman, 2002). According to this theory, aggressive thoughts form an associative network in memory. Once an aggressive thought is processed or stimulated, activation spreads along the links of the network to prime associated aggressive ideas and violent emotions, and may even provoke aggressive actions (Berkowitz, 1993). Thus, exposure to violent depictions in video games may activate the network of associated angry thoughts and emotions and potentially result in increased likelihood of aggressive behavior in real-life situations.

Observational Learning and Social Cognitive Theory.   Social cognitive theory postulates long-term effects of exposure to violence through the influence of exposure on the observational learning and imitation of violent acts, and ultimately the development of aggressive problem-solving scripts, hostile attribution biases, and normative beliefs approving of aggression (Bandura, 2001, 2002). Originally, the theory was developed by Bandura (1973) within the framework of behaviorism. Later, Bandura modeled symbolic representations of human responses to stimuli. The theory explains the psychological functioning of humans in terms of triadic reciprocal causation where "personal factors in the form of cognitive, affective, and biological events; behavioral patterns; and environmental events all operate as interactive determinants that influence each other bidirectionally" (Bandura, 2002, p. 121). It is assumed that the plasticity of human nature is fashioned by direct and observational experiences. Thus, individuals can not only learn while acting but also by observing others. Consistently, observing game characters behaving aggressively is assumed to increase the likelihood for the player to imitate the displayed behavior in real life. Furthermore, a player's identification with the game character should enhance observational learning as well. Frequent exposure to violent games may also lead to the formation of normative beliefs associated with the appropriateness of aggression and hostility in reallife social encounters (Huesmann et al., 2003).

Desensitization Theory.   Another alternative to explain long-term effects of exposure to violence in mass media is provided by desensitization theory. This theory is based on the assumption that most humans show an innate negative psychological and physiological reaction to observing violence. Desensitization may be defined as the habituation or attenuation of such distress-related cognitive, emotional, and behavioral reactivity to observations and thoughts of violence as a result of exposure to violence in real life as well as in the media (Rule & Ferguson, 1986). It is believed that the unpleasant physiological arousal usually associated with violence inhibits thinking about violence, disregarding violence, or behaving violently; however, as a result of continuous exposure to violent depictions, individuals are expected to no longer have such reactions. Violence becomes perceived as mundane and pervasive, and this may result in a heightened likelihood of violent thoughts and behaviors (Anderson et al., 2003; Huesmann et al., 2003; Sparks & Sparks, 2002).

Although desensitization theory has been extensively studied with regard to violent television viewing, there is only limited empirical support (Carnagey & Anderson, 2003). Regarding video games, one of the few attempts to directly test this hypothesis was made by Funk, Baldacci, Pasold, and Baumgardner (2004). From their survey of 150 fourth and fifth graders about exposure to violence in real life and in various media, the authors concluded that both video game and movie violence exposure are associated with stronger proviolence attitudes. Higher exposure to video game violence was also associated with lower empathy toward the victims.

Cultivation Theory.   Cultivation theory (Gerbner, Gross, Morgan, Signorielli, & Shanahan, 2002) concentrates on the enduring and common consequences of growing up and living with mass media—television in particular. Theories of the cultivation process attempt to understand and explain the dynamics of mass media as the distinctive and dominant cultural force in contemporary societies—"the source of the most broadly shared images and messages in history" (Gerbner et al., 2002, p. 43). The general hypothesis of the cultivation theory claims that heavy users of mass media are more likely to see the world in terms of the images, value systems, and ideologies propagated by the mass media. The result could be a social legitimization of the reality depicted in the mass media, which may again influence behavior. The heuristic processing model of cultivation effects offered by Shrum (2002) posits that frequent exposure to mainstream messages enhances the accessibility of mental schemata employed by individuals to process information and to evaluate them. The parallel to cognitive neoassociation theory (see above) is obvious.

In regard to applying cultivation theory to investigating negative effects of video games on aggression, several differences between comparable mass media, such as television, and video games seem to be problematic. On the one hand, the level of graphical quality or realism in video games is lower than in television. Therefore, we would expect weak or no cultivation effects as found in van Mierlo and van den Bulck's (2004) study, hereby reinforcing the assumption that a game is just a game. On the other hand, a video game usually requires complete attention and involvement, shows a higher frequency of violent scenes, and offers multiple opportunities to identify with violent characters (Anderson & Dill, 2000; Dill & Dill, 1998). According to this line of argument cultivation theory should predict larger cultivation effects of video games compared with other media.

Contemporary Theories Explicating Short-Term Effects

Any of the above outlined theories may also contribute to explaining short-term effects of playing violent video games on aggression. However, there are theoretical and methodological approaches, basically the notion of priming (Jo & Berkowitz, 1994) and excitation transfer (Zillmann, 2003), which are both primarily focused on rather short-term effects. At the same time, priming and excitation transfer may serve as conditions for any other negative effect including long-term effects. Furthermore, neuroscience methodology has recently been introduced into video game research (Koepp et al., 1998; Weber, Ritterfeld, & Mathiak, in press) in order to explore short-term arousal processes induced by violent video games more reliably than ever before.

Media Priming Theory.   According to media priming theory, exposure to media immediately increases the accessibility of information that is presented in media messages (Anderson & Huesmann, 2003; Jo & Berkowitz, 1994). Once specific information has been processed, it influences attitudes, norms, and even behavior. It is believed that exposure to aggression and violence in violent video games fosters the availability of aggressive schemata, which, subsequently, can be more easily used for processing information in other social contexts. In the long run, frequently primed angry cognitions, affects, and aggressive behaviors may become increasingly automatic in their invocation.

Excitation Transfer Theory.   Excitation transfer (Zillmann, 2003; Bryant & Miron, 2003) serves as another potential and widely discussed explanation as to why aggression may occur as an immediate effect of exposure to violence in video games. Media violence is believed to be exciting (arousing) for most people. Excitation transfer theory assumes that arousal from exposure to media messages facilitates the response of an individual to a certain stimulus of provocation (Tannenbaum & Zillmann, 1975). According to this theory, arousal from any source has the potential through misattribution processes to result "in inappropriate (disproportionate) responses to the current circumstances" (Bryant & Miron, 2003, p. 35); for instance, it can enhance the experience of anger and induce hostile or aggressive thoughts and behavior (Zillmann, 1978). Experiments testing this model with film and television stimuli (cf. Anderson et al., 2003) have demonstrated that aggressive behavior in response to provocations is more likely to occur immediately after exposure to violent media stimuli and often lasts no longer than a few minutes. Supporters of the excitation transfer model in video game research suggest that highly violent video games provide the arousal that is necessary for aggressive reactions in real life.

Psychophysiology and Neuroscience Research.   Excitation transfer theory in combination with long-term learning processes seems to provide a solid background to explain possible effects of playing violent video games on aggression and real-life violence. Nevertheless, it is questionable to what extent excitation can result from nonviolent media stimuli and if violent media stimuli can be considered as a direct cause of excitation linked to aggressive activation. We do know that playing violent video games show physiological effects that are different compared to playing less violent games or no games at all (Lynch, 1994, 1999) and that those effects may be even greater for children who already show more aggressive tendencies (Gentile, Lynch, Linder, & Walsh, 2004a; Gentile, Walsh, Ellison, Fox, & Cameron, 2004b). There is also evidence that violent content leads to a higher physiological arousal and therefore to a more intense stimulus when embedded in a narrative (Schneider, Lang, Shin, & Bradley, 2004). However, does this ultimately mean that violent content results in a higher excitement which transfers itself to aggressive activation or even behavior? This crucial question is still unanswered, but recent neuroscience research has the potential to provide new answers.

If individuals express aggressive thoughts or are exhibiting aggressive behavior, they seem to lack cognitive control of their emotions. Emotions are usually regulated in the human brain by a complex circuit consisting of the orbital frontal cortex, the amygdala, the anterior cingulate cortex, and some other interconnected regions. Through functional magnetic resonance imaging (fMRI), Davidson, Putnam, and Larson (2000) were able to show that unusual brain activity in the neural circuitry of emotion regulation appears as a result of impulsive violence and aggression. In particular, reduced activation in the cognitive and emotional subdivision of the anterior cingulate cortex and also amygdala seems to provide a reliable indicator of faulty emotion regulation and may be considered a neuronal correlate of aggression (Bush, Luu, & Posner, 2000; Sterzer, Stadler, Krebs, Kleinschmidt, & Poustka, 2003; Pietrini, Guazzelli, Basso, Jaffe, & Grafman, 2000).

Given these neurobiological correlates of aggression, neuroimaging studies can contribute significantly to identifying the direct impact of violent video games on aggressive cognitions. To our knowledge, there are only two pertinent neuroimaging studies that have focused on video game playing. By means of positron emission tomography (PET), Koepp et al. (1998) found evidence for dopamine release during video game playing that may play an additional role as a correlate of aggressive and violent cognitions (cf. Davidson et al., 2000). The distinctiveness of the correlation, however, is questionable as dopamine release indicates learning, attention, and sensorimotor coordination also. Mathews, Kronenberger, Wang, Lurito, Lowe, and Dunn (2005) explicitly studied the impact of violent media, including video games, on brain activity. The research team revealed similar brain activity patterns in children exposed to violent media within the last 12 months to children with a diagnosed aggressive behavior disorder. Unfortunately, the design of this study lacks internal validity because the findings are not only attributable to the media usage but also to many other confounded factors. Another fMRI study by Weber, Ritterfeld, and Mathiak (in press) overcomes some of the limitations of the aforementioned studies by controlling the treatment conditions while also considering the external validity of the game play. Thirteen volunteers (male, ages 18–26) played a violent first-person shooter game (*Tactical Operations*) while their brain activity within the neural circuitry of emotion regulation was measured. A high-resolution content analysis of the recorded game play provided information on the occurrence of violence in every moment. The study reveals that violent playing is highly associated with those brain activity patterns characteristic for aggression. As the authors used a within-subjects design, the findings may be interpreted as a short-term, causal effect that might be of importance in combination with long-term learning processes.

General Aggression Model

The general aggression model (GAM; Anderson & Bushman, 2002a; Carnagey & Anderson, 2003; Anderson, 2004; see also Buckley & Anderson, chap. 24, this volume) that integrates most theories of human aggression is a useful and often applied framework for understanding the effects of violent media. GAM describes a cyclical, dynamic pattern of interaction between a person and the environment or situation in which he or she lives. According to this theory, aggression is largely based on learning, activation, and application of aggression-related knowledge structures stored in memory (e.g., scripts, schemata). Two main models of violent media effects are offered: The single-episode GAM explicates the short-term effects of violence exposure whereas the multiple-episode GAM accounts for the long-term or cumulative effects of observing violence in real life and in mass media.

The single-episode model suggests that situational input variables such as recent exposure to violent media, aggressive cues, provocation or frustration, as well as personality factors such as traits, values, attitudes, or beliefs influence aggressive behavior through their impact on the person's present internal state, represented by cognitive, affective, and arousal variables. Violent media, for instance, may increase aggression by teaching observers how to behave aggressively, by priming aggressive cognitions (including previously learned aggressive scripts and aggressive perceptual schemata), by increasing arousal, and by creating an aggressive affective state.

Long-term effects explicated in the multiple-episode model also involve learning processes. The authors stated that humans learn from infancy how to perceive, interpret, judge, and respond to events in the physical and social environments. We all adopt various types of knowledge structures over time. These cognitive schemata are based on day-to-day observations of interactions with other people in real life and also imaginary scenarios. In this respect, violent video games may have an impact on real-life aggression because "each violent-media episode is essentially one more learning trial" (Anderson & Bushman, 2001, p. 355). Overall, the combination of single-episode and multiple-episode processes may result in the positive relation between exposure to media violence and aggressive or violent behavior.

EMPIRICAL RESEARCH FINDINGS

The fundamental question of this chapter remains still unanswered: Do violent video games cause aggression? To date, three meta-analytic reviews (Anderson & Bushman, 2001; Sherry, 2001; Anderson, 2004) complement numerous literature reviews (e.g., Anderson & Dill, 2000; Dill & Dill, 1998; Griffiths, 1999; Gunter, 1998).

Anderson and Bushman's (2001) meta-analysis cumulates empirical findings across 33 independent tests on the relation between video game violence and aggressive behavior, involving 3,033 participants. The average effect size of $r = 0.19$ proves to be positive and significant, however small. Effect sizes are greater if aggressive cognitions ($r = 0.27$) are investigated compared to studying aggressive affects ($r = 0.18$). Gender, age, or methodological design does not moderate the effects. Sherry (2001) considered 25 independent studies and arrived at a significant average effect size of $r = 0.15$ ($n = 2,722$) that is slightly smaller but still comparable to Anderson and Bushman's data. Two additional findings in Sherry's review need to be highlighted, however. A regression between effect sizes and moderator variables reveals the strongest positive relation between the year the game was released and the effect sizes ($\beta = 0.33$) as well as a notable *negative* relation between the playing time and aggression. Sherry concluded that on one side recent generation games increase the harmful effect while on the other side the time spent playing games reduces their negative impact. In contrast, Anderson and Bushman (2001) stated that the magnitude of effect does not depend on the amount of time spent playing violent games.

Recently, Anderson (2004) provided an update of his initial 2001 meta-analytic review while expanding the database to 32 independent samples with 5,240 participants. Data confirm a significant average effect size of $r = 0.20$ for aggressive behavior. Most interestingly, Anderson discriminates between 'best practice' and "non-best practice" studies by referring to the methodological rigor applied. Less rigorous studies show *smaller* effect sizes than more rigorous studies, suggesting that previous analyses on effect sizes may have underestimated the true magnitude of the negative consequences of playing violent video games. The average effect size of best practice studies regarding aggressive behavior is $r = 0.26$. Anderson (2004) summarized his findings by stating that exposure to violent video games is clearly and causally linked to increases in aggressive behavior, aggressive cognitions, aggressive affects, and cardiovascular arousal, and to decreases in helping behavior.

Overall, and according to Cohen's (1988) classification of effect sizes, there is a noticeable but small effect of playing violent video games on aggressive reactions (the effects are close to medium in size). Considering the increasingly realistic, elaborate, and involving design of recent game development and Sherry's findings of a positive and notable relation between release year of a violent game and effect one may expect considerably higher effect sizes in the future (see also Shapiro Pena-Herborn, & Hancock, chap. 19 this volume).

Furthermore, it is important to consider the dosage of a potential risk factor. Considering the amount of young people who are exposed to simulated violence, even a small effect can result in high societal costs or damage (Abelson, 1985; Prentice & Miller, 1992; Rosenthal, 1986). Anderson (2004) compared the effect sizes in the context of violent video games with effect sizes in other areas and stated that violent video game effect sizes are larger than the effects of studies that relate secondhand tobacco smoke to lung cancer, lead exposure to children's IQ, or condom use to the risk of HIV infection. However, Sherry (2001) pointed out that the effect sizes are still smaller than in the domain of violent TV and aggression. Translated into Cohen's d-metric, Sherry found a small effect of $d = 0.30$ for violent video games while Paik and Comstock (1994), for instance, yielded a medium effect of $d = 0.65$ for television violence and aggression. This statement is interesting, but certainly not comforting in a broader sense. Assuming that video games and TV are used complementarily and not as substitutes, and that effects may be mutually reinforcing (Slater, Henry, Swaim, & Anderson, 2003), the overall effect of media violence on aggression is clearly substantial (see also Anderson & Bushman, 2002b).

To answer the introducing question of whether violent mass media contribute to aggressive thoughts and behavior, we have critically examined the meta-analytic reviews, scrutinized the available literature reviews, and also considered studies that demonstrate negligible effects of playing violent video games on aggression (e.g., Collwell & Payne, 2000; Durkin, 1995; Durkin & Aisbett, 1999; Durkin & Barber, 2002; Kestenbaum & Weinstein, 1985; Scott, 1995). We are forced to conclude that within the bounds of the social science methodology, playing violent video games does increase aggressive reactions. However, whether the implied causality is justified will be explicated in the next section.

METHODOLOGICAL PITFALLS AND CHALLENGES

The crucial factor investigating the effect of violent video games is causality. Do violent computer games evoke aggression, or could it also be the reverse: Aggressive people use video games to cope with their problems? Roe and Muijs (1998) reported that boys who perform poorly in school tend to show higher motivation to achieve success in the world of video games—possibly to compensate for their failure. It is well-known that a single correlational study provides almost no information on causality. Even experimental studies may fail to prove causal relationships if control and experimental groups are not comparable or variables are confounded. If, for example, an experimental group plays a violent video game while the control group receives a racing or strategy game, not only the magnitude of violence but also numerous other aspects of content and game format are manipulated. Evidence of causality requires the combination of rigorous experimental studies with correlational studies using different operationalizations of both the independent (violent content of video games) and the dependent variable (aggression) but revealing consistent results. To date, the number and diversity of studies justify only careful causal interpretation of the relation between video games and aggressive reactions. Further exploration with longitudinal studies is still missing and badly needed. Cross-lagged panel designs (Kenny, 1973; Lazarsfeld, 1940) offer the best way to find strong evidence for a causal relationship.

Besides methodological critique, the common understanding of causality is also questionable. Traditionally, causality is defined as one or multiple variables affecting one or more other variables in a single direction. Understanding media effects requires a more complex model of causality: Slater et al. (2003), for example, stated that media violence research provides evidence that aggressive youth seek out media violence (selective exposure) and that media violence predicts aggression in youth. They argued that both relationships, when modeled over time, should be mutually reinforcing and generate exclusive effects. The authors introduced these processes as "downward spiral model" or "feedback loop model". Accordingly, aggressive tendencies may influence game players to expose themselves to violent content, which reinforces their existing aggressive predispositions. Bandura (2002) named similar ideas "triadic reciprocal causation" (see above).

Within the dynamic-transactional model of media effects (Früh, 1991; Früh & Schönbach, 1982), causes and effects can no longer be distinguished, because there is no stimulus or response. Instead, there are two basic transactions and multiple variables that can be simultaneously causes as well as effects and can be mutually intensifying or attenuating. The first transaction describes how audience members "read" and interpret messages in media or video games, which can be quite different from objective content (see below and Potter & Tomasello, 2003). Messages and stimuli are always subjective. The second transaction describes cognitive processes within the recipients, such as evaluating and learning messages. In summary, which video game a person plays and how this person interprets the messages of this video game affects what he or she learns; what he or she learns has an effect on which video games are subsequently selected and played and how the messages are interpreted; and so on. The exchanges between the two transactions build a dynamic process and an exclusive effect over time.

Vorderer and Weber (2003) pointed out that understanding interdependencies as dynamic, multicausal, transactional constructs is new to entertainment research but was introduced in other fields much earlier, for example, in economic research (e.g., Smith, 1776). The authors pointed to the necessity of modeling and analyzing such processes appropriately and suggested a connectionist approach in which artificial intelligence methodology is used for simulations.

Another important question relates to how best to operationalize the constructs "violent content" and "aggressive reaction" or "real-life violence". So far, various definitions of violent content had been used with possible effects being dependent on the applied definition (e.g., Anderson & Dill, 2000; Griffiths, 1999; Sherry, 2001). The same is true for "aggression" and "real-life violence" (e.g., Potter, 2003, pp. 87f). Griffiths (1999), for example, criticized that findings in the realm of video game violence are mainly based on observations of children's free play. This means that in fact aggressive *play* was measured and not aggressive *behavior* (Silvern & Williamson, 1987). Aggressive play or imitation does not imply the intent to harm. Interestingly, effects for aggressive play are considerably higher than for aggressive behavior (Cooper & Mackie, 1986; Hellendoorn & Harinck, 1997). The diversity in definitions and operationalizations may be another reason why some people in industry, politics, and even in academia draw inconsistent conclusions about the effects of video games on aggression. Additionally, this ambiguity makes it problematic to compare video game effect sizes to those in other fields of research. The operationalizations of lung cancer or an HIV infection are simply undisputed and, thus, the measurements more reliable than those of complex psychological variables like aggression or violent behavior.

Finally, a noteworthy methodological challenge stems from comparisons between interactive and noninteractive violent stimuli. It may be assumed that effects of violent state-of-the-art video games are higher than compared with television because the player is much more actively involved. However, this assumption is difficult to investigate, particularly in experimental designs. How can a noninteractive violent stimulus be created that is comparable to an interactive one, or vice versa? The control group could just watch a game that the experimental group played, but it may not look like the games played by the experimental groups. It is the nature of an interactive game that every player can play his or her "own" game. Klimmt, Vorderer, and Ritterfeld (2004) discussed possible solutions and proposed a design to compare interactive with noninteractive stimuli (see also Vorderer, 2000; Ritterfeld, Weber, Fernandes, & Vorderer, 2004).

CONCLUSION AND FUTURE RESEARCH

The current scientific evidence on negative consequences of playing violent video games, as well as the confirmed effect sizes, may be insufficient to ban violent video games. Nevertheless, we think that the pattern of significant, positive, rising, and notable effect sizes from diverse best practice studies ought to be sufficient to justify funding of national and international long-term studies. Such studies do not exist to date and are badly needed, both from academia as well as from the video game industry. Also, no study has ever investigated how players *understand* the video games they play. We should not expect violent video games (or other games for that matter) to affect players directly. Instead, it is more promising to expect possible effects to be moderated by the way users interpret the game; for example, they may interpret it as "just a game" and an opportunity to socialize with friends, or as a simulation of reality and a way to prepare for social life. Such interpretations of game play, and not the game itself, are the most fundamental cause of what the game can do to attitudes, affects, and behaviors of users (Potter & Tomasello, 2003).

We still know very little about players' interpretations and feelings while playing video games. We do not even know why exactly video games are so appealing, although we have learned what makes traditional entertainment so attractive (Vorderer, 2001; Zillmann & Vorderer, 2000). We certainly need to find out more about what children, adolescents, and adults do with a video game in order to understand what a video game, especially a violent game, can do to them.

REFERENCES

Abelson, R. P. (1985). A variance explanation paradox: When little is a lot. *Psychological Bulletin, 97,* 129–133.

Anderson, C. A. (2003). Violent video games: Myths, facts, and unanswered questions. *Psychological Science Agenda: Science Briefs, 16(5),* 1–3. Retrieved June 11, 2004, from http://www.apa.org/science/psa/sb-anderson.html.

Anderson, C. A. (2004). An update on the effects of playing violent video games. *Journal of Adolescence, 27,* 113–122.

Anderson, C. A., Berkowitz, L., Donnerstein, E., Huesmann, R. L., Johnson, J., Linz, D., Malamuth, N., & Wartella, E. (2003). The influence of media violence on youth. *Psychological Science in the Public Interest, 4,* 81–110.

Anderson, C. A., & Bushman, B. J. (2001). Effects of violent video games on aggressive behavior, aggressive cognition, aggressive affect, physiological arousal, and prosocial behavior: A meta-analytic review of the scientific literature. *Psychological Science, 12,* 353–359.

Anderson, C. A., & Bushman, B. J. (2002a). Human aggression. *Annual Review of Psychology, 53,* 27–51.

Anderson, C. A., & Bushman, B. J. (2002b). The effects of media violence on society. *Science, 295,* 2377–2379.

Anderson, C. A., & Dill, K. E. (2000). Video games and aggressive thoughts, feelings, and behavior in the laboratory and in life. *Journal of Personality and Social Psychology, 78,* 772–790.

Anderson, C. A., & Huesmann, L. R. (2003). Human aggression: A social-cognitive view. In M. A. Hogg & J. Cooper (Eds.), *Handbook of social psychology* (pp. 296–323). London: Sage.

Atkin, C. (1983). Effects of realistic TV violence vs. fictional violence on aggression. *Journalism Quarterly, 60,* 615–621.

Bandura, A. (1973). *Aggression: A social learning analysis.* Englewood Cliffs, NJ: Prentice-Hall.

Bandura, A. (2001). Social cognitive theory of mass communication. *Media Psychology, 3*, 265–299.

Bandura, A. (2002). Social cognitive theory of mass communication. In J. Bryant & D. Zillman (Eds.), *Media effects. Advances in theory and research* (pp. 121–154). Mahwah, NJ: Lawrence Erlbaum Associates.

Berkowitz, L. (1993). *Aggression: Its causes, consequences, and control.* New York: McGraw-Hill.

Bowman, R. P., & Rotter, J. C. (1983). Computer games: Friend or foe? *Elementary School Guidance and Counseling, 18*, 25–34.

Bryant, J., & Miron, D. (2003). Excitation-transfer theory and three-factor theory of emotion. In J. Bryant, D. Roskos-Ewoldsen, & J. Cantor (Eds.), *Communication and emotion: essays in honor of Dolf Zillmann* (pp. 31–59). Mahwah, NJ: Erlbaum.

Bush, G., Luu, P., & Posner, M. (2000). Cognitive and emotional influences in anterior congulate cortex. *Trends in Cognitive Sciences, 4*, 215–222.

Bushman, B. J. (2002). Does venting anger feed or extinguish the flame? Catharsis, rumination, distraction, anger, and aggressive responding. *Personality and Social Psychology Bulletin, 28*, 724–731.

Carnagey, N. L., & Anderson, C. A. (2003). Theory in the study of media violence: The general aggression model. In D. Gentile (Ed.), *Media violence and children* (pp. 87–106). Westport, CT: Praeger.

Calvert, S. L., & Tan, S. (1994). Impact of virtual reality on young adults' physiological arousal and aggressive thoughts: Interaction versus observation. *Journal of Applied Developmental Psychology, 15*, 125–139.

Cohen, J. (1988). *Statistical power analysis for the behavioral sciences.* Hillsdale, NJ: Lawrence Erlbaum Associates.

Collwell, J., & Payne, J. (2000). Negative correlates of computer game play in adolescents. *British Journal of Psychology, 91*, 295–310.

Cooper, J., & Mackie, D. (1986). Video games and aggression in children. *Journal of Applied Social Psychology, 16*, 726–744.

Davidson, R. J., Putnam, K. M., & Larson, C. L. (2000). Dysfunction in the neural circuitry of emotion regulation—a possible prelude to violence. *Science, 289*, 591–594.

Dill, K. E., & Dill, J. C. (1998). Video game violence: A review of the empirical literature. *Aggression and Violent Behavior: A Review Journal, 3*, 407–428.

Durkin, K. (1995). *Computer games: Their effects on young people.* Sydney, NSW: Office of Film and Literature Classification.

Durkin, K., & Aisbett, K. (1999). *Computer games and Australians today.* Sydney, NSW: Office of Film and Literature Classification.

Durkin, K., & Barber, B. (2002). Not so doomed: Computer game play and positive adolescent development. *Applied Developmental Psychology, 23*, 373–392.

Ellis, P. (2002). The research base on the impact of exposure to sexually explicit material: What theory and empirical studies offer. In D. Thornburgh & H. S. Lin (Eds.), *Youth, pornography, and the Internet* (pp. 143–160). Washington, DC: National Academies Press.

Entertainment Software Association. (2004). *Essential facts about the computer and video game industry.* Retrieved June 3, 2004, from http://www.theesa.com/EFBrochure.pdf.

Federal Trade Commission (2000). *Marketing violent entertainment to children: A review of self regulation and industry practices in the motion picture, music recording, and gaming industries.* Retrieved June 20, 2004, from http://www.ftc.gov/bcp/workshops/violence/.

Federal Trade Commission (2002). *Marketing violent entertainment to children: A twenty-one month follow-up review of industry practices in the motion picture, music recording and electronic game industries.* Retrieved June 20, 2004, from http://www.ftc.gov/bcp/workshops/violence/.

Feshbach, S. (1955). The drive-reducing function of fantasy behavior. *Journal of Abnormal and Social Psychology, 50*, 3–11.

Früh, W. (1991). *Medienwirkungen: Das dynamisch-transaktionale Modell [Media effects: The dynamic-transactional model].* Opladen, Germany: Westdeutscher Verlag.

Früh, W., & Schönbach, K. (1982). Der dynamisch-transaktionale Ansatz. Ein neues Paradigma der Medienwirkungen [The dynamic-transactional approach. A new paradigm of media effects]. *Publizistik, 27*, 74–88.

Funk, J. B., Baldacci, H., Pasold, T., & Baumgardner, J. (2004). Violence exposure in real-life, video games, television, movies, and the Internet: Is there desensitization? *Journal of Adolescence, 27*, 23–39.

Gentile, D. A., Lynch, P. J., Linder, J. R., & Walsh, D. A. (2004a). The effects of violent video game habits on adolescent hostility, aggressive behaviors, and school performance. *Journal of Adolescence, 27*, 5–22.

Gentile, D. A., Walsh, D. A., Ellison, P. R., Fox, M., & Cameron, J. (2004b). *Media violence as a risk factor for children: A longitudinal study.* Manuscript under review, Iowa State University at Ames.

Gerbner, G., Gross, L., Morgan, M., Signorielli, N., & Shanahan, J. (2002). Growing up with television: Cultivation processes. In J. Bryant & D. Zillman (Eds.), *Media effects. Advances in theory and research.* (2nd ed.), (pp. 43–68). Mahwah NJ: Lawrence Erlbaum Associates.

Golden, L. (1992). *Aristotle on tragic and comic mimesis.* Atlanta, G A: Scholars Press.

Griffiths, M. (1997). Video games and aggression. *Psychologist, 10*, 397–401.

Griffiths, M. (1999). Violent video games and aggression: A review of the literature. *Aggression and Violent Behavior, 4*, 203–212.

Grodal, T. (2000). Video games and the pleasures of control. In D. Zillman & P. Vorderer (Eds.), *Media entertainment: The psychology of its appeal* (pp. 197–212). Mahwah, NJ: Lawrence Erlbaum Associates.

Grossman, D. (2000). Violent video games are mass-murder simulators. *Executive Intelligence Review, 27*(22), 74–79.

Gunter, B. (1998). *The effect of video games on children: The myth unmasked.* Sheffield, UK: Sheffield Academic Press.

Gunter, B., Harrison, J., & Wykes, M. (2003). *Violence on television. Distribution, form, context, and themes.* Mahwah, NJ: Lawrence Erlbaum Associates.

Haninger, K., Ryan, M. S., & Thompson, K. M. (2004). Violence in Teen-rated video games. *Medscape General Medicine, 6*(1). Available at: http://www.medscape.com/viewarticle/468087.

Hearold, S. (1986). A synthesis of 1043 effects of television on social behavior. In G. Comstock (Ed.), *Public communication and behavior. Vol. 1* (pp. 65–133). San Diego, CA: Academic Press.

Hellendoorn, J., & Harinck. F. (1997). War toy play and aggression in Dutch kindergarten children. *Social Development, 6*, 340–354.

Huesmann, L. R., Moise-Titus, J., Podolski, C. L., & Eron, L. D. (2003). Longitudinal relations between children's exposure to TV violence and their aggressiveness and violent behavior in young adulthood: 1977–1992. *Developmental Psychology, 39*, 201–221.

Interactive Digital Software Association v. St. Louis County, *No. 02-3010 (8th United States Court of Appeals June 3rd, 2003).* Retrieved May 21, 2004, from http://www.ca8.uscourts.gov/tmp/023010.html.

Jo, E., & Berkowitz, L. (1994). A priming effect analysis of media influences: An update. In J. Bryant & D. Zillman (Eds.), *Media effects. Advances in theory and research* (2nd ed.), (pp. 43–60). Hillsdale, NJ: Lawrence Erlbaum Associates.

Kaiser Family Foundation. (2002). *Key facts: Children and video games.* Retrieved June 24, 2004, from http://www.kff.org/entmedia/3271-index.cfm.

Kenny, D. A. (1973). A quasi-experimental approach to assessing treatment effects in the nonequivalent control group design. *Psychological Bulletin, 82*, 345–362.

Kestenbaum, G. I., & Weinstein, L. (1985). Personality, psychopathology, and developmental issues in male adolescent video game use. *Journal of the American Academy of Child Psychiatry, 24*, 208–212.

Klimmt, C. (2004). Computer- und videospiele [Computer- and video games]. In R. Mangold, P. Vorderer, & G. Bente (Eds.), *Lehrbuch der Medienpsychologie [Media psychology textbook]* (pp. 695–716). Göttingen: Hogrefe.

Klimmt, C., Vorderer, P., & Ritterfeld, U. (2004). Experimentelle Medienforschung mit interaktiven Stimuli: Zum Umgang mit Wechselwirkungen zwischen, Reiz' und, Reaktion' [Experimental media research with interactive stimuli: Dealing with interaction between "stimulus" and „"response"]. In W. Wirth, E. Lauf, & A. Fahr (Eds.), *Forschungslogik und -design in der Kommunikationswissenschaft* [Reserach logic and design in communication research] (pp. 142–156). München, Germany: von Halem.

Koepp, M. J., Gunn, R. N., Lawrence, A. D., Cunningham, V. J., Dagher, A., Jones, T., Brooks, D. J., Bench, C. J., & Grasby, P. M. (1998). Evidence for striatal dopamine release during a video game. *Nature, 393*, 266–268.

Krahé, B., & Moeller, I. (2004). Playing violent electronic games, hostile attributional style, and aggression-related norms in German adolescents. *Journal of Adolescence, 27*, 53–69.

Kunczik, M. (2002). Gewaltforschung [Research on violence]. In M. Schenk (Ed.), *Medienwirkungsforschung [Media effects research].* (2nd ed., pp. 206–238). Tübingen, Germany: Mohr Siebeck.

Lazarsfeld, P. F. (1940). Panel studies. *Public Opinion Quarterly, 4*, 122–128.

Lynch, P. J. (1994). Type A behavior, hostility, and cardiovascular function at rest and after playing video games in teenagers. *Psychosomatic Medicine, 56*, 152.

Lynch, P. J. (1999). Hostility, Type A behavior, and stress hormones at rest and after playing violent video games in teenagers. *Psychosomatic Medicine, 61*, 113.

Mansfield, D. (2003, November 11). "Grand Theft Auto" makers fight $246M lawsuit. *USA Today.* Retrieved May 10, 2004, from http://www.usatoday.com/tech/news/2003-11-11-gta-lawsuit_x.htm.

Mathews, V. P., Kronenberger, W. G., Wang, Y., Lurito, J. T., Lowe, M. J., & Dunn, D. (2005). Media violence exposure and frontal lobe activation measured by functional magnetic resonance imaging in aggressive and non aggressive adolscents. *Journal of Computer Assisted Tomography, 29*, 287–293.

National Center for Injury Prevention and Control. (2001). *Injury fact book 2001–2002.* Atlanta, Centers for Disease Control and Prevention.

National Institute on Media and the Family. (2002). *MediaWise video game report card.* Retrieved June 24, 2004, from http://www.mediafamily.org/research/report_vgrc_index.shtml.

National Television Violence Study. (1999). *Technical report* (Vol. 3). Thousand Oaks, CA: Sage.

Paik, H., & Comstock, G. (1994). The effects of television violence on antisocial behavior: A meta-analysis. *Communication Research, 21*, 516–546.

Pietrini, P., Guazzelli, M., Basso, G., Jaffe, K., & Grafman, J. (2000). Neural correlates of imaginal aggressive behavior assessed by positron emission tomography in healthy subjects. *American Journal of Psychiatry, 157,* 1772–1781.

Potter, W. J. (2003). *The 11 myths of media violence.* Thousands Oaks, CA: Sage.

Potter, W. J., & Tomasello, T. K. (2003). Building upon the experimental design in media violence research: The importance of including receiver interpretations. *Journal of Communication, 53,* 315–329.

Prentice, D. A., & Miller, D. (1992). When small effects are impressive. *Psychological Bulletin, 112,* 160–164.

Ritterfeld, U., Weber, R., Fernandes, S., & Vorderer, P. (2004). Think science! Entertainment education in interactive theaters. *Computer in Entertainment, 1*(2). Retrieved May 30, 2004, from http://doi.acm.org/10.1145/973801.973819.

Roe, K., & Muijs, D. (1998). Children and computer games: A profile of heavy users. *European Journal of Communication, 13,* 181–200.

Rosenthal, R. (1986). Media violence, antisocial behavior, and the social consequences of small effects. *Journal of Social Issues, 42,* 141–154.

Rule, B. K., & Ferguson, T. J. (1986). The effects on media violence on attitude, emotions, and cognitions. *Journal of Social Issues, 42,* 29–50.

Scheele, B. (2001). Back from the grave: Reinstating the catharsis concept in the psychology of reception. In D. Schram & G. J. Steen (Eds.), *The psychology and sociology of literature: In honor of Elrud Ibsch* (pp. 201–224). Amsterdam: J. Benjamins.

Scheff, T. J., (1979). *Catharsis in healing, ritual, and drama.* Berkeley: University of California Press.

Schneider, E. F., Lang, A., Shin, M., & Bradley, S. D. (2004). Death with a story. How story impacts emotional, motivational, and physiological responses to first-person shooter video games. *Human Communication Research, 30,* 361–375.

Scott, D. (1995). The effect of video games on feelings of aggression. *The Journal of Psychology, 129,* 121–132.

Sherry, J. L. (2001). The effects of violent video games on aggression. A meta-analysis. *Human Communication Research, 27,* 409–431.

Shrum, L. J. (2002). Media consumption and perceptions of social reality: Effects and underlying processes. In J. Bryant & D. Zillmann (Eds.), *Media effects. Advances in theory and research* (2nd ed.), (pp. 69–95). Mahwah, NJ: Lawrence Erlbaum Associates.

Silvern, S. B., & Williamson, P. A. (1987). The effects of video game play on young children's aggression, fantasy, and prosocial behavior. *Journal of Applied Developmental Psychology, 8,* 453–462.

Slater, M. D. (2003). Alienation, aggression, and sensation-seeking as predictors of adolescent use of violent film, computer and website content. *Journal of Communication, 53,* 105–121.

Slater, M. D., Henry, K. L., Swaim, R., & Anderson, L. (2003). Violent media content and aggression in adolescents: A downward-spiral model. *Communication Research, 30,* 713–736.

Smith, A. (1776). *An inquiry into the nature and causes of the wealth of nations.* London: Methuen.

Smith, S. L., Lachlan, K., & Tamborini, R. (2003). Popular video games: Quantifying the presentation of violence and its context. *Journal of Broadcasting and Electronic Media, 47,* 58–76.

Sparks, G. G., & Sparks, C. W. (2000). Violence, mayhem, and horror. In D. Zillman & P. Vorderer (Eds.), *Media entertainment: The psychology of its appeal* (pp. 73–91). Mahwah, NJ: Lawrence Erlbaum Associates.

Sparks, G. G., & Sparks, C. W. (2002). Effects of media violence. In J. Bryant & D. Zillman (Eds.), *Media effects. Advances in theory and research* (2nd ed.), (pp. 269–285). Mahwah. Hillsdale, NJ: Lawrence Erlbaum Associates.

Sterzer, P., Stadler, C., Krebs, A., Kleinschmidt, A., & Poustka, F. (2003). Reduced anterior cingulate activity in adolescents with antisocial conduct disorder confronted with affective pictures, *NeuroImage, 19*(2), Supplement 1, 23.

Tannenbaum, P. H., & Zillman, D. (1975). Emotional arousal in the facilitation of aggression through communication. In L. Berkowitz (Ed.), *Advances in experimental social psychology* (pp. 149–192). New York: Academic Press.

Thompson, K. M., & Haninger, K. (2001). Violence in E-rated video games. *Journal of American Medical Association, 286,* 591–598.

van Mierlo, J., & van den Bulck, J. (2004). Benchmarking the cultivation approach to video game effects: A comparison of the correlates of TV viewing and game play. *Journal of Adolescence, 27,* 97–111.

Vorderer, P. (2000). Interactive entertainment and beyond. In D. Zillman & P. Vorderer (Eds.), *Media entertainment: The psychology of its appeal* (pp. 21–36). Mahwah, NJ: Lawrence Erlbaum Associates.

Vorderer, P. (2001). It's all entertainment—sure. But what exactly is entertainment? Communication research, media psychology, and the explanation of entertainment experiences. *Poetics, 29,* 247–261.

Vorderer, P., & Weber, R. (2003). Unterhaltung als kommunikationswissenschaftliches Problem. Ansaetze einer konnektionistischen Modellierung [Entertainment as a problem in communication sciences. First ideas of a connectionist approach.] In W. Früh & H. J. Stiehler (Eds.), *Theorie der Unterhaltung. Ein interdisziplinaerer Diskurs [Theory of entertainment. An interdisciplinary discourse]* (pp. 136–159). Köln, Germany: von Halem.

Weber, R., Ritterfeld, U., & Mathiak, K. (in press). Does playing violent video games induce aggression? Empirical evidence of a functional magnetic resonance imaging study. *Media Psychology.*

Zillmann, D. (1978). Attribution and misattribution of excitatory reactions. In J. H. Harvey, W. J. Ickes, & R. F. Kidd (Eds.), *New directions in attribution research* (Vol. 2, pp. 335–368). Hillsdale, NJ: Lawrence Erlbaum Associates.

Zillmann, D. (1994). Mechanisms of Emotional involvement with drama. *Poetics, 23*, 33–51.

Zillmann, D. (2000). The psychology of suspense in dramatic exposition. In P. Vorderer, H. J. Wulff, and M. Friedrichsen (Eds.), *Suspense: Conceptualizations, theoretical analyses, and empirical explorations* (pp. 199–231). Mahwah, NJ: Lawrence Erlbaum Associates.

Zillmann, D. (2003). Theory of affective dynamics: emotions and moods. In J. Bryant, D. Roskos-Ewoldsen, & J. Cantor (Eds.), *Communication and emotion: Essays in honor of Dolf Zillmann* (pp. 533–567). Mahwah, NJ: Lawrence Erlbaum Associates.

Zillman, D., & Vorderer, P. (Eds.). (2000). *Media entertainment: The psychology of its appeal.* Mahwah, NJ: Lawrence Erlbaum Associates.

A Theoretical Model of the Effects and Consequences of Playing Video Games

Katherine E. Buckley and Craig A. Anderson
Iowa State University

Though there has been considerable discussion of video game effects in several research literatures, theoretical integrations have been somewhat rare. Our own empirical work has focused primarily on the effects of violent video games on those who play them (e.g., Anderson et al., 2004; Anderson & Dill, 2000). That work has been framed in terms of the General Aggression Model, an integrative social-cognitive model designed to handle all influences on all types of human aggression (e.g., Anderson & Bushman, 2002; Anderson & Carnagey, 2004; Anderson & Huesmann, 2003). But that model itself can be further generalized to account for nonviolent effects of video games, many of which may well be quite beneficial to the individual as well as to larger society. Our primary goal in this chapter is to elucidate such a general model of video game effects.

WHO IS PLAYING WHAT KIND OF GAMES?

According to the Entertainment Software Association, 50 % of all U.S. Americans play video games, pushing entertainment software sales in the United States to $7 billion in 2003 (2004). This is more than double the $3.2 billion video games earned in domestic sales in 1995 (ESA, 2004). Although computers are not yet in every U.S. household, they are common and also are increasingly used in workplaces and schools. A 1997 U.S. census survey found that 36.6 % of households owned a home computer and that 47% of adults used computers daily either at home or work (Kominski & Newburger, 1999). In comparison, 75% of children 3–17 used computers daily either at home or school (Kominski & Newburger, 1999). In fact, 70% of children used computers *in schools* (Kominski & Newburger, 1999). More recent data suggest that computer ownership has grown substantially, so that by 2000, 70 % of homes with children 2–17 had computers (Woodard & Gridina, 2000). Similarly, 68% of homes with children 2–17 have video game equipment (Woodard & Gridina, 2000). A survey conducted in 2003 found that 87 % of children regularly play video games, but that these games are more popular with boys (96%) than girls (78%; Walsh, Gentile, Gieske, Walsh, & Chasco, 2003). In a recent survey of over 600 8th- and 9th-grade students, children averaged 9 hours per week of video game play overall, with boys averaging 13 hours per week and girls averaging 5 hours per week (Gentile, Lynch, Linder, & Walsh, 2004).

There are thousands of video games available. The content includes just about anything imaginable in both education and entertainment genres. Of much public concern has been the preponderance of games with violent content. A recent analysis found that about 89% of video games contain some violent content (Children Now, 2001). More than half of fourth- through eighth-grade children report preferences for games in which the main action is predominantly violent (Buchman & Funk, 1996; Funk, 1993). Other surveys of children and their parents found that: (a) about two thirds of children named violent games as their favorites; (b) about two thirds of parents are unable to correctly name their child's favorite game; and (c) 70% of the time that parents are incorrect, children describe their favorite game as violent (e.g., Funk, Flores, Buchman, & Germann, 1999). Shibuya and Sakamoto (2003) reported similar results in Japan, finding that 85% of the most popular video games of Japanese fifth graders contained violent content (for reviews on the content of video games, see B. P. Smith, chap. 4, this volume; S. Smith, chap. 5, this volume).

WHAT DO VIDEO GAMES TEACH?

Video games are used as teaching tools in situations ranging from elementary schools to military bases. McDonalds uses video games in employee training seminars. The Mayo Clinic uses video games like *Name That Congenital Abnormality* to train residents (Yaman, 2004). The Marcom Group, a company that specializes in workplace safety compliance training, uses interactive CDs, online services, and "jeopardy" style games to teach employees how to handle hazardous waste and similar tasks (Marcom, 2004).

Military Uses

One of the largest groups that has embraced the use of video games for training is the U.S. military. The United States Army seems to believe strongly in the ability of video games to train soldiers for combat, flying, driving tanks, and commanding troops. The Army Program Executive Office for Simulation, Training and Instrumentation (PEO STRI), the unit that oversees the implementation of video game training, has an operating budget of over $1 billion and "is dedicated to putting the power of simulation into the hands of our warfighters." PEO STRI has commissioned a family of simulations consisting of "live environment engagement capabilities" that will replicate "weapon effects of combat systems for collective training" (Public Affairs Office, US Army, 2004).

The Marines have even created a training game using a version of the commercially available Game *Doom* (Prensky, 2001). *Marine Doom* is played as a networked game by four-member teams on four separate computers in the same room. In the game, their goal is to coordinate their movements to eliminate an enemy bunker. According to Prensky (2001), the Marines are learning teamwork, communication, and concepts of command and control.

The military also uses games and online training to teach Joint Doctrine, the rules that govern how the branches of the military cooperate. In 2000, the U.S. military released *Joint Force Employment (JFE)*, a multimedia game designed to allow soldiers to "adjust friendly and enemy forces...to test varying military possibilities...and compete against state of the art computer artificial intelligence" (DTIC, 2004a). According to the US Department of

Defense, "*JFE* represents a true multi-media environment for joint doctrine education" (DTIC, 2004a).

The military has also released a series of courses that consist "of interactive multimedia presentations on key information contained in Joint Doctrine designed to enhance the Joint Doctrine learning experience" (DTIC, 2004b). The courses teach "primary lessons in planning, deploying, and employing joint military forces" (DTIC, 2004b). The goal of these online courses is to "bring Joint Doctrine to life" by presenting the information in a convenient format, employing varied instructional techniques, and taking advantage of the latest interactive multimedia technologies (DTIC, 2004b).

Finally, in June of 2004, elements of the United States military and Special Operations, as well as forces from the United Kingdom, Canada, the Netherlands, Norway, France, Germany, and Peru, participated in Combined Joint Task Force Exercise (CJTFEX) 04-2 (JFCOM, 2004). This was an exercise designed to coordinate a global network of live and virtual training components to train commanders to lead a global military. During this exercise, live participants took part from the waters off the U.S. East Coast, while virtual participants took part via computer networks from 20 different sites.

School and Other Intended Educational Uses

In schools, too, computer games have been effectively used to teach algebra and geometry (Corbett, Koedinger, & Hadley, 2001), biology (Ybarrondo, 1984), photography (Abrams, 1986), golfing skills (Fery & Ponserre, 2001), and computer programming (Kahn, 1999). Students with severe learning disabilities have learned life skills like grocery shopping in virtual reality environments (Standen & Cromby, 1996). The Pennsylvania Department of Migrant Education used a video game format to successfully teach migrant children math, reading, English fluency, and critical thinking skills (Winograd, 2001). A recent meta-analysis of the effectiveness of using educational software programs found positive effects on both early reading development ($d = 0.35$; $N = 26$) and math ($d = 0.45$; $N = 13$) (Murphy, et al., 2002).

Lieberman (1998) studied the effectiveness of a video game designed to teach diabetic children how to care for their disease. She found that children enjoyed the game as much as comparison children enjoyed playing pinball and spent the same amount of time playing over 6 months. The diabetes video game promoted self-esteem and social support, increased knowledge of diabetes self-care, and was related to positive health behaviors and positive health outcomes. Ultimately, this video game was able to successfully teach the attitudes, skills, and behaviors that it was designed to teach (Lieberman, 1997). However, video games can teach even when they are not intended to be educational (Lieberman, chap. 25, this volume).

Unintended Outcomes

Video games also appear to teach perceptual skills. For example, people who play video games show better attention to cues across the visual field and attend to more visual cues overall than people who do not play video games (Green & Bavelier, 2003). A recent study also found that surgeons who have some experience playing video games perform laparoscopic surgery faster and make fewer mistakes (Rosser et al., 2004). Laparoscopic surgery is a type of surgery in which a tiny camera is inserted into the abdomen through one small incision, and surgical instruments are inserted through other small incisions; once the camera and instruments are inserted, surgeons use keypads and joystick devices to operate while they watch their performance on a monitor.

Unfortunately, video games also are associated with a number of negative outcomes. For example, some research has linked high levels of video game playing with smoking (Kasper, Welsh, & Chambliss, 1999), obesity (Subrahmanyam, Kraut, Greenfield, & Gross, 2000), and poorer academic performance (e.g., Anderson & Dill, 2000; Anderson, Gentile, & Buckley, under review; Harris & Williams, 1985; Lieberman, Chaffee, & Roberts, 1988; Lynch, Gentile, Olson, & van Brederode, 2001; van Schie & Wiegman, 1997). Of particular concern are the potential negative effects of playing violent video games on aggression-related outcome variables.

Video games are designed to be entertaining, challenging, educational, and (frequently) violent (Children Now, 2001; Dietz, 1998; Dill, Gentile, Richter, & Dill, in press). Unfortunately, as children grow, they spend more time playing entertaining games and less time playing educational games (Scantlin, 2000). In addition to children spending more time with entertaining games, these games are becoming increasingly violent (Dill et al., in press). The potential problem with most violent video games is that they may well teach maladaptive ways of thinking, feeling, and behaving. That is, even though the manifest function (of game developers and players) of such games is to entertain or be entertained, the latent function is the teaching and learning of aggressive ways of responding to real situations outside of the virtual world.

Violent Game Effects. Just as educational video games can be used to teach school children, employees, and physicians, violent video games can teach aggression (for a review on the effects of playing violent video games, see Weber, Ritterfeld, & Kostygina, chap. 23, this volume; Lee and Peng, chap. 22, this volume). A comprehensive review of media violence effects on aggression and aggression-related variables found "unequivocal evidence that media violence increases the likelihood of aggressive and violent behavior in both immediate and long-term contexts" (Anderson et al., 2003, p. 81). This report also found that even though the effects tend to be smaller for more severe forms of violence than for milder forms, the magnitude on severe forms is sufficiently large to warrant concern. Of more direct relevance to this chapter, aggressive behavior has been positively associated with both real-life violent video game play and laboratory exposure to graphically violent video games (e.g., Anderson, 2004; Anderson & Dill, 2000).

Meta-analytic reviews of the video game research literature reveal that violent video games increase aggressive behavior in children and adults (Anderson, 2004; Anderson & Bushman, 2001; Sherry, 2001). Experimental and nonexperimental studies with males and females in laboratory and field settings support this conclusion. Analyses also reveal that exposure to violent video games increase physiological arousal and aggression-related thoughts and feelings. Playing violent video games also decreases prosocial behavior.

More generally, a wide range of studies have yielded five effects of violent media (Gentile & Anderson, 2003). People exposed to a lot of violent media: (a) tend to become meaner, more aggressive, and more violent; (b) tend to see the world as a scarier place; (c) tend to become more desensitized to violence (both in the media and in real life), more callous, and less sympathetic to victims of violence; (d) tend to get an increased appetite for seeing more violent entertainment; and (e) are less likely to behave prosocially. None of these effects, either of violent media in general or of violent video games in particular, are surprising to anyone who understands learning and social interaction processes.

Although the empirical effects of violent video games have been robust, they remain poorly accepted outside the research community. In this chapter we propose that video games teach whatever concepts are repeatedly rehearsed within them—algebra, diabetes care, or aggression. We show how video games teach through both long- and short-term exposure. Although our model is based on early aggression theories and on research on violent media, it is also based on modern social-cognitive models of learning, social information processing, and social interaction. We believe that our model is general enough to explain how video games teach and influence behavior. Although research into the teaching effects of video games has yielded consistent results, theories about why this is so have been neglected (Issroff & Scanlon, 2002).

HOW DO VIDEO GAMES TEACH?

There are many factors about video games that make them excellent teachers: they successfully get people's attention (Krendl & Lieberman, 1988), they teach attitudes necessary for successful behaviors (Anderson & Dill, 2000), they enable people to feel competent about performing a task (Krendl & Lieberman, 1988; Kozma, 1991), they are motivating (Krendl & Lieberman, 1988), they allow people to actively participate instead of passively watch (Krendl & Lieberman, 1988) they show all the steps necessary to perform a behavior or series of behaviors (Gentile & Anderson, 2003), and they allow repetitive practicing (Kozma, 1991).

There are at least three reasons why video games are motivating. First, players have control over the game; they can work at their own ability level and speed and repeat material as needed. Second, video games give immediate feedback (Lieberman, chap. 25, this volume). Video games often reward players for participating; these rewards (e.g., winning extra points and lives) increase the frequency of that behavior in that game and increase the player's motivation to persist at the game, in addition to teaching more positive attitudes toward the content of that game (Gentile & Anderson, 2003). Third, video games are motivating because they challenge players but remain doable (Jones, 1998; Lieberman, chap. 25, this volume). Video games begin easy, often teaching players the skills needed to continue, and become progressively more complex and challenging. Klimmt and Hartmann (chap. 10, this volume) call this balance of skill acquisition and increased challenge effectance motivation—players will continue to play games that give them feelings of self-efficacy.

Learning Theories

According to social learning and social cognitive theories, people learn either through direct experience or through observation (e.g., Bandura, 1973, 1983; Mischel, 1973; Mischel & Shoda, 1995). People learn by observing both themselves and others who may or may not be present. We can learn by observing people in the same room, on television, in movies, or in video games. When we observe people behave, we also see whether that behavior is rewarded or punished. People are more likely to imitate behaviors if they also witness that behavior being rewarded and less likely to imitate behaviors if they also witness that behavior being punished (e.g., Bandura, 1965; Bandura, Ross, & Ross, 1963).

A key distinction that emerged from early learning research with animals—a distinction that is frequently lost in modern discussions—is the difference between learning and performance. Briefly, learning entails the acquisition of new information regardless of whether it is ever used or displayed, whereas performance is the observable use of the newly acquired information. Thus, an organism may well acquire new information (e.g., learn how solve a type of algebra problem) but never actually use it. This distinction is important in many contexts, including social learning contexts. The child who watches a violent television episode may well learn a novel way of harming a provocateur but may be dissuaded from using the new aggressive behavior by observing that the person who used it was punished. Yet at a later point in time, perhaps when the threat of punishment has been removed, the novel aggressive behavior may be enacted.

People can learn many complicated behaviors, attitudes, expectations, beliefs, and perceptual schemata through observation and participation in video games. And, as they observe and perform these new behaviors, they are also learning scripts. Scripts are organized sets of knowledge that define situations and guide behavior, for example, knowing what to expect and how to act in a restaurant or at a movie. Scripts are well-learned and well-rehearsed concepts often involving an understanding of causation, goals, and plans of action (Abelson, 1981; Anderson, Benjamin, & Bartholow, 1998; Schank & Abelson, 1977).

Once a script is learned, it can guide how we perceive and interpret similar situations, and can tell us how to behave appropriately. The more familiar the current situation is and the more similarities the current situation has with a previously learned script, the more likely that script will be activated. Many aspects of the initial situation can become part of the script. When these aspects are found in subsequent situations, they can become cues that activate the script. Overall, people's behavior is guided by learning, internalizing, and applying scripts (Huesmann, 1986, 1998).

Social Problem Solving

It is important to understand that video games frequently teach social problem-solving skills (Goldsworthy, Barab, & Goldsworthy, 2000), positive or negative, whether intended or not. Social problem solving refers to the many cognitive processes involved in interpreting a social situation and behaving in ways that are congruent with one's own goals. There are four processes used in social problem solving: (a) encoding and interpreting environmental cues, (b) generating and selecting goals, behaviors, or scripts to guide behavior; (c) evaluating how appropriate the selected script is, and (d) behaving followed by an interpretation of the reactions to that behavior (Anderson & Huesmann, 2003).

Cognitive Theories

Theories about scripts are based on theories of cognition and memory (Anderson et al., 1998; Berkowitz, 1990; Collins & Loftus, 1975). According to these theories, memory is a network of concepts and their links. Individual cognitive concepts are known as nodes; as we learn more about a concept, links are developed connecting that node with others. Or, links may be developed to connect concepts that have been experienced together. For example, "furry animal" might conjure images of dogs and cats or allergic reactions depending on one's experiences. Some people will have the furry animal node closely linked to the nodes for various types of furry animals, but for others, furry animals will be closely linked to allergies.

When a set of nodes become strongly linked together, they are known as knowledge structures. Well-known scripts are knowledge structures and as such they can include emotions, behavioral responses, and beliefs. Knowledge structures (a) develop from experience; (b) influence perception; (c) can become automated; (d) can contain affective states, behavioral scripts, and beliefs; and (e) are used to guide interpretations and responses to the environment (Anderson & Bushman, 2002).

Some concepts are more easily linked together, for example, humans appear to easily develop links between frustration, pain, and anger with aggression (Berkowitz, 1990). Highly similar concepts, and concepts that frequently are simultaneously activated, become strongly linked. Activating a node or knowledge structure depends on the number and strength of links to that node or knowledge structure. Some stimuli are not strong enough to completely activate the target knowledge structure; instead, they may only activate one or more nodes. This impartial activation is called priming the knowledge structure.

A primed knowledge structure is more likely to be used than other wholly inactive knowledge structures. For example, research has found that pictures of weapons can automatically prime aggressive thoughts (Anderson et al., 1998). However, the specific thoughts primed by pictures of weapons depends on the type of weapon (hunting vs. assault) *and* the individual's learning history (and hence, his knowledge structures) involving weapons. In other words, aggressive thoughts and subsequent aggressive behavior were interactively influenced by both the content of the weapon pictures and the past learning histories of the individual participants (Bartholow, Anderson, Carnagey, & Benjamin, 2004).

Priming also works on behavioral scripts. Bargh, Chen, and Burrows (1996) and Carver, Ganellen, Froming, and Chambers (1983) have shown that solving anagram problems whose solutions are aggressive words increases the chances of aggressive (or rude) behavior immediately afterwards, if participants are provoked.

A MODEL TO ORGANIZE AND UNDERSTAND THE EFFECTS OF VIDEO GAMES

Anderson and colleagues proposed the General Aggression Model (GAM) to explain how violent video games increase aggressive thoughts, feelings, and behaviors (for a short review of GAM, see Weber et al., chap. 24, this volume). GAM describes a continuous cycle of interaction between a person and the environment and combines elements from many theories including: social learning and social cognitive theories (e.g., Bandura, 1971, 1973; Bandura, Ross, & Ross, 1961, 1963; Mischel, 1973; Mischel & Shoda, 1995), Cognitive Neoassociationist Model (Berkowitz, 1986, 1990), social information-processing model (Anderson, Anderson, & Deuser, 1996; Anderson & Bushman, 2002; Anderson & Dill, 2000; Anderson & Huesmann, 2003; Crick & Dodge, 1994; Dodge & Crick, 1990), affective aggression model (Geen, 1990), script theory (Huesmann, 1986), and excitation transfer model (Zillmann & Bryant, 1983). Most recently, GAM has incorporated the developmental approaches of risk and resilience and the cumulative risk model (see Anderson et al., under review).

In this chapter we will expand this model of aggression into our General Learning Model (GLM). Although our model is based on early aggression theories and on research on violent media, it is also based on modern social-cognitive models and developmental approaches. We believe that our model is general enough to explain how video games teach and influence behavior.

Input Variables

A person's behavior is based on two types of input variables: personal and situational (see Fig. 24.1). Personal variables include what a person brings to the current situation, specifically: attitudes, beliefs, goals, behavioral tendencies, previous experience, and emotions. These internal variables show substantial consistency over time and situations, the result of consistent use of knowledge structures such as scripts, beliefs and schemas, and affective components (Anderson & Bushman, 2002; Anderson & Huesmann, 2003). Situational variables are the features of the environment around the individual; they include media, objects, settings, and other people. Situational variables may vary widely but may also be fairly consistent over time, as individuals tend to be in the same or similar situations repeatedly.

Personal variables that may influence people's ability to learn from video games include many variables that are relevant to learning in general: age, grade level, ability level (including

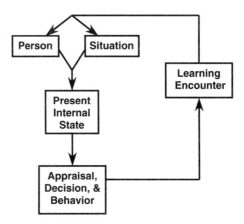

FIG. 24.1. The general learning model: simplified view.

learning disabled and low-performing students), income level, and self-esteem (Lieberman, 1998). Other variables that may influence people's ability to learn from video games include variables that are more specifically relevant to video games, including an individual's history of media exposure and how much a person's comprehension of information is affected by the surrounding field of information (field-dependence, field-independence; Ghinea & Chen, 2003). There are also personal variables that may be relevant to the effects of *violent* video games: player sex; age; bullies; victims of bullies; having poor social problem-solving skills; having poor emotion regulation abilities; being more hostile in personality; having a history of aggressive behavior; or having parents who do not monitor or limit video game play; and having aggressive beliefs, attitudes, values, long-term goals, and scripts (Anderson & Bushman, 2002).

Situational variables also influence people's ability to learn from video games, and perhaps the most important situational variables include aspect of the games themselves, such as: the extent to which a game keeps players interested and playing, game content (violent, nonviolent, educational), and current game exposure (number of minutes per exposure, number of exposures per week, number of weeks). Other variables inherent in video games include whether the game focuses on drill-and-practice for factual recall (*Reader Rabbit* or *Knowledge Munchers*), or uses simulations or role-playing to model reality (*The Sims* or *MS Flight Simulator;* Murphy et al., 2002; Squire, 2003). Situational variables that are relevant to the effects of violent video games include aggressive cues (weapons and aggressive words), provocation, frustration, pain, drugs, and other incentives (rewards) (e.g., Anderson & Bushman, 2002).

Interactions of Input Variables

Learning is the result of the complex combination of personal and situational variables. These variables can interact to increase or inhibit learning. For example, some children have low self-esteems, but by playing games that: (a) have a character similar to themselves and (b) are challenging without being too difficult, children can experience increased self-esteem (Lieberman, 1998). In fact, many researchers now view learning as a process through which personal predispositions are modified through situational influences (Huesmann, 1997; Tremblay, 2000). Similarly, all of the social-cognitive models of aggression agree that person variables and current situational variables combine to influence the individual's present internal state, for example, pain and trait hostility combine interactively to affect aggressive cognitions, so that a person who is high in trait hostility will react disproportionately to pain (Anderson, Anderson, Dill, & Deuser, 1998).

Routes

Input variables affect how a person responds by influencing one's internal state. Three types of interrelated states are especially important: cognition, affect, and arousal.

Cognition. Some input variables make constructs more readily accessible in memory, resulting in an increase in related thoughts, or scripts, or both. As script theory has contended, situational variables may activate scripts that can bias the interpretation of a situation and the possible responses to that situation (Huesmann, 1986). Research has shown that the process by which knowledge structures are activated is cognitive but can with practice become completely automatic and operate without awareness (Schneider & Shiffrin, 1977; Todorov & Bargh, 2002). That is, many cognitive processes are unconscious, and some that initially have conscious elements become automatized to the point of occurring outside of consciousness.

Cognitive variables that are influenced by personal and situational input variables include: thoughts, attributions, beliefs, attitudes, perceptual schema, expectation schema, and behavior scripts. Aggressive personal and situational input variables, for instance, can increase aggressive thoughts and attributional biases, aggressive beliefs and attitudes, aggressive perceptual and expectation schema, and aggressive behavior scripts. For example, several studies found that exposure to violent video games increases aggressive thoughts (e.g., Anderson & Bushman, 2001). Of course, nonviolent situational variables, such as playing a prosocial game, can increase many forms of nonviolent outcome variables, such as the accessibility of prosocial thoughts.

Affect. Input variables can directly influence mood and emotion, which can later influence behavior. Two examples of how affect can influence learning and behavior include mood-congruent cognition and mood-dependent memory. Mood-congruent cognition is the phenomenon in which a person's mood increases the processing of information that is affectively similar. Mood-dependent memory is the finding that information learned in a particular mood is best retrieved in that mood, regardless of the information's affect. This means that people will pay more attention to information that matches their mood, think about this information longer, and are more likely to remember this information later when in the same mood. For example, depressed people recall more negative information than positive (Berry, 1997), and any aversive stimulation, such as heat, can create negative affect, which can prime a network of cognitive structures that increases a person's aggressive cognitions and behavior (Berkowitz, 1990).

Another way affect can influence learning and behavior is through the mere-exposure effect; the repeated exposure of an object increases its attractiveness. This is true even when people are not aware of the exposure (Kunst-Wilson & Zajonc, 1980). Players often develop emotional responses to the characters and stories within video games; these emotions engage and motivate players (Lieberman, chap. 25, this volume). A third way involves systematic desensitization processes, in which repeated exposure to initially fearful stimuli in a positive context (e.g., violent stimuli in a fun video game) leads to a reduction in fearful emotions to that type of stimuli (e.g., Carnagey, Bushman, & Anderson, under review).

Arousal. Entertaining video games are generally arousing, as are many educational video games. However, too much or too little arousal can have a strong impact on learning by inhibiting the learning of new material or leaving the learner too bored to pay attention, respectively (Deshpande & Kawane, 1982; Yerkes & Dodson, 1908). If the material has already been well learned, increases in arousal are less likely to inhibit the retrieval and use of that information (Berkowitz, 1990). However, if the material is not well learned, increases in arousal are more likely to inhibit the learning and use of that information.

Interactions of Routes. As we have seen, input variables can influence cognition, affect, and arousal, but these three routes are also highly connected to one another; cognition and arousal influence affect (Schachter & Singer, 1962) and affect influences cognition and arousal (Bower, 1978). Too much arousal may inhibit ability to think about and learn new information; too little arousal may lead to lack of motivation and passive instead of active learners. Or, hostile cognitions and angry affect may bias which cognitive scripts and knowledge structures people use.

Outcomes

Outcomes from playing video games include learning facts, from drill and practice routines; learning specific behaviors, from playing simulation and role-playing games that model reality; learning new perceptual and decision schemata, from many types of games; and generating personality changes that occur when a person's habitual thought and behavior patterns begin to change, as a result of repeated video game play.

Long-Term Effects of Video Games: Factual Learning. Most educational software has focused on using drill and practice routines to enhance the learning of specific, concrete facts. This type of software is easily incorporated into classrooms as it allows students to practice individually as much or as little as needed (Squire, 2003). Students generally receive quick feedback about their performance and are able to repeat tasks until they are successful.

Long-Term Effects of Video Games: Learning Behaviors. Simulation and role-playing games attempt to model physical or social systems in a manner that is consistent with reality. However, unlike reality, simulations allow learners to manipulate variables that are normally unalterable (create a town's power supply), view phenomena from new perspectives (a town planner, a pilot), observe the system's behavior over time (How do time, growth, and natural disasters affect a town?), and pose hypothetical questions to a system (How does a limited power supply affect a town's growth?) (e.g., Squire, 2003). Thus, video games can enhance learning of many types of complex behaviors.

Long-Term Effects of Video Games: Changes in Personality. Personality results from the development and construction of knowledge structures. Influences include biological factors, of course, but are based primarily on life experiences, or "biosocial interactions." How people perceive the world and react to it depends upon the particular situational factors in their world and on the knowledge structures they have learned and habitually use (see Fig. 24.2). Creating long-lasting changes in people's knowledge structures also changes their personalities (Mischel & Shoda, 1995; Sedikides & Skowronski, 1990).

Learning—and ultimately, performance— depends on several complex internal processes— input variables affecting cognition, affect, and arousal, which in turn influence both processes of appraising the current situation and making appropriate decisions. The outcome of appraisal and decision processes also depends on a person's available attention. If a person has sufficient time and cognitive capacity, and if the immediate appraisal outcome is both important and *unsatisfying*, then the person will engage an effortful set of reappraisals (see Fig. 24.3). Reappraisals consist of searching for additional information in order to view the situation differently, perhaps in search of clarity or of a preferred outcome or interpretation. For example, if a likeable character in a mystery novel appears to have committed a murder,

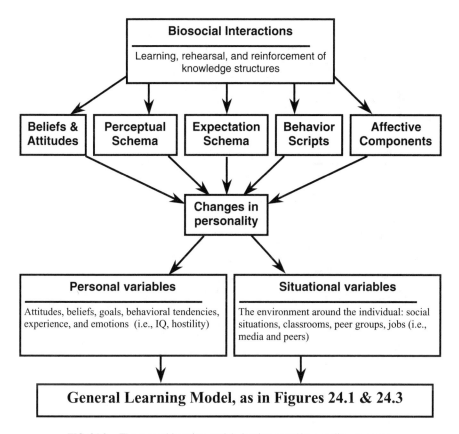

FIG. 24.2. The general learning model: developmental/personality processes.

one might reread earlier passages to learn whether this character is not so likeable, or to discover possible reinterpretations of the evidence that implied the character's misdeed. If resources are insufficient, or if the outcome of immediate appraisal is either unimportant *or* satisfying, then action and understanding will be dictated by the immediate appraisal and the knowledge structure accessed in that appraisal. Of course, one of the key elements required for reappraisal is time, and time is less likely to be available while watching television and films than reading novels, and is almost entirely absent while playing many video games.

The general learning model is well founded in psychological theories of learning. From those theories we understand that people can learn complicated behaviors through observation and that by observing and performing behaviors people are also learning scripts; overall, people's behavior is guided by learning, internalizing, and applying scripts. We also know that well-known scripts or knowledge structures can include emotions, behavioral responses, and beliefs. These knowledge structures (a) develop from experience; (b) influence perception; (c) can become automated; (d) can contain affective states, behavioral scripts, and beliefs; and (e) are used to guide interpretations and responses to the environment. These theories all adopt similar premises about the processing of information in social problem solving, the social-cognitive structures involved, the interacting role of emotions and cognitions, and the interaction of person and situation. The General Learning Model is itself merely an attempt to unify these many strands into a common framework.

We also know that video games use drill and practice techniques or simulate reality to allow players the opportunity to solve problems. Simulation games allow players to (a) encode

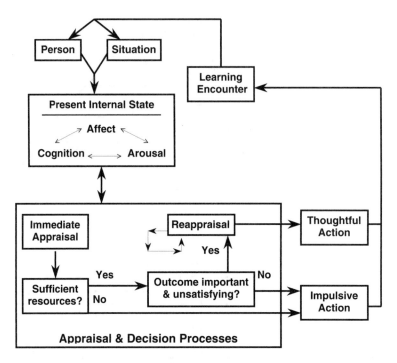

FIG. 24.3.   The general learning model: expanded causes and processes.

and interpret environmental cues; (b) generate and select goals, behaviors, or scripts to guide behavior; (c) evaluate how appropriate the selected script is; and (d) behave followed by an interpretation of the reactions to that behavior.

Furthermore, the cyclical process of GLM allows researchers to apply it to long-term effects of exposure to video games. We can view each episodic cycle of GLM as a learning trial, leading (with practice) to the development of well-rehearsed (and eventually automatized) knowledge structures of various kinds. Repeated exposure to certain stimuli makes related knowledge structures more readily accessible. Over time, these knowledge structures are activated more automatically and are more likely to be used in later situations. The development, automatization, and reinforcement of these knowledge structures change the individual's personality.

There are two final primary conclusions that should be drawn about the educational use of video games: games teach and content matters. We see in a variety of contexts that people learn by playing video games. Children are able to enhance their reading and math skills by playing games like *Reader Rabbit* and *Math Blaster*, and adults learn about heart disease and military concepts of teamwork, communication, and command and control from games like *Name That Congenital Abnormality* and *Marine Doom*. At this time, the idea that video games teach should not be in doubt, but researchers, educators, and parents need to pay more attention to the idea that the content of video games does matter. Video games teach regardless of their content, regardless of the intended outcome of the game developer, and regardless of the intended use of the game by the player. In brief, games teach whatever they inspire the game player to rehearse mentally and behaviorally. Enactment of what is learned depends on other factors, of course. If the individual never encounters a situation relevant to some particular fact or script, then that fact or script will not be used. Many people may never encounter a situation in which to apply the Pythagorean theorem (outside of a classroom), and so may never perform the actions and decisions that would make use of that knowledge. Similarly, many people may never be in a sufficiently provoking situation to trigger use of certain violent behavior scripts. However, the better learned that the underlying knowledge structure has become (i.e., the more contexts with which it is associated, the more easily and automatically it comes to mind . . .), the more likely it is to be used, whether in designing one's dream home (Pythagorean theorem) or lashing out at a provoking spouse.

As we have seen through the cyclical process of the GLM, each cycle is a learning trial, leading eventually to the development of well-rehearsed knowledge structures of various kinds and to the development, automatization, and reinforcement of these knowledge structures, thereby changing the individual's personality. The kind of knowledge structures a person develops depends on the content of the games played—content matters.

REFERENCES

Abelson, R. P. (1981). Psychological status of the script concept. *American Psychologist, 36*, 715–729.

Abrams, A. (1986, January). *Effectiveness of interactive video in teaching basic photography Skills.* Paper presented at the Annual Convention of the Association for Educational Communication and Technology, Las Vegas, NV.

Anderson, C. A. (2004). An update on the effects of violent video games. *Journal of Adolescence, 27*, 113–122.

Anderson, C. A., Anderson, K. B., & Deuser, W. E. (1996). Examining an affective aggression framework: Weapon and temperature effects on aggressive thoughts, affect, and attitudes. *Personality and Social Psychology Bulletin, 22*, 366–376.

Anderson, C. A., Benjamin, A. J., & Bartholow, B. D. (1998). Does the gun pull the trigger? Automatic priming effects of weapon pictures and weapon names. *Psychological Science, 9*, 308–314.

Anderson, C. A., Berkowitz, L., Donnerstein, E., Huesmann, L. R., Johnson, J. D., Linz, D., Malamuth, N. M., & Wartella, E. (2003). The influences of media violence on youth. *Psychological Science in the Public Interest, 4*, 81–110.

Anderson, C. A., & Bushman, B. J. (2001). Effects of violent video games on aggressive behavior, aggressive cognition, aggressive affect, physiological arousal, and prosocial behavior: A meta-analytic review of the scientific literature. *Psychological Science, 12*, 353–359.

Anderson, C. A., & Bushman, B. J. (2002). Human aggression. *Annual Review of Psychology, 53*, 27–51.

Anderson, C. A., & Carnagey, N. L. (2004). Violent evil and the general aggression model. In A. Miller (Ed.), *The social psychology of good and evil* (pp. 168–192). New York: Guilford.

Anderson, C. A., Carnagey, N. L., Flanagan, M., Benjamin, A. J., Eubanks, J., & Valentine, J. C. (2004). Violent video games: Specific effects of violent content on aggressive thoughts and behavior. *Advances in Experimental Social Psychology, 36*, 199–249.

Anderson, C. A., & Dill, K. E. (2000). Video games and aggressive thoughts, feelings, and behavior in the laboratory and in life. *Journal of Personality and Social Psychology, 78*, 772–790.

Anderson, C. A., Gentile, D. A., & Buckley, K. E. (2004). *violent video game effects on children and adolescents: Further developments and tests of the general aggression model. Manuscript submitted for publication.*

Anderson, C. A., & Huesmann, L. R. (2003). Human aggression: A social-cognitive view. In M. A. Hogg & J. Cooper (Eds.), *Handbook of social psychology.* (pp. 296–323). London: Sage.

Anderson, K. B., Anderson, C. A., Dill, K. E., & Deuser, W. E. (1998). The interactive relations between trait hostility, pain, and aggressive thoughts. *Aggressive Behavior, 24*, 161–171.

Bandura, A. (1965). Influence of models' reinforcement contingencies on the acquisition of imitative responses. *Journal of Personality & Social Psychology, 1*, 589–595.

Bandura, A. (1971). Social learning of moral judgments. *Journal of Personality & Social Psychology, 11*, 275–279.

Bandura, A. (1973). *Aggression: A social learning analysis.* Oxford, England: Prentice-Hall.

Bandura, A. (1983). Self-efficacy determinants of anticipated fears and calamities. *Journal of Personality & Social Psychology, 45*, 464–468.

Bandura, A., Ross, D., & Ross, S. A. (1961). Imitation of film-mediated aggressive models. *Journal of Abnormal & Social Psychology, 66*, 3–11.

Bandura, A., Ross, D., & Ross, S. A. (1963). Transmission of aggression through imitation of aggressive models. *Journal of Abnormal & Social Psychology, 63*, 575–582.

Bargh, J. A., Chen, M., & Burrows, L. (1996). Automaticity of social behavior: Direct effects of trait construct and stereotype activation on action. *Journal of Personality & Social Psychology, 71*, 230–244.

Bartholow, B. D., Anderson, C. A., Carnagey, N. L., & Benjamin, Jr., A. J. (2005). Interactive effects of life experience and situational cues on aggression: The weapons priming effect in hunters and nonhunters. *Journal of Experimental Social Psychology, 41*, 48–60.

Berkowitz, L. (1986). Situational influences on reactions to observed violence. *Journal of Social Issues, 42*, 93–106.

Berkowitz, L. (1990). On the formation and regulation of anger and aggression: A cognitive-neoassociationistic analysis. *American Psychologist, 45*, 494-503.

Berry, G. E. (1997). Information processing in anxiety and depression: Attention responses to mood congruent stimuli. *Dissertation Abstracts International, 58*(2-B), 967.

Bower, G. H. (1978). Emotional mood as a context for learning and recall. *Journal of Verbal Learning and Verbal Behavior, 17*, 573–1585.

Buchman, D. D., & Funk, J. B. (1996). Video and computer games in the 90s: Children's time commitment and game preference. *Children Today, 24*, 12–16.

Carnagey, N. L., Bushman, B. J., & Anderson, C. A. (2003). Video game violence desensitizes players to real world violence. Under review.

Carver, C. S., Ganellen, R. J., Froming, W. J., & Chambers, W. (1983). Modeling: An analysis in terms of category accessibility. *Journal of Experimental Social Psychology, 19*, 403–421.

Children Now (2001, December). *Fair play? Violence, gender and race in video games.* Sacramento, CA: Author.

Collins, A. M, & Loftus, E. F. (1975). A spreading-activation theory of semantic processing. *Psychological Review, 82*, 407–28.

Corbett, A. T., Koedinger, K. R., & Hadley W. (2001). Cognitive tutors: From the research classroom to all classrooms. In P. S. Goodman (Ed.), *Technology enhanced learning* (235–263). Mahwah, NJ: Lawrence Erlbaum Associates.

Crick, N. R., & Dodge, K. A., (1994). Social information-processing mechanisms in reactive and proactive aggression. *Child Development, 67*, 993–1002.

Deshpande, S. W., & Kawane, S. D. (1982). Anxiety and serial verbal learning: A test of the Yerkes-Dodson Law. *Asian Journal of Psychology and Education, 9*, 18–23.

Dietz, T. L. (1998). An examination of violence and gender role portrayals in video Games: Implications for gender socialization and aggressive behavior. *Sex Roles: A Journal of Research, 38*, 425–442.

Dill, K. E., Gentile, D. A., Richter, W. A., & Dill, J. C. (in press). Violence, sex, race, and age in popular video games: A content analysis. In E. Cole & D. J. Henderson (Eds.), *Featuring females: Feminist analyses of the media.* Washington, DC: American Psychological Association.

Dodge, K. A., & Crick, N. R. (1990). Social information-processing bases of aggressive behavior in children. *Personality & Social Psychology Bulletin. Special Issue: Illustrating the value of basic research, 16*, 8–22.

DTIC (Defense Technical Information Center). (2004a). *The central facility for the collection and dissemination of scientific and technical information for the Department of Defense.* Retrieved June 8, 2004, from http:// www.dtic.mil/.

DTIC (Defense Technical Information Center). (2004b). *The central facility for the collection and dissemination of scientific and technical information for the Department of Defense.* Retrieved June 8, 2004, from http:// www.dtic.mil/doctrine/history/historical.htm.

ESA (Entertainment Software Association). (2004). Retrieved May 4, 2004, from http://www.theesa.com/ pressroom.html.

Fery, Y. A., & Ponserre, S. (2001). Enhancing the control of force in putting by video game training. *Ergonomics, 44*, 1025–1037.

Funk, J. B. (1993). Reevaluating the impact of video games. *Clinical Pediatrics, 32*, 86–90.

Funk, J. B., Flores, G., Buchman, D. D., & Germann, J. N. (1999). Rating electronic games: Violence is in the eye of the beholder. *Youth & Society, 30*, 283–312.

Geen, R. G. (1990). *Human aggression. Mapping social psychology series.* Belmont, CA: Brooks/Cole.

Gentile, D. A., & Anderson, C. A. (2003). Violent video games: The newest media violence hazard. In D. Gentile (Ed.), *Media violence and children* (pp. 131–152). Westport, CT: Praeger.

Gentile, D. A., Lynch, P. L., Linder, J. R., & Walsh, D. A. (2004). The effects of violent video game habits on adolescent hostility, aggressive behaviors, and school performance. *Journal of Adolescence, 27*, 5–22.

Ghinea, G., & Chen, S. Y. (2003). The impact of cognitive styles on perceptual distributed multimedia quality. *British Journal of Educational Technology, 34*, 393–406.

Goldsworthy, R. C., Barab, S. A., & Goldsworthy, E. L. (2000). The STAR Project: Enhancing adolescents' social understanding through video-based, multimedia scenarios. *Journal of Special Education Technology, 15*, 13–26.

Green, C. S., & Bavelier, D. (2003). Action video game modifies visual selective attention. *Nature, 423*, 534–537.

Harris, M. B., & Williams, R. (1985). Video games and school performance. *Education, 105*, 306–309.

Huesmann, L. R. (1986). Psychological processes promoting the relation between exposure to media violence and aggressive behavior by the viewer. *Journal of Social Issues, 42*, 125–139.

Huesmann, L. R. (1997). Observational learning of violent behavior: Social and biosocial processes. In A. Raine & P. A. Brennan (Eds.), *Biosocial bases of violence. NATO ASI series: Series A: Life sciences, Vol. 292* (pp. 69–88). New York, : Plenum.

Huesmann, L. R. (1998). The role of social information processing and cognitive schema in the acquisition and maintenance of habitual aggressive behavior. In R. G. Geen, & E. Donnerstein (Eds.), *Human aggression: Theories, research, and implications for social policy* (pp. 73–109). San Diego, CA: Academic Press.

Issroff, K., & Scanlon, E. (2002). Educational technology: The influence of theory. *Journal of Interactive Media in education, 6*, 1–13.

JFCOM (United States Joint Forces Command). (2004). Retrieved June 8, 2004, from http://www.jfcom.mil/about/exercises/cjtfex04.htm.

Jones, M. G. (1998, February). *Creating electronic learning environments: Games, flow, and the user interface.* Paper presented at the meeting of the Association for Educational Communications and Technology, St. Louis, MO.

Kahn, K. (1999, June). *A computer game to teach programming.* Paper presented at the National Educational Computing Conference, Atlantic City, NJ.

Kasper, D., Welsh, S., & Chambliss, C. (1999). Educating students about the risks of excessive videogame usage. pennsylvania. (ERIC Document Reproduction Service No. ED426315).

Kominski, R., & Newburger, E. (1999, August). *Access denied: Changes in computer ownership and use: 1984–1997.* Paper presented at the Annual Meeting of the American Sociological Association, Chicago, Illinois.

Kozma, R. B. (1991). Learning with media. *Review of Educational Research, 61*, 179–211.

Krendl, K. A., & Lieberman, D. A. (1988). Computers and learning: A review of recent research. *Journal of Educational Computing Research, 4*, 367–389.

Kunst-Wilson, W. R., & Zajonc, R. B. (1980). Affective discrimination of stimuli that cannot be recognized. *Science, New Series, 207*, 557–558.

Lieberman, D. A. (1997). Interactive video games for health promotion: Effects on knowledge, self-efficacy, social support, and health. In R. L. Street Jr. & W. R. Gold (Eds.), *Health promotion and interactive technology: Theoretical applications and future directions. LEA's communication series* (pp. 103–120). Mahwah, NJ: Lawrence Erlbaum Associates.

Lieberman, D. A. (1998, July). *Health education video games for children and adolescents: Theory, design, and research findings.* Paper presented at the Annual Meeting of the International Communication Association, Jerusalem, Israel.

Lieberman, D. A., Chaffee, S. H., & Roberts, D. F. (1988). Computers, mass media, and schooling: Functional equivalence in uses of new media. *Social Science Computer Review, 6*, 224–241.

Lynch, P. J., Gentile, D. A., Olson, A. A., & van Brederode, T. M. (2001, April). *The effects of violent video game habits on adolescent aggressive attitudes and behaviors.* Paper presented at the Biennial Meeting of the Society for Research in Child Development, Minneapolis, MN.

Marcom (The Marcom Group, Ltd.). (2004). Quality safety and health training products for Today... and tomorrow. Retrieved May 2, 2004, from http://www.marcomltd.com/products.php#safe.html.

Mischel, W. (1973). Toward a cognitive social learning reconceptualization of personality. *Psychological Review, 80*, 252–283.

Mischel, W., & Shoda, Y. (1995). A cognitive-affective system theory of Personality: Reconceptualizing situations, dispositions, dynamics, and invariance in personality structure. *Psychological Review, 102*, 246–268.

Murphy, R. F., Penuel, W. R., Means, B., Korbak, C., Whaley, A., & Allen, J. E. (2002). *E-DESK: A review of recent evidence on the effectiveness of discrete educational software.* (Prepared for: Planning and Evaluation Service, U. S. Department of Education.) SRI International.

Prensky, M. (2001). *Digital game-based learning.* New York: McGraw-Hill.

Public Affairs Office, US Army. (2004). Army awards one Tactical Engagement Simulation System (OneTESS). Retrieved May 2, 2004, from http://www.peostri.army.mil.PAO/pressrelease/OneTESS.jsp.

Rosser, J. C. Jr., Lynch, P. J., Haskamp, L. A., Yalif, A., Gentile, D. A., & Giammaria, L. (2004, January). Are video game players better at laparoscopic surgery? Paper presented at the Medicine Meets Virtual Reality Conference, Newport Beach, CA.

Scantlin, R. M. (2000). Interactive media: An analysis of children's computer and video game use. *Dissertation Abstracts International, 60* (12-B), 6400.

Schachter, S., & Singer, J. (1962). Cognitive, social, and physiological determinants of emotional state. *Psychological Review, 69*, 379–399.

Schank, R. C., & Abelson, R. P. (1977). *Scripts, plans, goals and understanding: An inquiry into human knowledge structures.* Oxford, England: Lawrence Erlbaum Associates.

Schneider, W., & Shiffrin, R. M. (1977). Controlled and automatic human information processing: I. Detection, search, and attention. *Psychological Review, 84*, 1–66.

Sedikides, C., & Skowronski, J. J. (1990). Towards reconciling personality and social psychology: A construct accessibility approach. *Journal of Social Behavior & Personality, 5*, 531–546.

Sherry, J. L. (2001). The effects of violent video games on aggression: A meta-analysis. *Human Communication Research, 27*, 409–431.

Shibuya, A., & Sakamoto, A. (2003). The quantity and context of video game violence in Japan: Toward creating an ethical standard. In K. Arai (Ed.), *Social contributions and responsibilities of simulation and gaming* (pp. 305–314). Tokyo, Japan: Association of Simulation and Gaming.

Squire, K. (2003). Video games in education. *International Journal of Intelligent Simulations and Gaming, 2*.

Standen P. J., & Cromby, J. J. (1996). Can students with developmental disability use virtual reality to learn skills which will transfer to the real world? In H. J. Murphy (Ed.), *Proceedings of the Third International Conference on Virtual Reality and Persons with Disabilities*. California State University Center on Disabilities, Northridge, CA.

Subrahmanyam, K., Kraut, R. E., Greenfield, P. M., & Gross, E. F. (2000). The impact of home computer use on children's activities and development. *Future of Children, 10*, 123–44.

Todorov, A., & Bargh, J. A. (2002). Automatic sources of aggression. *Aggression & Violent Behavior, 7*, 53–68.

Tremblay, R. E. (2000). The development of aggressive behaviour during childhood: What have we learned in the past century? *International Journal of Behavioral Development, 24*, 129–141.

van Schie, E. G. M., & Wiegman, O. (1997). Children and videogames: Leisure activities, aggression, social integration, and school performance. *Journal of Applied Social Psychology, 27*, 1175–1194.

Walsh, D., Gentile, D., Gieske, J., Walsh, M., and Chasco, E. (2003, December). *Eighth Annual Video Game Report Card*. National Institute on Media and the Family.

Winograd, K. (2001). Migrant families: Moving up with technology. *Converge, 4*, 16–18.

Woodard, E. H., & Gridina, N. (2000). *Media in the home: The fifth annual survey of parents and children*. The Annenberg Public Policy Center of the University of Pennsylvania.

Yaman, D. (2004). *Why games work*. Retrieved May 2, 2004 from http://learningware.com/clients/ gameswork.html.

Ybarrondo, B. A. (1984). A study of the effectiveness of computer-assisted instruction in the high school biology classroom. Idaho. (ERIC Document Reproduction Service No. ED265015).

Yerkes, R. M., & Dodson, J. D. (1908). The relation of strength of stimulus to rapidity of habit formation. *Journal of Comparative Neurology & Psychology, 18*, 459–482.

Zillmann, D., & Bryant, J. (1983). Pornography and social science research: . . . higher moralities. *Journal of Communication, 33*, 111–114.

25

What Can We Learn From Playing Interactive Games?

Debra A. Lieberman

University of California, Santa Barbara

Interactive games are powerful environments for learning. Research consistently finds that players learn new skills, knowledge, insights, attitudes, or even behaviors in games that challenge them to think, explore, and respond (see, for example, Betz, 1995; Gee, 2003; Dempsey et al., 1996; Jayakanthan, 2002; Kirriemuir, 2002; Lieberman, 1997, 2001a; Potter, K. R. 1999; Prensky, 2001).

How do games stimulate and support learning? Consider the following features of well-designed games, found also in the best nongame learning environments (Choi & Hannafin, 1995; Reigeluth & Squire, 1998; Schank et al., 1994; Tennyson & Breuer, 2002). Typically, interactive games challenge players to solve compelling problems. Players learn by doing, in a virtual setting that responds to every move or decision or input they make. They interact with the game environment, develop skills to succeed in that environment, and rehearse those skills repeatedly. They have opportunities to experiment, fail, and try again until they succeed, and they receive help when needed. Games usually adapt to players' abilities and keep the level of difficulty in a range that is challenging but not impossible for each individual. Players receive feedback on their progress and they are able to see how their choices enhance or hinder their advancement toward their goal. They learn what is valued by receiving rewards (e.g., gaining points or status) or punishments (e.g., losing points or status) for their decisions and performance. They may also observe role-model characters experiencing positive or negative consequences for their behaviors. And, players often collaborate with other people, learning information, skills, and strategies from each other.

These well-established approaches to teaching and learning occur with skillful tutors and classroom teachers, and also with interactive games. It is important to note that the capacity of games to teach does not guarantee that their lessons will be desirable ones. For example, the entertainment industry has produced a variety of popular games that promote fear, hate, and violence. Most studies investigating their effects on players' emotions, attitudes, and behavior conclude that players learn these lessons well, sometimes to the point of antisocial behavior (see Buckley & Anderson, this volume; Calvert et al., 2002; Cassell & Jenkins, 1998; Cooper & Mackie, 1986; Kinder, 1996, 2000; Lee & Peng, this volume; Schutte et al., 1998; S. Smith, this volume; Weber, Ritterfeld, & Kostygina, this volume). On the other hand, games designed to teach more valuable lessons can also be effective (see Amory et al., 1999; Bosworth, 1994; Lieberman, 2001a; Subrahmanyam & Greenfield, 1994; Thomas, Cahill, & Santilli, 1997; Walshe et al., 2003), and the curriculum of games has been expanding into new topic areas and applications (Gee, 2003; Goldstein, 2003; Prensky, 2001; Stewart, 1997; Wolf, 2002). Almost any message could be conveyed, supported, and rehearsed in an interactive game. To paraphrase former FCC Commissioner Nicholas Johnson's famous quotation made decades ago about the effects of television, and substituting interactive "games" for "television," it is fair to say today that "All games are educational games. The question is: What are they teaching?"

To begin to answer that question, and to consider implications for future game design, this chapter cites research that has identified many kinds of learning that take place with games and, in some cases, how this learning happens. It organizes current research on interactive games and learning into nine areas:

- Motivation to learn
- Perception and coordination
- Thinking and problem solving
- Knowledge
- Skills and behaviors
- Self-regulation and therapy
- Self-concepts
- Social relationships
- Attitudes and values

MOTIVATION TO LEARN

Interactive games can motivate people to learn, including those who at first are not particularly interested in the subject matter (Reigeluth & Squire, 1998; Lepper & Henderlong, 2000; Lieberman & Linn, 1991). The incentive of trying to win a game, which is an extrinsic, external motivation for learning, can draw players into the learning activity, which can then stimulate an intrinsic, internalized interest in the subject matter itself (Cordova & Lepper, 1996; Millians, 1999). For example, quiz games and simulation games have been used successfully both to attract adolescents to online health information and to increase their interest in health topics (Bosworth, 1994).

People who play interactive games in the United States are diverse (see Raney, Smith, & Baker, this volume; Weber this volume) and more than 50% of the population plays them regularly (ESA, 2003). Enthusiasts come from all socioeconomic backgrounds, genders (43% of game players are female), and age groups (65% of game players are age 18 or older), and the amount of leisure time spent playing interactive games continues to grow (Buchanan & Funk, 1996; ESA, 2005; Ipsos-Insight, 2003). In a survey of 1,500 representative U.S. households (ESA, 2003), respondents ranked interactive games as the most fun form of entertainment, ahead of watching television, surfing the Internet, and going to the movies. The top reasons they selected for playing interactive games were, "It's fun," "They're challenging," and "I like to play with friends and family." More than three quarters of all U.S. children play interactive games, and those who play games spend an average of about 1 to 1.5 hours per day playing them during leisure time (ESA, 2005; Huston et al., 1999; Ipsos-Insight, 2003; Roberts et al., 1999).

More than half of the U.S. population has grown up with games and other computer-based media, and many in this group are comfortable with the technology, enjoy responding to interactive content, and often want and expect to be able to *do* something with a screen in addition to watching it (Harel, 2002; Papert, 1993; Rieber, 1996; Wartella & Jennings, 2000). Following are some of the reasons why interactive games are so attractive and motivating to the younger half of the population and, increasingly, to the older half as well.

Appeal

Research on the appeal of interactive games has found that players especially like to experience challenging goals, stimulation of curiosity, control over the action, and fantasy themes (Cordova & Lepper, 1996; Ermi, Helio, & Mayra, 2004; Malone & Lepper, 1987); verisimilitude, which is authenticity in graphics, sound, and other sensory cues of the presentation (Reid, 1997); problem solving (Reid, 1997); competition (Cordova & Lepper, 1996; Reid, 1997); and collaboration (Cordova & Lepper, 1996). Also, children are attracted to games that support them in developing the skills needed to succeed in the game, and they like explorations and construction activities that help them achieve game-related goals (Ermi, Helio, & Mayra, 2004).

Challenge to Reach a Goal

The central motivating feature of an interactive game is the goal it sets for the player. It is rewarding to succeed in a game and to experience a sense of efficacy and control, and players are often motivated to strive intensely to achieve these rewards (Klimmt & Hartmann, this volume). The challenge to perform and win makes a game highly involving, especially when the challenge is difficult, it is possible to lose, and there is more than one pathway to winning (Malone & Lepper, 1987; Reid, 1997). Players become deeply immersed in games that challenge them—for instance to solve a problem, explore an environment or system, share strategies with others, make something happen, complete a puzzle, prevail over an obstacle, compete against time, or find an answer. In all these examples players have a goal. Learning that occurs in the process of attempting to reach a goal can increase players' interest, motivation to learn, and learning outcomes (Cordova & Lepper, 1996; Schank et al., 1994).

Emotions and Fun

Players enjoy the entertaining, playful aspects of interactive games, which could include an enthralling story; appealing characters; lush production values; a sense of social presence; making choices that affect the direction of the game; assuming the role of a character and playing with a new personality or identity; the extreme emotions that come with failure and success; and the pleasure of interacting with other characters and players (see Gee, 2003; Prensky, 2001; Tamborini & Skalski, this volume; Vorderer, 2000; Vorderer et al., this volume; Wolf, 2002; Wolf & Perron, 2003). These experiences can heighten players' emotional responses to an interactive game and motivate their effort to learn (Garris, Ahlers, & Driskell, 2002; Lepper & Henderlong, 2000).

Interactivity

Games are interactive when their responses take into account previous actions or messages generated by both the player and the game. A game's response to the player at any given point depends to a great extent on the player's previous choices and inputs, in much the same way a good conversation is based thoughtfully on the previous messages two people have exchanged. (See Rafaeli, 1988, for a discussion of interactivity; also see Vorderer, 2000.)

Interactivity can involve feedback and help messages that are tailored to the individual player and that adapt to the player's changing abilities, and these elements can be educationally effective (Betz, 1995; Kafai et al., 1998; Millians, 1999), providing the kind of learning support that expert tutors give to students (Choi & Hannafin, 1995; Dempsey et al., 1996; Lepper & Henderlong, 2000; Reigeluth & Squire, 1998). Players typically enjoy interactive, experiential learning that gives them a great deal of control, involves them in active decision making, and provides continuous feedback that lets them know how well they are doing and gives hints and support if they are having problems with the material (Amory et al., 1999; Garris, Ahlers, & Driskell, 2002; Harel, 2002; Stewart, 1997). Learning outcomes improve significantly when players receive a flexible and appropriate level of assistance based on recent performance (Davis, 1999; Luckin, 2001); creativity and problem-solving skills can also improve with this kind of adaptive help (Tennyson & Breuer, 2002).

Engagement

Interactive games can be extremely demanding, often requiring deep levels of attention, managing more than one cognitive task at a time, or making quick responses. Players actively participate in a game—applying knowledge, devising strategies, making decisions, using skills, and reviewing the outcomes. The demands of a game engage players in what has been called "productive play," the learning and accomplishment that can occur when playful activities lead, for instance, to building virtual worlds, manipulating simulations, or solving problems (Rieber, 1996). Other terms for this process are "serious play" (Rieber, 1998) with games, and "hard fun" with computer-based interactive media in general (Papert, 1993, 1996). In many cases, the people who are resistant to the more traditional didactic methods of teaching and learning are happy to engage in the productive play and hard fun of interactive games (Betz, 1995; Ermi, Helio, & Mayra, 2004; Papert, 1993).

Flow

Participation in a challenging activity bounded by rules often brings a sense of pure concentration and immersion, or "flow" (Csikszentmihalyi, 1990, 2000; Weisler & McCall, 1976). It is a state of pleasure, well-being, and increased cognitive efficiency that occurs during an absorbing task. People experience flow when they are challenged enough to do their best, yet not challenged beyond their abilities. Flow can occur whenever an activity involves intense focus and a sense of control—during work, creative endeavors, sports, or play. People in a state of flow often lose their sense of time and place while they are completely absorbed in the concentrated effort. Children and adults enjoy being in a state of flow when playing a challenging yet achievable interactive game. Here again, the components of interactive games that make them attractive and compelling are the same components that motivate engagement and learning.

Story Line

Intriguing stories can attract and involve an audience (Gee, 2003; Sood, Marard, & Witte, 2004). Some games are designed so that the plot will move forward only after the player has completed a task. In these cases, the goals of the game are interrelated with the unfolding of the story, requiring players to solve problems in order to advance the plot. The desire to see what happens next in a story is a powerfully motivating force (Amory et al., 1999; Gee, 2003; Goldstein, 2003; Vorderer, 2000; Wolf, 2002).

PERCEPTION AND COORDINATION

Most of the research on effects of games on perceptual skills and coordination has been done with children and adolescents, and some has been done with college students. Many studies ask whether players learn and improve the cognitive and motor skills they use while playing interactive games.

Spatial Perception

Research has found that some visually oriented interactive games improve players' spatial perception and visualization skills, such as the ability to recognize 3-dimensional shapes and mentally rotate them (Dorval & Pepin, 1986; Gunter, 1998; Lowery & Knirk, 1983; Okagaki & Frensch, 1994; Subrahmanyam & Greenfield, 1994). For example, in one study students in Grades 5, 7, and 9 played interactive games that required the use of spatial skills, and students in each grade subsequently improved the spatial skill of mental rotation and became more adept at using mental maps as an aid to memory (McClurg & Chaille, 1987).

Visual Style of Communication

Games often rely heavily on animation, pictures, and diagrams, and not so much on words, to convey information. It makes sense that players of visually oriented games would become more proficient at reading images and might even develop a preference for using images to express themselves. A cross-cultural experiment, designed to show causality, and conducted with college students in the United States and Italy, found that playing a visually oriented interactive game shifted players' representational styles from verbal to iconic and spatial. Participants who spent time playing the interactive game later used more diagrams to convey information, whereas those who played a board game with content similar to the interactive game continued to use a verbal style and did not shift toward a more visual style (Greenfield et al., 1994a).

Cognitive Processing, Visual Processing, and Eye–Hand Coordination

There is also evidence that game playing can improve cognitive processing skills such as visual discernment, which involves the ability to divide visual attention and allocate it to two or more simultaneous events (Greenfield et al., 1994b); parallel processing, the ability to engage in multiple cognitive tasks simultaneously (Gunter, 1998); and other forms of visual discrimination including the ability to process cluttered visual scenes and rapid sequences of images (Riesenhuber, 2004). Experiments have also found improvements in eye-hand coordination after playing video games (Rosenberg et al., 2005).

Correlational studies have found that children who are frequent video game players have greater capacity than infrequent players and nonplayers for deep concentration, are more adept at making quick decisions and responses, and have exceptional eye–hand coordination (Gunter, 1998). Unlike the controlled experiments discussed in this section so far, correlational studies examining the skills of frequent game players are snapshots in time and are usually not designed to show causality. They cannot tell us if frequent video game playing causes the skills to emerge; an alternative explanation would be that young people who already have the skills are the ones who enjoy and seek out interactive games the most.

Manual Dexterity and Speed

Action-based interactive games, where the player pushes buttons on a game controller device, require players to develop an expert level of manual dexterity and the ability to make quick decisions—skills that are also needed to conduct laparoscopic surgery. In one study, surgeons who usually spent at least 3 hours a week playing action video games during leisure time, made about 37% fewer mistakes in laparoscopic surgery practice tasks and performed them 27% faster than their colleagues who did not play video games (Dobnik, 2004; also see Rosenberg et al., 2005).

Technological Skills and Career Interests

Some have noted (e.g., Gee, 2003; Prensky, 2001; Subrahmanyam & Greenfield, 1994) that the perceptual and coordination skills that players develop with visually oriented interactive games, such as the ability to perceive and apply fast-paced multichannel information on computer screens, are skills also needed for technological literacy. The enjoyment and skill development in video game playing may stimulate interest in a career involving advanced technological work, scientific research, or computer programming. Games may serve as training grounds for future technical careers.

THINKING AND PROBLEM SOLVING

Some theorists have noted that interactive games require players to apply and rehearse sophisticated thinking and problem-solving skills, such as close observation, inferring of the rules and structure of a game, logical thinking, hypothesis testing, experimentation, and strategy development (Gee, 2003; Hogle, 1996). An examination of the tasks involved in playing various game genres concludes that games with puzzles and complex questions may improve players' ability to think logically and tactically; simulation games may enhance scientific thinking, such as the ability to control for a single a variable; and adventure games may increase players' skills in observation, analysis of systems, and coaching of others (Prensky, 2001). Following are some of the skills that have been tested in research.

Self-Directed Learning

Studies have found that games challenge players to pay attention, monitor and evaluate their own actions, use strategies such as grouping and the use of imagery as aids to memory, reason inductively and deductively, apply new knowledge to novel situations, and use affective strategies such as anxiety reduction and self-encouragement (Chaika, 1996; Goldstein, 2003; Hogle, 1996; Oyen & Bebko, 1996). One study, for example, found that college students improved in cognitive flexibility and creativity after they participated in interactive games that demanded and rehearsed those skills (Doolittle, 1995).

Game players may learn new ways to learn in order to win at games. As players become more aware of and adept at their own learning strategies, they begin to apply them more appropriately. The strategies may include, for example, learning through trial and error, finding patterns that lead to discovering an answer, or using an inductive style of learning (Lepper & Henderlong, 2000; Lieberman & Linn, 1991).

Components of Learning

Children who are expert players of interactive games tend to be highly skilled at self-monitoring, pattern recognition, iconic representation, principle-based decision making, qualitative thinking, and memory (Vandeventer, 1998), and these skills are components of some of the broader cognitive tasks of learning, such as classifying, planning, critical thinking, scientific and mathematical thinking, decision making, and problem solving (Greenfield, 1993; Kafai et al., 1998; Subrahmanyam et al., 2001). As mentioned earlier, studies of expert players often provide no causal information. It is possible that either game playing causes the skills to develop or alternatively the preexistence of these skills leads the person to play interactive games.

Transfer of Skills

One study provides an example of skill transfer without any coaching or support. It found that arcade-style action games increased players' abilities in inductive discovery, and they transferred their newly acquired skills from the game environment to a non–game environment that required them to use scientific–technical representation (Greenfield et al., 1994b).

KNOWLEDGE

Interactive games often require players to take in new information, apply it toward solving a problem, receive feedback on their performance, and then apply the information again until they are successful. This combination of obtaining information and using it repeatedly helps players learn and retain the information, and applying it in a meaningful context helps them develop deeper understanding (Heinich et al., 1996; Kafai et al., 1998; Potter, K.R., 1999; Reigeluth & Squire, 1998).

Learning Outcomes

Hundreds of studies find that people learn with appropriately designed interactive media, such as software and the Internet (see Fletcher-Finn, 1995; Kozma, 1994; Kulik & Kulik, 1991; Mayer & Moreno, 2002; Mayer, Schustack, & Blanton, 1999; Subrahmanyam et al., 2001; Wartella & Jennings, 2000). Learning is especially well supported when learners proceed at their own ability level and pace, receive individualized and constructive performance feedback, receive help when needed, and review material until they understand it thoroughly (Jayakanthan, 2002; Jonassen & Land, 2000; Reigeluth & Squire, 1998; Tennyson & Breuer, 2002). These features are commonly built into interactive games, including games designed purely for entertainment and games intended for learning and behavior change (Lieberman, 1997). To a greater or lesser extent, most games support players' learning of the content needed to succeed in the game. Following are examples of research on learning effects with interactive games:

Situated Learning and Mindfulness

Games provide situations that stimulate players to learn and apply knowledge. Players learn-by-doing in contexts that afford them some control over environments that change as a result of their decisions. Not only can players gain information when their learning is situated this way, but they also gain other kinds of knowledge, such as insights into how their decisions affect the physical or social environment portrayed in the game and how the components of systems are interrelated. They also learn firsthand about values, social relationships, and socially acceptable strategies for approaching and solving problems. It is important to note that games contain their own logic and assumptions about cause and effect, which may be biased or inaccurate, intentionally or unintentionally, on the part of the designer. The game environment responds to the player according to those built-in assumptions and can convey persuasive lessons about appropriate ways to behave (Fogg, 2003). For example, many violent entertainment games demonstrate that crime and murder bring status and power to the perpetrator; health games may show medication use as the only way to become healthy, without including alternative therapies or healthy lifestyle changes as viable routes to health; and AIDS prevention games may portray abstinence as the only way to prevent infection.

Situated learning in interactive games can motivate deep cognitive engagement. To succeed in the situation, players become especially mindful of the content they need to know in order to win the game and mindful of the skills they must apply, and they use effortful and strategic processes of thinking (Lieberman & Linn, 1991; Potter, K.R, 1999). This close attention and intense mental effort lead to deeper understanding, learning, and retention of the material (Choi & Hannafin, 1995). There is some evidence that involvement in learning activities tends to be stronger when learning is game-based, compared to more traditional approaches to teaching and learning (Cordova & Lepper, 1996; Dempsey et al., 1996; Hogle, 1996), and this leads to more learning and retention of the content. One study found that children's enjoyment and mental effort were more intense while playing interactive games than while watching animations or solving visual puzzles (Yamada, 1998).

Pedagogical Agents

A character in an interactive game can be designed to interact with the player as a pedagogical agent, serving as a role model, tutor, or guide to support learning, either as a participant in the game environment or as a helper on the side (Amory et al., 1999; Jayakanthan, 2002). One study provides evidence that such agents might be effective in games. It found that seventh-grade students who learned biology with assistance from a software-based interactive pedagogical agent were more interested in the lesson and were better able to transfer the knowledge and skills they learned in the lesson, compared to students who learned biology but did not interact with the agent (Moreno et al., 2001).

Simulations

A simulation is a representation of a physical or social system that lets the user change its parameters and observe its dynamics (Aldrich, 2003; Heinich et al., 1996). With interactive media a simulation can be an algorithm-based artificial world that has some properties of the real world. For example, there are simulations that enable users to learn how ecosystems work, how to lead a country and deal with international conflict, how chemicals interact, how to fly an airplane, how to use food and insulin to keep a diabetic character's blood glucose in the normal range, how to manage a city, how to build a business, how to keep a family happy and thriving, and so on. Simulations can simplify a view of a system by eliminating some of the variables; they can speed up or slow down time so that processes and outcomes are easier to observe; they allow the user to manipulate variables that are not immediately alterable in the real world (such as raising and lowering the Earth's temperature to observe the impact of global warming); and they are safe because any dangerous outcomes are depicted but not physically experienced.

When a simulation challenges the user to achieve a goal within the simulated world, then it becomes a game. Some simulation games incorporate interactive features that enhance the learning experience, such as calculation tools, multiple modalities to represent events (for example, using the modalities of sound, text, mathematical formulas, and graphs to represent the current state of the simulation); performance feedback; characters that interact socially with the player to teach, guide, or help; logbooks to keep track of the player's decisions and their effects; and debriefing scenarios that help learners reflect on what they achieved and how they got there.

Research has discovered a broad range of learning outcomes that can occur with simulation games, such as, multidisciplinary learning across the curriculum, where students see how academic subjects are interrelated when they try to solve real-world problems (Betz, 1995); insight into cause and effect within complex systems, where learners make decisions and immediately see the consequences (Corbeil, 1999); development of skills in logic and decision making (Aldrich, 2003; Goldstein, 2003); and moral and ethical development as learners see how their decisions can affect others (Aldrich, 2003; Millians, 1999; Reigeluth & Squire, 1998).

SKILLS AND BEHAVIORS

Interactive games can challenge players to apply a wide variety of skills. There are games, for instance, that develop players' skills in reading, math, business management, crisis management, military combat, committing crime, fistfighting, singing, dancing, playing chess or basketball, negotiating safe sex, and taking care of one's health. Direct participation in virtual tasks with interactive media such as games has in some cases developed learners' skills better than traditional didactic, less participatory, forms of learning (Betz, 1995; Kozma, 1994; Moreno et al., 2001).

Example: Health Behavior

Interactive health games can improve players' health-related skills and behaviors. Field studies and clinical trials of a series of console-based (Nintendo) interactive health games found, for example, that players with chronic health conditions (diabetes or asthma) improved their self-care skills and their prevention and self-care behaviors, and this led to improved diabetes or asthma outcomes (Lieberman, 2001a).

In a controlled trial, diabetic children and adolescents were randomly assigned to take home either a diabetes self-management video game or an entertainment video game with no health content, and were told they could play their game as much or as little as they wished, as long as they followed all household rules about when and for how long they were allowed to play interactive games. The study found that participants in both groups played their game about 1.5 hours per week on average over the course of 6 months, but only the group that received the diabetes game increased their communication about diabetes with family and peers, increased their diabetes knowledge and perceived self-efficacy for diabetes self-care, and improved their diabetes-related skills and self-care behaviors. As a result their urgent care and emergency visits related to diabetes decreased by 77%, dropping from an average of about 2.4 visits per child per year down to about 0.5 visits per child per year. The control group that received the entertainment video game experienced no significant changes in diabetes-related skills, behaviors, or outcomes (Brown et al., 1997; Lieberman, 2001a).

Similar randomized controlled trials of an asthma self-management game, with asthmatic children and adolescents, found improvements in their asthma-related knowledge, self-efficacy, self-care skills, and behaviors, and 35 to 40% reductions in urgent care and emergency visits related to asthma, and in missed school days due to asthma. Other studies also found increases in players' asthma knowledge and perceived self-efficacy for asthma self-care (Lieberman, 1997, 2001a).

Another example of behavior change with an interactive health game is a nutrition game that was designed for the elementary school curriculum to increase children's consumption of fruits and vegetables. A randomized study in 26 elementary schools found that fourth graders who participated in the game-based curriculum for 5 weeks ate 1.0 servings more per day of fruits and vegetables, during the week after the curriculum ended, than did the control group that did not participate (Baranowski et al., 2003).

Example: Aggressive Behavior

Effects of violent interactive games on aggressive behavior have been studied extensively (for reviews of the research and discussion of theory in this area, see Buckley & Anderson, chap. 5, this volume, Kinder, 1996, 2000; Lee & Penge, this volume; W. J. Potter, 1999; S. Smith, this volume; Weber, Ritterfeld, & Kostygina, this volume), especially with children and adolescents, who are considered to be particularly vulnerable. Several studies have found increases in aggressive behaviors and in fearful and hostile emotional states after playing violent games, but not after playing nonviolent games (Cooper & Mackie, 1986; Lee & Peng, chap. 22, this volume, Schutte et al., 1988; Weber, Ritterfeld, & Kostygina, chap. 23, this volume).

SELF-REGULATION AND THERAPY

Interactive games have been used to teach people how to regulate physiological processes such as brain waves and relaxation, cognitive processes involved in allocating attention, emotional reactions to events and the environment, and phobias.

Regulating Brain Waves and Attention

Biofeedback games challenge players to keep their brain in a particular wave state in order to progress in the game. Images, sounds, and events in the games provide biofeedback to the player and show the current brain wave state. In addition to self-control of brain waves, interactive games can motivate players to self-monitor and regulate their attention. In one study, players with attention deficit disorder (ADD) improved their ability to sustain their attention by playing an interactive game that detects the player's brain waves (Pope & Bogart, 1996). The game is based on a biofeedback system that was developed to assess the mental engagement of airplane pilots. When the system detects, from the player's brain waves, that attention is waning, the game becomes more difficult to play. The player can only succeed in the game by maintaining an adequate level of attention, and is motivated to attend in order to win the game.

Therapy, Social Skills, and Pain Management

In a clinical context, interactive games have led to positive therapeutic outcomes for children and adolescents (see Griffiths, 2003). They have helped young people undergoing chemotherapy and psychotherapy, children with emotional and behavioral problems, and youngsters with communication and social skill problems related to impulsivity, ADD, and autism. In addition to teaching young people how to regulate attention, manage emotions, and interact socially, interactive games have also helped distract patients to reduce their perception of pain during physical therapy and during medical procedures for conditions such as Erb's palsy, muscular dystrophy, and burns.

Phobias

Virtual environments have been used very successfully in exposure therapy, for patients who have phobias such as fear of snakes, spiders, public speaking, elevators, and flying (Wiederhold, 2003). Under the guidance of a therapist, patients learn to approach the object of their fears in small, incremental steps. In the past this was done through direct experience, but now virtual environments are providing a more economical way to achieve the same outcomes. In one study, an interactive game provided effective therapy for auto accident victims who wanted to reduce their fear of driving (Walshe et al., 2003).

SELF-CONCEPTS

Interactive games provide performance feedback that can call attention to a player's skills and accomplishments to such an extent that they influence players' sense of self. Following are a few of the ways games teach players about themselves, especially in the areas of self-esteem and self-efficacy.

Self-Esteem and Pride of Achievement

Players enjoy having control over the action in an interactive game. Control helps make the experience immersive and involving, allowing players to explore their own pathways through the material, make choices, and experience the resulting rewards (Klimmt & Hartmann, this volume). When they have a high level of control and then succeed, they feel pride and self-esteem because the success was based to a great extent on their own decisions and skills (Colwell, Grady, & Rhaiti, 1995; Corbeil, 1999; Luckin, 2001; Stewart, 1997; Tennyson & Breuer, 2002).

Combining Interactivity and Privacy

Computer-based media such as interactive games offer users the unique experience of receiving interactive and individualized performance feedback *without* the presence or surveillance of another person. This combination of interactivity and privacy can be especially effective for game players who are not yet confident of their skills and would be embarrassed to try them out in front of others. Working alone, a player can rehearse new skills in a game—and receive feedback and help—without fear of publicly exposing their weaknesses. After the skills improve, the player will be eager to show others, and both the sense of pride in one's own achievement and the social approval received from others can increase self-esteem (Amory et al., 1999; Lieberman, 2004; Millians, 1999).

Another advantage of privacy plus interactivity can occur when players want to explore a topic without others knowing they are interested in it, such as sex, alcohol, drugs, or other high-risk behaviors. People can save face and avoid the embarrassment of admitting an interest in forbidden, taboo, or illegal behaviors, while still being able to assess their own risks, develop prevention and self-care skills, and obtain information, in an interactive environment (Bosworth, 1994).

Self-Efficacy

Social cognitive theory (Bandura, 1997, 2004; Klimmt & Hartmann, this volume) points to the importance of the individual's sense of self-efficacy as a mediator of behavior change. Self-efficacy is the belief that one is capable of carrying out a particular activity or behavior. People with high self-efficacy for an activity that they consider to be desirable are more likely to engage in that activity than people whose self-efficacy is low. Encouragement, positive performance feedback, vicarious experience, and the actual experience of success can raise a person's sense of self-efficacy and can increase the likelihood of future behavior.

Some interactive games are designed to increase players' self-efficacy by giving them vicarious experiences in which they can succeed. Players apply specific skills and the game supports them with encouragement, feedback, help, and rehearsal and application of skills until they are successful. Research has found that interactive games can help players improve their self-efficacy for skills involved in HIV/AIDS prevention (Thomas, Cahill, & Santilli, 1997), diabetes self-management (Brown et al., 1997; Lieberman, 2001a), and asthma self-management (Lieberman, 1997, 1999).

SOCIAL RELATIONSHIPS

A discussion of interactive games and learning should consider not only the solo activity of a person playing a game but also the social environment in which game playing occurs. Research finds that for most people interactive game playing is essentially a social activity. Game players are not social isolates; instead they tend to meet friends outside of work or school more often than occasional players or nonplayers do (Colwell, Grady, & Rhaiti, 1995; Funk, Germann, & Buchman, 1997; Orleans & Laney, 2000), and they often play interactive games with family and friends instead of playing alone (ESA, 2003). Although most research finds that game players are more socially active than nonplayers, one study found no differences in the amount of social interaction for game players and nonplayers (Phillips et al., 1995).

In addition to playing games with others, interactive game players like to talk about games when they are not playing (ESA, 2003). Players often help each other with game strategies and this develops their knowledge about topics presented in the game, and also helps improve social and communication skills (Goldstein, 2003; Vandeventer, 1998).

Social Recognition for Game Skills

The desire to demonstrate expertise to others and gain social approval motivates players to invest considerable effort so that they will ultimately perform well in games (Griffiths, 1997). When they gain the admiration and approval of peers, they are then motivated to continue striving to succeed in order to stay in the spotlight (Sakamoto, 1994). Video game arcades, for example, are meeting places for adolescents, where they can socialize without parental control and can show off their skills to others (Michaels, 1993).

Using Social Interaction to Enhance Learning

Games have been created to encourage social interaction as a route to learning (Choi & Hannafin, 1995; Griffiths, 1997; Mayer, Schustack, & Blanton, 1999; Reid, 1997). The health games discussed earlier (Lieberman, 2001a) were designed to stimulate discussion about health, by including two-player options that required both players to communicate with each other and cooperate in order to win the game. Clinical trials found that young people who were randomly assigned to take home a health game for six months gained more health knowledge, shared more information about the health topic with family and friends, and discussed their personal feelings about the health topic with others more often than those who were randomly assigned to take home a nonhealth entertainment game for the same period of time (Lieberman, 1997, 2001a). When people have opportunities to talk about their health with others this way, they receive more social support and this is associated with better self-care, better health, and more effective coping strategies when problems arise (Peterson & Stunkard, 1989).

Game Communities as Learning Environments

Communities develop around games, offline and online, and they often involve extensive social interaction and sharing of knowledge. With offline games (standalone console, computer, or arcade games), players meet to play together face-to-face and enjoy both competitive and collaborative experiences. Online games (via consoles or computers) bring people together from diverse geographical locations. Some of the most commercially successful interactive games each have dozens of fan sites where players create, share, and trade game objects, maps, levels, scenarios, game codes, and stories. They develop personal relationships and social structures, use their own language and jargon related to games, and collaborate in a group effort to beat the game, and these interactions are an important part of the game playing experience (Amory et al., 1999; Bruckman, 1998).

Some games are designed especially for collaboration and group participation, activities that foster learning. Playing in a MUD (Multi User Domain—or Dungeon or Dimension) or MOO (MUD, Object Oriented) allows players to collaborate in groups to complete quests, solve puzzles, or defeat enemies. Each character has unique strengths and weaknesses, and usually cannot survive without working cooperatively with other characters who have different, complementary abilities. This builds a sense of community, where players design and construct personally meaningful projects, and are self-motivated and peer-supported (Bruckman, 1998). Participation in online game communities is growing and the more popular role-playing games have many thousands of players participating online at any given time.

ATTITUDES AND VALUES

Games can potentially affect players' attitudes and values related to learning, social roles, and behaviors such as violence or nurturing. More research is needed in this area.

Attitudes About Learning: Enjoyment and Rewards

Players may develop more positive attitudes about learning after they learn in an interactive game. There is evidence that players who have enjoyable and productive learning experiences with a game will develop and sustain positive attitudes about learning that subject outside the game environment, and will develop more positive attitudes toward learning in general. This is especially likely to happen if the rewards of the game focus on a job well done and on the value of the learning itself, and not on extrinsic rewards such as points and prizes (Lepper & Henderlong, 2000; Papert, 1993; Schunk & Zimmerman, 1994).

Attitudes About Others: Role Playing and Role Modeling

Role playing and role modeling in games can teach attitudes and values (Cassell & Jenkins, 1998; Gunter, 1998; Kinder, 2000; Lee & Peng, this volume; Lieberman, 2001b). Social cognitive theory (Bandura, 1997, 2004) explains how attractive role model characters that are similar to the player can teach by example. Players may observe role model characters or interact directly with them. They may even play the role of the character, make behavioral choices, and virtually experience the rewards and punishments that occur as a result of the character's actions. Further research is needed on the attitudinal effects of having a first-person experience of virtual rewards and punishments in the context of an interactive game.

Research on effects of violent games offers some evidence that role playing and role modeling can influence attitudes and values. For example, playing violent games can desensitize players to the horrors of real-world violence, increase their hostility and mistrust of others, and teach them to accept violence as a legitimate way to solve problems (Buckley & Anderson, this volume; Potter WJ, 1999). Violent games often portray repetitive, sometimes constant, use of weapons or fistfights, and this violent behavior is rewarded with flashy graphics, sound effects, game progress, points, and other affirmations. Violent role modeling and role playing like this may significantly influence players' attitudes about the appropriateness of violence in society (Kinder, 1991, 1996; Wartella & Jannings, 2000; Wartella, O'Keefe, & Scantlin, 2000; Weber, Ritterfeld, & Kostygina, this volume).

CONCLUSION

Interactive games are dynamic learning environments that can motivate players to achieve, and can instill confidence, stimulate thinking and problem solving, and successfully support the development of new knowledge, skills, and behavior. Current research has found a broad and diverse range of learning outcomes with interactive games, both desirable and undesirable.

The most effective interactive games intended for learning are designed on the basis of well-established theories and learning principles from the fields of education, psychology, communication, human–computer interaction, and the arts and humanities. Game designers should know their intended players well, should have a clear idea of what they want the game to achieve and why, and should understand how to craft a game so it will lead to the intended outcomes. Theory and research can contribute significantly to help identify players' needs and characteristics, determine the learning outcomes the game will bring about, shape the goals the player will be challenged to achieve, guide user testing before the game is completed, and develop evaluation studies to test hypothesized outcomes and to explain how the game helped make them happen (see Hanna et al., 1999; Lieberman, 1999).

Along with knowing the kinds of learning outcomes games are capable of achieving, designers should know the limitations of interactive games as learning environments. More research is needed to identify the kinds of learning that are—and are not—well supported by interactive games. Future research should also investigate the strengths and weaknesses of instructional design approaches in games, processes of learning with games in general, variation in players' cognitive needs and abilities and the implications for design, and the role of social interaction and emotional responses in learning.

All age groups, including very young children, play interactive games and their abilities and interests vary. Children have particularly special needs because they select, attend to, and cognitively and even physically process media differently than adults do. In general, they are ready and eager for interactive media at extremely young ages but they need content and formats designed for their developmental capabilities and learning needs. From birth to age 18, children progress from concrete to abstract thinking; from an egocentric view of events to an ability to take the other's perspective; from holding very few to holding many schemas, or mental models, about the way events occur and how social and physical environments function; from low interest in learning rules and following them to high; from low reading skills and media literacy skills to high; from focusing on content regardless of its relevance to the main message to focusing almost entirely on content likely to be most relevant to the main message; and from "centration" on one attribute in a presentation to perceiving multiple attributes simultaneously (Blumberg, 1998; Calvert, 1999; Calvert, Jordan, & Cocking, 2002; Druin et al., 1999; Vorderer & Ritterfeld, 2003; Wartella & Jennings, 2000). These and other developmental shifts should be addressed in the design of interactive games so that the material is appropriate for specific age groups and ability levels and, whenever feasible, supports and enhances children's cognitive, emotional, and social development (Luckin, 2001; Wartella, O'Keefe, & Scantlin, 2000).

Many of the interactive games tested in research so far are screen-based and use a console game controller or a computer keyboard as an input device, but this is changing. Now interactive games are moving to other platforms (such as, cell phones, robots, virtual environments, interactive TV, smart toys, DVDs, and handheld mobile devices with geographical positioning systems) and some of them provide new kinds of input devices (such as, dance pad on the floor, camera pointed at the player, microphone for voice input, sensors that detect brain waves and other physiological states, haptic devices that detect movement and pressure, movable objects that can be sensed by a computer system). These new ways of playing interactive games will raise new questions about effects on learning, and what can be learned, in each environment, while at the same time the fundamental questions about processes of human learning will continue to be relevant.

As the design strategies and technological capabilities of interactive games evolve, research should continue to investigate both intended and unintended learning, so that designers of entertainment games can avoid teaching undesirable lessons and designers of games for learning can improve players' learning experiences and outcomes. There is much more research and theory development yet to be done. In the meantime, current research in this nascent field has already shown us that indeed we can learn a great deal from playing interactive games.

REFERENCES

Aldrich, C. (2003). *Simulations and the future of learning: An innovative (and perhaps revolutionary) approach to e-learning.* New York: Jossey-Bass/ Pfeiffer.

Amory, A., Naicker, K., Vincent, J., & Adams, C. (1999). The use of computer games as an educational tool: Identification of appropriate game types and game elements. *British Journal of Educational Technology, 30*(4), 311–321.

Bandura, A. (1997). *Self-efficacy: The exercise of control.* New York: Freeman.

Bandura, A. (2004). Social cognitive theory for personal and social change by enabling media. In A. Singal, M. J. Cody, E. M. Rogers, & M. Sabido (Eds.), *Entertainment-education and social change: History, research, and practice* pp. 75–96. Mahwah, NJ: Laurence Erlbaum Associates.

Baranowski, T., Baranowski, J., Cullen, K.W., Marsh, T., Islam, N., Zakeri, I., Honess-Morreale, L., & deMoor, C. (2003). Squire's Quest! Dietary outcome evaluation of a multimedia game. *American Journal of Preventive Medicine, 24*(1), 52–61.

Betz, A. Joseph. (1995). Computer games: Increased learning in an interactive multidisciplinary environment. *Journal of Educational Technology Systems, 24*(2), 195–205.

Blanchard, J., & Stock, W. (1999). Meta-analysis of research on a multimedia elementary school curriculum using personal and video-game computers. *Perceptual and Motor Skills, 88*(1), 329–336.

Blanton, W. E., Moorman, G. B., Hayes, B. A., & Warner, M. L. (1997). Effects of participation in the Fifth Dimension on far transfer. *Journal of Educational Computing Research, 16*, 371–396.

Blumberg, F. C. (1998). Developmental differences at play: Children's selective attention and performance in video games. *Journal of Applied Developmental Psychology, 19*(4), 615–624.

Bosworth, K. (1994). Computer games and simulations as tools to reach and engage adolescents in health promotion activities. *Computers in Human Services, 11*(1–2), 109–119.

Brown, S. J., Lieberman, D. A., Gemeny, B. A., Fan, Y. C., Wilson, D. M., & Pasta, D. J. (1997). Educational video game for juvenile diabetes: Results of a controlled trial. *Medical Informatics, 22*(1), 77–89.

Bruckman, A. (1998). Community support for constructionist learning. *Computer Supported Cooperative Work, 7,* 47–86.

Buchanan, D. D., & Funk, J. B. (1996). Video and computer games in the 90's: Children's time commitment and game preference. *Children Today, 24,* 2–15.

Calvert, S. (1999). *Children's journeys through the information age.* Boston: McGraw-Hill.

Calvert, S. L., Jordan, A. B., & Cocking, R. R. (Eds.). (2002). *Children in the digital age: Influences of electronic media on development.* Westport, CT: Praeger.

Cassell, J., & Jenkins, H. (Eds.). (1998). *From Barbie to Mortal Kombat: Gender and computer games.* Cambridge, MA: MIT Press.

Chaika, G. V. (1996). Computer games as a powerful tool for development of memory and attention. *International Journal of Psychology, 31*(3–4), 84156–84165.

Choi, J., & Hannafin, M. J. (1995). Situated cognition and learning environments: Roles, structures, and implications for design. *Educational Technology Research and Development, 43*(2), 53–69.

Colwell, J., Grady, C., & Rhaiti, S. (1995). Computer games, self-esteem, and gratification of needs in adolescents. *Journal of Community and Applied Social Psychology, 5,* 195–206.

Cooper, J., & Mackie, D. (1986). Video games and aggression in children. *Journal of Applied Social Psychology, 16*(8), 726–744.

Corbeil, P. (1999). Learning from the children: Practical and theoretical reflections on playing and learning. *Simulation & Gaming, 30*(2), 163–180.

Cordova, D. L., & Lepper, M. R. (1996). Intrinsic motivation and the process of learning: Beneficial effects of contextualization, personalization, and choice. *Journal of Educational Psychology, 88,* 715–730.

Csikszentmihalyi, M. (1990). *Flow: The psychology of optimal experience.* New York: Harper & Row.

Csikszentmihalyi, M. (2000). *Beyond boredom and anxiety: Experiencing flow in work and play.* New York: Jossey-Bass.

Davis, N. (1999). Young children, videos and computer games. *Journal of Computer Assisted Learning, 15*(4), 334–334.

Dempsey, J. V., Lucassen, B. A., Haynes, L. L., & Casey, M. S. (1996). *Instructional applications of computer games.* New York: American Educational Research Association. (ERIC Document Reproduction Service No. ED 394 500).

Dobnik, V. (April 7, 2004). *Video game playing surgeons make fewer mistakes.* Associated Press.

Doolittle, J. (1995). Using riddles and interactive computer games to teach problem-solving skills. *Teaching of Psychology, 22*(1), 33–36.

Dorval, M., & Pepin, M. (1986). Effect of playing a video game on a measure of spatial visualization. *Perceptual Motor Skills, 62,* 159–162.

Druin, A., Bederson, B., Boltman, A., Miura, A., Knotts-Callahan, D., & Platt, M. (1999). Children as our technology design partners. In A. Druin (Ed.), *The design of children's technology.* San Francisco: Morgan Kaufmann pp. 51–72.

Ermi, L., Helio, S., & Mayra, F. (2004). *The power of games and control of playing: Children as the actors of game cultures.* Report from Hypermedia Laboratory Net Series 6. University of Tampere, Finland.

ESA. (2003). *ESA 2003 consumer survey.* Washington, DC: Entertainment Software Association.

ESA. (2005). *2005 sales, demographics and usage data: Essential facts about the computer and video game industry.* Retrieved August 2, 2005, from http://www. theesa.com/files/2005EssentialFacts.pdf.

Fletcher-Finn, C. M. (1995). The efficacy of computer-assisted instruction (CAI): A meta-analysis. *Journal of Educational Computing Research, 12,* 219–241.

Fogg, B. J. (2003). *Persuasive technology: Using computers to change what we think and do.* San Francisco: Morgan Kaufmann.

Funk, J. B., Germann, J. N., & Buchman, D. D. (1997). Children and electronic games in the United States. *Trends in Communication, 2,* 111–126.

Garris, R., Ahlers, R., & Driskell, J. E. (2002). Games, motivation & learning: A research and practice, model. *Simulation & Gaming, 33*(4), 441–467.

Gee, J. P. (2003). *What video games have to teach us about learning and literacy.* New York: Palgrave Macmillan.

Goldstein, J. (2003). People @ play: Electronic games. In H. van Oostendorp (Ed.), *Cognition in a digital world* (pp. 25–45). Mahwah, NJ: Lawrence Erlbaum Associates.

Greenfield, P. M. (1993). Representational competence in shared symbol systems: Electronic media from radio to video games. In R. R. Cocking & K. A. Renninger (Eds.), *The development and meaning of psychological distance* (pp. 161–183). Hillsdale, NJ: Lawrence Erlbaum Associates.

Greenfield, P. M., Camaioni, L., Ercolani, P., Weiss, L., Lauber, B., & Perucchini, P. (1994a). Cognitive socialization by computer games in two cultures: Inductive discovery or mastery of an iconic code? *Journal of Applied Developmental Psychology, 15*(1), 59–85.

Greenfield, P. M., de Winstanley, P., Kilpatrick, H., & Kaye, D. (1994b). Action video games and informal education: Effects on strategies for dividing visual attention. *Journal of Applied Developmental Psychology, 15*(1), 105–124.

Griffiths, M. (1997). Friendship and social development in children and adolescents: The impact of electronic technology. *Educational and Child Psychology, 14*, 25–37.

Griffiths, M. (2003). The therapeutic use of videogames in childhood and adolescence. *Clinical Child Psychology & Psychiatry,8*(4), 547–554.

Gunter, B. (1998). *The effects of video games on children: The myth unmasked.* Sheffield, England: Sheffield Academic Press.

Hanna, L., Risden, K., Czerwinski, M., & Alexander, K. (1999). The role of usability research in designing children's computer products. In A. Druin (Ed.), *The design of children's technology.* San Francisco: Morgan Kaufmann pp. 3–26.

Harel, I. (2002). Learning new-media literacy: A new necessity for the young clickerati generation. *Telemedium, The Journal of Media Literacy, 48*(1), 17–26.

Heinich, R., Molenda, M., Russell, J. D., & Smaldino, S. E. (1996). *Instructional media and technologies for learning* (5th ed.). Englewood Cliffs, NJ: Prentice-Hall.

Hogle, J. G. (1996). *Considering games as cognitive tools: In search of effective "edutainment."* University of Georgia Department of Instructional Technology. (ERIC Document ED 425 737).

Huston, A. C., Wright, J. C., Marquis, J., & Green, S. B. (1999). How young children spend their time: Television and other activities. *Developmental Psychology, 35*(4) 912–925.

Ipsos-Insight Research. (2003). *Video game consumer survey, 2003.* New York.

Jayakanthan, R. (2002). Application of computer games in the field of education. *The Electronic Library, 20*(2), 98–105.

Jonassen, D. H., & Land, S. (2000). *The theoretical foundations of learning environments.* Mahwah, NJ: Lawrence Erlbaum Associates.

Kafai, Y. B. (1995). *Minds in play: Computer game design as a context for children's learning.* Mahwah, NJ: Lawrence Erlbaum Associates.

Kafai, Y. B., Franke, M., Ching, C., & Shih, J. (1998). Games as interactive learning environments fostering teachers' and students' mathematical thinking. *International Journal of Computers for Mathematical Learning, 3*(2), 149–193.

Kinder, M. (1991). *Playing with power in movies, television and video games.* Berkeley: University of California Press.

Kinder, M. (1996). Contextualising video game violence: From Teenage Mutant Ninja Turtles 1 to Mortal Kombat 2. In P. M. Greenfield & R. R. Cocking (Eds.), *Interacting with video.* (pp. 25–38) Norwood, NJ: Ablex.

Kinder, M. (Ed.). (2000). *Kids' media culture (console-ing passions).* Durham, NC: Duke University Press.

Kirriemuir, J. K. (2002). *The relevance of video games and gaming consoles to the higher and further education learning experience.* JISC TechWatch Report, at http://www.jisc.ac.uk/index.cfm?name=techwatch_report_0201.

Kozma, R. B. (1994). Will media influence learning? Reframing the debate. *Educational Technology Research & Development, 42*, 7–19.

Kulik, C. C., & Kulik, J. A. (1991). Effectiveness of computer-based instruction: An updated analysis. *Computers in Human Behavior, 7*, 75–94.

Lepper, M. R., & Henderlong, J. (2000). Turning "play" into "work" and "work" into "play": 25 years of research on intrinsic versus extrinsic motivation. In C. Sansone & J. Harackiewicz (Eds.), *Intrinsic and extrinsic motivation: The search for optimal motivation and performance* (pp. 257–307). San Diego: Academic Press.

Lieberman, D. A. (1997). Interactive video games for health promotion: Effects on knowledge, self-efficacy, social support, and health. In R. L. Street, W. R. Gold, & T. Manning (Eds), *Health promotion and interactive technology: Theoretical applications and future direction* pp. 103–120. Mahwah, NJ: Lawrence Erlbaum Associates.

Lieberman, D. A. (1999). The researcher's role in the design of children's media and technology. In A. Druin (Ed.), *The design of children's technology* pp. 73–97. San Francisco: Morgan Kaufmann.

Lieberman, D. A. (2001a). Management of chronic pediatric diseases with interactive health games: Theory and research findings. *Journal of Ambulatory Care Management, 24*(1), 26–38.

Lieberman, D. A. (2001b). Using interactive media in communication campaigns for children and adolescents. In R. Rice & C. Atkin (Eds.), *Public communication campaigns (3rd ed.,* pp. 373–388). Newbury Park, CA: Sage.

Lieberman, D. A. (2004). *Uses and gratifications of the Dance Dance Revolution video game: How players use it for fun, competition, dance performance, social interaction, and working out.* Unpublished report: University of California, Santa Barbara.

Lieberman, D. A., & Linn, M. C. (1991). Learning to learn revisited: Computers and the development of self-directed learning skills. *Journal of Research on Computing in Education, 23*(3) 373–395.

Lowery, B. R., & Knirk, F. G. (1983). Micro-computer video games and spatial visualization acquisition. *Journal of Educational Technology System, 11*(2), 155–166.

Luckin, R. (2001). Designing children's software to ensure productive interactivity through collaboration in the Zone of Proximal Development (ZPD). *Information Technology in Childhood Education Annual, 12*, 57–85.

Malone, T. W., & Lepper, M. R. (1987). Making learning fun: A taxonomy of intrinsic motivations for learning. In R. E. Snow & M. J. Farr (Eds.), *Aptitude, learning and instruction III, Conative and affective process analyses* pp. 223–253. Hillsdale, NJ: Lawrence Erlbaum Associates.

Mayer, R. E., & Moreno, R. (2002). Aids to computer-based multimedia learning. *Learning and Instruction, 12*(1), 107–119.

Mayer, R. E., Schustack, M. W., & Blanton, W. E. (1999). What do children learn from using computers in an informal, collaborative setting? *Educational Technology, 39*, 27–31.

McClurg, P. A., & Chaille, C. (1987). Computer games: Environments for developing spatial cognition? *Journal of Educational Computing Research, 3*, 95–111.

Michaels, J. W. (1993). Patterns of video game play in parlors as a function of endogenous and exogenous factors. *Youth and Society, 25*(2), 272–289.

Millians, D. (1999). Simulations and young people: Developmental issues and game development. *Simulation & Gaming, 30*(2), 199–226.

Moreno, R., & Mayer, R. E. (2004). Personalized messages that promote science learning in virtual environments. *Journal of Educational Psychology, 96*(1), 165–173.

Moreno, R., Mayer, R. E., Spires, H. A., & James, L. (2001). The case for social agency in computer-based teaching: Do students learn more deeply when they interact with animated pedagogical agents? *Cognition & Instruction, 19*(2), 177–213.

Okagaki, L., & Frensch, P. A. (1994). Effects of video game playing on measures of spatial performance: Gender effects in late adolescence. *Journal of Applied Development Psychology, 15*, 33–58.

Orleans, M., & Laney, M. C. (2000). Children's computer use in the home: Isolation or sociation? *Social Science Computer Review, 18*, 56–72.

Oyen, A. S., & Bebko, J. M. (1996). The effects of computer games and lesson contexts on children's mnemonic strategies. *Journal of Experimental Child Psychology, 62*(2), 173–189.

Papert, S. (1993). *The children's machine: Rethinking school in the age of the computer.* New York: Basic Books.

Papert, Seymour. (1996). *The connected family.* Athens, GA: Longstreet Press.

Peterson, C., & Stunkard, A. J. (1989). Personal control and health promotion. *Social Science and Medicine, 28*, 819–828.

Phillips, C. A., Rolls, S., Rouse, A., & Griffiths, M. D. (1995). Home video game playing in schoolchildren: A study of incidence and patterns of play. *Journal of Adolescence, 18*(6), 687–691.

Pope, A. T., & Bogart, E.H. (1996). Extended attention span training system: Video game neurotherapy for attention deficit disorder. *Child Study Journal, 26*(1), 39–50.

Potter, K. R. (1999). Learning by doing: A case for interactive contextual learning environments. *Journal of Instruction Delivery Systems, 13*(1), 29–33.

Potter, W. J. (1999). *On media violence.* Thousand Oaks, CA: Sage Publications.

Prensky, Marc. (2001). *Digital game-based learning.* New York: McGraw-Hill.

Rafaeli, S. (1988). Interactivity: From new media to communication. In R. Hawkins, J. Wiemann, & S. Pingree (Eds.), *Advancing communication science: Merging mass and interpersonal processes* (pp. 110–134). Newbury Park, CA: Sage.

Reid, M. (1997). *An exploration of motives for video game use: Implications for the study of an interactive medium.* Dissertation Abstracts International section A: Humanities & Social Sciences, 58(1-A), July.

Reigeluth, C. M., & Squire, K. D. (1998). Emerging work on the new paradigm of instructional theories. *Educational Technology, 38*(4), 41–47.

Rieber, L. P. (1996). Seriously considering play: Designing interactive learning environments based on the blending of microworlds, simulations, and games. *Educational Technology Research & Development, 44*, 43–58.

Rieber, L. P. (1998). The value of serious play. *Educational Technology, 38*(6), 29–37.

Riesenhuber, M. (2004). *An action video game modifies visual processing. Trends in Neurosciences, 27*(2), 72–74.

Roberts, D. F., Foehr, U. G., Rideout, V. J., & Brodie, M. (1999, November). *Kids and media @ the new millenium: A comprehensive national analysis of children's media use.* Menlo Park, CA: Kaiser Family Foundation.

Rosenberg, B. H., Landsittel, D., & Averch, T. D. (2005) Can video games be used to predict or improve laparoscopic skills? *Journal of Endourology, 19*(3), 372–376.

Sakamoto, A. (1994). Video game use and the development of sociocognitive abilities in children: Three surveys of elementary school students. *Journal of Applied Social Psychology, 24*(1), 21–42.

Schank, R. C., Fano, A., Bell, B., & Jona, M. (1994). The design of goal-based scenarios. *Journal of the Learning Sciences, 3,* 305–345.

Schunk, D. H., & Zimmerman, B. J. (Eds.). (1994). *Self-regulation of learning and performance: Issues and educational applications.* Hillsdale, NJ: Lawrence Erlbaum Associates.

Schutte, N. S., Malouff, J. M., Post-Gorden, J. C., & Rodasta, A. L. (1988). Effects of playing videogames on children's aggressive and other behaviors. *Journal of Applied Social Psychology, 18*(5), 454–460.

Sood, S., Marard, T., & Witte, K. (2004). The theory behind entertainment-education. In, A. Singal, M. J. Cody, E. M. Rogers, & M. Sabido (Eds.), *Entertainment-education and social change: History, research, and practice* (pp. 117–149). Mahwah, NJ: Laurence Erlbaum Associates.

Stewart, K. M. (1997). Beyond entertainment: Using interactive games in Web-based instruction. *Journal of Instruction Delivery Systems, 11*(2), 18–20.

Subrahmanyam, K., & Greenfield, P. M. (1994). Effect of video game practice on spatial skills in girls and boys. *Journal of Applied Developmental Psychology, 15,* 13–32.

Subrahmanyam, K., Greenfield, P., Kraut, R., & Gross, E. (2001). The impact of computer use on children's and adolescents' development. *Journal of Applied Developmental Psychology, 22*(1), 7–30.

Tennyson, R. D., & Breuer, K. (2002). Improving problem solving and creativity through use of complex-dynamic simulations. *Computers in Human Behavior, 18*(6), 650–668.

Thomas, R., Cahill, J., & Santilli, L. (1997). Using an interactive computer game to increase skill and self-efficacy regarding safer sex negotiation: Field test result. *Health Education and Behavior, 24*(1), 71–86.

Vandeventer, S. S. (1998). *Expert behavior among outstanding videogame-playing children.* Dissertation Abstracts International Section A: Humanities & Social Sciences, 58(11-A), May.

Vorderer, P. (2000). Interactive entertainment and beyond. In D. Zillmann & P. Vorderer (Eds.), *Media entertainment: The psychology of its appeal* (pp. 21–36). Mahwah, NJ: Lawrence Erlbaum Associates.

Vorderer, P., & Ritterfeld, U. (2003). Children's future programming and media use between entertainment and education. In E. L. Palmer & B. Young (Eds.), *The faces of televisual media: Teaching, violence, selling to children* (2nd ed., pp. 241–262). Mahwah, NJ: Lawrence Erlbaum Associates.

Walshe, D. G., Lewis, E. J., Kim, S. I., O'Sullivan, K., & Wiederhold, B. K. (2003). Exploring the use of computer games and virtual reality in exposure therapy for fear of driving following a motor vehicle accident. *CyberPsychology & Behavior, 6*(3), 329–334.

Wartella, E. A., & Jennings, N. (2000). Children and computers: New technology—old concerns. *Future of Children, 10*(2), 31–43.

Wartella, E., O'Keefe, B. O., & Scantlin R. M. (2000). *Children and interactive media—a compendium of current research and directions for future.* A Report to The Markle Foundation.

Weisler, A., & McCall, R. R. (1976). Exploration and play: Resume and redirection. *American Psychologist, 31*(7), 492–508.

Wiederhold, B. K. (2003). The impact of the Internet, multimedia and virtual reality on behavior and society. *CyberPsychology & Behavior, 6*(3), 225–227.

Wolf, M. (2002). *The medium of the video game.* Austin: University of Texas Press.

Wolf, M., & Perron, B. (2003). *The video game theory reader.* New York: Routledge.

Yamada, F. (1998). Frontal midline theta rhythm and eyeblinking activity during a VDT task and a video game: useful tools for psychophysiology in ergonomics. *Ergonomics, 41*(5), 678–688.

26

Video Games for Entertainment and Education

Ute Ritterfeld
University of Southern California

René Weber
Michigan State University

While the morality of games and their ethical implications in an educational context have been questioned from the very beginning of game technology evolution (e.g., McLean, 1978), video games[1] have become not only increasingly attractive for players of both genders (Burke, 2000), various ethnicities (Bickham et al., 2003), and ages (IDSA, 2003), but are also utilized more and more for educational purposes (see, for example, the annual serious games summit, http://www.seriousgamessummit.com/). Thus, studying the education potential for this new and controversial medium is of tremendous importance.

Although comprehensive effect theories on the specific impact of video game playing are still missing, more than a decade of research provides us with an impressive body of literature demonstrating mostly negative, but also significant positive effects (see, for introduction, Mitchell & Savill-Smith, 2004). The majority of research has focused on the potential negative effects of video game playing; however, this is due to an abundance of studies on violent games and does not reflect the potential of video games as a medium. While Weber, Ritterfeld, and Kostygina (chap. 24) discuss the findings on violent video game playing, hostility, and aggression, Lee and Peng (chap. 22) as well as Lieberman (chap. 25) give a comprehensive overview on effect studies with an emphasis on learning, and Durkin (chap. 21) further elaborates the benefits of video game playing for adolescents. This chapter focuses on the potential of video game playing to facilitate *developmental processes* through the unique combination of interactive entertainment and learning, while at the same time taking a rather theoretical perspective.

The term "developmental processes" refers to an understanding of media usage and effects in the context of human development. Human development is the result of continuous transactions between a person's biological constitution and his or her physical, social, and media environment over the life span. According to this perspective, media usage is not random but already reflects the developmental processes of a user who is selecting some media or content over others, and processing it according to his or her developmental capacities, previous experiences, and current developmental needs. At the same time, the media usage influences the developmental processes of the user.

Taking a developmental approach seriously results in proposing distinctly different challenges for media usage and effect studies usually undertaken in the field of communication: (1) Impact studies should reflect evolving mental representations of the game content. Individuals may interpret the same video game play in completely different ways. For instance, one person takes the video game world as a realistic simulation, whereas another reads it in a metaphoric or even ironic way (Potter & Tomasello, 2003). (2) The study of play (Ohler & Nieding, chap. 8) demonstrates clearly that play is not a random activity selected to overcome boredom but rather a highly rational choice that serves the need of developmental processes. From that functionalist point of view, the selection of video games, as well as the video game play, serves the player's developmental needs (Havighurst, 1971; see also Durkin, chap. 21). Accordingly, analyses of video game playing should take the developmental status of the player into account. (3) For a developmentalist, changes over time are crucial. Consequently, media impact studies should overtake a longitudinal perspective, considering that media effects may vary substantially in the realm of a person's development.

Vygotsky (1978) introduced the "zone of proximal development" to explain which input characteristics are most likely to influence human development. According to his view, an input is most influential if it matches the developmental stage of the person, that is, if it connects to the established mental structure and extends it. This principle can be easily adopted to explain media effects, especially those of interactive media: If the user chooses the challenges he or she can manage successfully, his or her mental organization is most likely influenced through the video game play. Vygotsky's concept is compatible with the notion of developmental tasks (Havighurst, 1971), which claims that individuals actively seek out challenges that are appropriate to master the improvement of cognitive, emotional, social, or behavioral skills. Accordingly, a player is expected to lose interest in a specific game if the game no longer provides support in resolving developmental tasks.

Similar to a child's play in the first few years but in contrast to other media activities, video game play offers the potential of utilizing interactive entertainment to improve developmental processes in an age far beyond preschool years. Moreover, video game play accounts for highly intrinsic activity leading to intense experiences of presence (Tamborini & Skalski, chapter 16), thus allowing for maximum focus on the activity and related information processing (Biocca, 2002).

As early as 1981, Malone introduced three factors of intrinsic motivation derived from video game play: challenge, fantasy, and curiosity. According to Malone, "challenge" occurs in a situation of uncertain outcome. Prototypically, challenge is imposed through time constraints, competition with other players or agents (social norm of reference) or previous results (individual norm of reference), and hence creates winners and losers. Winning a game is supposed to enhance motivation to play the game again if the games posed an adequate challenge (Malone, 1981). However, recent findings indicate that losing a game can be intrinsically motivating as well, if the player receives positive feedback for his or her effort (Vansteenkiste & Deci, 2003). Nevertheless, the feedback does not take away negative feelings of disappointment. "Fantasy" is described as a state of cognitive and emotional involvement with the video game play facilitating game play skills. Most important, fantasy points to the virtual character of the experience that allows overcoming boundaries real-life experiences would impose. Finally, "curiosity" is divided into sensory and cognitive components, with sensory curiosity being dependent on the game surface and cognitive curiosity referring to the narrative of game play.

Although video games are most commonly used by older children, adolescents, and young adults, the often-made assumption that the media format is less suitable for older or younger audiences is questionable. Rather, the current user profiles reflect an industry that produces video game primarily for teenagers. The potential for other demographic groups to engage in video game play more has yet to be acknowledged. For very young children (e.g., Din & Calao, 2001), special need children (e.g., Rizzo et al., 2001), disabled individuals (e.g., Hasdai, Jessel, & Weiss, 1998) or geriatric clienteles (e.g., Shapiro, 1995) in particular, developmentally appropriate technology and content may be the most important tool to facilitate developmental processes. Because intrinsic motivation is heavily influenced by hope for success and fear of failure (Lepper & Henderlong, in press), appropriate challenges will most likely result in appreciation of the technology. Rightfully, the increase of computer literacy evolved as one of the main efforts in adult education (see Askov, Maclay, & Meenan, 1987).

Thus, video games offer an opportunity for intrinsically motivated, high-involvement experiences freed from real-life constraints that allow performances in the zone of proximal development facilitating individualized format developmental processes while reaching out to all members of society. If the developmental outcomes are considered valuable, video game play would be the ultimate paradigm for entertainment–education.

In the following section we provide support for this thesis, first addressing more closely the unique *characteristics of video game play* to facilitate developmental processes, then briefly outlining *enjoyment* and *learning* as the two kinds of game-related experiences blended together in the *entertainment–education* paradigm.

CHARACTERISTICS OF VIDEO GAME PLAY

From a historical perspective, books, television, and video games represent the evolution of make-belief: Books provide the reader with a *narrative* yet communicate through written language alone; television introduces simulation of the world through spoken language and images; and video games allow for *interactivity* and—potentially—*intelligent* reactions of the system (see Table 26.1). Compared to television, the primary disadvantage of video games, being less sophisticated and believable in *simulation*, is about to vanish (see Anderson, Funk, & Griffiths, 2004). The newest video game graphics already blur the boundaries between video documentation and animation.

Although video game play differs significantly from television watching, a theory on its differential impact is still lacking. Instead, researchers tend to apply models from television research to playing video games, despite being well aware that this generalization is flawed. The interactive—and even more, the intelligent—nature of a video game results in the incomparability of game play. In fact, every game play experience is unique.

The interactivity of the medium allows a player to choose settings or the unfolding of a narrative, to participate in the narrative, pursue goals, accept challenges, and experience

TABLE 26.1
The Evolution of Make-Belief

Characteristics of Medium	Books	Television	Video Games
Narrativity	Yes	Yes	Yes
Simulation	No	Yes	Yes
Interactivity	No	No	Yes
Intelligence	No	No	Yes

self-efficacy. Moreover, the user can interact with the system and other players. The term "intelligence of the medium" refers to the fact that the video takes the player's play history into account. In a rudimentary way, it could mean that the program "remembers" the player's name. A highly sophisticated program adapts game play to player achievements, failures, or even preferences. Thus, the embedded challenges can be tailored on an individual basis. Moreover, the system may give feedback to the player, tutor him or her, or make recommendations to improve game play.

One popular understanding of video games is that there will be an increased impact on its users compared to television *because* of the interactive possibilities (Markle Foundation, 2003). However, the potential for interacting with or through the medium is neither equivalent with the perceived interactivity nor with the interaction that takes place. Factual interactivity may enhance the user's involvement (Rockwell & Bryant, 1999) and hence his or her deeper processing of content (e.g., Ritterfeld, Weber, Fernandes, & Vorderer, 2004). Calvert and Tan (1994) provided evidence that this thesis holds true for aggressive thoughts. A recent study by Lee, Jin, Park, and Kang (2004) demonstrated that elaboration of a video game narrative led to greater presence, but research on the specific impact of interactivity of video games is still lacking. One reason for the shortage of scientific investigations may be methodological constraints. A study aiming to control the impact of interactivity on players' experiences and play outcomes would require an experimental setting in which interactive video game play is compared to non—interactive video game play. The latter would merely consist of watching video game play. However, as mentioned above, each game play experience is unique, differing in content, experience, and subsequently, impact. A noninteractive version of the video game play must therefore be identical to the interactive video game play, excepting the possibility to interact. The only—although compromised—possibility to realize such a comparison is to produce recorded versions of individual sessions of video game play and compare matched pairs of individuals who either play or watch them. But, even watching a game played by another person does not reflect the game play the viewer would have conducted him or herself. Moreover, matched pair samples designs depend heavily on the comparability of the paired individuals. Variables relevant for video game play such as gender, age, video game literacy, previous video game play experiences, topic-related interest, knowledge, and so on would have to be controlled carefully, altogether still allowing only limited approximation to comparability.

For instance, a recent study about epilepsy in children (Pellouchoud, Smith, McEvoy, & Gevins, 1999) compared a group of children playing a game with another group of children watching the game, although without matching individuals according to control variables. The authors used electroencephalographic recording (EEG) to observe mental activity in playing versus watching a video game demonstrating equally high activation. To our knowledge, no study to date has applied a matched pair design to control for the effect of interactivity.

Another interesting difference between television and video playing has been identified in the context of the uses and gratification approach (see Sherry, Lucas, Greenberg, & Lachlan, chapter 15): Whereas television has always been used for entertainment *and* educational purposes, video games are primarily played for entertainment. Purposeful education, it seems, is not significant in playing such video games. The use of "game" as a label that connotes pleasurable, enjoyable activities reflects this distinctive characteristic. Moreover, the games industry initially adopted the notion of video games as merely entertaining media, introducing so-called educational or learning video games as a supplemental category. The differentiation between entertaining and educational video games is still widely accepted (Lee & Peng, chap. 22). This not only implies that most video games have no educational potential, but also takes for granted that educational video games are significantly less fun than "real" video games.

LEARNING

Yet, the distinction between entertaining and educational video games simplifies the understanding of educational value. If education is defined through explicit learning goals embedded in the video game play, the distinction may be justified. However, educational value derives from a much broader variety of learning opportunities (see Nathan & Robinson, 2001). In fact, every change in the mental organization of a person is considered development, respectively learning (Bjorklund, 2000). Strictly speaking, every input processed influences the user and thus, results in learning. The question is *what* we learn and for *how long* the effect persists. With this understanding of learning, we may even consider learning to play a specific video game an educational outcome. The fact that such an outcome is usually not considered educational only reflects moral standards that distinguish between desired and undesired effects.

From a pedagogical perspective, enhancement in cognitive and metacognitive skills is the most desired outcome. Cognitive skills include, for instance, spatial abilities (see the avant-garde work by Greenfield and colleagues presented in the special issue of the *Journal of Applied Developmental Psychology*, 1994; but also more recently De Lisi & Wolfrod, 2002), linguistic competence (e.g., Din & Calao, 2001; Veale, 1999), knowledge acquisition, decision making (e.g., Ko, 2002), or problem solving (e.g., Ritterfeld et al., 2004). Metacognition refers to the fact that humans are conscious about their thinking: One can select, evaluate, and modify strategies for knowledge acquisition, problem solving, or other learning processes (Schneider & Lockl, 2002). For example, Oyen and Bebko (1996) successfully applied video games for the development of memory-enhancing strategies. Metacognitive skills play a major role in education because they help individuals learn how to learn. Research suggests that teaching metacognitive skills results in increased learning, especially in unfamiliar situations in which habitual responses are not successful (Scruggs, Mastropieri, Monson, & Jorgenson, 1985). Computer simulated worlds are designed to confront the user with a broad variety of unfamiliar situations in which he or she is supposed to act. It seems plausible that video game players benefit from metacognitive strategies in situations in which a challenge cannot be mastered. Metacognitive strategies unfold even at the very moment a child begins to play a video game. As Ko (2002) observed, video game play patterns vary between random guessers and good problem solvers, indicating the use of metacognitive strategies in the latter.

Pillay (2003) investigated the possible transfer of computer game literacy to performance on computer-based instructional tasks. Results indicate that transfer depends on the similarity between the designs of the computer game design and the instructional software. Moreover, linear-cause-and-effect computer games were found to encourage means–end analysis strategies, whereas adventure games experiences influenced inferential and proactive thinking in the usage of the instructional software.

Giving children the opportunity to learn how to program their own games is another way that video games contribute educational value. It enhances not only their technological skills, but also problem solving, creativity, and more (Kafai, 1995, 1998). The development of technological skills seems especially important for overcoming gender differences, as girls are usually less interested in videos than boys are (Gallup Poll, 1997). However, if girls are provided with the opportunity to use computers successfully, most gender differences vanish (Ching, Kafai, & Marshall, 2000).

One possible undesired impact of video game playing on sensomotoric control was primarily addressed through the criticism on ego-shooters (see Weber, Ritterfeld, & Kostygina, chap. 24) and racing games. Ego-shooters are assumed to cultivate the usage of firearms, and a player's driving habits in racing games are supposed to transfer into real-life driving styles. Although they are currently being investigated, empirical evidence for any of these assumptions is meager (Klimmt, 2003).

Comparably negative are assumptions regarding attitude formation through violent video games (Weber, Ritterfeld, & Kostygina, chap. 24). Sherer (1998) used simulation games to enhance moral development, with only partial success. The discussion of other possible contributions of video games to the development of positive attitudes such as altruism or social acceptance is still neglected. Theoretically, the potential is immense if one considers the global experiences of online gamers communicating with each other. Such communication experiences are supposed to cultivate attitudes as the most sensitive reflector for social norms. More generally, the study of video game-mediated content should be connected with impact studies to investigate the potential of video games (compared to other media) to lead to attitude formation.

Additionally, if video games are considered cultivators, the development of socioemotional skills should be influenced, too (see also Subrahmanyam, Greenfield, Kraut, & Gross, 2001). First, it is assumed that computer game-mediated communication enhances communication skills. But identity formation, autonomy, tolerance for frustration, or coping behavior may also improve extensively through video game play (for details, see Durkin, chap. 21).

Finally, we would like to emphasize the adaptability of video game technology to specifically improve handicaps or disabilities. Rizzo et al. (2001) developed and evaluated computer game-based educational technology for children with attention deficits. Scientific Learning developed games specifically for language and reading-impaired children, which resulted in significant improvement (Tallal, Miller, Jenkins, & Merzenich, 1997; Veale, 1999). Hasdai, Jessel, and Weiss (1998) used a driving simulator to help children with either progressive muscular dystrophy or cerebral palsy improve their ability to operate a wheelchair safely. To secure transfer effects, the authors designed the game environment according to the school environment in which the children were living. Results demonstrate that simulation games are efficient, though not fully compensating for, real-life practice. Olney (1997) successfully implemented computer games in training programs to scaffold communication in children with autism, Down syndrome, mental retardation, cerebral palsy, or pervasive developmental disorder. Finally, games can contribute to facing the increasing social challenge of multilingual environments, through requesting technologies for children to independently learn a second (or third) language that is not used at home (see Baltra, 1990).

In sum, video games can be specifically tailored to enhance cognitive, metacognitive, socioemotional, or behavioral skills, even addressing various user needs. With the development of more sophisticated intelligent systems, the potential for education in formal and informal contexts will rise dramatically. Still, there is concern that the time spent playing a video game may be harmful. In fact, several studies report that the amount of video game playing correlates negatively with school performance (for an overview, see Gentile, Lynch, Linder, & Walsh, 2004). However, most of the video game play considered in the various studies contained violence. Lieberman, Chaffee, and Roberts (1988) demonstrated an interaction effect between the amount of time spent on playing video games and the content of the games played: Children frequently playing violent video games performed more poorly in school than children who rarely used those games. On the other hand, school performance increased in children who frequently played educational games. Durkin and Barber (2002) reported data suggesting a curvilinear relationship between the amount of playing and positive outcomes.

Most interestingly, parents are somehow negligent in controlling their children's video playing. In a study by Gentile et al. (2004), 43% of adolescents who play video games said their parents did not control their video game usage at all. Also, children who reported that their parents set limits for video game playing performed better in school than children who experienced no control. However, available studies on the association between the time invested in video game play and the differential impact of various games report only short-term effects. To date, we are still lacking longitudinal studies on the educational impact of video game play.

Finally, educational value can be defined through the pathway selected for learning. Explicit, instructional learning is considered the major goal of education. However, developmental psychologists illuminate the fact that the younger children are, the more their learning is incidental and implicit[2] (Bjorklund, 2000, p. 18). Accordingly, implicit learning seems to be the main pathway for children and adolescents to learn video game play (Sala & Boyer, 1994). As mentioned above, video game playing may offer the possibility of an experience that is very similar to early learning in childhood, in which developmental tasks are performed without being aware of the developmental goal. Moreover, playing video games may result in a mental activity integrating intense cognitive, metacognitive—and, depending on the game—sometimes even social experiences while providing enjoyment at the same time.

ENJOYMENT

At this point, we will briefly outline the concept of enjoyment in the context of media studies. Here, enjoyment is considered the core experience of entertainment (Vorderer, Klimmt, & Ritterfeld, 2004), defining the sum of positive reactions toward media experiences being cognitive, affective, or conative. This view is consistent with empirical evidence in psychology indicating that the motivational basis of human activity relies on two rather independent systems: a so-called approach system and an avoidance system (Elliot & Thrash, 2002). Activation of the approach system results in enjoyment, whereas activation of the avoidance system leads to pain (Berridge, 2003). Enjoyment in game play may result from (a) sensory delight; (b) suspense, thrill, and relief; or (c) achievement, control, and self-efficacy (Vorderer et al., 2004).

Over 3 decades of game production, technological improvements in game design resulted in a dramatic increase of sensory delight, mostly visual and sonic, though possibly even kinesthetic and olfactory, in interactive theaters (e.g., Universal Studios' show "Shrek II"). However, the development of narrative aspects lags behind. Most games offer similar and quite simple stories (Clanton, 2000). But, it is the narrative that ultimately provides entertainment value beyond the pleasure of senses (see Lee et al., 2004). Suspense, thrill, and relief are enjoyable reactions that result from narratives (Brock, Strange, & Green, 2002). Moreover, avatars and agents allow for parasocial relationships, a phenomena that is widely acknowledged for its contribution to entertainment (Vorderer et al., 2004).

There is no doubt that enjoyment plays a key role in achievement (Sansone & Harackiewicz, 2000). However, authors disagree about whether enjoyment and intrinsic motivation are distinct qualities (e.g., Sansone & Smith, 2000) or one and the same (e.g., Linnenbrink & Pintrich, 2000). The intrinsic motivation to play a game results from promoting enjoyable experiences (approach state) and from preventing boredom or failure (avoidance state). For instance, a study by Lee, Sheldon, and Turban (2003) demonstrates how enjoyment and performance in achievement settings depend on the individual's capability to focus mentally on the task given. Persons high in achievement motivation in general enjoy challenges much more than individuals low in achievement motivation (Durik & Harackiewicz, 2003). Low achievers are more task-dependent and tend to enjoy tasks aligned with their interests more. In general, enjoyment through achievement can be a result of either mastery or performance (Linnenbrink & Pintrich, 2000). Mastery orientation is based on the need for a deeper understanding of a task, whereas performance orientation results from the wish to be superior and to win. Most interestingly, research on sport games demonstrates that enjoyment seems not to differ significantly between cooperative and competitive tasks (Tauer & Harackiewicz, 2004).

In sum, entertainment in video game play reflects enjoyable experiences while playing the game. Thus, the concept of enjoyment is closely related—if not identical—to intrinsic motivation, which again is crucial for developmental processes, as we address further in the following section.

ENTERTAINMENT–EDUCATION

The abovementioned distinction between educational and entertainment games (for more details, see Lee & Peng, chap. 22) is based on the assumption of a single dimension with education and entertainment defining the poles. However, entertainment and education should be conceptualized as two different dimensions. A distinction between educational and entertainment games neglects the possibility that the two may coexist. The construction of a 2-dimensional space with entertainment and education as axes allows the deduction of a variety of hypotheses regarding the relationship between the two.

Whether and to which amount the entertainment–education paradigm is applicable to interactive technology remains still unclear. In theory, one can argue for a general relationship between entertainment and learning that is (a) either linear positive (entertainment as facilitator), (b) linear negative (distraction hypothesis), or (c) inverse u-shaped (moderate enjoyment hypothesis) (see Fig. 26.1).

Overcoming the abovementioned simplifying dichotomy of media entertainment and education was the—at first implicit, increasingly more explicit—goal of a paradigm called entertainment–education. The general idea to utilize enjoyment for learning is undoubtedly not new, but the purposeful use of media—especially video games—to facilitate these means is still incipient. Mostly applied in the area of health communication, entertainment–education programming in radio and television have recently proven to be tremendously effective for knowledge acquisition, attitude, and even behavior change (see Singhal & Rogers, 2002; Slater, 2002).

There are basically three pathways for entertaiment–education (see Fig. 26.2): First, an entertainment experience is used as a motivational facilitator to process educational information (motivation paradigm). The function of entertainment in this experience can be described as a "door opener" to allocate attention to educational content, to develop interest in the content, and finally process the educational information delivered (Ritterfeld et al., 2004; Vorderer & Ritterfeld, 2003). It is hereby implied that the content alone would not be a strong enough attractor to ensure processing, and it requires enrichment through entertainment. Accordingly, entertainment media can be enriched with educational games. For instance, the addition of

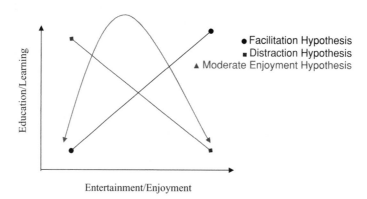

FIG. 26.1.   Theoretical assumptions on the entertainment–education link.

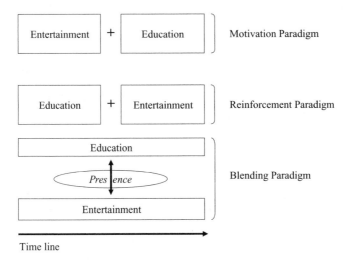

FIG. 26.2. Paradigms of entertainment–education.

information to entertainment can be found in some video games providing hyperlinks to research on specific issues (e.g., Sesame Street Web-based games). Thus, entertainment serves as a mediator between content and the educational impact. This approach has been proven successful for attentiveness toward linguistic information (Ritterfeld, Klimmt, Vorderer, & Steinhilper, 2005), science education (e.g., Ritterfeld et al., 2004), and health education (e.g., Ritterfeld & Jin, in press).

The second paradigm (reinforcement paradigm) argues from the contrary perspective with utilizing entertainment for reinforcement purposes. The promise of such an approach is to enhance motivation to process educational content. Reinforcement can come as a surprise after succeeding or be expected (Malone, 1981). A reinforcement strategy is applied in most educational games through scores, virtual money, fun animations, or the reward of progress in the video game play. The effect of repeated usage of educational content (e.g., Vorderer, Böcking, Klimmt, & Ritterfeld, in press) on phonological awareness or phoneme distinction (Veale, 1999), for example, demonstrates the effectiveness of entertainment as a reinforcement strategy at first glance. But, the studies do not reveal how motivated subjects were while participating in the training. Reinforcement may result in an unintended and undesired motivational decrease. If individuals are intrinsically motivated, they do not have to be preached to. Nor do they need incentives in order to perform a task. In these individuals, reinforcement may even reduce intrinsic motivation on the long run (Lepper & Henderlong, 2000).

Both the motivation and the reinforcement paradigm are based on an additive concept of entertainment and education, in which educational information is added to an entertaining program or vice versa. Educational goals are mostly explicit and primarily intentional learning is aspired. A combined game format is used to teach cognitive skills, such as in the popular *Mathematic* or *Reading Blasters* that are now available for all performance levels. The game format consists of appealing settings, characters, feedback, and reinforcement. The user knows exactly what he or she is expected to learn. The surplus of gaming compared to instruction derives from the hope that the fun elements provide additional pleasure. However, the most pleasure in educational settings is obtained from achieving a goal (Shah & Kruglanski, 2000). Entertaining byproducts may increase pleasure of the senses but cannot replace lacking success. Strictly speaking, even a highly entertaining format cannot compensate for a user who fails to achieve the learning goal. On the other side, if a user indeed succeeds, he or she might not need the entertainment surplus at all. The assumption that entertainment contexts facilitate learning in general is plausible at first glance, but loses suggestive power if applied to all players.

From a developmental perspective, changes in mental organization most likely occur through (a) repeated processing, which not only establishes networks but also reinforces new pathways; (b) problem solving; and (c) generalization to other domains. There is no evidence that reinforcement can actually facilitate developmental processes other than the first. For instance, why should a child improve his or her mathematic problem solving if a correct result is reinforced with attractive animation? It is much more likely that already acquired skills can be practiced in such an entertaining context.

Practicing skills does not allow solving new tasks that require cogitating. But, it enhances the fluency and speed of the performance and reduces mistakes. Expertise research states that practice is crucial for academic performance (Ericsson, 1998). Deliberate practice means vast and extensive intrinsically motivated training. As Schneider and Stefanek (2004) summarized, deliberate practice and an IQ that is slightly above average guarantee the best results for children's GPAs. Similarly crucial for performance is an early start with practice that allows stabilizing acquired competences that are necessary to build on for further developments. Longitudinal data reveal that a combination of intense practice, highly developed interest, and intrinsic motivation predicts above average performance (Schneider & Stefanek, 2004). Video games provide a terrific opportunity to realize the described triad (practice, interest, intrinsic motivation).

However, intrinsic motivation and interest are expected to be much higher in video games selected for entertainment purposes than those chosen for educational reasons. Consequently, practice will be lower in educational games. Blumberg (2000) drew this conclusion by investigating the role of learning goals prior to playing a computer game (*Sonic the Hedgehog 2*). His findings reveal an age and skill dependent pattern: Older children improved their performance when they were trying to be especially attentive while playing the video game whereas the effect reversed for younger children. The latter improved their video game play if they were pursuing personal goals for mastering the game.

Yet, as Slater (2002) pointed out, entertainment–education unfolds its greatest potential strategy if the information provided becomes an essential part of the entertainment experience. The prototype for a blend between entertainment experience and education (blending paradigm) is game play in early childhood before the child becomes conscious about educational goals and settings (see Ohler & Nieding, chap. 8). Around kindergarten age, children learn to distinguish between educational and play purposes while at the same time shaping their attitudes toward educational activities (Oerter, 1999). If they enjoy educational settings and develop interest in instructional learning, the distinction between primarily entertainment and primarily learning activities is unnecessary. Although being of different qualities, humans may enjoy both activities and learn in both settings. Children whose attitudes toward learning tend to become negatively formed will eventually develop a distinction between enjoyable entertainment activities and unattractive learning activities. These children should profit most from entertainment–education strategy. Entertainment–education intends to adopt the unified approach of early game play for older children who are less motivated to learn, but also for adolescents and even adults.

As mentioned above, blending an entertaining narrative with educational content became a popular entertainment–education strategy in health-related TV and radio programming. Most interestingly, this strategy has been proven to be tremendously successful for audiences that otherwise could not be reached with persuasive messages (Singhal, Cody, Rogers, & Sabido, 2004). Although Lee et al. (2004) most recently provided empirical evidence for the effectiveness of a narrative in video game play for the experience of presence, the educational relevance still needs to be investigated.

Recently, so-called *pedagogical dramata* (e.g., Marsella, Johnson, & LaBore, 2003), which utilize interactive game play to teach successful communication strategies, were introduced. For example, Clarke and Schoech (1994) developed a simulation game for adolescents to learn impulse control, and Oakley (1994) pioneered a video game play for at-risk teens displaying consequences of drug abuse. Marsella and his team introduced an interactive technology in which parents of children suffering from cancer learn to communicate with their sick child and cope with crisis in the family and work place (Marsella et al., 2003). Miller and Read (in press) developed a similar technology to teach homosexual men safe-sex negotiations. Evaluation results demonstrate the acceptance of the technology in the target groups, respectively, and its impact on communication skills and behavior (Miller & Read, in press). Similar to educational games, pedagogical dramata pursue an explicit learning goal. In contrast to educational games, they offer simulated situations that otherwise cannot be experienced for training purposes. The user dives into the simulation in order to improve communication or behavioral skills in a protected, safe environment. Because the development of pedagogical dramata is still in its very beginning, the technology has not yet been applied to less mature audiences, but it is easy to imagine how the principle can be applied to all age groups and specifically tailored for different target groups.

In sum, most video games currently available follow the motivation and/or the reinforcement paradigm to combine entertainment with education. The role of game play in the motivation paradigm can be described as seduction with the goal of entertainment to allocate attention and direct it to educational content (see Table 26.2). It is implied that the entertaining experience of game play provides enough intrinsic motivation for continuation and ultimately learning. Educational goals can hereby be implicit or explicit, facilitating intentional and/or incidental learning that may include complex problem solving. The reinforcement paradigm differs from the motivation paradigm; in using entertainment as reward, it provides a means to improve extrinsic motivation. A reward system implies that the educational goal has to be explicit, and learning is intentional. As mentioned above, game play as reinforcement may facilitate practice but does not enable a person to solve problems.

Both pathways combine entertainment and education in sequence. In contrast, the blending paradigm claims a parallel experience of entertainment and education, facilitating intrinsic motivation to resolve developmental tasks. The educational goal is implicit, and learning is primarily incidental. Incidental learning is not predictable; education relies on the fact that a gamer chooses games and plays them to support his or her developmental tasks. Thus, this kind of game play mimics the most effective play observed in young children.

Children play various games with differential impact on their developmental processes. One predominant game format to resolve developmental tasks is role-play (Oerter, 1999), which simulates real life, engages fantasies, and involves new formats for communication and exploration of behavior. Most importantly, children are highly immersed and intrinsically

TABLE 26.2
Characteristics of Motivation, Reinforcement, and Blending Paradigm

	Role of Game Play	Focus of Motivation	Educational Goal	Goal Communication	Model of Learning
Motivation Paradigm	Seduction	Intrinsic	Allocation of Attention	Explicit	Either or
Reinforcement Paradigm	Reward	Extrinsic	Practice	Explicit	Intentional
Blending Paradigm	Simulation	Intrinsic	Developmental tasks	Implicit	Incidental

motivated while role-playing. Consequently, one expects a positive linear correlation between entertainment and education for role-play. Generalizing this to video game play, we assume that the mimic of role-play is most sufficient in providing highly immersive experiences with entertainment and education combined. We consider the experience of nonmediation (presence) an indicator for a successful blend, which is illustrated in Fig. 26.2.

Finally, research on the improvement of intrinsic motivation points to social context, which has long been underestimated. Not only are values or attitudes that build interest, set learning goals, and enhance achievement motivation heavily influenced (Ryan & Deci, 2000; Jacobs & Eccles, 2000) by social contexts, but social contexts also serve as a direct motivator. Multiplayer games, especially multiplayer online games where individuals meet in a virtual environment, provide a social context freed from real-life constraints.

Taken together, we propose that the impact of entertainment on the educational experience differs according to which paradigm of entertainment–education is applied. Simulation games should allow for a variety of incidental learning effects, whereas prototypical educational video games mostly enhance intentional learning by using entertainment as motivator or reinforcement strategy. Given that the game provides content via a socially valued format, desired educational outcomes are practice and higher attention toward the content.

Because the entertainment and the educational experience are not identical, one may argue that they compete for attention, with the risk of too much entertainment reducing the educational value. Due to the abovementioned possible negative consequences of reinforcement and the here-described potential for distraction through high entertainment, it may be expected that the relationship between entertainment and education (as described in Fig. 26.1) is not linear. Consequently, we propose a curvilinear correlation between intentional learning and enjoyment.

In conclusion, from a pedagogical perspective, games that use entertainment for motivation and reinforcement are undoubtedly valuable. If, as elaborated above, high entertainment may reduce the educational impact in these games, game development faces the challenge of identifying the optimal balance between entertainment and educational content. But, even perfectly balanced "entertainment+education" games provide only limited surplus to resolve developmental tasks. Thus, from a developmental perspective, the ultimate pathway of game-based entertainment–education is offered through online multiplayer simulation.

NOTES

[1]To stay consistent with the title of this volume, we use the term "video game" as a generic category including all electronic and interactive games.

[2] Incidental and intentional learning are antagonists with intentional learning being goal-directed and incidental learning being unplanned (Kerka, 2000). The term "implicit learning" refers to the learning process as being not conscious, opposed to explicit learning (Perrig, 1996).

REFERENCES

Anderson, C. A, Funk, J. B., & Griffiths, M. D. (2004). Contemporary issues in adolescent video game playing: Brief overview and introduction to the special issue. *Journal of Adolescence, 1*, 1–3.

Askov, E., Maclay, C., & Meenan, A. (1987). Using videos for adult literacy instruction. In W. M. Rivera & S. M. Walker (Eds.), *Lifelong Learning Research Conference Proceedings* (pp. 1–26). College Park: University of Maryland.

Baltra, A. (1990). Language learning through video adventure games. *Simulation & Gaming, 4*, 445–452.

Berridge, K. C. (2003). Pleasures of the brain. *Brain & Cognition, 52*, 106–128.

Bickham, D. S., Vandewater, E. A., Huston, A. C., Lee, J. H., Caplovitz, A. G., & Wright, J. C. (2003). Predictors of children's electronic media use: An examination of three ethnic groups. *Media Psychology, 2*, 107–137.

Biocca, F. (2002). The evolution of interactive media. In M. C. Green, J. J. Strange, & T. C. Brock (Eds.), *Narrative impact. Social and cognitive foundations* (pp. 97–130). Mahwah, NJ: Lawrence Erlbaum Associates.

Bjorklund, D. F. (2000). *Children's thinking. Developmental function and individual differences.* Belmont, CA: Wadsworth.

Blumberg, F. C. (2000). The effects of children's goals for learning on video game performance. *Journal of Applied Developmental Psychology, 6,* 641–653.

Brock, T. C., Strange, J. J., & Green, M. C. (2002). Power beyond reckoning. In M. C. Green, J. J. Strange, & T. C. Brock (Eds.), *Narrative impact. Social and cognitive foundations* (pp. 1–16). Mahwah, NJ: Lawrence Erlbaum Associates.

Burke, K. (2000). *Sixty percent of all Americans play video games, contributing to the fourth straight year of double-digit growth for the interactive entertainment industry.* Retrieved June 1, 2004, from Interactive Digital Software Association Web site: http://www.isda.com/releases/4-21-2000.html.

Calvert, S. L., & Tan, S. (1994). Impact of virtual reality on young adults' physiological arousal and aggressive thoughts: Interaction versus observation. *Journal of Applied Developmental Psychology, 15,* 125–139.

Ching, C. C., Kafai, Y. B., & Marshall, S. (2000). Spaces for change: Gender and technology access in collaborative software design projects. *Journal for Science Education and Technology, 1,* 45–56.

Clanton, C. (2000). Lessons from game design. In E. Bergman (Ed.), *Information appliances and beyond* (pp. 300–334). New York: Harcourt Academic.

Clarke, B., & Schoech, D. (1994). A video-assisted therapeutic game for adolescents: Initial development and comments. *Videos in Human Services, 1–2,* 121–140.

De Lisi, R., & Wolfrod, J. L. (2002). Improving children's mental rotation accuracy with video game playing. *Journal of Genetic Psychology, 3,* 272–282.

Din, F. S., & Calao, J. (2001). The effects of playing educational video games on kindergarten achievement. *Child Study Journal, 2,* 95–102.

Durik, A. M., & Harackiewicz, J. M. (2003). Achievement goals and intrinsic motivation: Coherence, concordance, and achievement orientation. *Journal of Experimental Social Psychology, 4,* 378–385.

Durkin, K., & Barber, B. (2002). Not so doomed: Video game play and positive adolescent development. *Applied Journal of Developmental Psychology, 23,* 373–392.

Elliot, A. J., & Thrash, T. M. (2002). Approach–avoidance motivation in personality: Approach and avoidance temperament and goals. *Journal of Personality and Social Psychology, 82,* 804–818.

Ericsson, K. A. (1998). The scientific study of expert levels of performance: General implications for optimal learning and creativity. *High Ability Studies, 1,* 75–100.

Gallup Poll. (1997). U.S. teens and technology. Retrieved July 22, 2004, from http://www.nsf.gov/od/lpa/nstw/teenov.htm.

Gentile, D. A., Lynch, P. J., Linder, J. R., & Walsh, D. A. (2004). The effects of violent video game habits on adolescent hostility, aggressive behaviors, and school performance. *Journal of Adolescence, 1,* 5–22.

Greenfield, P., & Cocking, R. R. (Eds.). (1994). Effects of interactive entertainment technologies on development. Special issue of *Journal of Applied Developmental Psychology, 1.*

Hasdai, A., Jessel, A. S., & Weiss, P. L. (1998). Use of a video simulator for training children with disabilities in the operation of a powered wheelchair. *American Journal of Occupational Therapy, 3,* 215–220.

Havighurst, R. J. (1971). *Developmental tasks and education.* (3rd ed.). New York: Longman.

IDSA (2003). *Top ten industry facts.* Interactive Digital Software Association. Retrieved June 14, 2004, from Interactive Digital Software Association Web site: http://www.idsa.com/pressroom_main.html.

Jacobs, J. E., & Eccles, J. S. (2000). Parents, task values, and real-life achievement-realted choices. In C. Sansone & J. M. Harackiewicz (Eds.), *Intrinsic and extrinsic motivation. The search for optimal motivation and performance* (pp. 408–443). New York: Academic Press.

Kafai, Y. B. (1995). *Minds in play: Video game design as a context for children's learning.* Hillsdale, NJ: Lawrence Erlbaum Associates.

Kafai, Y. B. (1998). Children as software users, designers and evaluators. In A. Druin (Ed.), *The design of children's interactive technologies* (pp. 123–145). San Francisco: Morgan Kaufman.

Kerka, S. (2000). Incidental learning. Trends and issues alert No. 18. Retrieved November 11, 2003, from http://www.cete.org/acve/docgen.asp?tbl=tia&ID=140.

Klimmt, C. (2003). Racing games and driving behavior: An effects model of interactive entertainment. Paper presented at the 53rd Annual Conference of the International Communication Association (ICA), San Diego.

Ko, S. (2002). An empirical analysis of children's thinking and learning in a video game context. *Educational Psychology, 2,* 219–233.

Lee, F. K., Sheldon, K. M., & Turban, D. B. (2003). Personality and the goal-striving process: The influence of achievement goal patterns, goal level, and mental focus on performance and enjoyment. *Journal of Applied Psychology, 2,* 256–265.

Lee, K. M., Jin, S., Park, N., & Kang, S. (2004, October). Effects of narratives on feelings of presence in video-game playing. Paper presented at the Games @ USC Summit, Los Angeles.

Lepper, M. R., & Henderlong, J. (2000). Turning "play" into "work" and "work" into "play": 25 years of research on intrinsic versus extrinsic motivation. In C. Sansone & J. M. Harackiewicz (Eds.), *Intrinsic and extrinsic motivation. The search for optimal motivation and performance* (pp. 257–310). New York: Academic Press.

Lepper, M. R., & Henderlong, J. (in press). Motivation and instruction. In J. W. Guthrie (Ed.), *Encyclopedia of education.* New York: MacMillan.

Lieberman, D. A., Chaffee, S. H., & Roberts, D. F. (1988). Videos, mass media, and schooling: Functional equivalence in uses of new media. *Social Science Video Review, 6*, 224–241.

Linnenbrink, E. A., & Pintrich, P. R. (2000). Multiple pathways to learning and achievement: The role of goal orientation in fostering adaptive motivation, affect, and cognition. In C. Sansone & J. M. Harackiewicz (Eds.), *Intrinsic and extrinsic motivation. The search for optimal motivation and performance* (pp. 196–230). New York: Academic Press.

Malone, T. W. (1981). Toward a theory of intrinsically motivating instruction. *Cognitive Science, 4*, 333–369.

Markle Foundation. (2003). *Children and interactive media research compendium update.* Retrieved May 25, 2004, from http://www.markle.org/news/interactive_media_update.pdf.

Marsella, S. C., Johnson, L. W., & LaBore, C. M. (2003). *Interactive pedagogical drama for health interventions.* Paper presented at the 11th International Conference on Artificial Intelligence in Education, Australia.

McLean, H. W. (1978). Are simulations and games really legitimate? *Audiovisual Instruction, 7*, 12–13.

Miller, L. C., & Read, S. J. (in press). *Virtual sex: Creating environments for reducing risky sex.* In K. Portnoy and S. Cohen (Eds.), *Virtual decisions: Digital simulations for teaching reasoning in the social sciences and humanities.* Mahwah, N.J.: Lawrence Erlbaum Associates.

Mitchell, A., & Savill-Smith, C. (2004). *The use of video and video games for learning.* London: Learning and Skills Development Agency.

Nathan, M., & Robinson, C. (2001). Considerations of learning and learning research: Revisiting the "media effects" debate. *Journal of Interactive Learning Research, 1*, 69–88.

Oakley, C. (1994). SMACK: A video driven game for at-risk teens. *Videos in Human Services, 1–2*, 97–99.

Oerter, R. (1999). *Psycholgie des Spiels.* Weinheim: Beltz.

Olney, M. (1997). A controlled study of facilitated communication using video games. In D. Biklen & D. N. Cardinal (Eds.), *Contested words, contested science: Unraveling the facilitated communication controversy.* Special education series (pp. 96–114). New York: Teachers College Press.

Oyen, A.-S., & Bebko, J. M. (1996). The effects of video games and lesson contexts on children's mnemonic strategies. *Journal of Experimental Child Psychology, 2*, 173–189.

Pellouchoud, E., Smith, M. E., McEvoy, L., & Givens, A. (1999). Mental effort-related EEG modulation during videogame play: Comparison between juvenile subjects with epilepsy and normal control subjects. *Epilepsia, 4*, 38–43.

Perrig, W. J. (1996). Implizites Lernen [Implicit learning]. In J. Hoffmann & W. Kintsch (Eds.), *Lernen* [Learning] (pp. 203–234). Enzyklopädie für Psychologie CII/7. Göttingen: Hogrefe.

Pillay, H. (2003). An investigation of cognitive processes engaged in by recreational video game players: Implications for skills of the future. *Journal of Research on Technology in Education, 3*, 336–350.

Potter, W. J., & Tomasello, T. K. (2003). Building upon the experimental design in media violence research: The importance of including receiver interpretations. *Journal of Communication, 53*, 315–329.

Ritterfeld, U., & Jin, S.-A. (in press). Fighting stigma attached to people suffering from mental illness using entertainment–education strategy. Journal of Health Psychology.

Ritterfeld, U., Klimmt, C., Vorderer, P., & Steinhilper, L. (2005). The effects of a narrative audio tape on preschoolers' attention and entertainment experience. *Media Psychology, 7*, 47–72.

Ritterfeld, U., Weber, R., Fernandes, S., & Vorderer, P. (2004). Think science! Entertainment education in interactive theaters. *Videos in Entertainment: Educating Children through Entertainment, 2/1*, http://doi.acm.org/10.1145/973801.973819.

Rizzo, A. A., Buckwalter, J. G., McGee, J. S., Bowerly, T., van der Zaag, C., Neumann, U., et al. (2001). Virtual environments for assessing and rehabilitating cognitive/functional performance. *Presence, 4*, 359–374.

Rockwell, S. C., & Bryant, J. (1999). Interactivity and enjoyment in a multimedia entertainment application for children. *Media Psychology, 4*, 244–259.

Ryan, R. M., & Deci, E. L. (2000). When rewards compete with nature: The undermining of intrinsic motivation and self-regulation. In C. Sansone & J. M. Harackiewicz (Eds.), *Intrinsic and extrinsic motivation. The search for optimal motivation and performance* (pp. 14–56). New York: Academic Press.

Sala, T. E., & Boyer, H. P. (1994). Implicit learning and representation of knowledge in tasks of interactive system control. *Cognitiva, 1*, 47–65.

Sansone, C., & Harackiewicz, J. M. (2000). Controversies and new directions—is it déjà vu all over again? In C. Sansone & J. M. Harackiewicz (Eds.), *Intrinsic and extrinsic motivation. The search for optimal motivation and performance* (pp. 444–454). New York: Academic Press.

Sansone, C., & Smith, J. L. (2000). Interest and self-regulation: The relation between having to do and wanting to. In C. Sansone & J. M. Harackiewicz (Eds.), *Intrinsic and extrinsic motivation. The search for optimal motivation and performance* (pp. 343–374). New York: Academic Press.

Schneider, W., & Lockl, K. (2002). The development of metacognition and knowledge in children and adolescents. In T. Perfect & B. Schwartz (Eds.), *Applied metacognition* (pp. 224–247). Cambridge, UK: Cambridge University Press.

Schneider, W., & Stefanek, J. (2004). Entwicklungsveränderungen allgemeiner kognitiver Fähigkeiten und schulbezogener Fertigkeiten: Evidenz für einen Schereneffekt? [Developmental changes of cognitive skills and school performance: Evidence for an increasing gap?] *Zeitschrift für Entwicklungspsychologie und Pädagogische Psychologie, 36*, 147–159.

Scruggs, T. E., Mastropieri, M. A., Monson, J., & Jorgenson, C. (1985). Maximizing what gifted students can learn: Recent findings of learning strategy research. *Gifted Child Quarterly, 4*, 181–185.

Shah, J. Y., & Kruglanski, A. W. (2000). The structure and substance of intrinsic motivation. In C. Sansone & J. M. Harackiewicz (Eds.), *Intrinsic and extrinsic motivation. The search for optimal motivation and performance* (pp. 106–130). New York: Academic Press.

Shapiro, P. (1995). *Video use and the elderly.* Retrieved July 2, 2004, from http://www.his.com/~pshapiro/videos.and.elderly.html.

Sherer, M. (1998). The effect of videoized simulation games on the moral development of junior and senior high-school students. *Videos in Human Behavior, 2*, 375–386.

Singhal, A., Cody, M. J., Rogers, M. E., & Sabido, M. (2004). *Entertainment–education and social change: History, research, and practice.* Mahway, NJ: Lawrence Erlbaum Associates.

Singhal, A., & Rogers, E. M. (2002). A theoretical agenda for entertainment–education. *Communication Theory, 2*, 117–135.

Slater, M. (2002). Entertainment education and the persuasive impact of narratives. In M. C. Green, J. J. Strange, & T. C. Brock (Eds.), *Narrative impact. Social and cognitive foundations* (pp. 157–181). Mahwah, NJ: Lawrence Erlbaum Associates.

Subrahmanyam, K., Greenfield, P., Kraut, R., & Gross, E. (2001). The impact of video use on children's and adolescents' development. *Journal of Applied Developmental Psychology, 1*, 7–30.

Tallal, P., Miller, S. L., Jenkins, W. M., & Merzenich, M. M. (1997). The role of temporal processing in developmental language-based learning disorders: Research and clinical implications. In B. A. Blachman (Ed), *Foundations of reading acquisition and dyslexia: Implications for early intervention* (pp. 49–66). Mahwah, NJ: Lawrence Erlbaum Associates.

Tauer, J. M., & Harackiewicz, J. M. (2004). The effects of cooperation and competition on intrinsic motivation and performance. *Journal of Personality Social Psychology, 6*, 849–861.

Vansteenkiste, M., & Deci, E. L. (2003). Competitively contingent rewards and intrinsic motivation: Can losers remain motivated? *Motivation & Emotion, 4*, 273–299.

Veale, T. K. (1999). Targeting temporal processing deficits through Fast ForWord(R): Language therapy with a new twist. *Language, Speech, & Hearing Services in Schools, 4*, 353–362.

Vorderer, P., Böcking, S., Klimmt, C., & Ritterfeld, U. (in press). What makes preschoolers listen to narrative audio tapes? Zeitschrift für medienpsychologie.

Vorderer, P., Klimmt, C., & Ritterfeld, U. (2004). Enjoyment: At the heart of media entertainment.*Communication Theory, 4*, 388–408.

Vorderer, P., & Ritterfeld, U. (2003). Children's future programming and media use between entertainment and education. In E. Palmer & B. Young (Eds.), *The faces of televisual media: Teaching, violent, selling to children* (pp. 241–264). Mahwah NJ: Lawrence Erlbaum Associates.

Vygotsky, L. S. (1978). *Mind in society: The development of higher psychological processes.* Cambridge, MA: Harvard University Press. Published originally in Russian in 1930.

27

Game Playing and Adolescents' Development

Kevin Durkin
University of Strathclyde

The absence of video game play in the life of a contemporary early adolescent is a risk indicator. Indeed, it is possible—though it remains to be tested empirically—that the absence of video games is an impediment to what has become normal development, and a more serious one than the absence of television.

Television is usually excluded from the home for reasons of cultural protection, intellectual elitism, or religious or political ideology. It is likely that the very small number of households that commit to this exceptional self-deprivation are motivated to generate alternative, quite possibly richer, activities. In contrast, the nonavailability of video/computer games in Western homes is more likely to be due to poverty, parental technological anxiety/Luddism, or children's lack of interest—and each of these signals risk. Furthermore, because absent video games are not as immediately obvious a problem as absent TV, the risk may be overlooked and no compensatory activities instigated.

It is intriguing that we have reached this stage of evolution in just a generation or two. Many adult readers of this book will have passed through their entire adolescence without playing video games, and most will not have suffered from the loss (or so we like to think). But, it is now nonnormative to spend one's teens without computer or video[1] games—they are new cultural artifacts (Greenfield, 1994, 1998) of widespread acceptance. Households that cannot afford to purchase games and game equipment are clearly disadvantaged—perhaps not irredeemably, but the odds are hostile. Parents who refuse to allow their children access to computer games are taking risks, perhaps not enormous but certainly worth reviewing. And, young people who do not want to play the games are different, which is always risky. Being different is not necessarily a bad thing—for example, if one prefers ballet training or cross-country skiing then the hedonic, sensorimotor, spatial, and social benefits accessible to the masses through computer games might be obtained in other ways—but difference can reflect adjustment problems and it can hinder normal peer relations.

As the great majority of young people have already decided, life is better with video games (Barnett et al., 1997; Colwell, Grady, & Rhaiti, 1995; Durkin & Aisbett, 1999; Roberts, Foehr, Rideout, & Brodie, 1999; Vorderer, Bryant, Pieper, & Weber, this volume). This chapter considers some of the reasons why this is the case. The broad aim is to show that game use is a psychologically healthy part of the developmental period of adolescence (which is not to deny that computer games are also important to people at other stages of the life span, nor to deny that they are sometimes played by adolescents who are not experiencing psychologically healthy development). The chapter will necessarily be speculative in parts, drawing implications from broader developmental research that have not yet been tested exhaustively in relation to computer games, because of the enduring preoccupation of social scientists with violent fantasy content.

The chapter begins with a brief overview of the nature of adolescent development, noting the important physiological, cognitive, and social changes that occur during this part of life. It is stressed that all of these changes have emotional concomitants. We turn then to the place of computer games in this developmental context, considering in turn the place of computer games in relation to cognitive, social, and emotional development.

It should be noted that this chapter is not centrally concerned with effects. Effects models generally assume a largely passive audience who is somehow shaped by media input and illustration. Issues relating to effects are discussed in greater detail in other contributions to this volume (Buckley & Anderson; Lee & Peng; Weber, Ritterfeld, & Kostygina). The present chapter draws on the field of developmental psychology, which assumes in contrast to linear effects models that developing persons are active and resilient beings, engaged in a constant exploration of their environments, informed by the available culture and technologies, manifesting individual differences, and changing over time (see also Ritterfeld & Weber, this volume). From a developmental perspective, while "effects" of playing computer games are certainly possible and worthy of investigation, they are not expected invariably to be strong and, even where they are demonstrable, need to be understood in relation to a host of other factors that bear more profoundly and more extensively on human behavior and development. This does not mean that the processes involved in adolescents' uses of computer games are trivial but, rather, that those processes need to be understood in relation to the developing person's needs and contexts.

THE NATURE OF ADOLESCENCE

Adolescence is a lengthy period of pre-adult life, arguably becoming lengthier in Western societies. Because of individual differences in the onset of puberty and social/cultural variations in the organization of young people's roles, it is difficult to define exactly when adolescence begins or ends (Durkin, 1995; Spear, 2000). Rather than debate definition, the focus here will be on the period between about 12 and 20 years. It is important to bear in mind that this encompasses a wide span of physical, intellectual, and interpersonal growth. Adolescence is generally acknowledged as a period of radical change in terms of physical (Archibald, Graber & Brooks-Gunn, 2003), cognitive (Byrnes, 2003; Granic, Dishion, & Hollenstein, 2003; Piaget & Inhelder, 1966), and social (Durkin, 1995; Granic et al., 2003) domains. Even a cross-sectional inspection of a sample of adolescents will reveal enormous heterogeneity. This is obvious but often overlooked in popular accounts of teenagers as the victims of new technologies.

The important cognitive changes of adolescence include an increasing ability to handle complex and abstract problems, to entertain and even test competing hypotheses about causal relations, and to manipulate images and conjecture alternative possible states of affairs (Byrnes, 2003; Inhelder & Piaget, 1958). For many developmentalists, adolescent cognitive development is regarded as ushering in a revolution in the ways in which the young person is able to think about and respond to the world. The radical cognitive changes that begin in early adolescence are not isolated cerebral activity but they relate directly to other important developments. They reflect massive neural reorganization and hormonal instability (Spear, 2000, 2003). The adolescent brain, Spear pointed out, "is a brain in transition, and differs anatomically and neurochemically from that of the adult" (2000, p. 446). Reorganization and growth are accompanied by heightened emotional arousal and volatility of mood (Granic et al., 2003; Rosenblum & Lewis, 2003; Spear, 2000).

Some of the cognitive developments are themselves emotionally unsettling. For example, during adolescence young people become more sensitive to the opinions of others and more relativistic in their appraisals of reality (Byrnes, 2003; Klaczynski, 2000). It can be disturbing to discover that the world one had previously thought of in rather concrete terms is actually amenable to a plurality of interpretations and standpoints.

All of this takes place in micro- and macrosocial contexts, where other people and institutions are communicating new expectations and role demands (Durkin, 1995). In dealing with these issues, some of the most prominent concerns of adolescents include identity formation (Erikson, 1968; Kroger, 2003) and achieving autonomy from parents (Granic et al., 2003; Zimmer-Gembeck & Collins, 2003). The young person has to establish who she or he is, or wishes to become, and how she or he relates to the rest of the social world. The emergence of autonomy in adolescence involves behavioral, cognitive, and emotional dimensions, and these are interdependent (Eccles et al., 1997; Steinberg & Silverberg, 1986; Zimmer-Gembeck & Collins, 2003). During this period of life, there is an increasing need to organize one's own actions and to experience the agentic self as a source of those actions (Deci & Ryan, 2000).

Leisure is central to the lives of contemporary adolescents (Csikszentmihalyi & Larson, 1984; Raymore, Barber, Eccles, & Godbey, 1999). Importantly, leisure is an arena in which young people deal with the kinds of developmental tasks noted above. These include developing a sense of self, handling peer relations, and managing emotions, including conflictual emotions and aggression. Again, it is important to recognize that leisure behavior is not homogenous: That is, it is not the case that all young people do the same things at the same time for the same reasons (Raymore et al., 1999).

The above merely outlines a few of the fundamental developmental and contextual issues that bear on the psychology of adolescence. However, this sketch establishes that adolescents are much more than the passive recipients of whatever environmental stimuli (including media) surround them. They are complex, active, and varied beings engaging with the complex, multidimensional ecologies in which they live. How do computer games fit into their lives? We consider in turn cognitive, social, and emotional aspects of the relationships between young people and this technology, though it will already be clear that these factors overlap.

COMPUTER GAMES AND ADOLESCENT COGNITIVE DEVELOPMENT

Cognitive development in adolescence is by no means straightforward. Certainly, on several measures, there are reasonably reliable age-related increases in performance (Byrnes, 2003). But, large sample studies of adolescents' performances in formal operational and other scientific reasoning tasks often yield dismal results. Byrne (2003) noted that, while older adolescents tend to have richer metacognitive abilities than younger adolescents, many demonstrate only mediocre performance in scholastically related tasks and reasoning about controversial topics. Educational performance is multidetermined, but there is little doubt that part of the story is due to motivation. In contrast, in the cognitively demanding, high-speed decision-making environment of computer games, motivation is rarely a problem (Bryant & Davies, this volume; Lee & Peng, this volume; Vorderer, Klimmt, & Ritterfeld, 2004).

The frontal lobes are still maturing during adolescence and a range of executive functions, working memory, and metacognitive strategies continue to develop. Computer games engage these capacities. We still know relatively little about how cognitive capacities are used/ stretched/ reinforced via computer game play, but we do know that performance on cognitively demanding tasks is superior in young people (and adults) when the level of motivation is high (Byrnes, 2003; Klimmt & Hartmann, this volume; Ritterfeld & Weber, this volume).

Computer games seem to be well designed to meet some of the cognitive developmental needs of adolescents. Goals are becoming particularly salient to the young person, but many important goals are often long-term. Computer games offer microworlds in which goals can be pursued in much faster time, with much more readily measurable consequences. As Klimmt and Hartmann (this volume) show in some detail, these environments call for intentional, alert and analytical participation, requiring a range of skills that are frequently tested to their limits. Furthermore, in the computer world, goals can be reiterated, failures redeemed, and past triumphs relived or exceeded by starting a new game (see also Bryant & Davies, this volume).

There is experimental evidence of cognitive and/or perceptual skill gains through computer game play (Buckley & Anderson, this volume; Greenfield, Brannon, & Lohr, 1994; Greenfield, Camaioni, et al., 1994; Greenfield, deWinstanley, Kilpatrick, & Kaye, 1994; Lee & Peng, this volume; Okagaki & French, 1994; Subrahmanyam & Greenfield, 1994). This does not confirm that all computer game play is invariably beneficial in all respects for all players (Greenfield, 1994), but it does indicate that positive outcomes are possible in at least some domains. Importantly, cognitive and spatial skills, involvement in technologies increasingly pivotal to contemporary educational and occupational demands, interaction with the products of artificial intelligence, and positive self-concepts of computer-related skills are all facilitated through computer game play (Greenfield, 1994). It has also been speculated that young people's exposure to the imagery and task demands of computer games may help explain the increases in recent decades on some tests of nonverbal intelligence (Greenfield, 1998; Subrahmanyam, Greenfield, Kraut, & Gross, 2001). More research is needed to chart the cognitive developmental course of game usage during adolescence; as noted by Ritterfeld and Weber (this volume), the same game or game elements may be interpreted in different ways at different levels of development, and the same skills requirements may engage quite different processes according to the young person's broader cognitive competencies.

Other cognitive developmental issues that call for future research concern the implications for adolescents of engagement with the fantasy dimensions of games. Fantasizing is a cognitive skill (Gunter, 1980; Singer, 1966) and therefore likely to increase with development and to be particularly intense in adolescence. Computer games offer rich fantasy environments, which are likely to draw upon the relevant cognitive activities. It is possible that this may stimulate imaginative development (players are drawn into vivid and extraordinary worlds where events can become unpredictable) or hinder it (the games are so lavish that lay players may find it hard, or unnecessary, to elaborate spontaneously). Once more, individual and developmental differences are likely in these respects but have been scarcely investigated to date.

COMPUTER GAMES AND ADOLESCENT SOCIAL DEVELOPMENT

Early concern about computer games included speculation that they were addictive, solitary pursuits that would absorb the vulnerable young and inhibit their social development. In fact, very little evidence has emerged to confirm fears of addiction (but see Lee & Peng, this volume for a discussion of dependency). Many players do go through phases of what some observers and perhaps the individuals themselves might view as excessive use (Creasey & Myers, 1986; Durkin & Aisbett, 1999; Mitchell, 1985), but interest and involvement typically fluctuate. Even among males aged 8 to 18 years, only 21% report playing more than an hour per day (Roberts et al., 1999). It remains to be seen whether computer games will overtake television as the main leisure activity for the majority of young people (Funk, 1993; Kubey & Larson, 1990; Roberts et al., 1999), though their widespread popularity among the young is not in dispute (Vorderer et al., this volume).

Less than 1% of children are rated by teachers or by themselves as game dependent or "addicted" (Shotton, 1989), and these figures may be overestimates as they rely on lay judgments of dependency/ addiction and are unlikely to entail reliable assessments of changes over time. A 1% addiction rate is very small compared to addictions among adolescents to dangerous substances like nicotine. There may well be some young people for whom involvement in computer game play becomes problematic, but we lack careful studies of the social, developmental, and clinical contexts in which these instances occur. It is very likely that where there are serious problems associated with overuse of games, the young person is dealing with other serious problems. Research rather than newspaper headlines is needed on this topic.

The Social Nature of Game Use

More importantly, the early anxieties presumed what has since become very clear was an inaccurate image of how young people integrate computer games into their lives. In general, children and adolescents enjoy spending computer game time in the company of family members or friends (Cupitt & Stockbridge, 1996; Durkin & Aisbett, 1999; Kubey & Larson, 1990; Phillips et al., 1995; Vorderer & Ritterfeld, 2003). In focus group interviews with enthusiastic young players, we found that the majority claimed to prefer to play with others rather than alone (Durkin & Aisbett, 1999). When we put it to them that game playing could turn them into isolates, they rejected the idea:

> "I don't reckon it's anti-social. A big bunch of us played computer games at school and we all shared knowledge. . . . It was a shared experience." (male, mid-teens)

> "With computer games, it's educational and you're bonding with other people who are helping you play." (female, mid-teens)

These findings are consistent with many other reports. For example, in an interview study conducted in Californian video game arcades, adolescents said that they found little interference with family life due to video game play (Egli & Myers, 1984). Sherry, Lucas, Greenberg, and Lachlan (this volume) show that social interaction is a primary motivation for game play throughout adolescence, especially for males.

As in other aspects of life, game players are socially selective: They prefer partners of about the same skill level. Again, competition and challenge are clearly important considerations here (Raney et al., this volume; Sherry et al., this volume; Vorderer, 2000). Our interviewees explained that they did not want to beat an opponent too easily and did not want to be beaten continually. Just as games may be selected to match one's mood (see Bryant & Davies, this volume), so are partners:

> "My friend Chris is good to play with. He's fun. He mucks around." (male, preteens)

> "Friends for fun—Matt, my brother, for a challenge." (male, preteens).

Focus groups and interviews can provide useful indicators of social behavior and reasoning but they are not necessarily representative, and they can be compromised by particularly voluble respondents or unwitting researcher influence. However, in a nationally representative survey of over 400 Australian adolescents (aged 12 to 17 years), we found that 87% reported playing with others at least once a month and 76% of frequent players played with others at least once a week (Durkin & Aisbett, 1999).

One third of the Australian sample in Durkin and Aisbett (1999) reported that they play games with others over the Internet. The predominant favorite reasons given by adolescents and young adults for participation in multiplayer online games are social (contact with others, being able to assist others, being a member of the relevant club, etc.). In UK research, these were nominated by 44% of adolescents and 55% of young adults (Griffiths, Davies, & Chappell, 2004). Next most popular features were game-specific (character role-playing, casting magic, game longevity). Chan and Vorderer (this volume) show that not only do the technologies of multiple player games meet strong social motivations in many players, but they may also serve as a stepping-stone to other forms of contact (e-mail, telephone, or meetings).

In a sample of 120 11- to 17-year-olds, Colwell et al. (1995) found that, among boys, amount of play was correlated with preferring computer games to friends. However, the more boys played, the more they saw their friends out of school. These seemingly anomalous findings call for further research and they could be explained in different ways. A "third factor" account is possible: Higher levels of game play and higher levels of unsupervised peer socialization could each be associated with lower levels of parental control. Alternatively, increasing peer interaction during adolescence, while normative, can be taxing at times. Perhaps, if one's friends are adolescent males, increased exposure to them renders *Castle Wolfenstein* all the more congenial a place to hang out.

GAMES AND POSITIVE SOCIAL DEVELOPMENT

Durkin and Barber (2002) investigated the relationships between amount of play and several measures of social behavior and adjustment in a sample of over 1,000 sixteen-year-old Americans. One advantage of this study was that it drew on data from the Michigan Study of Adolescent Life Transitions (MSALT), an ongoing longitudinal study that was designed primarily to address other developmental questions (Eccles et al., 1989) but did include a measure of computer game use. While many experimental studies are vulnerable to the risk of demand characteristics communicating to the participants that they are expected to become more antisocial, the MSALT study is unlikely to have transmitted expectations about specific relations among a large battery of measures.

Comparing participants who never played computer games, low players (play occasionally), and high players (play daily or almost daily), we found several advantages to the player groups. Low and high game players had more positive self-concepts in respect of intelligence and computer skills than did the never players. High players rated their mechanical skills higher than did never players. (There were no differences among the groups on self-concepts of interpersonal skills or leadership ability.) Low and high players reported less substance use than did never players. Depressed mood was significantly lower in the low play group (the other two groups did not differ from each other). These measures depend on self-ratings but, interestingly, consistent results obtained on more objective measures, such as academic achievement. Low players had higher grade point averages (GPAs) than each of the other groups. This finding is broadly compatible with the point made by Ritterfeld and Weber (this volume) that higher GPAs are predicted on the basis of higher IQ and deliberate practice; we do not have independent measures of the IQ of adolescents in this sample but it is at least plausible that some young people of above average intelligence manifest in computer game play, as elsewhere, a readiness to practice skills.

In contrast to the popular assumption that computer game play inhibits sociability—but consistent with the evidence reported in this and several other contributions to this volume that game play often has a social motivation—both low and high play groups reported higher levels of family closeness and attachment to school than did never players. Low and high play participants scored lower on a measure of association with "risky" peers than did never players. Low and high players were significantly more likely to be involved in organized extracurricular activities and clubs than were the never players, and there was a marginal effect in the same direction for participation in sports teams.

Durkin and Barber (2002) interpreted these results as supporting the view that young people who engage regularly in voluntary, structured, and challenging activities will tend to enjoy relatively positive development (Eccles & Barber, 1999; Larson, 2000). It is unlikely that playing computer games in itself "causes" healthy psychological development, but it may be one manifestation of the natural, active, and constructive exploration and exploitation of the environment that is assumed by most contemporary developmental theories.

Games and Autonomy

We saw earlier that the achievement of autonomy is one of the fundamental tasks of a healthy adolescence. One cannot become autonomous from parents by being with them all of the time and by accepting every choice they make on one's behalf. At the behavioral level, young people begin to spend more time with their peers, and they take up activities that are not necessarily disapproved of by parents but are less directly regulated by them. At the cognitive level, they are increasingly likely to formulate their own perspectives, opinions, and strategies. At the emotional level, they are experiencing powerful new feelings about themselves, their bodies and their social lives, concomitant with tensions over parental authority. Reflecting the interweaving behavioral, cognitive, and emotional aspects of this major life transition, parent–child conflict tends to increase during early to mid-adolescence (Collins & Laursen, 1999; Granic et al., 2003; Laursen, Coy, & Collins, 1998).

In this context, computer games have considerable appeal, especially for younger adolescents not yet ready for dating and drinking. There are at least two levels in which computer games offer attractions to young people seeking to develop personal autonomy.

First, changes in the patterns of media use are part of a broader readjustment of personal lives as young people move away from parental control and increase their peer-related activities (Arnett, 2003; Durkin, 1995; Roe, 1989). Recent generations have tended to enjoy the additional liberty provided by their considerable superiority in the game domain relative to their parents, though it is interesting to speculate whether this pattern of generational discrepancy will continue indefinitely. (What happens when today's highly skilled players, who will most likely maintain their enthusiasm for games into adulthood, become parents of young adolescents over whom they will have the advantage of many years of game expertise?).

Second, computer games allow an escape into a world where one has symbolic agentic autonomy—the great thing about games is that the player makes the choices (Klimt & Hartmann, this volume; Vorderer, 2000). Games involve cognitive activity and challenges, but answerable largely to the machine, to oneself, or to like-minded peers rather than controlling adults. Games can be emotionally arousing, and there is the scope to choose the emotions—for example, whether one plays for fun, for the alleviation of real-world stresses and anxieties, for adrenalin rushes, for self-aggrandizement, or for vengeance (see also Bryant & Davies, this volume; Raney et al., this volume).

Games and Social Identity

Establishing one's sense of identity is a protracted achievement, taking place over more than a decade or longer (Erikson, 1968; Kroger, 2003; Meilman, 1979). In the quest for self-determination, most adolescents explore possible identities, directly (through introspection, career fantasies, activities, fashions, peer selections) and vicariously, through identification with role models, including media figures (Giles & Maltby, 2004; Taveras et al., 2004). Computer games offer an exciting new medium for these traditional preoccupations, allowing the young person a vast array of potential identification figures and milieux ranging from the familiar to the fantastic. We know relatively little about how salient computer game characters and environments are to young people exploring identity issues, though it is reasonable to suppose that there are individual and developmental differences in this respect. As Chan and Vorderer (this volume) document, some games (especially avatar-based and MMOGs) afford opportunities variously to adopt, enhance, hide, and disguise identities. This feature appears to be attractive to many players (not only adolescents); future research may reveal how choices and emotional investments vary as a function of progress and setbacks in the search for identity.

Being known to be playing computer games per se may be an important part of social identity during some phases of adolescence: It makes a public statement about who one is and thereby forms part of one's social capital (Chan & Vorderer, this volume; Raney et al., this volume). Colwell et al. (1995) suggested that, as computer game play has become normative for males, it has become integrated with peer relations such that increasing amount of play is associated with increasing amount of peer interaction. As suggested earlier, at certain stages of adolescence, it may be risky not to play; it marks one out as not sharing popular peer interests.

Adolescents have traditionally engaged in pastimes and media uses that facilitate the establishment of social identity and group membership (Arnett, 2003; Christenson, 2003). Adolescents are also dealing with changes in their relations with institutions beyond the family, most notably the school. Young people who find it difficult to achieve and enhance self-esteem and social reputations via the conventional system tend to find alternative routes, often via socially disapproved activities (Emler & Reicher, 1995). One such route is involvement in controversial media (Roe, 1989). Roe argued that VCR use (which, like computer games more recently, was the focus of a moral panic in the early 1980s) was appropriated by some low-achieving adolescents as a subcultural medium for the expression of their disaffection with authority. Nevertheless, it is important to keep in mind that the popularity of computer games is now almost universal among adolescent males and increasing among adolescent females.

Developing identity is closely interwoven with developing one's gender role. Part of the complex and sometimes conflicting demands on the contemporary young male is that he demonstrate physical autonomy and toughness, and boys become increasingly sensitive to this expectation during adolescence (Wichstrom, 1999). In some contexts, this is manifest in actual physical engagement (street fighting, etc.), but for many young people in Western societies overt aggression is not a routine activity and it is proscribed by competing aspects of the modern male role, whereby intellect and manners are expected to prevail over testosterone and brawn. It is also rather dangerous, and not all young males wish to push their physical risk taking quite that far. Hence, "violent" activity via computer screens provides a safe means of being tough. It also offers a means of demonstrating to one's peers that one is tough (Goldstein, 1998) and appropriately anti-establishment—there is the added bonus that you can be confident that some adults will disapprove of some game contents. Games are produced primarily by males, for males and, in respect to gender roles, content tends to be highly traditionally stereotyped (Dietz, 1998; Lee & Peng, this volume). The competitive nature of some game play also lends itself to (and is most likely a reflection of) the maintenance and constant review of social hierarchies among young males (see also Raney et al., this volume). As Klimmt and Hartmann (this volume) speculate, there may also be experiential and reward consequences of game play that are more readily complementary to masculine than feminine social identities.

Zillmann (1998) made the persuasive case that the appeal of horror movies to adolescents needs to be understood in relation to gender role demands during this phase of life. Boys affect to be underwhelmed or amused by horror content, in keeping with their aspirations, while girls are happy to provide public displays of distress and alarm in the face of such materials. The public nature of the reactions are important, because the reflective company of peers is crucial to role enactment. Zillmann and colleagues' work shows that young people report greater enjoyment of horror movies in peer company, and that the nature of the reaction interacts with gender. Males find greater enjoyment when in the company of others who are squeamish (masculine superiority). Females find the company of tough males more appealing (Zillmann, Weaver, Mundorf, & Aust, 1986). Very similar processes appear to occur in relation to computer game play. Boys like to share games with their male peers, and girls' reactions vary from lack of interest, through spectator roles, to enthusiasm. Vivid illustrations of these phenomena are readily available at your local video arcade.

COMPUTER GAMES AND ADOLESCENT EMOTIONAL DEVELOPMENT

It has been stressed here that the cognitive, social, and physiological changes of adolescence are interwoven with emotional experience. Byrnes (2003) pointed out that effective decision making involves being able to manage one's emotions. Emotional management is problematic for adolescents: They find it difficult to control their own displays and to adjust their self-presentations to accommodate the behavior and expectations of others (Rosenblum & Lewis, 2003; Saarni, 1989). Boys in particular have problems in organizing their emotions, partly because they are expected to repress feminine feelings and partly because the peer culture often values "cool" or "tough" images.

In this context, it would not be surprising to find that favored leisure activities have strong emotional associations and functions. Self-evidently, young people like computer games. When observed during play, whether in homes, arcades or laboratories, most young participants are clearly enjoying themselves, and the activity is accompanied by laughter and intense engagement (Durkin & Aisbett, 1999; Kubey & Larson, 1990; Michaels, 1993). In a survey of over 400 Australian adolescents, Durkin and Aisbett asked participants to indicate what words best describe how they feel when they are playing the game that they play most often. Consistent with independent findings reviewed by Raney et al. and Sherry et al. (this volume), the most frequently mentioned emotions were happy, having fun, enjoyment (reported by 54% of the sample), excitement, exhilaration (39%), relaxed, peaceful, calm, quiet (14%), and challenged (14%). Fewer participants reported feeling frustrated (8%), bored (5%), angry, aggressive, warlike (5%). Subjectively, as noted also by Raney et al., the activity is predominantly perceived as emotionally positive.

The low levels of reported boredom while playing favorite computer games contrasts dramatically with adolescents' reports of their experiences in other environments, such as school. Many teenagers find the latter environment varies from dull to chronically boring (Csikszentmihalyi, Rathunde, & Whalen, 1993; Larson & Richards, 1991; Papert, 1998). Pervasive feelings of boredom are signs of psychological health risk in adolescents (Hunter & Csikszentmihalyi,

2003). The relationship between computer game play and boredom calls for further research to determine the extent to which game play is chosen as a relatively easily accessed means of coping with boredom and the extent to which games are appealing to young people who are already developing positively. Given the breadth of popularity of computer games, it is likely that they serve different functions in different lives and, again, these may vary within individuals over the course of development.

One way in which computer games appear to be used by at least some adolescents is in the form of emotional self-regulation (Bryant & Davies, this volume; Sherry et al., this volume). This is a basic function of all play—we turn to leisure for relaxation, arousal, diversion, and other emotional needs. Among other regulatory functions, one benefit of computer games for at least some players is that it may be cathartic. Catharsis has tended to be dismissed in media psychology, largely as an outcome of debates about the impact of television and movies. But, computer games offer different media experiences. In particular, the "intervention potential" (Bryant & Zillmann, 1997; Vorderer, 2000) of this medium is high. As pointed out by Bryant and Davies (this volume) and Ritterfeld and Weber (this volume), this feature is often associated in popular and some scientific discussions with the possibility that games might instigate or exacerbate aggressive tendencies in players. However, there are also arguments for a possible cathartic role and these call for future research.

Zillmann (1998) noted that catharsis theories of media use (especially, uses of aggressive media content) presume that certain anxieties, tensions, and other negative emotions exist as preconditions for whatever relief the media experiences may provide. That is, one has to be suffering something before finding catharsis. This, he points out, is not an established fact in adult audiences. However, it is plausible that such conditions obtain for quite a lot of the time in adolescents. To what extent these negative states prompt game play has not been tested extensively, though certainly in interview studies young players report that they sometimes use games purposefully in this way (see below).

Zillmann (1998) made the further point that for a catharsis theory to be borne out fully we would need evidence that the media experiences are indeed sufficiently powerful to afford the relief sought. The rather small amount of evidence in respect of other media is controversial but generally interpreted as failing to support the catharsis hypothesis, and quite possibly demonstrating the opposite effect (see Zillmann, 1998; for a review).

However, there could be more than one possibility. For example, satisfaction of the catharsis hypothesis according to the scientific criteria Zillmann (and most other psychologists) would advocate may be more stringent than the lay phenomenological test. That is, some players may believe that they are obtaining cathartic relief, even if the validity of their beliefs is hard to demonstrate objectively.

There may also be individual differences in propensity for cathartic relief (or in the conditions required): Some people may obtain cathartic relief from aggressive game play, and others may not. Gunter (1980) argued that a critical variable may be the ability to fantasize. In a discussion of aggressive material on television, he proposed that individuals who are high on fantasizing skill can use media contents to moderate their own affective states, including reduction of aggressive feelings. Although this argument remains to be tested in respect of computer game play, there is some evidence from some (though by no means all) laboratory studies that children exposed to aggressive content score lower on aggression measures shortly after play (Graybill, Kirsch, & Esselman, 1985; Silvern & Williamson, 1987). Also consistent with Gunter's arguments is Funk and Bachman's finding (1996) of a positive relationship between proportion of violent fantasy games among teenage girls' game preferences and the girls' self-concepts of job competence. Young women who have positive views of their employability may also have the cognitive resources to exploit computer games constructively.

Self-reports elicited in an interview study (Durkin & Aisbett, 1999) indicate that many young people do believe that they can use game play, including aggressive games, to resolve pent-up tensions and improve their mood:

"If I don't like him, I'll bash the hell out of him—like if it's my science teacher, it's getting rid of irritations." (female, mid-teens)

"When people I'm working for get ratty, I play some mindless games and get rid of my irritations. The characters in Wolfenstein look like my customers." (female, mid-teens)

It is very likely also that there is intraindividual variation: Sometimes catharsis is obtained, and sometimes it is not (indeed, emotional responses of varied sorts can be experienced within a game or game session; see Grodal, 2000; Vorderer, 2000). Some young people interviewed by Durkin and Aisbett (1999) said that the venting benefits of computer game play were variable and much depends on how the game goes: If one is angry and then frustrated by poor performance in a game, the negative emotions can increase rather than dissipate.

Catharsis is clearly not "the" explanation for the popularity of computer games. The duration of any cathartic outcomes is doubtful. However, it may be a factor for some players, under some circumstances, some of the time, and it is premature to write it off on the basis of assumptions that all media have equivalent effects. (Other therapeutic uses of games are discussed by Lee & Peng and Ohler & Nieding, this volume).

CONCLUSIONS

It was suggested at the beginning of this chapter that the absence of computer game play in the life of a contemporary adolescent signals risk. Mounting evidence is consistent with this claim: Young people invest and gain a lot from computer game play, and those who do not play are more likely to be at risk of social adjustment problems. However, by definition, risk is not a guarantee: Some young people will lead perfectly healthy and well-adjusted lives without computer games. Furthermore, there are risks and there are risks. Missing out on computer games may be unfortunate, but it is not as problematic as exposure to dysfunctional family life, bullying, poor schooling, unemployment, poverty, racism, substance abuse, pollution, crime, or warfare.

This chapter has argued the need for a developmental approach to the understanding of computer games in the lives of contemporary adolescents. Such an approach focuses on the complex psychological phenomena of developing people in multidimensional ecologies. In contrast to approaches that largely ignore development and emphasize instead the presumed damage caused by external stimuli, a developmental perspective leads to the more optimistic possibility that several hundred million adolescents around the world are proactively and successfully integrating computer games into their lives and profiting from the experiences that ensue.

NOTE

[1]Because this chapter is not concerned with issues of format difference, the terms *video games* and *computer games* are used interchangeably.

REFERENCES

Archibald, A. B., Graber, J, A., & Brooks-Gunn, J. (2003). Pubertal processes and physiological growth in adolescence. In G. R. Adams & M. D. Berzonsky (Eds.), *Blackwell handbook of adolescence* (pp. 24–47). Oxford, UK: Blackwell.

Arnett, J. J. (2003). Music at the edge: The attraction and effects of controversial music on young people. In D. Ravitch & J. P. Viteritti (Eds.), *Kid stuff: Marketing sex and violence to America's children* (pp. 125–142). Baltimore: Johns Hopkins University Press.

Barnett, M. A., Vitaglione, G. D., Harper, K. K., Quackenbush, S. W., Steadman, L. A., & Valdez, B. S. (1997). Late adolescents' experiences with and attitudes toward videogames. *Journal of Applied Social Psychology, 27*, 1316–1334.

Bryant, J., & Zillmann, D. (1977). The mediating effect of the intervention potential of communications on displaced aggressiveness and retaliatory behavior. In B. D. Ruben (Ed.), *Communication Yearbook 1* (pp. 291–306). New Brunswick, NJ: Transaction.

Byrnes, J. P. (2003). Cognitive development during adolescence. In G. R. Adams & M. D. Berzonsky (Eds.), *Blackwell handbook of adolescence* (pp. 227–246). Oxford, UK: Blackwell.

Christenson, P. G. (2003). Equipment for living: How popular music fits in the lives of youth. In D. Ravitch & J. P. Viteritti (Eds.), *Kid stuff: Marketing sex and violence to America's children* (pp. 96–124). Baltimore: Johns Hopkins University Press.

Collins, W. A., & Laursen, B. (1999). *Relationships as developmental contexts*. Mahwah, NJ: Lawrence Erlbaum Associates.

Colwell, J., Grady, C., & Rhaiti, S. (1995). Computer games, self-esteem and gratification of needs in adolescents. *Journal of Community and Applied Social Psychology, 5*, 195–206.

Creasey, G. L., & Myers, B. J. (1986). Video games and children: Effects on leisure activities, schoolwork, and peer involvement. *Merrill-Palmer Quarterly, 32*, 251–262.

Csikszentmihalyi, M., & Larson, R. (1984). *Being adolescent: Conflict and growth in the teenage years*. New York: Basic Books.

Csikszentmihalyi, M., Rathunde, K., & Whalen, S. (1993). *Talented teenagers: The roots of success and failure*. New York: Cambridge University Press

Cupitt, M., & Stockbridge, S. (1996). *Families and electronic entertainment*. Sydney: Australian Broadcasting Authority/ Office of Film and Literature Classification.

Deci, E. L., & Ryan, R. (2000). The "what" and "why" of goal pursuits: Human needs and the self-determination of behavior. *Psychological Inquiry, 11*, 227–268.

Dietz, T. L. (1998). An examination of violence and gender role portrayals in video games: Implications for gender socialization and aggressive behavior. *Sex Roles, 38*, 425–442.

Durkin, K. (1995). *Developmental social psychology. From infancy to old age*. Oxford, UK: Blackwell.

Durkin, K., & Aisbett, K. (1999). *Computer games and Australians today*. Sydney: Office of Film and Literature Classification.

Durkin, K., & Barber, B. (2002). Not so doomed: Computer game play and positive adolescent development. *Journal of Applied Developmental Psychology, 23*, 373–392.

Eccles, J. S., & Barber, B. L. (1999). Student council, volunteering, basketball, or marching band: What kind of extracurricular involvement matters? *Journal of Adolescent Research, 14*, 10–43.

Eccles, J. S., Early, D. Frasier, K., Belansky E., & McCarthy, K. (1997). The relation of connection, regulation, and support for authority to adolescents' functioning. *Journal of Adolescent Research, 12*, 263–186.

Eccles, J. S., Wigfield, A., Flanagan, C. A., Miller, C., Reuman, D. A., & Yee, D. (1989). Self-concepts, domain values, and self-esteem: Relations and changes at early adolescence. *Journal of Personality, 57*, 283–310.

Egli, E. A., & Myers, L. S. (1984). The role of video game playing in adolescent life: Is there reason to be concerned? *Bulletin of the Psychonomic Society, 22*, 309–312.

Emler, N., Reicher, S. (1995). *Adolescence and delinquency: The collective management of reputation*. Oxford, UK: Blackwell.

Erikson, E. H. (1968). *Identity: Youth and crisis*. New York: Norton.

Funk, J. B. (1993). Reevaluating the impact of video games. *Clinical Pediatrics, 32*, 86–90.

Funk, J. B., & Bachman, D. D. (1996). Playing violent video and computer games and adolescent self-concept. *Journal of Communication, 46*, 19–32.

Giles, D. C., & Maltby, J. (2004). The role of media figures in adolescent development: Relations between autonomy, attachment, and interest in celebrities. *Personality and Individual Differences, 36*, 813–822.

Goldstein, J. H. (1998). Immortal kombat: War toys and violent video games. In J. H. Goldstein (Ed.), *Why we watch. The attractions of violent entertainment* (pp. 53–68). New York: Oxford University Press.

Granic, I., Dishion, T. J., & Hollenstein, T. (2003). The family ecology of adolescence: A dynamic systems perspective on normative development. In G. R. Adams & M. D. Berzonsky (Eds.), *Blackwell handbook of adolescence* (pp. 60–91). Oxford, UK: Blackwell.

Graybill, D., Kirsch, K. R., & Esselman, E. D. (1985). Effects of playing violent versus nonviolent video games on the aggressive ideation of aggressive and nonaggressive children. *Child Study Journal, 15*, 199–205.

Greenfield, P. M. (1994). Video games as cultural artifacts. *Journal of Applied Developmental Psychology, 15*, 3–12.

Greenfield, P. M. (1998). The cultural evolution of IQ. In U. Neisser (Ed.), *The rising curve: Long-term gains in IQ and related measures* (pp. 81–123). Washington, DC: American Psychological Association.

Greenfield, P. M., Brannon, C., & Lohr, D. (1994). Two-dimensional representation of movement through three-dimensional space: The role of video game experience. *Journal of Applied Developmental Psychology, 15*, 87–103.

Greenfield, P. M., Camaioni, L., Ercolani, P., Weiss, L., Lauber, B. A., & Perucchini, P. (1994). Cognitive socialization by computer games in two cultures: Inductive discovery or mastery of an iconic code? *Journal of Applied Developmental Psychology, 15*, 59–85.

Greenfield, P. M., deWinstanley, P., Kilpatrick, H., & Kaye, D. (1994). Action video games as informal education: Effects on strategies for dividing visual attention. *Journal of Applied Developmental Psychology, 15*, 59–85.

Griffiths, M. D., Davies, M. N. O., & Chappell, D. (2003). Breaking the stereotype: The case of online gaming. *CyberPsychology and Behavior, 6*, 81–91.

Griffiths, M. D., Davies, M. N. O., & Chappell, D. (2004). Online computer gaming: A comparison of adolescent and adult gamers. *Journal of Adolescence, 27*, 87–96.

Grodal, T. (2000). Video games and the pleasures of control. In D. Zillmann & P. Vorderer (Eds.), *Media entertainment. The psychology of its appeal* (pp. 197–212). Mahwah, NJ: Lawrence Erlbaum Associates.

Gunter, B. (1980). The cathartic potential of television drama. *Bulletin of the British Psychological Society, 33*, 448–450.

Hunter, J. P., & Csikszentmihalyi, M. (2003). The positive psychology of interested adolescents. *Journal of Youth and Adolescence, 32*, 27–35.

Inhelder, B., & Piaget, J. (1958). *The growth of logical thinking from childhood to adolescence*. New York: Basic Books.

Klaczynski, P. A. (2000). Motivated scientific reasoning biases, epistemological beliefs, and theory polarization: A two-process approach to adolescent cognition. *Child Development, 71*, 1347–1366.

Kroger, J. (2003). Identity in adolescence. In G. R. Adams & M. D. Berzonsky (Eds.), *Blackwell handbook of adolescence* (pp. 205–226). Oxford, UK: Blackwell.

Kubey, R., & Larson, R. (1990). The use and experience of the new video media among children and young adolescents. *Communication Research, 17*, 107–130.

Larson, R., & Richards, M. (1991). Boredom in the middle school years: Blaming schools versus blaming students. *American Journal of Education, 99*, 418–443.

Larson, R. W. (2000). Toward a psychology of positive youth development. *American Psychology, 55*, 170–183

Laursen, B., Coy, K. C., & Collins, W. A. (1998). Reconsidering changes in parent–child conflict across adolescence: A meta-analysis. *Child Development, 69*, 817–832.

Meilman, P. W. (1979). Cross-sectional age changes in ego identity status during adolescence. *Developmental Psychology, 15*, 230–231.

Michaels, J. W. (1993). Patterns of video game play in parlors as a function of endogenous and exogenous factors. *Youth and Society, 25*, 272–289.

Mitchell, E. (1985). The dynamics of family interacation around home video games. *Marriage and Family Review, 8*, 121–135.

Okagaki, L., & French, P. A. (1994). Effects of video game playing on measures of spatial performance: Gender effects in late adolescence. *Journal of Applied Developmental Psychology, 15*, 33–58.

Papert, S. (1998). Does easy do it? *Games Developer*, 88.

Piaget, J., & Inhelder, B. (1966). *The psychology of the child*. New York: Basic Books.

Phillips, C. A., Rolls, S., Rouse, A., & Griffiths, M. D. (1995). Home video game playing in schoolchildren: A study of incidence and patterns of play. *Journal of Adolescence, 18*, 687–691.

Raymore, L. A., Barber, B. L., Eccles, J. S., & Godbey, G. C. (1999). Leisure behavior pattern stability during the transition from adolescence to young adulthood. *Journal of Youth and Adolescence, 28*, 79–103.

Roberts, D. F., Foehr, U. G., Rideout, V. J., & Brodie, M. (1999). *Kids and media @ the new millennium*. Menlo Park, CA: Henry J. Kaiser Family Foundation.

Roe, K. (1989). School achievement, self-esteem, and adolescents' video use. In M. R. Levy (Ed.), *The VCR age: Home video and mass communication* (pp. 168–189), Newbury Park, CA: Sage.

Roe, K. (1995). Adolescents' use of socially disvalued media: Towards a theory of media delinquency. *Journal of Youth and Adolescence, 24*, 617–631.

Rosenblum, G. D., & Lewis, M. (2003). Emotional development in adolescence. In G. R. Adams & M. D. Berzonsky (Eds.), *Blackwell handbook of adolescence* (pp. 269–289). Oxford, UK: Blackwell.

Saarni, C. (1989). Children's understanding of strategic control of emotional expression in social transactions. In C. Saarni & P. L. Harris (Eds.), *Children's understanding of emotion* (pp. 181–208). Cambridge, UK: Cambridge University Press.

Shotton, M. A. (1989). *Computer addiction? A study of computer dependency.* London: Taylor & Francis.

Silvern, S. B., & Williamson, P. A. (1987). The effects of video game play on young children's aggression, fantasy, and prosocial behavior. *Journal of Applied Developmental Psychology, 8,* 453–462.

Singer, J. L. (1966). *Daydreaming: An introduction to the experimental study of inner experience.* New York: Random House.

Spear, L. P. (2000). The adolescent brain and age-related behavioral manifestations. *Neuroscience and Biobehavioral Reviews, 24,* 417–463.

Spear, L. P. (2003). Neurodevelopment during adolescence. In D. Cicchetti & E. Walker, (Eds.), *Neurodevelopmental mechanisms in psychopathology* (pp. 62–83). New York: Cambridge University Press.

Steinberg, L., & Silverberg, S. (1986). The vicissitudes of autonomy in early adolescence. *Child Development, 57,* 841–851.

Subrahmanyam, K., Greenfield, P. M., Kraut, R. E., & Gross, E. (2001). The impact of computer use on children's and adolescents' development. *Journal of Applied Developmental Psychology, 22,* 7–30.

Subrahmanyam, K., & Greenfield, P. M. (1994). Effect of video game practice on spatial skills in girls and boys. *Journal of Applied Developmental Psychology, 15,* 13–32.

Taveras, E. M., Rifas-Shima, S. L., Field, A. E., Frazier, A. L., Colditz, G. A., & Gillman, M. W. (2004). The influence of wanting to look like media figures on adolescent physical activity. *Journal of Adolescent Health, 35,* 41–50.

Vorderer, P. (2000). Interactive entertainment and beyond. In D. Zillmann & P. Vorderer (Eds.), *Media entertainment: The psychology of its appeal* (pp. 21–36). Mahwah, NJ: Lawrence Erlbaum Associates.

Vorderer, P., Klimmt, C., Ritterfeld, U. (2004). Enjoyment: At the heart of media entertainment. *Communication Theory, 14,* 388–408.

Vorderer, P., & Ritterfeld, U. (2003). Children's future programming and media use between entertainment and education. In E. L. Palmer & B. Young (Eds.), *The faces of televisual media: Teaching, violence, selling to children* (2nd ed., pp. 241–262). Mahwah, NJ: Lawrence Erlbaum Associates.

Wichstrom, L. (1999). The emergence of gender difference in depressed mood during adolescence: The role of intensified gender socialization. *Developmental Psychology, 35,* 232–245.

Zillmann, D. (1998). The psychology of the appeal of portrayals of violence. In J. H. Goldstein (Ed.), *Why we watch. The attractions of violent entertainment* (pp. 179–211). New York: Oxford University Press.

Zillmann, D., Weaver, J. B., Mundorf, N. & Aust, C. F. (1986). Effects of an opposite-gender companion's affect to horror on distress, delight, and attraction, *Journal of Personality & Social Psychology. 51,* 586–594.

Zimmer-Gembeck, M. J., & Collins, W. A. (2003). Autonomy development during adolescence. In G. R. Adams & M. D. Berzonsky (Eds.), *Blackwell handbook of adolescence* (pp. 175–204). Oxford: Blackwell.

Author Index

L

T

Subject Index